NATO ASI Series

Advanced Science Institutes Series

A series presenting the results of activities sponsored by the NATO Science Committee, which aims at the dissemination of advanced scientific and technological knowledge, with a view to strengthening links between scientific communities

The Series is published by an international board of publishers in conjunction with the NATO Scientific Affairs Division

A Life Sciences B Physics	Plenum Publishing Corporation London and New York
C Mathematical and Physical Sciences D Behavioural and Social Sciences E Applied Sciences	Kluwer Academic Publishers Dordrecht, Boston and London
F Computer and Systems Sciences G Ecological Sciences H Cell Biology	Springer-Verlag Berlin Heidelberg New York London Paris Tokyo Hong Kong

Series H: Cell Biology Vol. 41

The ASI Series Books Published as a Result of Activities of the Special Programme on CELL TO CELL SIGNALS IN PLANTS AND ANIMALS

This book contains the proceedings of a NATO Advanched Research Workshop held within the activities of the NATO Special Programme on Cell to Cell Signals in Plants and Animals, running from 1984 to 1989 under the auspices of the NATO Science Committee.

The books published as a result of the activities of the Special Programme are

Vol 1: **Biology and Molecular Biology of Plant-Pathogen Interactions.** Edited by J A Bailey 1986

Vol 2: **Glial-Neuronal Communication in Development and Regeneration.** Edited by H H Althaus and W. Seifert 1987

Vol 3: **Nicotinic Acetylcholine Receptor: Structure and Function.** Edited by A Maelicke 1986

Vol 4: **Recognition in Microbe-Plant Symbiotic and Pathogenic Interactions.** Edited by B Lugtenberg 1986.

Vol. 5. **Mesenchymal-Epithelial Interactions in Neural Development.** Edited by J R. Wolff, J. Sievers, and M. Berry 1987.

Vol 6: **Molecular Mechanisms of Desensitization to Signal Molecules.** Edited by T.M Konjin, P.J M Van Haastert, H Van der Starre, H Van der Wel, and M D. Houslay 1987

Vol 7: **Gangliosides and Modulation of Neuronal Functions.** Edited by H. Rahmann 1987

Vol 9: **Modification of Cell to Cell Signals During Normal and Pathological Aging.** Edited By S Govoni and F Battaini 1987

Vol 10. **Plant Hormone Receptors.** Edited by D Klämbt. 1987

Vol 11: **Host-Parasite Cellular and Molecular Interactions in Protozoal Infections.** Edited by K -P Chang and D Snary 1987.

Vol 12: **The Cell Surface in Signal Transduction.** Edited by E Wagner, H Greppin, and B Millet 1987

Vol 19: **Modulation of Synaptic Transmission and Plasticity in Nervous Systems.** Edited by G. Hertting and H.-C Spatz 1988

Vol. 20 **Amino Acid Availability and Brain Function in Health and Disease.** Edited by G Huether 1988

Vol 21 **Cellular and Molecular Basis of Synaptic Transmission.** Edited by H Zimmermann 1988

Vol. 23: **The Semiotics of Cellular Communication in the Immune System.** Edited by E E Sercarz, F Celada, N A. Mitchison, and T Tada. 1988

Vol 24: **Bacteria, Complement and the Phagocytic Cell.** Edited by F C Cabello and C Pruzzo 1988

Vol 25 **Nicotinic Acetylcholine Receptors in the Nervous System.** Edited by F. Celementi, C Gotti, and E Sher 1988

Vol 26. **Cell to Cell Signals in Mammalian Development.** Edited by S W de Laat, J.G Bluemink, and C.L. Mummery 1989

Vol 27: **Phytotoxins and Plant Pathogenesis.** Edited by A Graniti, R D Durbin, and A Ballio 1989

Vol 31: **Neurobiology of the Inner Retina.** Edited by R. Weiler and N.N Osborne 1989

Vol 32 **Molecular Biology of Neuroreceptors and Ion Channels.** Edited by A Maelicke 1989

Vol 33: **Regulatory Mechanisms of Neuron to Vessel Communication in the Brain.** Edited by F Battaini, S Govoni, M S Magnoni, and M Trabucchi 1989.

Vol 35 **Cell Separation in Plants: Physiology, Biochemistry and Molecular Biology.** Edited by D J Osborne and M B Jackson 1989

Vol 36 **Signal Molecules in Plants and Plant-Microbe Interactions.** Edited by B J J Lugtenberg 1989

Vol 39 **Chemosensory Information Processing.** Edited by D Schild 1990

Vol 41 **Recognition and Response in Plant-Virus Interactions.** Edited by R S S Fraser 1990

Recognition and Response in Plant-Virus Interactions

Edited by

R.S.S. Fraser

AFRC Institute of Horticultural Research,
Worthington Road, Littlehampton, West Sussex
BN17 6LP, U.K

Springer-Verlag Berlin Heidelberg New York
London Paris Tokyo Hong Kong
Published in cooperation with NATO Scientific Affairs Division

Proceedings of the NATO Advanced Research Workshop on Recognition and Response in Plant-Virus Interactions, held at Chichester, West Sussex, U K , April 12–16, 1989

Library of Congress Cataloging-in-Publication Data NATO Advanced Research Workshop on Recognition and Response in Plant-Virus Interactions (1989 Chichester, England) Recognition and response in plant-virus interactions/edited by R S S Fraser p cm —(NATO ASI series Series H, Cell biology, vol 41) "Proceedings of the NATO Advanced Research Workshop on Recognition and Response in Plant-Virus Interactions, held at Chichester, West Sussex, U K , April 12–16, 1989"—T p verso
ISBN-13: 978-3-642-74166-1 e-ISBN-13: 978-3-642-74164-7
DOI: 10.1007/ 978-3-642-74164-7
1 Plant viruses—Congresses 2 Plant-microbe relationships—Congresses 3 Plant viruses—Host plants—Congresses I Fraser, R S S II Title III Series QR351 N39 1989 581 2'34—dc20 90-9565

Softcover reprint of the hardcover 1st edition 1990

2131/3140-543210 – Printed on acid-free-paper

PREFACE

Signalling between plants and viruses forms a special model system within the NATO Programme on "Cell-to-cell Signals in Plants and Animals". Clearly the pathogen side of the interaction, the virus, is not a cell. But the recognition events between plant and virus, and the subsequent signalling and response mechanisms, involve host- and pathogen-specified molecules, just as is thought to occur in plant-fungus or plant-bacterium interactions.

The particular attraction of viruses - and of viroids - is that they can now be completely defined in molecular terms. For several of these minimal pathogens, the complete sequences of the nucleic acids and the proteins specified are known. There is much information from X-ray diffraction and related techniques on the three-dimensional structure of the particle. In contrast, the molecular biology of the host reaction to virus infection is less well understood; this area is now receiving greater attention, and there have been several recent advances in our knowledge. For this reason, it was timely to review progress by holding a NATO Advanced Research Workshop on "Recognition and Response in Plant-Virus Interactions".

The Workshop took place from 12 - 16 April 1989, at Bishop Otter College, Chichester, West Sussex, U.K. It was attended by more than 50 participants, including representatives from eleven NATO countries. The objective was to examine signalling events, defined in the broadest sense, at all stages of the viral replicative cycle, and to follow the subsequent response of the plant or consequences for virus replication. The Workshop consisted of a series of major lectures, shorter oral presentations, and poster and discussion sessions. This book consists of chapters which are based on the major lectures, but which have been written with a broad approach, to encompass much of the other material discussed at the Workshop, with the aim of providing a comprehensive coverage of the subject area.

Organizing a NATO Advanced Research Workshop, and editing the resulting book, are enjoyable and stimulating tasks. They are made thus by the willing help one receives from many quarters. I am grateful to John Friend and Alain Jubier of the NATO Cell-to-cell Signals Programme, and to the Workshop Organizing Committee - Bryan Harrison, Kees van Loon, Michael Wilson and Milt Zaitlin - for advice during the planning phase. The smooth running of the Workshop was ensured by the efforts of Jill Rogers and her staff at Bishop Otter College, and by Peter Grimbly, Jane Mepham,

Ruthanne Smith and Andrew Smith from IHR Littlehampton. I would like to thank the participants in the Workshop for their presentations and contributions to the discussions, and for preparing their chapters for publication. Finally, I am most grateful for assistance in the production of this book, to Mike Ledieu for electronic text processing, to Brenda Barnard for preparing the diagrams, and to my secretary Jane Mepham for much-valued help in preparation of the text.

Ron Fraser
Workshop Director
December 1989

CONTENTS

Preface

SECTION 1. INTRODUCTION AND COMPARATIVE STUDIES

R S S Fraser Recognition and response in plant-virus interactions: some underlying concepts 1

A Vivian Recognition in resistance to bacteria 17

J Mansfield Recognition and response in plant-fungus interactions 31

SECTION 2. PLANT-VIRUS INTERACTIONS: TRANSMISSION, HOST RANGE AND GENETICS

A F Murant Specificity and recognition events in the transmission of plant viruses by vectors 53

R S S Fraser The genetics of plant-virus interactions: mechanisms controlling host range, resistance and virulence 71

J M Thresh Plant virus epidemiology: the battle of the genes 93

SECTION 3. MOLECULAR MECHANISMS OF PLANT-VIRUS INTERACTION. I. INFECTION, REPLICATION AND SPREAD

T M A Wilson, K A Plaskitt, J W Watts, J K Osbourn and P A C Watkins Signals and structures involved in early interactions between plants and viruses or pseudoviruses 123

R Goldbach, R Eggen, C de Jager, A van Kammen, J van Lent, G Rezelman and J Wellink Genetic organization, evolution and expression of plant viral RNA genomes 147

R Gargouri and A-L Haenni Replication of plant RNA viruses *in vivo* in relation to virus-host interactions 163

M Tabler and M Tsagris Viroid replication mechanisms 185

T Godefroy-Colburn, F Schoumacher, C Erny, A Berna, O Moser, M-J Gagey and C Stussi-Garaud The movement protein of some plant viruses 207

SECTION 4. MOLECULAR MECHANISMS OF INTERACTION. II. PATHOGENESIS AND SYMPTOM FORMATION

V Conejero, J M Bellés, F García-Breijo, R Garro, J Hernández-Yago, I Rodrigo and P Vera Signalling in viroid pathogenesis 233

D Baulcombe, M Devic and M Jaegle The molecular biology of satellite RNA from cucumber mosaic virus 263

J F Antoniw, R F White and W Xie Cryptic viruses of beet and other plants 273

R J Whenham and R S S Fraser Plant growth regulators, viruses and plant growth 287

L C van Loon, O van Kooten, E G A Linders, C Meurs and M M G Wijdeveld Recognition and disease development in the tobacco-TMV system 311

M Zaitlin, J E Schoelz, R Grzesik and J O Gaspar Entry of tobacco mosaic virus RNA into tobacco chloroplasts *in vivo* and *in vitro* 329

SECTION 5. RECOGNITION AND RESPONSE IN RESISTANCE AND VIRULENCE

J N Culver and W O Dawson Modifications of the coat protein gene of tobacco mosaic virus resulting in the induction of necrosis 337

K-W Mundry, W Schaible, M Ellwart-Tschürtz, H Nitschko, C Hapke Hypersensitivity to tobacco mosaic virus in *N'*-gene hosts: which viral genes are involved? 345

H J M Linthorst, B J C Cornelissen, J A L van Kan, M van de Rhee, R L J Meuwissen, M T González Jaén and J F Bol Induction of plant genes by compatible and incompatible virus-plant interactions 361

B Fritig, S Kauffmann, J Rouster, B Dumas, P Geoffroy, M Kopp and M Legrand Defence proteins, glycanhydrolases and oligosaccharide signals in plant-virus interactions 375

G Loebenstein and A Gera Inhibitor of virus replication associated with resistance responses 395

B D Harrison, H Barker and P M Derrick Intercellular spread of potato leafroll luteovirus: effects of co-infection and plant resistance 405

SECTION 6. EXPLOITING SIGNALLING IN CROP PROTECTION

L A Urban, J L Sherwood, J A M Rezende and U Melcher Examination of mechanisms of cross protection with non-transgenic plants 415

R S Nelson, P A Powell and R N Beachy Coat protein-mediated protection against virus infection 427

R Hull The use and misuse of viruses in cloning and expression in plants 443

Index 459

RECOGNITION AND RESPONSE IN PLANT-VIRUS INTERACTIONS: SOME UNDERLYING CONCEPTS

R S S Fraser
AFRC Institute of Horticultural Research
Worthing Road
Littlehampton
West Sussex BN17 6LP
UK

ABSTRACT

Plant-virus interactions form a model system within the NATO Special Programme on cell-to-cell signalling in plants and animals. Recognition events between plants and viruses, and subsequent signalling, lead to diverse responses which affect many aspects of pathogenesis. This paper reviews the importance of plant-virus signalling processes in five main areas: virus transmission, host range and genetic interactions; molecular mechanisms of virus replication and spread; molecular mechanisms of pathogenesis and symptom formation; recognition in resistance and virulence; and exploiting signalling in crop protection.

INTRODUCTION

The 1960s and 70s were decades in which there was an explosion of knowledge of gene structure, and of the mechanisms of expression of the genetic information. The 1980s saw more emphasis being placed on the mechanisms of control of gene activity. But there was also a significant move towards investigation of how living organisms are co-ordinated at more complex levels than those of the individual operon or single cell. This led to increasing interest in mechanisms of cell-to-cell signalling, whereby different parts of a single organism affect each other, organisms react to changing signals from the environment, and different organisms interact with each other. Cell-to-cell signalling processes are important in, for example, the control of differentiation and plasticity of developmental processes, and in recognition and response processes in parasitic and

NATO ASI Series, Vol H 41
Recognition and Response in Plant-Virus Interactions
Edited by R S S Fraser

symbiotic interactions. Clearly, some cell-to-cell signalling processes are likely to lead to alterations in the patterns of gene expression in the receiver cells. But these changes in pattern of gene expression will be more complex than the single-operon controls studied previously, and may also initiate a cascade of subsequent changes by transmission of further signals. In these respects, cell-to-cell signalling studies may be regarded as the second generation of investigations of control of gene activity.

To mark the importance of cell-to-cell signals in plant and animal development, a NATO Special Programme was established by the NATO Science Committee in 1984, and a series of Advanced Research Workshops has been held over the succeeding five years.

The interactions between plants and their microbial pathogens present some excellent opportunities for studying signalling between two organisms. These interactions have been the subject of earlier Workshops in the Programme (Bailey, 1986; Lugtenberg 1986; Scannerini *et al.,* 1988). Depending on the types of host- and pathogen-specified molecules interacting, and on the nature of the recognition events and ensuing further signals, the outcome may be full susceptibility and pathogen establishment, or the induction of host resistance with rejection of the pathogen. Although the initial recognition systems in host and pathogen are often under extremely simple genetic controls, the consequences of the first recognition event between the two organisms can signal diverse and complex changes in future metabolism of both partners, close to and distant from the site of initial infection, and varying with whether the outcome is resistance or susceptibility.

Plant-virus signalling and dependent events form a special model system within the Programme. Clearly, the pathogen half of the interaction is not a cell, and might not be expected to be capable of the same degree of complexity of signals as those emanating from a bacterium or fungus. But recognition and signalling events between plants and viruses do nevertheless involve the universal principle, of interactions between host- and pathogen-specified molecules, just as happens with microbial pathogens and plants. The particular attraction of viruses, and indeed of their simpler cousins the viroids, is that they can now be completely defined in molecular terms. For several viruses, there is now complete information on the nucleic acid and protein sequences, and a significant body of information on secondary and tertiary structure of the virions and other virus-specified molecules. This depth of knowledge, and the comparative simplicity of viral pathogens, together offer excellent opportunities for studying the molecular biology of signalling: how a minimal, defined pathogen can send signals which affect the response of the host, and how signals returned from the host can affect the pattern of gene expression and gene product activity of the pathogen. Indeed, despite the basic simplicity of the viral *pathogen*, the processes of viral *pathogenesis* in the plant are very complex, and involve signalling at several levels. The overall objective of this book is to examine these layers of interaction and response; to identify recognition events between plant- and virus-specified molecules, and to follow the responses to these interactions, and how they affect further development of the pathogenesis.

To a limited extent, viruses may serve as a useful model system for studying plant-pathogen recognition phenomena, and may provide information and ideas which can assist in understanding the more complex signalling events in plant-bacterium and plant-fungus interaction. But the reverse is also true, and the history of this branch of science has shown how studies of plant-virus interactions can benefit from application of concepts developed in studies of plant-microbe interactions. For this reason, the introductory section of this book contains chapters summarizing the current state of knowledge of recognition and response in plant-bacterium and plant-fungus interactions, by Vivian and Mansfield respectively.

WHERE DO RECOGNITION AND RESPONSE EVENTS OCCUR IN PLANT-VIRUS INTERACTIONS?

Transmission, host range and genetics

Events which affect the success or otherwise of pathogenesis by a particular virus with a particular plant species are not restricted to those occurring once the plant and virus have been brought together. The second section of this book explores some areas where recognition - or lack of it - may occur outside the plant, or at a higher level of complexity than the within-plant interaction between plant and virus.

One factor is whether the virus is successfully transmitted to the potential host. Many viruses have evolved complex and, normally, rather specific relationships with particular animal or fungal vectors. Interaction with the vector can involve a number of different types of recognition events which affect the success and features of transmission. Some of the mechanisms are explored in the chapter by Murant.

The interaction - or lack of interaction - of a particular virus with a plant species is also controlled by genetic factors in the virus and plant. A spectrum of relationships is controlled, from non-host immunity/non-pathogenesis, through to the genetic determination of resistance or susceptibility in species which are normally hosts to a particular virus. Phenomena in the latter group are, of course, directly relevant to a later section of this book on signalling in resistance, but are considered at this stage because the genetic controls may overlap with those involved in the broader concept of host range control and non-host immunity. Genetical aspects of plant-virus interactions are reviewed in the chapter by Fraser.

The third concept operating above the level of the individual infected plant is time, and the co-evolution of plants and their viruses. This section therefore ends with a chapter by Thresh on epidemiology, or how recognition and response processes can affect evolution at the population level.

Molecular mechanisms of plant-virus interactions: replication and spread

When the virus and potential host do come together, whether with the assistance of a vector or by chance mechanical transmission, there are several stages in the development of pathogenesis where plant- and virus-specified molecules interact. In doing so, they may further the cause of virus multiplication, and they may provoke responses in the plant which we recognize as disease symptoms. It is convenient to treat the two areas separately. The third section of the book deals with events related to virus multiplication and spread. Put simply, a virus contains information which, in a suitable host, demands that the virus be replicated. But the virus itself does not contain, and does not specify, all of the biochemical equipment required for this replication. It therefore requires to make use of various items of host synthetic equipment, and this entails a number of recognition events, and the resultant consequences.

When a virus has, by mechanical inoculation or by the assistance of a vector, reached a location in the potential host where replicative processes may begin, an early requirement is to expose the genetic information so that it may be expressed (Wilson, 1985). Most virus particles are rather stable - some highly so - and require quite extreme conditions to ensure disassembly *in vitro*. Such conditions clearly do not exist within the host cell. The first chapter in this section, by Wilson *et al.*, summarizes some intriguing recent work showing how host components are involved in exposure, and subsequent expression, of the viral genome. The best-understood case, that of tobacco mosaic virus

(TMV) is perhaps the classical example of a recognition event leading to a profound series of consequences for both plant and virus. It appears that host 40S ribosomal subunits recognize a special RNA sequence at the 5' end of TMV RNA, which is unencapsidated or easily exposed *in vivo*, and that the 80S ribosomes which then assemble are actively involved in the translation of the "early" proteins required for virus replication. Much of the evidence comes from studies *in vitro*, but there is increasing support, and much further information, from studies using leaves and protoplasts.

Having exposed the genetic information for translation - although there are still many aspects to be clarified - the next topic is to consider how the genomic information is organized and expressed in different viruses. Here, a particularly fruitful approach has been to compare different virus groups, both within the plant kingdom, and between plant and animal kingdoms (Goldbach, 1987). From the considerable amount of sequence data now available, some intriguing comparisons can be made, and these are discussed in the chapter by Goldbach *et al.* These comparisons have implications for the understanding of particular functions, and for the evolution of viruses. On the latter point, the evidence for recombination events in the evolution of DNA and RNA viruses is becoming much stronger.

Replication of the viral genome is clearly a critical stage in virus multiplication, and one which has received a considerable amount of attention from researchers. By analogy with viruses of organisms other than plants, and from evidence from a limited number of plant viruses, it is argued that the replicase may involve both virus- and plant-coded proteins (Mouchés *et al.,* 1984). There should be some form of recognition between the two sets, in forming either a multi-subunit replicase, or components which function sucesssively and coordinately in time. There is progress towards understanding these systems, which is reviewed in the chapter by Gargouri and Haenni, but the amount of information varies considerably between virus groups. In the special case of viroids, there is generally a fuller understanding of the involvement of host enzymes: the replication of these minimal pathogens is therefore treated in a separate chapter by Tabler and Tsagris.

Assembly of progeny virus particles from the replicated nucleic acids and coat protein(s) is generally regarded - at least for the simpler plant viruses - as a non-enzymatic process, and in a number of cases has been performed *in vitro* using only virus-coded components (Davies, 1985). The topic therefore does not merit a special chapter in this volume, but the possible role of host components and organelles in the *in planta* production processes of progeny virus particles are considered in the chapter on genomic replication.

All of the above processes refer to replication within the initially infected cell, or protoplast. But it is obvious that viruses spread to lesser or greater extents within the plant: the mechanisms involved have long posed some of the major unanswered questions of plant virology. In what form does the virus move - as virion, nucleic acid or "informosome" (Hull, 1989)? Are there separate mechanisms for cell-to-cell spread, and long-distance spread, presumably in the vascular system? What virus-encoded proteins are involved, and what do they do? Are host membranes and plasmodesmata modified by the virus-encoded movement proteins? Evidence relevant to some of these questions is assessed in the chapter by Godefroy-Colburn *et al.* on virus movement. However, this is clearly an exciting growth area in plant virology, and some other chapters, such as those by Goldbach *et al.*, Harrison *et al.*, and Wilson *et al.* also contain new information and speculations about movement mechanisms.

Molecular mechanisms of interaction: pathogenesis and symptom formation

In comparison with the large research effort on the molecular biology of virus structure and replication in recent years, studies of the host have received much less attention (Fraser, 1987). Yet there are numerous intriguing questions about the host response. How do minimal pathogens such as viroids and viruses - the former specifying no proteins, and the latter containing the genetic information to specify up to about a dozen - cause such a diversity and intensity of pathogenic effects in the host, which is genetically so much more complex? Are the pathogenic effects of benefit in any way to the pathogen in its replicative cycle; do they represent a passive consequence of the presence of a significant component of "foreign" metabolism; or do they represent measures taken by the host to ameliorate or even resist the pathogenesis? Are the pathogenic responses truly representative of a natural co-evolution of host and virus, or has intensive breeding of crop species - which are used for most plant virus studies - caused a distortion of some more harmonious, naturally-evolved interaction?

Whatever the answer to these speculations, there can be little doubt that most effects of pathogenesis are triggered by interactions between plant- and virus-specified molecules. The objective in this section of the book is to consider some of the possible mechanisms of interaction, and how these are transduced to affect various aspects of cell metabolism, and organ or whole plant development.

The treatment begins with the mechanisms of viroid pathogenicity in the chapter by Conejero *et al.* There is considerable information on the relationship between viroid sequence, and the effects of small changes in sequence, and the pathogenic effects produced. Some of these aspects were touched on in the chapter by Tabler and Tsagris in the preceeding section. There is also a little information, and some speculation, on the means by which the viroid molecule may interact with host molecules to trigger symptom formation. Then there is a gap in our understanding, until we come to the downstream mechanisms which may cause visible disease symptoms directly. In these, viroid-induced increase in ethylene synthesis appears to be a major factor.

Satellite RNAs occur with some isolates of a small number of viruses, such as cucumber mosaic virus. They can modify the symptoms produced by the virus, making them either milder or more severe (Jacquemond and Leroux, 1982). This is clearly an interesting special case of signalling, in which there may be a three-way interaction between virus, satellite and host, or in which the satellite may interact directly and only with the virus to modulate its pathogenic propensities. Some of the molecular mechanisms involved and the potential for exploiting this alteration of signalling in crop protection are considered in the chapter by Baulcombe *et al.*

In discussing how viruses cause visible pathogenic effects, it can be instructive to look at the growing number of cases known where plants have cryptic viruses, which do not form visible symptoms, and which cannot be transmitted other than through the plant's own seed and pollen. The main features of cryptic virus infections are reviewed in the chapter by Antoniw *et al.* Despite the lack of visible effects, some cryptic viruses do cause significant losses of crop yield, so there are still interactions and responses which are detrimental to the host, and which require biochemical explanations. Other intriguing questions include whether there is any advantage to the virus in adopting, or being forced to adopt, a strategy of not producing visible symptoms, and whether this is related in any mechanistic way to the inability of these viruses to be transmitted by vectors or by mechanical means. Finally, is the disease syndrome associated with cryptic viruses a consequence of their apparent relationship to double-stranded RNA mycoviruses and if so, what are the biochemical mechanisms?

Many studies of viral pathogenesis and symptom formation have involved tobacco mosaic virus (TMV) and tobacco and tomato hosts. Partly, this has been because the virus is stable and readily produced in large amounts, and has quite pronounced effects on the host. Partly, it reflects the diversity of symptoms produced by different strains of TMV. This has been re-inforced by the interest from early times in using this virus for genetic studies. Many of the mutants produced were detected and characterized in terms of the different symptoms they caused. These studies have given a wealth of research tools - and have probably made the problem more complicated by introducing pathways of symptom formation which may never have evolved or survived in nature! Be that as it may, the three final chapters in this section, by Whenham and Fraser; Van Loon *et al.* and Zaitlin *et al.*, discuss a number of mechanisms by which TMV may interact with the plant, and create a series of pathways for symptom development.

Recognition in resistance and virulence

Mechanisms of resistance to viruses are diverse, and probably involve different types of recognition events. Generally, the plant component of the early interaction is poorly understood; some possible models are considered in the chapter on genetics of plant virus interactions in an earlier section. But recently, the use of recombinant DNA techniques has allowed fine genetic analysis of the genomes of RNA viruses, and determinants which affect interactions with host resistance genes can now be located with great precision (Saito *et al.*, 1987; Meshi *et al.*, 1988a,b). Early results have confirmed the diversity of virus-coded molecules likely to be involved in different types of recognition event. The first two chapters in this section, by Culver and Dawson, and by Mundry *et al.* describe recent work locating the viral determinant of virulence against the *N'* gene of *Nicotiana sylvestris* in the coat protein, with a possible modulating influence of the 30 kDa transport protein.

What does seem clear on the plant side is that the resistance mechanism may be either constitutive - already present in the healthy plant - or may be induced after the plant-virus recognition event (Fraser, 1988). These two types of mechanism must involve fundamentally different types of signalling. The latter is thought to involve changes in the pattern of gene expression, and there is now detailed experimental evidence on some of the steps involved. The pathogenesis-related (PR) proteins have provided an excellent model system for studies of the induction of host genes after infection; the necrotic reaction to TMV in tobacco especially has been used as an effective inducer. Two aspects are considered in this book: the molecular structure of the PR protein genes and their promoters, in the chapter by Linthorst *et al.*, and the cascade controls and various signals which may induce not only PR proteins, but also some other types of defence-related proteins, in the chapter by Fritig *et al.* Paradoxically, although

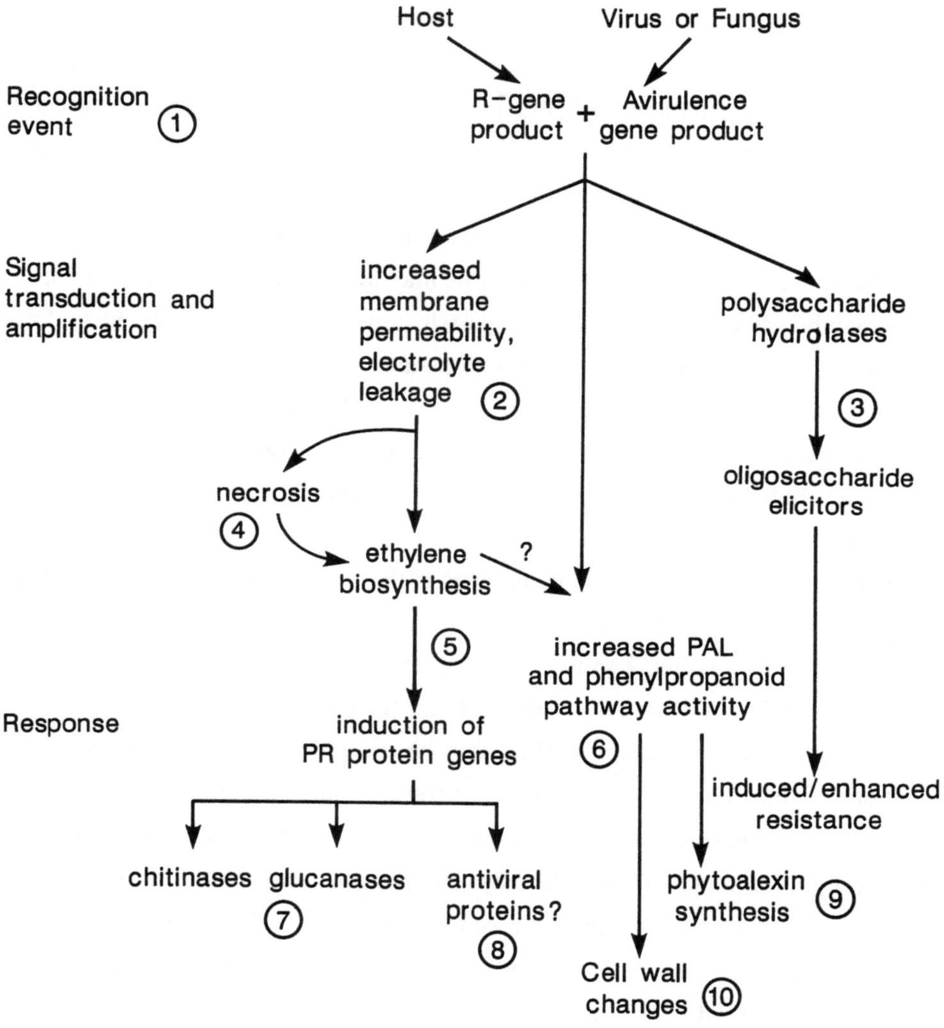

Figure 1. Cascade control of metabolic changes after an initial recognition event between a resistant plant host and an avirulent strain of virus or fungus. Representative references are indicated by numbers in circles. 1. Ellingboe, 1982; 2. Pennazio and Sapetti, 1982; 3. Fritig *et al.*, 1987 and Kopp *et al.*, 1989; 4. De Laat *et al.*, 1981; 5. Van Loon, 1985; 6. Collendavelloo *et al.*, 1983; 7. Kauffmann *et al.*, 1987; 8. Kassanis *et al.*, 1974; 9. Bailey and Burden, 1973; 10. Favali *et al.*, 1978.

research on PR proteins has focussed on a possible role in virus resistance in the two decades since their discovery (van Loon and Van Kammen, 1970; Gianinazzi *et al.*, 1970), no antiviral activity has been convincingly demonstrated (Fraser, 1982), whereas other functions such as resistance to fungal or bacterial invaders have now been established for several members of the group.

One general point is perhaps worth making in relation to cascade controls. A single pathogen can enter into a recognition event; generally this seems to be controlled in a genetically simple manner by both host and virus. This triggers the cascade of responses, some of which are involved in the operation of the actual resistance mechanism or mechanisms against the virus, and some of which now appear to be involved in defence mechanisms which would be useful against secondary invaders. The battery of defences against the initiating virus may involve a series of measures, some of which may be effective early in pathogenesis, and some later (reviewed by Fraser, 1985). Indeed, there is no reason *a priori* to assume that there should be any single critical defence mechanism against the virus. In a small number of cases, the links in the cascade have been established, as shown in Fig. 1. However, the intermediate signalling steps between the initial recognition event and the observed response have still to be discovered.

A second point to make about cascades is that a series of inverted cascades can be superimposed. Thus taking the induction of PR proteins as a typical response in the cascade shown in Fig. 1, Fig. 2 shows how they can be induced by an inverted cascade from a number of quite disparate biotic and abiotic agencies. It is to be expected that lower down in this inverted cascade, there may be common signalling pathways, and the involvement of ethylene as a signal might represent one such. But it is likely that the initial recognition and response mechanisms for the different inducers may involve quite separate recognition mechanisms, and may send signals down quite different pathways. Superimposing Figs 1 and 2 would give a complex *network* of signalling pathways rather than a straightforward cascade. As the evidence presented by Fritig *et al.* in this volume suggests, the outcome of signalling after an initial recognition response may have to be traced by zig-zagging though a network, rather than simply being propagated downwards in a cascade.

In contrast to the cascade approach, some investigators have tried to isolate single or small numbers of compounds which may be involved in virus resistance, and which may be shown to have an antiviral activity in assay systems (reviewed by Verma and Prasad, 1989). This section of the book contains a chapter by Loebenstein *et al.* on an antiviral protein which may be involved in the local lesion reaction to tobacco mosaic virus in *N* gene tobacco.

This section ends with a description by Harrison *et al.* of a completely different type of resistance, which appears to operate against virus movement. This harks back

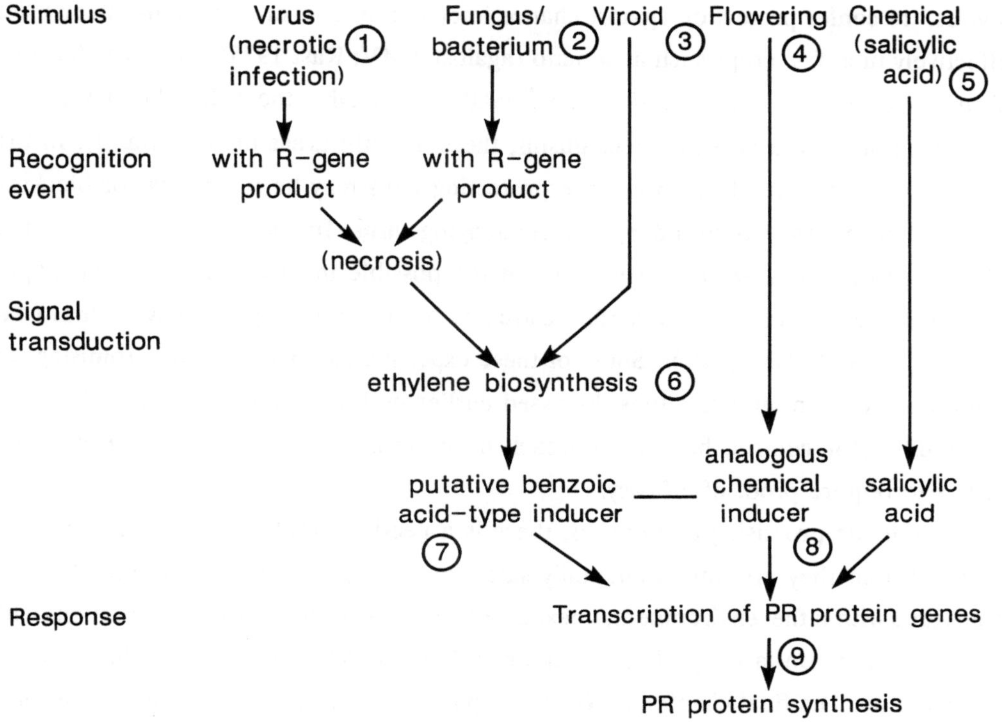

Figure 2. "Inverted cascade" control of metabolic changes, showing how a variety of biotic and abiotic agents can lead to a particular response, in this case the induction of PR protein synthesis. Some of the initial stimuli involve clearly defined recognition events, others may or may not. Representative references are indicated by numbers in circles. 1. Van Loon and Van Kammen, 1970; Gianinazzi *et al.,* 1970; 2. Ahl *et al.,* 1981; 3. Conejero *et al.,* 1979; 4. Fraser, 1981; 5. White, 1979; 6. De Laat and Van Loon, 1983; 7. Van Loon, 1983; 8. Cleland and Ajani, 1974; 9. Hooft van Huijsduijnen *et al.,* 1985.

to the earlier discussion of virus movement, and emphasizes the value of broadening the number of types of resistance mechanism studied.

Exploiting signalling in crop protection

An ultimate practical aim of being able to understand signalling in plant-virus interactions is to exploit this knowledge in crop protection (Fraser, 1989). If the nature of the recognition and signalling pathways is understood, altered or false signals may be sent to induce resistance mechanisms or to interfere with virus replication at some stage. The final section of this book discusses some of the established mechanisms, and some new opportunities offered by modern methods of genetic manipulation.

Cross protection, in which a crop is deliberately inoculated with a mild strain of a virus, to confer protection against chance infection by a severe strain, has been used effectively in a few crops such as tomato (against TMV; Rast 1975) and citrus (against citrus tristeza virus; Costa and Müller, 1980). Although there has been vigorous discussion about the possible mechanisms, there are still areas of uncertainty. In the context of the theme of this book, the protecting virus may be saturating or blocking a signalling pathway required by the challenging virus. In the first chapter on the phenomenon, Urban *et al.* review some of the possible mechanisms. The next paper deals with recent attempts to confer resistance by transforming plants with viral coat protein genes (Nelson *et al.*). Some of these experiments appear highly promising for field use: work on viral satellites discussed earlier by Baulcombe *et al.* may also offer possibilities for control, but the approach using antisense RNAs to viral RNAs still seems to require proof of efficacy.

For the promising approaches, there is a need to conduct parallel experiments to check that they are environmentally safe, from the point of view of spread of the transgene within the environment, unexpected and possibly deleterious interactions with the target and non-target pathogens, and possible long-term effects on pathogen/plant co-evolution. The final chapter, by Hull, couples a discussion of these issues with some wider aspects of the use of plant virus components as gene vectors and promoters. These may be considered as further examples of virus-plant signalling.

EPILOGUE: WHERE NEXT?

The Workshop, and the papers in this book, record advances on a broad front in understanding signalling, recognition and response as key processes in plant-virus interactions. In his summing up at the end of the Workshop, Bryan Harrison highlighted a number of areas where particularly exciting progress has been made. These included:

1. the analysis of viral gene function by mutation;
2. work on movement functions specified by the virus, and how these may interact with the host;
3. studies of resistance-breaking isolates of viruses;
4. the move towards giving greater weight to studies of the plant rather than the virus alone, and
5. the impact of transgenic plants, not only in crop protection, but also as tools

for investigating plant-virus interactions in new ways.

It is to be hoped that there will be continuing developments in these areas. But one of the major gaps in this panoply of progress is the failure to date to isolate specific host resistance genes and thus to go on to determine their modes of action. This, it must be said, is a problem also facing those who are investigating genes for resistance to bacteria and fungi. However, although the attendant problems are great, it is likely that resistance genes will be isolated in the not-too-distant future, given the scale of resource being devoted to such work. Fine genetic mapping of viral virulence, and an understanding of the relationship between acquisition of resistance-breaking behaviour, and the alteration in the function of the viral protein (or RNA) involved, may offer clues in the hunt for the product coded by the host resistance gene. It is to be hoped that some future NATO Workshop on plant-virus interactions will include detailed treatment of how host resistance gene products and viruses interact, and a full explanation of the signalling of the subsequent resistance response.

REFERENCES

Ahl P, Benjama A, Samson R, Gianinazzi S (1981) New host proteins (b-proteins) induced together with resistance to a secondary infection following a bacterial infection in tobacco. Phytopathol Z 102:201-212

Bailey JA (ed) (1986) Biology and molecular biology of plant pathogen interactions. Springer Verlag, Berlin

Bailey JA, Burden RS (1973) Biochemical changes and phytoalexin accumulation in *Phaseolus vulgaris* following cellular browning caused by tobacco necrosis virus. Physiol Plant Pathol 3:171-177

Cleland CR, Ajami A (1974) Identification of the flower-inducing factor isolated from aphid honeydew as being salicylic acid. Plant Physiol 54:904-906

Collendavelloo J, Legrand M, Fritig B (1983) Plant disease and the regulation of enzymes involved in lignification. Increased rate of de novo synthesis of the three tobacco O-methyltransferases during the hypersensitive response to infection by tobacco mosaic virus. Plant Physiol 315:601-603

Conejero V, Picazo I, Segado P (1979) Citrus exocortis viroid (CEV): protein alterations in different hosts following viroid infection. Virology 97:454-456

Costa AS, Müller GW (1980) Tristeza control by cross protection:- A US-Brazil cooperative success. Plant Dis 64:538-541

Davies JW (ed) (1985) Molecular plant virology. Vol. 1. CRC Press, Boca Raton.

De Laat AMM, Van Loon LC, Vonk CR (1981) Regulation of ethylene biosynthesis in virus-infected tobacco leaves. I. Determination of the role of methionine as the precursor of ethylene. Plant Physiol 68:256-261

De Laat AMM, Van Loon LC (1983) The relationship between stimulated ethylene production and symptom expression in virus-infected tobacco leaves. Physiol Plant Pathol 22:261-273

Ellingboe AH (1982) Genetic aspects of active defense. In: Wood RKS (ed) Active defense mechanisms in plants. Plenum Press, New York, London, p 179

Favali MA, Conti GG, Bassi M (1978) Modifications of the vascular bundle ultrastructure in the "resistance zone" around necrotic lesions induced by tobacco mosaic virus. Physiol Plant Pathol 13:247-251

Fraser RSS (1981) Evidence for the occurrence of the 'pathogenesis related' proteins in leaves of healthy tobacco plants during flowering. Physiol Plant Pathol 19:69-76

Fraser RSS (1982) Are 'pathogenesis-related' proteins involved in acquired systemic resistance of tobacco plants to tobacco mosaic virus? J Gen Virol 58:305-313

Fraser RSS (1985) Mechanisms involved in genetically controlled resistance and virulence: virus diseases. In: Fraser RSS (ed) Mechanisms of resistance to plant diseases. Nijhoff, Dordrecht, p 143

Fraser RSS (1987) Biochemistry of virus-infected plants. Research Studies Press, Letchworth; Wiley & Sons, New York

Fraser RSS (1988) Virus recognition and pathogenicity: implications for resistance mechanisms and breeding. Pestic Sci 23:267-275

Fraser RSS (1989) Control of plant viruses. Plants Today 2:100-105

Fritig B, Kauffmann S, Dumas B, Geoffroy P, Kopp M, Legrand M (1987) Mechanism of the hypersensitivity reaction of plants. In: Evered D, Harnett S (eds) Plant resistance to viruses. Ciba Foundation Symposium 133. Wiley, Chichester, p 92

Gianinazzi S, Martin C, Valee J-C (1970) Hypersensibilité aux virus, temperature et proteines solubles chez le *Nicotiana* Xanthi n.c. Apparition de nouvelles macromolecules lors de la repression de la synthèse virale. C R Hebd Seances Acad Sci 270D:2383-2386

Goldbach RW (1987) Genome similarities between plant and animal RNA viruses. Microbiol Rev 4:197-202

Hooft van Huijsduijnen RAM, Cornelissen BJC, van Loon LC, van Boom JH, Tromp M, Bol JF (1985) Virus-induced synthesis of messenger RNAs for precursors of pathogenesis-related proteins in tobacco. EMBO J 4:2167-2171

Hull R (1989) The movement of viruses in plants. Annu Rev Phytopathol 27:213-40

Jacquemand M, Leraux JP (1982) Cucumber mosaic virus-associated RNA-5. 2. Virus-satellite RNA relationships in various host plants. Agronomie 2:55-62

Kassanis B, Gianinazzi S, White RF (1974) A possible explanation of the resistance of virus-infected tobacco plants to second infection. J Gen Virol 23:11-16

Kauffmann S, Legrand M, Geoffroy P, Fritig B (1987) Biological function of 'pathogenesis-related' proteins: four PR proteins of tobacco have 1,3- -glucanase activity. EMBO J 6:3209-3212

Kopp M, Rouster J, Fritig B, Darvill A, Albersheim P (1989) Host-pathogen interactions XXXII. A fungal glucan preparation protects Nicotianae against infection by viruses. Plant Physiol, in press

Legrand M, Kauffmann S, Geoffroy P, Fritig B (1987) Biological function of "pathogenesis-related" proteins: four tobacco PR-proteins are chitinases. Proc Natl Acad Sci USA 84:6750-6754

Lugtenberg B (ed) (1986) Recognition in microbe-plant symbiotic and pathogenic interactions. Springer Verlag, Berlin

Meshi T, Motoyoshi F, Adachi A, Watanabe Y, Takamatsu N, Okada Y. Two concomitant base substitutions in the putative replicase genes of tobacco mosaic virus confer the ability to overcome the effects of a tomato resistance gene, Tm-1. EMBO J 7:1575-1580

Mouchés C, Candresse T, Bove JM (1984) Turnip yellow mosaic virus RNA-replicase

contains host and virus-encoded subunits. Virology 134:78-90
Pennazio S, Sapetti C (1982) Electrolyte leakage in relation to viral and abiotic stresses inducing necrosis in cowpea leaves. Biol Plant 24:218-255
Rast ATB (1975) Variability of tobacco mosaic virus in relation to control of tomato mosaic in glasshouse crops by resistance breeding and cross protection. Agric Res Rep (Neth) 834
Saito T, Meshi T, Takamatsu N, Okada Y (1987) Coat protein gene sequence of tobacco mosaic virus encodes a host response determinant. Proc Natl Acad Sci USA 84:6074-6079
Scannerini S, Smith D, Bonfaute-Fasolo, P, Gianinazzi-Pearson V (eds) (1988) Cell to cell signals in plant, animal and microbial symbiosis. Springer Verlag, Berlin
Van Loon LC (1983) The induction of pathogenesis-related proteins by pathogens and specific chemicals. Neth J Plant Pathol 89:265-273
Van Loon LC (1985) Pathogenesis-related proteins. Plant Mol Biol 4:111-116
Van Loon LC, Van Kammen A (1970) Polyacrylamide disc electrophoresis of the soluble leaf proteins from *N tabacum* var Samsun and Samsun NN. II. Changes in protein constitution after infection with TMV. Virology 40:199-211
Verma HN, Prasad V (1989) Virus inhibitors and inducers of resistance: potential avenues for biological control of viral diseases, In: Mukerji KG, Tewari JP (eds) Biocontrol of plant diseases, Vol. III. CRC Press, Boca Raton, in press
White RF (1979) Acetylsalicylic acid (aspirin) induces resistance to tobacco mosaic virus in tobacco. Virology 99:410-412
Wilson TMA (1985) Nucleocapsid disassembly and early gene expression by positive-strand RNA viruses. J Gen Virol 66:1201-1207

RECOGNITION IN RESISTANCE TO BACTERIA

Alan Vivian
Molecular Genetics Group
Science Department
Bristol Polytechnic
Coldharbour Lane
Bristol
BS16 1QY
UK

ABSTRACT

Recognition in resistance to plant-pathogenic bacteria involves two classes of genes in the pathogen: avirulence (*avr*) genes and harp (*hrp*) genes. Avirulence genes are matched with specific resistance genes in the host, resulting in incompatibility, the outcome being recognized as a hypersensitive response in the host. Harp genes are involved in the production of a hypersensitive response in incompatible interactions and in pathogenicity in compatible interactions.

INTRODUCTION

Bacteria that cause disease in plants do not appear to have developed any overt specialized means of gaining entry to their plant hosts (in contrast to many phytopathogenic fungi). Rather, entry occurs by way of natural openings, such as stomata, or through wounds. There is some evidence to suggest that in many instances, involving pathogens that are otherwise epiphytic, considerable populations can develop at sub-stomatal locations within the plant (Mansvelt and Hattingh, 1989). The subsequent progress of these invasions depends upon the recognition (or lack of it) of the pathogen by the host. In the case of most phytopathogenic bacteria, disease results from the failure of the host to recognize the pathogen. However, non-pathogenic bacteria such as *Escherichia coli* or *Pseudomonas fluorescens*, which cause no visible symptoms after infiltration into the leaves of tobacco (*Nicotiana tabacum*), are also "not recognized". Presumably in the latter case this is due to the absence of specific genes

NATO ASI Series, Vol. H 41
Recognition and Response in Plant-Virus Interactions
Edited by R.S S Fraser

for pathogenicity, which are present in all phytopathogenic bacteria.

This paper will include examples from a number of groups of phytopathogenic bacteria, and here it may be useful to describe these briefly. Many studies have involved pathogens belonging to the *Pseudomonas syringae* group. Most of these bacteria are epiphytic, and cause a variety of leaf-spotting diseases. This bacterial "species" is subdivided into a number of recognized pathovars, based on their host of origin and host range (Dye *et al.*, 1980; Young *et al.*, 1978). *P. s.* pv. *syringae* embraces a wide range of strains, causing disease symptoms in a wide range of host species. Other pathovars are restricted to a much narrower host range. Those which have been the subject of studies mentioned here include pv. *glycinea* which attacks soybean (*Glycine max*); pv. *phaseolicola* which attacks common bean (*Phaseolus vulgaris*); pv. *pisi* which attacks pea (*Pisum sativum*); pv. *tabaci* which causes wildfire on tobacco (*Nicotiana tabacum*) and pv. *tomato* which attacks tomato (*Lycopersicon esculentum*).

Xanthomonas campestris includes a very extensive group of pathovars, two of which have been used for studies of avirulence genes. These are *X. c.* pv. *malvacearum* which causes angular leaf spot on cotton (*Gossypium hirsutum*), and pv. *vesicatoria* which causes spot disease of pepper (*Capsicum annuum*). The possibility of positive factors determining host range has been investigated in *X. c.* pv. *translucens*, which causes watersoaking in grasses and cereals.

Classified as a separate species, *P. solanacearum* causes bacterial wilt and is subdivided into three races: race 1 attacks a wide range of hosts, while the other two races are more restricted in their host ranges (Lozano and Sequeira, 1970). Pathovars of *P. syringae* also frequently exhibit further subdivision of isolates into races, distinguished on the basis of differential interaction with a range of cultivars of a single host species. Thus *P. s.* pv. *glycinea* comprises nine recognized races (Cross *et al.*, 1966; Thomas and Leary, 1980; Fett and Sequeira, 1981), *P. s.* pv. *phaseolicola* three races (Taylor *et al.*, 1987) and *P. s.* pv. *pisi* currently seven (Taylor *et al.*, 1989; JD Taylor, personal communication).

Incompatibility between pathogen and host may involve specific recognition at two levels: race/cultivar-specific interactions or pathovar/host species-specific interactions. The outcome of this incompatibility is usually accompanied by a similar reaction in terms of visible symptoms in the affected host plant, namely as a hypersensitive response (HR) reaction (Klement, 1982). HR is a localized defence mechanism occurring at or close to the site of inoculation or invasion of the pathogen.

Rapid necrosis of plant cells is followed by the accumulation of phytoalexins and other factors, which together result in the containment and death of the invading bacteria.

There is now considerable experimental evidence to show that in many cases the basis for the recognition event which triggers the HR is due to matching dominant avirulence (*avr*) genes in the pathogen and dominant resistance (R) genes in the host. Disease occurs if this genetically-determined (gene-for-gene) recognition is absent. The section about avirulence genes in this paper will review the experimental evidence for the activity of these genes in race/cultivar and also pathovar/host species interactions.

Several investigators have reported the production of HR-deficient bacterial mutants, following transposon and other types of mutagenesis (Boucher *et al.,* 1985; Malik *et al.,* 1987; Panopoulos *et al.,* 1985). Often these mutants were initially isolated as non-pathogenic on their normal hosts, and were subsequently found to be also HR-deficient on non-host plants. Since the ability to induce HR is strictly confined to a fairly wide range of plant pathogenic bacteria, it was always likely that the basis of HR was closely related to the pathogenic nature of the bacterium. Further work, by two groups in particular, has led to the identification of clusters of genes involved in both HR and pathogenicity: these genes are designated *hrp* or phonetically "harp" genes. The second section of this paper will consider the current state of knowledge of harp genes and their likely relationship to the function of avirulence genes in both *Pseudomonas* and *Xanthomonas.*

The final section will consider the evidence for a totally opposite system of recognition, which seems to operate in some strains of both *Pseudomonas* and *Xanthomonas*, involving recognition which leads to disease rather than to resistance.

AVIRULENCE (*avr*) GENES

The first clear demonstration of a gene-for-gene relationship in the specificity of bacterial plant pathogens was provided by Stall *et al.* (1986) and Swanson *et al.* (1988), who showed that an avirulence gene ($avrBs_1$) cloned from a plasmid in a race 2 strain of *X. campestris* pv. *vesicatoria* interacts specifically with a single resistance gene (Bs_1) in the pepper cultivar ECW10R. This was rapidly followed by the cloning of $avrB_1$ from *P. s.* pv. *glycinea* race 1, which was shown to interact specifically with a single resistance gene *Rpg1* in soybean cultivars Harosoy and Norchief (Staskawicz *et al.,* 1987). A third example has been the identification of *avrAspi1* in *P. s.* pv. *pisi* race 2, which has been shown to interact with resistance gene R2 in pea cvs Early Onward and Fortune (Vivian *et al.,* 1989).

Although precise information on the corresponding host genotypes is still lacking for a number of other *avr* genes, studies of *avr* genes are nevertheless contributing valuable DNA sequence data, which may enable some clues to emerge about the nature and mode of action of the bacterial gene products they specify. Considerable variation appears to exist, if the examples published to date are indicative. Sequencing of the first avirulence gene (*avrA*), cloned from a race 6 strain of *P. s.* pv. *glycinea*, has revealed that it occupies an open reading frame (ORF) of 2721 bp, and specifies a protein of 100 kDa which was expressed in maxicells (Napoli and Staskawicz, 1987). However, since *avr* genes clearly are involved in recognition, it was supposed that their gene products might well be located at or near the cell surface, thus permitting direct contact with the plant. There has therefore been interest to examine "computer-translated" amino-acid sequences to look for regions of hydrophobicity, which might indicate a role for the gene product as either an outer or inner membrane protein. Cloning from *P. s.* pv. *glycinea* and sequencing of two further genes, *avrB* and *avrC*, both smaller than *avrA*, has failed to detect significant hydrophobic regions or sequences that resemble known signal sequences for transport of the gene product across the membrane (Tamaki *et al.,* 1988).

The avirulence gene *avrBs*$_1$ from *X. c.* pv. *vesicatoria* was identified and sequenced and found to comprise two ORFs, the second of which (1334 bp) encoded a protein calculated to be about 50 kDa and which showed considerable amino-acid sequence homology with *avrA* from *P. s.* pv. *glycinea* (Ronald and Staskawicz, 1988). ORF 2 was apparently dependent on ORF 1, which probably provides transcriptional readthrough to permit expression of ORF 2.

The location of avirulence genes within the bacterium, and also whether plasmids (which occur frequently in many plant pathogenic bacteria) play a significant role in specificity, have been matters of speculation for some time. The discovery that *avrBs*$_1$ was located on a plasmid, which also determined resistance to copper in *X. c.* pv. *vesicatoria*, has indicated the complex relationships that may result from accepted control measures, such as the application of copper sprays (Stall *et al.,* 1986). It is already clear that avirulence genes can be either chromosome or plasmid-borne (Table 1). Studies with the "cryptic" plasmid, pMC7105 in *P. s.* pv. *phaseolicola* have demonstrated the potential for complex integration/excision events of plasmid and chromosome (Szabo

Table 1. *Avirulence genes in bacterial pathogens*

Gene	Bacterium	Reference
Chromosomal		
avrA	*P. syringae* pv. *glycinea*	Napoli and Staskawicz (1987)
avrB	*P. syringae* pv. *glycinea*	Tamaki *et al.* (1988)
avrAsph1	*P. syringae* pv. *phaseolicola*	Hitchin *et al.* (1989)
avrAspi1	*P. syringae* pv. *pisi*	Vivian *et al.* (1989)
$avrB_2$		
$avrB_3$		
$avrB_6$	*X. c.* pv. *malvacearum*	Gabriel *et al.* (1986)
$avrB_N$		
$avrB_{In}$		
avrRxv	*X. c.* pv. *vesicatoria*	Whalen *et al.* (1988)
$avrBs_2$	*X. c.* pv. *vesicatoria*	Whalen *et al.* (1988)
Plasmid-borne		
avrC	*P. s.* pv. *glycinea*	Tamaki *et al.* (1988)
avrBspi2	*P. s.* pv. *pisi*	A Bavage and A Vivian, unpublished results
$avrBs_1$	*X. c.* pv. *vesicatoria*	Swanson *et al.* (1988)
$avrBs_3$	*X. c.* pv. *vesicatoria*	Whalen *et al.* (1988)
Location not reported		
avr I		
avr II	*P. s.* pv. *tomato*	Kobayashi *et al.* (1989)
avr III		

and Mills, 1984a). These have been associated with repetitive DNA sequences, some of which may resemble insertion sequences (Szabo and Mills, 1984b). A recently reported insertion sequence, IS476, has been shown to be responsible for spontaneous race-change mutants involving *avrBs*$_1$ (Kearney *et al.,* 1988).

The cloning from race 0 of *P. s.* pv. *glycinea* of two avirulence genes, *avrB* from the chromosome and *avrC* from a plasmid, showed that there was up to 80% homology of amino-acid sequence in the carboxyl ends of the proteins, together with an overall homology of 46%, yet both genes showed distinct HR phenotypes (Tamaki *et al.,* 1988). The homology raises the possibility that these genes may have arisen by duplication of an ancestral sequence, and significantly, both genes were flanked by repeated DNA sequences (Staskawicz *et al.,* 1987). Normally there is a strong position-3 bias for G+C in *Pseudomonas*, but both genes had relatively low overall G+C contents (*avrB* 46%, and 40% at position 3; *avrC* 47%, and 40% at position 3), which might indicate the recent acquisition of both of these genes in *P. s.* pv. *glycinea* (Tamaki *et al.,* 1988).

In *P. s.* pv. *pisi*, it seems likely that avirulence genes are located both chromosomally and on plasmids. The race 2 gene *avrAspi1* is located on the chromosome in strain 203 (Vivian *et al.,* 1989), but a recently discovered gene *avrBspi2* in a race 3 strain 870A is located on a plasmid (AD Bavage and A Vivian, unpublished data).

Hitchin *et al.,* (1989) have recently reported the cloning of an avirulence gene from *P. s.* pv. *phaseolicola* race 3 which confers avirulence on races 1 and 2 towards the bean cv. Tendergreen. Introduction of the *avr* library clone into either *P. s.* pv. *syringae* or *X. c.* pv. *phaseoli* did not affect lesion formation in cv. Tendergreen. These authors also quote a report (Patil and Shintaku, 1986) of the cloning of a similar gene from race 1 conferring avirulence towards bean cv. Red Mexican.

The cloning of five avirulence genes from a single strain of *X. c.* pv. *malvacearum*, a pathogen of cotton, indicated that all were chromosomal, three being linked on a 32 kb DNA fragment, and the presence for at least some of these avirulence genes of colinear DNA sequences in a virulent strain (lacking all avirulence genes). This raises the possibility that some avirulence genes may have corresponding virulence alleles (Gabriel *et al.,* 1986). Evidence from other systems (*avrC* and race 6 in *P. s.* pv. *glycinea*; Staskawicz *et al.,* 1987; *avrAspi1* in races 3, 4, 5 and 7 or *P. s.* pv. *pisi;* L Mur and A Vivian, unpublished) tends to confirm this, although *avrA, B* apparently have no alleles (i.e. conserved homologous sequences in races that do not express the respective

Avr phenotype) in *P. s.* pv. *glycinea* (Staskawicz *et al.,* 1984; Kobayashi *et al.,* 1989).

The question of whether pathovar/host species incompatibility is determined by matching avirulence and resistance genes in a similar fashion to race/cultivar incompatibility was addressed by Whalen *et al.* (1988), working with *X. c.* pv. *vesicatoria.* This bacterium is the cause of leaf spot disease on pepper and tomato, and a clone was isolated from a tomato race 1 gene library that converted *X. c.* pv. *phaseoli* to avirulence on bean cv. Sprite, but not on cv. Bush Blue Lake. The gene was designated *avrRxv*, and was shown to inhibit disease production by some *X. campestris* pathovars on their normally susceptible hosts. Resistance to *avrRxv* segregated in beans as a single incompletely-dominant gene, designated *Rxv*. Significantly, homologous DNA sequences were only detected in other races of *X. c.* pv. *vesicatoria*, and not in other pathovars of *X. campestris*.

An analogous type of gene was also found by Kobayashi *et al.* (1989), who introduced cosmid clones from a gene library of *P. s.* pv. *tomato* into *P. s.* pv. *glycinea* and tested the transconjugants on soybean cultivars that were compatible with the original *P. s.* pv. *glycinea* strain. The aim was to seek to identify genes from the tomato pathogen, which might be responsible for eliciting HR in soybean. Three classes of avirulence gene were detected. Whereas *P. s.* pv. *tomato* elicits HR on all soybean cultivars, when transconjugants carrying individual avirulence determinants from *P. s.* pv. tomato were inoculated into a range of soybean cultivars, some cultivars produced an HR, while others were susceptible. The class I avirulence gene produced a differential cultivar response that was identical to that obtained for race 6 of *P. s.* pv. *glycinea.* Collectively the three classes of avirulence determinant accounted for the necessary span to confer avirulence towards all soybean cultivars. The class I avirulence gene was homologous to the previously cloned race/cultivar gene *avrA* from race 6 of *P. s.* pv. *glycinea*, suggesting that single avirulence genes may function at different levels of specificity in different pathovars (Kobayashi *et al.,* 1989).

Further evidence in support of this potential dual role for some avirulence genes comes from the work at Bristol and Wye, where *avrAspi1* has been introduced into *P. s.* pv. *phaseolicola* and shown to confer avirulence towards all bean cultivars. A single Tn5 insertion in *avrAspi1* which abolishes its race/cultivar avirulence in *P. s.* pv. *pisi*, also abolishes the pathovar/host species avirulence in *P. s.* pv. *phaseolicola* (J Fillingham, JW Mansfield, LAJ Mur and A Vivian, unpublished results).

HARP (*hrp*) GENES

Eight mutants of *P. s.* pv. *phaseolicola* obtained after mutagenesis with Tn5 were found to have lost the ability to induce HR on tobacco and other host plants. Six of these mutants were non-pathogenic on a susceptible bean cultivar, and failed to induce HR in resistant bean cultivars. A single cosmid from a *P. s.* pv. *phaseolicola* gene library restored wild type phenotype to seven of the eight mutants, indicating a clustering of genes involved in HR (Lindgren *et al.*, 1986). These genes, designated *hrp*, showed the presence of identical sequences in *P. s.* pv. *glycinea* races 0, 1, and 6, and pv. *tabaci* strain BR-2; and strong cross-hybridization with pv. *glycinea* races 4 and 5, with two further strains of pv. *tabaci* and with pv. *angulata*. Weaker homology was detected in pvs *coronafaciens*, *lachrymans*, *pisi*, *syringae* and *tomato*, but no homology was detected with *X. c.* pv. *glycinea* and pv. *phaseoli* (Lindgren *et al.*, 1988). This contrasts with the results of Boucher *et al.* (1987), using *hrp* genes from *P. solanacearum*, which are clustered in a similarly-sized region (approx 17-20 kb) to the *hrp* cluster from *P. s.* pv. *phaseolicola*. Boucher *et al.* (1987) showed that these genes have sequence homologies with eight pathovars of *X. campestris*.

Lindgren *et al.* (1988) transferred a Tn5-induced Hrp^- mutation from *P. s.* pv. *phaseolicola* into the genome of *P. s.* pv. *tabaci* and three races of *P. s.* pv. *glycinea* by marker-exchange. The resulting progeny were no longer pathogenic on their respective hosts, nor did they elicit HR on non-host plants; they were however restored to Hrp^+ by acquisition of a recombinant plasmid carrying the appropriate DNA from *P. s.* pv. *phaseolicola*. The marker-exchange mutants for races 0 and 5 of *P. s.* pv. *glycinea* carried different avirulence genes, but these were unable to elicit HR on incompatible soybean cultivars, indicating that the expression of *avr* genes is dependent upon a fully functional *hrp* region (Lindgren *et al.*, 1988).

Mukhopadhyay *et al.* (1988) have sequenced an *hrp* locus in *P. s.* pv. *syringae* contained on a 3.9 kb DNA fragment. This comprised two ORFs and the deduced amino acid sequences of an 83 kDa polypeptide contained features characteristic of known integral membrane proteins.

The *hrp* region in *P. solanacearum* can be deleted following growth in the presence of acridine orange (Boucher *et al.*, 1986). Homology of the *hrp* region was found in 52 strains of *P. solanacearum* tested, and most of these strains were able to give rise to acridine orange-resistant (Acr-r) mutants which showed the pleiotropic Hrp^-

phenotype (Boucher *et al.*, 1988). This includes auxotrophy for methionine and excretion of a brown pigment during growth in rich culture medium (Message *et al.*, 1978) and changes in the structures of exopolysaccharides and lipopolysaccharides (Drigues *et al.*, 1985).

A cosmid, pVir2 from a gene library of the wild type *P. solanacearum* comprising about 25 kb of DNA inserted into pLAFR3, was found to restore all the Tn5-induced mutants, but not the Acr-r mutants (Boucher *et al.*, 1987). In contrast to the *hrp* genes in *P. syringae* pathovars, these mutations lie within an 85 kb deletion in a megaplasmid of greater than 1000 kb (Boucher *et al.*, 1986).

An *hrp* gene cluster has been identified and characterized in a strain of *P. s.* pv. *syringae* isolated from wheat (*Triticum aestivum*). Mutants lacking the ability to induce HR and which have concomitantly lost the ability to cause rapid K^+ ion efflux/H^+ ion influx in suspension-cultured tobacco cells (designated XR^-) were isolated by Tn5-mutagenesis (Baker *et al.*, 1987). A recombinant plasmid, pHIR11, containing a 31 kb insert of *P. s.* pv. *syringae* DNA restored wild type phenotype to all HR^-, XR^- mutants and conferred the ability to induce HR in *P. s.* pv. *tabaci* and in a tomato strain of *P. s.* pv. *syringae* on their normal hosts (tobacco and tomato, respectively). Introduction of pHIR11 into the non-pathogen *P. fluorescens* conferred the ability to elicit a typical HR and XR on tobacco. *E. coli* carrying pHIR11 inoculated at high concentrations in tobacco elicited an HR-like response. These results imply that this plasmid contains virtually all the genetic information necessary for the bacterium to elicit HR in certain host plants. A single Tn*phoA* insertion in pHIR11, which failed to confer HR on *P. fluorescens*, inactivated the HR in both the wheat and tomato strains of *P. s.* pv. *syringae* when marker-exchanged (Huang *et al.*, 1988).

POSITIVE DETERMINANTS OF HOST SPECIFICITY

All of the examples discussed above provide evidence for the existence of active genetic determinants of recognition, which result in the induction of plant defence systems. In other words, the bacteria cause substances to be produced that, directly or indirectly, limit their host range as pathogens. Disease results from an absence of (negative) recognition. It might be supposed, however, that it could be favourable to a pathogen's survival to extend its host range, wherever possible by the acquisition of functional determinants to enable it to evade the resistance of its host(s). Both Hitchin *et al.* (1989) and Staskawicz *et al.* (1984), in the course of detecting avirulence genes, have sought and failed to find evidence for such positively-acting genes which might extend host range. However, recent papers by Mellano and Cooksey (1988) and by Ma *et al.* (1988) have indicated that such positively-acting genes do exist. Clearly, it will be important to discover the relationship between these contrary systems of host range determination.

Mellano and Cooksey (1988) investigated *X. c.* pv. *translucens*, a pathogen of cereals and grasses. Individual strains within the pathovar were found to differ in their host range and co-inoculation of wide- and narrow-host-range strains resulted in a wide-host-range reaction. Mutants were obtained from a wide-host-range strain which had lost pathogenicity towards barley, wheat, triticale or rye, without any reduction of pathogenicity towards other hosts. The authors concluded that these results indicate the presence of independent factors whose function is required for pathogenicity on a particular host or hosts. Clearly it will be important to attempt to isolate such genes from gene libraries of *X. c.* pv. *translucens*.

Further evidence comes from the work of Ma *et al.* (1988) on *P. solanacearum* strains. Cosmid clones from a gene library of a race 3 strain, pathogenic on potato, but non-pathogenic on peanut, were introduced into a race 1 strain which was pathogenic on peanut. The aim was to identify *avr* determinants for avirulence towards peanut: none were found. However, when the converse experiment was undertaken, introducing a gene library from the race 1 strain into the race 3 strain, a cosmid was found that conferred the ability to cause wilting on peanut in the race 3 strain. Ma *et al.* (1988) have experiments underway to attempt to identify and characterize the gene or genes involved, and the outcome cannot fail to be of great interest to all of us engaged in unravelling the genetic basis of bacterial diseases in plants.

CONCLUSIONS

It can be seen that there is diversity in both avirulence genes and harp genes, between different strains or groups of plant pathogenic bacteria. At the same time there is also evidence of homology, sometimes extending across relatively unrelated groups. The evidence that some avirulence genes may be relatively recent acquisitions by their bacterial hosts, argues for a high degree of potential genetic mobility of these determinants, with important consequences for disease control. The potential ecological and evolutionary significance of these findings pose many questions, and should provide a new impetus to many aspects of practical plant pathology.

While avirulence genes may provide the basis of both race/cultivar and pathovar/host-species specificity, these genes seem to be dependent upon a fully functional set of harp genes. Whether some groups such as *P. solanacearum* have a fundamentally different and possibly mutually exclusive system of recognition for determining host range, based on positively-acting factors, must remain a subject for speculation and further experimentation.

ACKNOWLEDGEMENTS

I am indebted to Adrian Bavage and John Taylor for helpful comments on the manuscript.

REFERENCES

Baker CJ, Atkinson MM, Collmer A (1987) Concurrent loss in Tn5 mutants of *Pseudomonas syringae* pv. *syringae* of the ability to induce the hypersensitive response and host plasma membrane K^+/H^+ exchange mechanism. Phytopathology 77:1268-1272

Boucher C, Barberis P, Trigalet A, Demery D (1985) Transposon mutagenesis of *Pseudomonas solanacearum*: isolation of Tn5-induced avirulent mutants. J Gen Microbiol 131:2449-2457

Boucher CA, Martinel A, Barberis P, Allgoing G, Zischek C (1986) Virulence genes are carried by a megaplasmid of the plant pathogen *Pseudomonas solanacearum*. Mol Gen Genet 205: 270-275

Boucher C, Van Gijsegem F, Barberis PA, Arlat M, Zischek C (1987) *Pseudomonas solanacearum* genes controlling both pathogenicity on tomato and hypersensitivity on tobacco are clustered. J Bacteriol 169:5626-5632

Boucher CA, Barberis PA, Arlat M (1988) Acridine orange selects for deletion of *hrp* genes in all races of *Pseudomonas solanacearum*. Mol Plant-Microbe Interact 1:282-288

Cross JE, Kennedy BW, Lambert JW, Cooper RL (1966) Pathogenic races of the bacterial blight pathogen of soybeans, *Pseudomonas glycinea*. Plant Dis Rep 50:557-560

Drigues P, Demery-Lafforgue D, Trigalet A, Dupin P, Samain D, Asselineau J (1985) Comparative studies of lipopolysaccharide and exopolysaccharide from a virulent strain of *Pseudomonas solanacearum* and from three avirulent mutants. J Bacteriol 162:504-509

Dye DW, Bradbury JF, Goto M, Hayward AC, Lelliott RA, Schroth MN (1980) International standards for naming pathovars of phytopathogenic bacteria and a list of pathovar names and pathotype strains. Rev Plant Pathol 59:153-168

Fett WF, Sequeira L (1981) Further characterization of the physiologic races of *Pseudomonas glycinea*. Can J Bot 59: 283-287

Gabriel DW, Burges A, Lazo GR (1986) Gene-for-gene interactions of five cloned avirulence genes from *Xanthomonas campestris* pv. *malvacearum* with specific resistance genes in cotton. Proc Natl Acad Sci USA 83:6415-6419

Hitchin FE, Jenner CE, Harper S, Mansfield JW, Barber CE, Daniels MJ (1989) Determinant of cultivar specific avirulence cloned from *Pseudomonas syringae* pv. *phaseolicola* race 3. Physiol Mol Plant Pathol 34:309-322

Huang H-C, Schuurink R, Denny TP, Atkinson MM, Baker CJ, Yucel I, Hutcheson SW, Collmer A (1988) Molecular cloning of a *Pseudomonas syringae* pv. *syringae* gene cluster that enables *Pseudomonas fluorescens* to elicit the hypersensitive response in tobacco plants. J Bacteriol 170:4748-4756

Kearney B, Ronald PC, Dahlbeck D, Staskawicz BJ (1988) Molecular basis for evasion of plant host defence in bacterial spot disease of pepper. Nature 332:541-543

Klement Z (1982) Hypersensitivity. In: Mount MS Lacy GH (eds) Phytopathogenic prokaryotes. Vol. 2. Academic Press, New York p 149

Kobayashi DY, Tamaki SJ, Keen NT (1989) Cloned avirulence genes from the tomato pathogen *Pseudomonas syringae* pv. *tomato* confer cultivar specificity on soybean. Proc Natl Acad Sci USA 86:157-161

Lindgren PB, Peet RC, Panopoulos NJ (1986) Gene cluster of *Pseudomonas syringae* pv. "*phaseolicola*" controls pathogenicity of bean plants and hypersensitivity on nonhost plants. J Bacteriol 168:512-522

Lindgren PB, Panopoulos NJ, Staskawicz BJ, Dahlbeck D (1988) Genes required for pathogenicity and hypersensitivity are conserved and interchangeable among pathovars of *Pseudomonas syringae*. Mol Gen Genet 211:499-506

Lozano JC, Sequeira L (1970) Differentiation of races of *Pseudomonas solanacearum* by a leaf infiltration technique. Phytopathology 60:833-838

Ma Q-S, Chang M-F, Tang J-L, Feng J-X, Fan M-J, Han B, Liu T (1988) Identification of DNA sequences involved in host specificity in the pathogenesis of *Pseudomonas solanacearum* strain T2005. Mol Plant-Microbe Interact 1:169-174

Malik AN, Vivian A, Taylor JD (1987) Isolation and partial characterization of three classes of mutant in *Pseudomonas syringae* pathovar *pisi* with altered behaviour towards their host, *Pisum sativum*. J Gen Microbiol 133:2393-2399

Mansvelt EL, Hattingh MJ (1989) Scanning electron microscopy of invasion of apple leaves and blossoms by *Pseudomonas syringae* pv. *syringae*. Appl Environ Microbiol 55:533-538

Mellano VJ, Cooksey DA (1988) Development of host range mutants of *Xanthomonas campestris* pv. *translucens*. Appl Environ Microbiol 54:884-889

Message B, Boistard P, Pitrat M, Schmit J, Boucher CA (1978) A new class of fluidal avirulent mutants of *Pseudomonas solanacearum* unable to induce a hypersensitive reaction. In: Proc 4th Int Conf Plant Pathog Bact, Station de pathologie vegetale INRA, Beaucouze, France p 823

Mukhopadhyay P, Williams J, Mills D (1988) Molecular analysis of a pathogenicity locus in *Pseudomonas syringae* pv. *syringae*. J Bacteriol 170:5479-5488

Napoli C, Staskawicz BJ (1987) Molecular characterization and nucleic acid sequence of an avirulence gene from race 6 of *Pseudomonas syringae* pv. *glycinea*. J

Bacteriol 169:572-578
Panopoulos NJ, Lindgren PB, Willis DK, Peet RC (1985) Clustering and conservation of genes controlling the interactions of *Pseudomonas syringae* pathovars with plants. In: Sussex I, Ellingboe AH, Crouch M, Malmberg R (eds) Plant cell/cell interactions. Cold Spring Harbor Laboratory Press, New York, p 69
Patil SS, Shintaku MH (1986) Cloning and characterization of an avirulence determinant from *Pseudomonas syringae* pv. *phaseolicola*. In: Genetic manipulation of *Pseudomonads*: applications in biotechnology and medicine. Abstracts EMBO Workshop, University of Geneva
Ronald PC, Staskawicz BJ (1988) The avirulence gene *avrBs*$_1$ from *Xanthomonas campestris* pv. *vesicatoria* encodes a 50 kD protein. Mol Plant-Microbe Interact 1:191-198
Stall RE, Loschke DC, Jones JB (1986) Linkage of copper resistance and avirulence loci on a self-transmissible plasmid in *Xanthomonas campestris* pv *vesicatoria*. Phytopathology 76:240-243
Staskawicz BJ, Dahlbeck D, Keen NT (1984) Cloned avirulence gene of *Pseudomonas syringae* pv. *glycinea* determines race-specific incompatibility on *Glycine max* (L). Merr. Proc Natl Acad Sci USA 81:6024-6028
Staskawicz BJ, Dahlbeck D, Keen N, Napoli C (1987) Molecular characterization of cloned avirulence genes from race 0 and race 1 of *Pseudomonas syringae* pv. *glycinea*. J Bacteriol 169:5789-5794
Swanson J, Kearney B, Dahlbeck D, Staskawicz B (1988) Cloned avirulence gene of *Xanthomonas campestris* pv. *vesicatoria* complements spontaneous race-change mutants. Mol Plant-Microbe Interact 1:5-9
Szabo LJ, Mills D (1984a) Characterization of eight excision plasmids of *Pseudomonas syringae* pv. *phaseolicola*. Mol Gen Genet 195:90-95
Szabo LJ, Mills D (1984b) Integration and excision of pMC7105 in *Pseudomonas syringae* pv. *phaseolicola*: involvement of repetitive sequences. J Bacteriol 157:821-827
Tamaki S, Dahlbeck D, Staskawicz B, Keen NT (1988) Characterization and expression of two avirulence genes cloned from *Pseudomonas syringae* pv. *glycinea*. J Bacteriol 170:4846-4854
Taylor JD, Teverson DM, Davis JHC (1987) Halo-blight of *Phaseolus* bean. Rep Natl Veg Res Stn for 1986 p 63
Taylor JD, Bevan JR, Crute IR, Reader SL (1989) Genetic relationship between races of *Pseudomonas syringae* pathovar *pisi* and cultivars of *Pisum sativum*. Plant Pathol, 38: in press
Thomas MD, Leary JV (1980) A new race of *Pseudomonas glycinea*. Phytopathology 70:310-312
Vivian A, Atherton GT, Bevan JR, Crute IR, Mur LAJ, Taylor JD (1989) Isolation and characterization of cloned DNA conferring specific avirulence in *Pseudomonas syringae* pv. pisi to pea (*Pisum sativum*) cultivars, which possess the resistance allele, R2. Physiol Mol Plant Pathol 34:335-344
Whalen MC, Stall RE, Staskawicz BJ (1988) Characterization of a gene from a tomato pathogen determining hypersensitive resistance in non-host species and genetic analysis of this resistance in bean. Proc Natl Acad Sci USA 85:6743-6747
Young JM, Dye DW, Bradbury JF, Panagopoulos CG, Robbs CF (1978) A proposed nomenclature and classification for plant pathogenic bacteria. N Z J Agric Res 21:153-177

RECOGNITION AND RESPONSE IN PLANT/FUNGUS INTERACTIONS

John Mansfield
Department of Biochemistry and Biological Sciences
Wye College (University of London)
Ashford
Kent
TN25 5AH
UK

ABSTRACT

Many different physical routes are taken by fungi during their colonization of plants and this variability applies important constraints on the type of signalling which is possible between host and pathogen. Thus, some fungi grow only within the intercellular spaces whereas others (notably the obligate parasites producing haustoria) make direct contact with the plant cell membrane. Mechanisms of resistance effective against the various strategies for parasitism adopted by fungi will be considered.

Varietal resistance is often expressed by a hypersensitive reaction (HR) to avirulent pathotypes but the more general resistance of non-host plants is usually associated with less dramatic and localized changes in plant cell walls, including deposition of lignin, silica and hydroxyproline-rich glycoproteins. Carbohydrates have been shown to have the capacity to act as important signalling molecules, eliciting plant responses such as phytoalexin biosynthesis but the activity of oligosaccharides during infection is unclear. Elicitors with well defined cultivar and race-specific activity remain elusive.

Lettuce downy mildew disease caused by the obligate parasite *Bremia lactucae* has been used as a model system for studying the processes of recognition underlying race-specific resistance. Experiments will be described which will allow discussion of the identification of primary responses, the role of plasma membrane damage in the HR, the identification of elicitors of the HR and possible routes for the isolation of gene-specific receptor proteins in lettuce.

NATO ASI Series, Vol H 41
Recognition and Response in Plant-Virus Interactions
Edited by R S S Fraser

INTRODUCTION

This article is concerned primarily with signalling which controls the activation of processes of resistance to fungal pathogens at the site of infection. These processes usually determine the success or failure of the fungus to colonize an infected plant and therefore control host specificity. It is, however, important to remember that some of the more striking examples of specificity in fungal pathogens do not concern host range but the physical routes taken by the fungus during attempted, but not always successful, colonization of plant tissue. Development of characteristic infection structures by phytopathogenic fungi appears to be controlled by signals from the invaded plant but, apart from the location of stomata by rust fungi on certain susceptible hosts, there are no confirmed examples of host-specific control of fungal differentiation (Heath, 1974; Wynn and Staples, 1981; Staples *et al.,* 1986). Fungal pathogens attempt to follow their programmed stages of development on susceptible hosts, resistant cultivars or non-host plants and must, therefore, respond to a series of non-specific plant signals. The signals that control this basic, pathogenesis-related differentiation are poorly understood, but they may provide valuable targets for the development of strategies for disease control.

Spatial constraints and modes of nutrition

The different routes taken by fungi during colonization of plants were discussed by Heath (1986) and I have illustrated specific examples in Fig. 1. The location of the fungus applies important constraints on the type of signalling which is possible between potential pathogen and host. Thus, the dikaryotic form of a rust fungus (a in Fig. 1, e.g. *Uromyces phaseoli* var. *vignae)* develops haustoria in host cells. The haustorium is in direct contact with the plant plasma membrane which is invaginated within the penetrated cell (Heath, 1974). Such direct contact would allow the activity of the widest possible range of signalling molecules including components which might not be released from the surface of the fungal wall.

By contrast, for a plant to respond to a pathogen which grows intercellularly (e.g. *Cladosporium fulvum,* the tomato leaf mould fungus, b in Fig. 1) a signal molecule must be released from fungal hyphae and cross the plant cell wall before reaching the plant cell membrane (De Wit *et al.,* 1988). An extreme form of developmental specialization is displayed by the vascular wilt fungi whose growth is preferentially within xylem vessels (e.g. *Verticillium albo-atrum*, Bishop and Cooper, 1983). The types of interface occurring

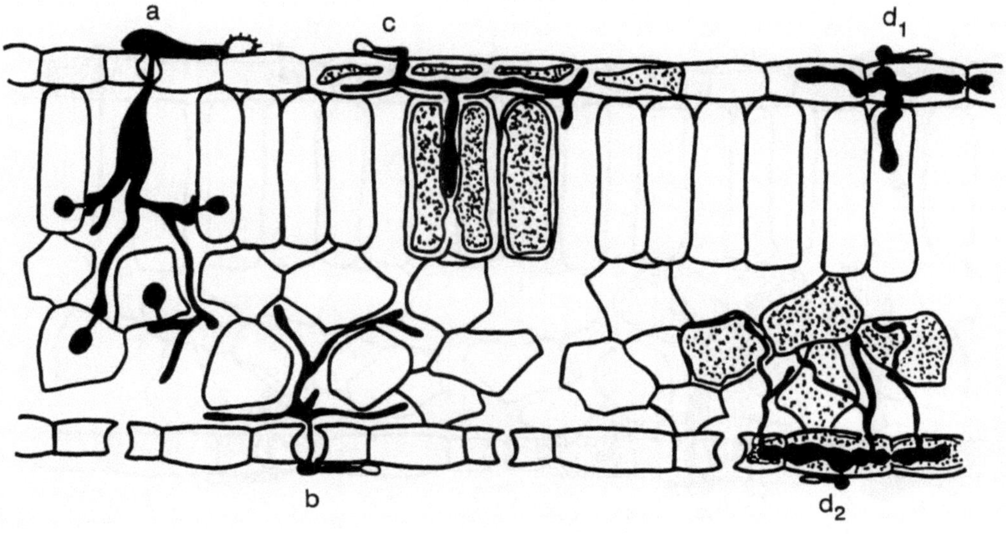

Figure 1. Routes taken by pathogenic fungi: a) biotrophic rust fungus, stomatal penetration and production of intercellular hyphae and intracellular haustoria, b) biotrophic intercellular growth of *Cladosporium fulvum*, c) necrotrophic *Botrytis* sp., direct penetration into epidermal cells, growth of intramural and intercellular hyphae within necrotic tissue, d) the temporary biotroph *Colletotrichum lindemuthianum*; the initial phase of biotrophic intracellular growth of large primary hyphae (d_1) is followed in the necrotrophic phase (d_2) by the rapid intra- and intercellular growth of secondary hyphae. Dead plant cells are shaded.

at the cellular level between plants and fungi were described in detail by Bracker and Littlefield (1973).

The morphological development of fungal pathogens is closely related to their mode of nutrition. Thus, taxonomically unrelated rusts (basidiomycetes), powdery mildews (ascomycetes) and downy mildews (oomycetes) are all fastidious biotrophs restricted to feeding from living plant cells and adapted to do so by producing haustoria from intercellular hyphae. The parallel evolution of the haustorium in these groups of fungi is becoming more apparent through detailed physiological and ultrastructural studies (Gay *et al.*, 1988; Woods *et al.*, 1988a).

The infection routes followed by many necrotrophic fungi, which feed on nutrients released from killed cells, may also be similar amongst unrelated species. Thus, *Botrytis cinerea* (ascomycete) and *Rhizopus stolonifer* (zygomycete) can both be rampant necrotrophs and both have an initial period of growth through plant cell walls which swell following the action of pectic enzymes (c in Fig. 1; Mansfield and Richardson,

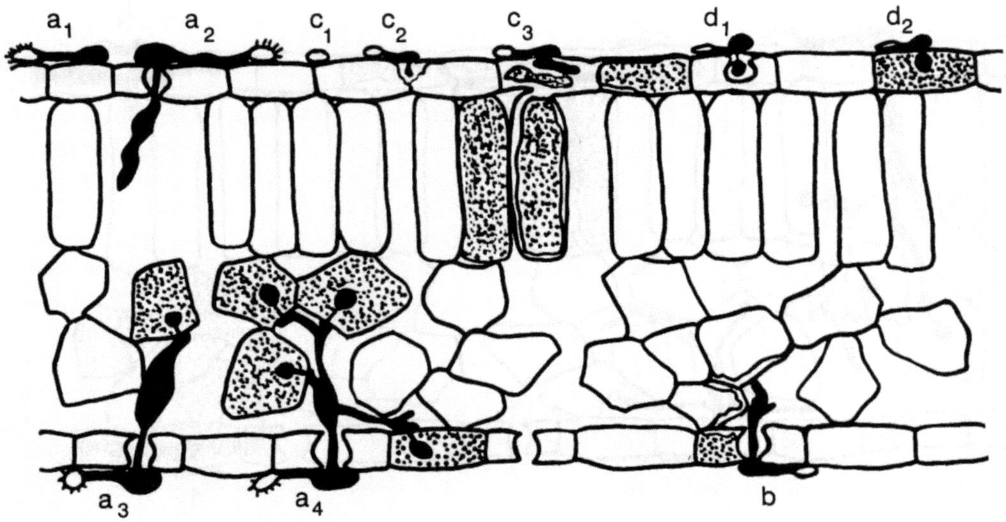

Figure 2. Sites of restriction of fungal development: a) inhibition of rust fungus on the leaf surface (a_1), during formation of a sub-stomatal vesicle (a_2), and following the HR after production of the first rudimentary haustorium (a_3) or at a later stage of infection (a_4), b) induction of the HR and cell wall alterations restricts *Cladosporium fulvum*, c) inhibition of *Botrytis* sp., failure to germinate (c_1), prevention of penetration by cell wall alterations (c_2) or cessation of growth in a limited lesion (c_3), d) colonization by *Colletotrichum lindemuthianum* may be restricted by papilla formation (d_1) or the HR (d_2). Dead plant cells are shaded.

1981).

The third type of nutrition displayed by pathogenic fungi may be described as temporary biotrophy (Bailey, 1982). Biotrophic, often intracellular growth is followed by development of a necrotic lesion. Critically, successful colonization is never achieved without the initial period of biotrophy. Well-studied examples of this type of parasite are *Colletotrichum lindemuthianum* (ascomycete), as illustrated by d in Fig. 1, and *Phytophthora infestans* (oomycete).

MECHANISMS OF RESISTANCE AND PATHOGENICITY

Some sites at which fungal growth may be restricted in resistant plants are indicated in Fig. 2. Many factors have been identified which may prevent colonization of plants by fungi. They fall into two categories: constitutive factors which are present

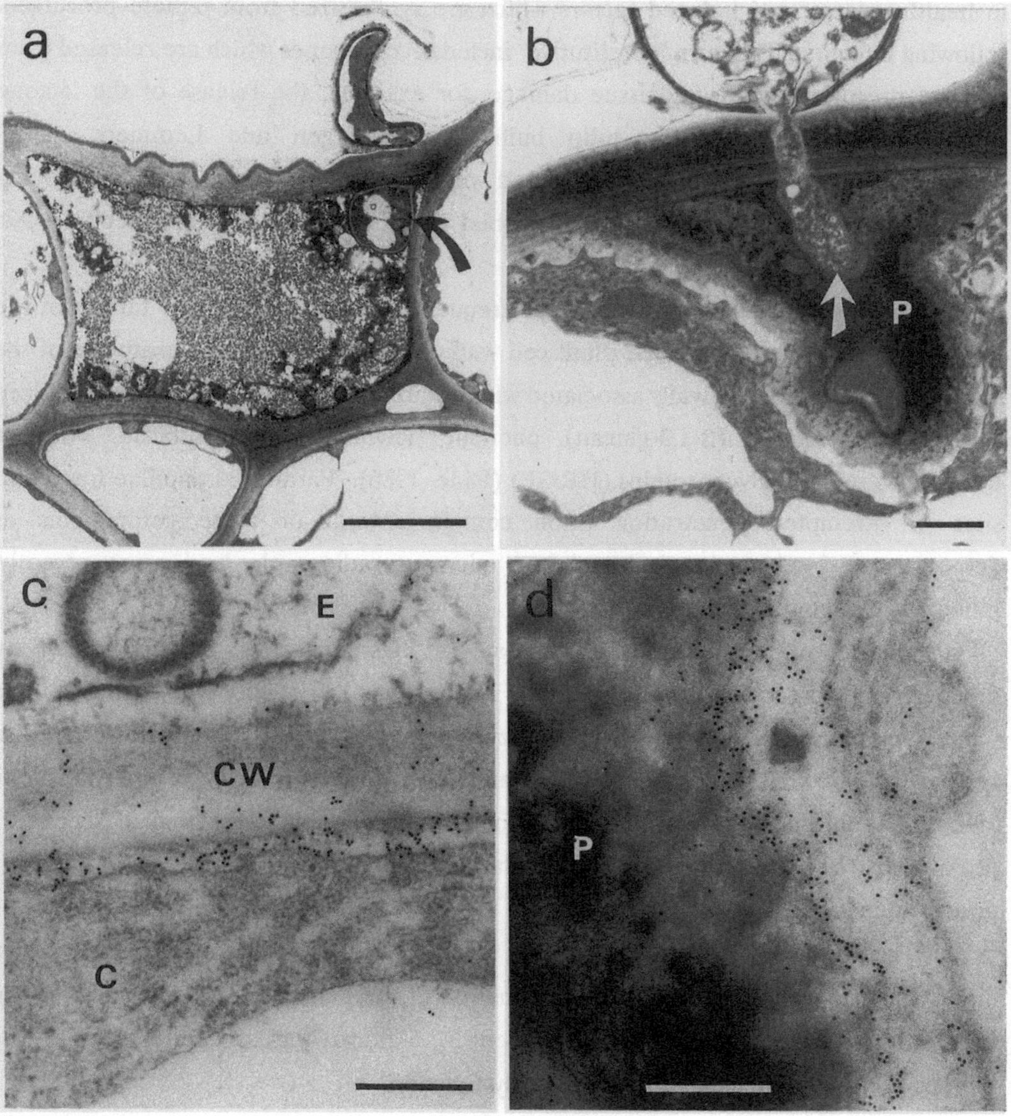

Figure 3. Electron micrographs of resistant interactions between hypocotyls of *Phaseolus vulgaris* and *Colletotrichum lindemuthianum* 96h after inoculation. a) HR, the epidermal cell penetrated by a small infection hypha (arrow) has disorganized cytoplasm, bar = 5μm. b) Papilla reaction, encasement of a small infection hypha (arrow) by a multi-layered papilla (P), bar = 1μm. c) and d) Immunogold labelling of hydroxyproline-rich glycoproteins using antiserum to $HRGP_{2b}$ from melon callus (Mazau *et al.,* 1988). In c), which shows an epidermal cell (E) undergoing the HR and an adjacent uninfected cortical cell (C), note that most gold particles (10nm) are associated with the cortical cell wall (CW). d) Immunogold labelling of part of a papilla (P). Plates kindly provided by Dr RJ O'Connell, Long Ashton Research Station, Bristol, UK.

in healthy plants, and induced factors which are synthesized from remote precursors following infection. The term "constitutive" includes compounds which are released from inactive precursors following tissue damage, for example, the release of the lactone tulipalin from tuliposides in tulip bulbs (Beijersbergen and Lemmers, 1972). Constitutive factors are typically associated with non-host resistance, pathogens being adapted to tolerate constitutive antimicrobial factors present in their specific hosts (Mansfield, 1983).

The resistance of non-hosts is also frequently associated with the failure of the invading fungus to penetrate the plant cell wall. There are numerous examples of the modification of plant cell walls associated with failure to penetrate, such as the localized deposition of callose (ß-1,3-glucan), phenolic polymers, lignin, suberin, silica or hydroxyproline-rich glycoproteins (HRGP) (Ride, 1986). Paramural papillae formed at sites of attempted penetration often contain several of these components as demonstrated by histochemical staining, microautoradiography and more recently, immunocytochemical techniques (Fig. 3).

The deposition of papillae as illustrated in Figs 2 and 3 has in many cases been demonstrated to require protein synthesis by the plant. It seems that challenge by a fungus stimulates *de novo* synthesis of enzymes controlling the modification and also synthesis of cell wall components (Vance and Sherwood, 1976; Stewart and Mansfield, 1985). Papilla deposition is an example of localized directed secretion of secondary products by the activated plant. The regulation of such a localized secretory activity is poorly understood.

The localized formation of callose may not require protein synthesis. Kauss (1985) has proposed that localized perturbation of the plasma membrane may lead to Ca^{2+} influx which activates calcium dependent ß-1,3-glucan synthase, and thereby callose synthesis, close to the site of membrane dysfunction.

Resistance at the varietal level and in particular race specific resistance, is usually expressed by the hypersensitive reaction (HR) of penetrated cells (see Figs 2 and 3). The HR is also often observed in non-host plants and may be defined as the rapid and localized death of challenged plant cells associated with the restriction of microbial colonization. Plant cell death is an essential feature of the HR but, in different plant species, there is variation in the biochemical events associated with the reaction. Thus, in some plants, low molecular weight phytoalexins may accumulate (*Vicia faba*; Mansfield *et al.,* 1980), whereas in others different antimicrobial systems such as

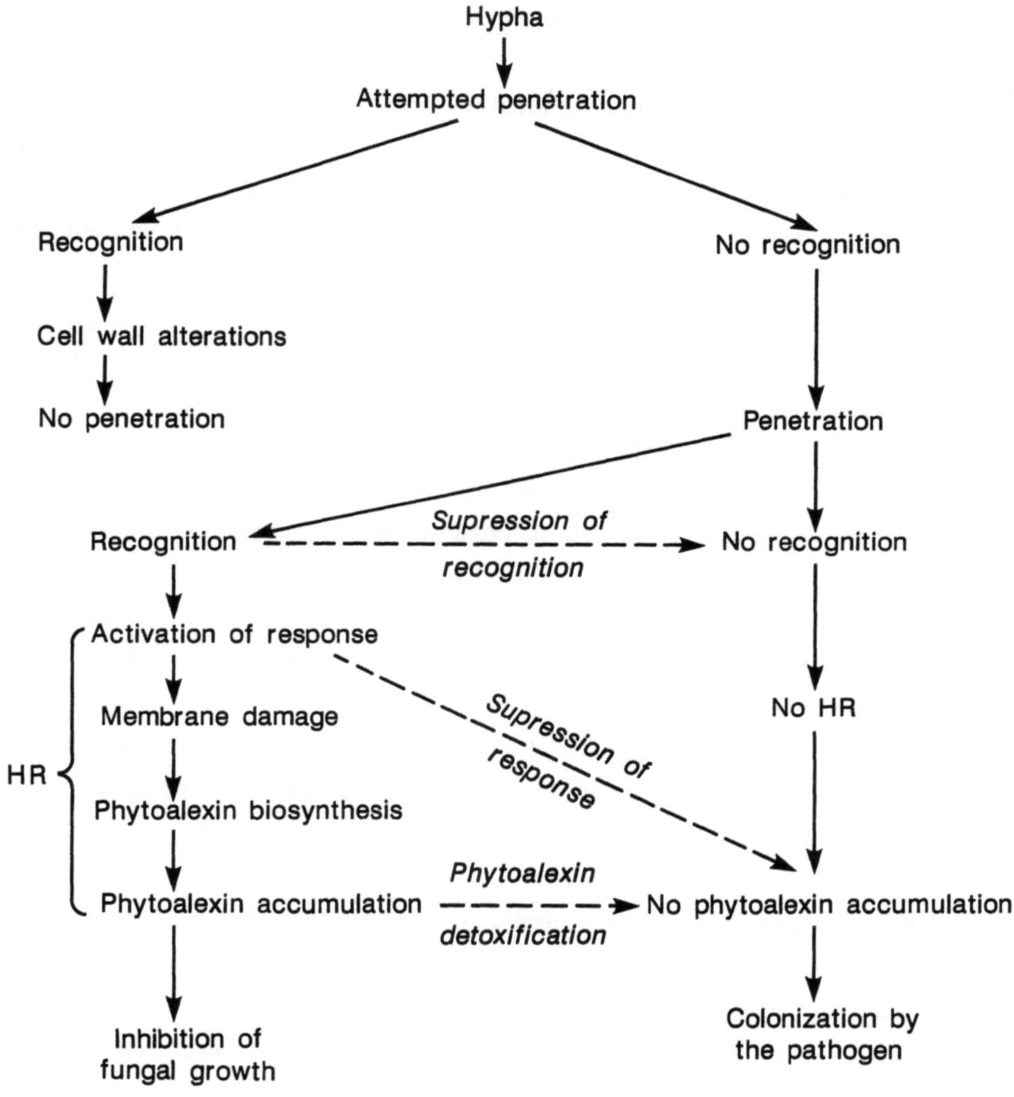

Figure 4. Recognition leading to resistance during plant/fungus interactions.

widespread lignification (wheat, Beardmore *et al.*, 1983) suberization (potato, Kolattukudy *et al.*, 1989) and HRGP deposition (cucumber; Hammerschmidt *et al.*, 1984) may be activated. A more general response is the accumulation of ß-1,3-glucanase and chitinase (Mauch *et al.*, 1988; Mauch and Staehelin, 1989). Paradoxically, the generation of an antimicrobial environment need not be implicated in the interaction which led

Stackman (1915) to propose the term *hypersensitive response* (equivalent to HR), namely the death of penetrated cells in plants resistant to rust fungi. Plant cell death would immediately terminate the intimate nutritional relationship between potential host and biotrophic rust and, in theory, cause inhibition of fungal development through starvation.

Some features of papilla deposition and the HR in bean hypocotyl cells challenged by *C. lindemuthianum* are illustrated in Fig. 3. Immunocytochemical localization of HRGPs using a polyclonal antibody raised to melon $HRGP_{2b}$ (Mazau *et al.,* 1988) illustrates the localized accumulation of the glycoprotein within the papilla surrounding the infection hypha and in the wall surrounding the epidermal cell undergoing the HR.

In order for a fungus to colonize a plant it must overcome potential mechanisms of resistance. This may be achieved by failing to provoke a response or by detoxifying antimicrobial factors that may be produced. Possible routes to successful parasitism and the critical stages at which signalling associated with recognition events may determine the outcome of the plant/fungus interaction are outlined in Fig. 4. Phytoalexin accumulation has been emphasized as the end point of the response as it is the best studied process of resistance.

Perhaps the simplest hypothesis to explain the establishment of biotrophic parasites (metaphorically described as "confidence men"; Keen, 1986) is that the plant fails to recognize signals from invading hyphae. Barriers to penetration and the HR are not induced. The possible involvement of suppressors (particularly race-specific suppressors) of recognition is not well established (Heath, 1980; Bushnell and Rowell, 1981; Crucefix *et al.,* 1984).

Plants must be aware of the presence of necrotrophic parasites (described as "thugs" by Keen, 1986) but may fail to produce an antimicrobial environment because tissues may be killed or inactivated before individual cells are able to synthesize resistance factors such as phytoalexins. The importance of the ability of *Botrytis fabae* to kill cells of *V. faba* during the early stages of infection has been described in detail (Hutson and Mansfield, 1980; Mansfield, 1984). The suppression of the resistance response by *B. fabae* is thought to be due to the secretion of pectic enzymes adapted to act on host cell walls.

Accumulation of phytoalexins may also be prevented if the pathogen is adapted to detoxify its host's phytoalexins. The most detailed analysis of the role of phytoalexin detoxification is in the interaction between isolates of *Nectria haematococca* and pisatin, the pterocarpanoid phytoalexin from pea (Van Etten *et al.*, 1982). Biochemical and genetical analyses have demonstrated a close correlation between virulence on pea and the ability of *N. haematococca* to detoxify pisatin by a cytochrome P450-dependent demethylase. The demethylase gene, which is induced by pisatin and related phenolics, has been cloned in *Aspergillus* (Kistler and Van Etten, 1984; Weltring *et al.*, 1988). Induction of pisatin demethylase is a good example of a fungus responding to a signal from the plant.

Elicitors and primary responses

The nature of the signal molecules produced by fungi, and the receptors and primary responses involved in plants is the subject of continued research and debate (see recent reviews by De Wit, 1986; Callow *et al.*, 1988 and Lamb *et al.*, 1989). The activity of putative signals and receptors must fit within the *structural* and *genetic framework* which defines the biology of the host/parasite interaction. In race specific resistance a *gene-for-gene relationship* in which resistance alleles specifically interact with complementary avirulence alleles has been suggested for numerous interactions and effectively proven for a few host/fungus combinations (Ellingboe, 1985; Crute, 1985; Michelmore *et al.*, 1988).

The model most generally accepted for control of recognition during race specific resistance to fungal pathogens involves interactions between the surface architecture of the fungus, and receptor proteins in the plasma membrane of the host plant. This recognition event usually leads to a hypersensitive reaction characterized by occurrence of irreversible membrane damage in the challenged cell and the *de novo* synthesis of defence-related proteins and secondary metabolites in responding tissue. Receptor proteins in the plant are thought to be the products of resistance genes, and surface features in the pathogen are thought to be controlled by genes for avirulence in the pathogen. This hypothesis awaits proof through the isolation of components of the cognon/cognor interaction and characterization of genes for avirulence and disease resistance. Although considerable progress has been made with the isolation of genes controlling avirulence in bacteria (Staskawicz *et al.*, 1988; Vivian, this volume), we are comparatively ignorant about the control of recognition in diseases caused by the major fungal pathogens; the rusts, powdery and downy mildews, all of which are biotrophic parasites. The relationship between recognition and the activity of defence-related mRNAs is also unclear.

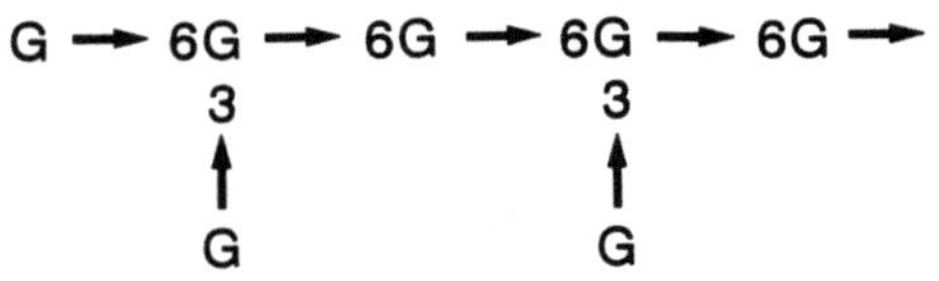

Figure 5. Structure of the hepta-ß-glucoside which is a potent elicitor of phytoalexin accumulation in soybean (Sharp *et al.,* 1984).

Recent analysis suggests that gene-for-gene interactions may be extended to species specificity (Valent, 1986; Niks, 1988; Newton and Crute, 1989). However, the more generally accepted view is that non-host resistance may be under polygenic control (Heath, 1980) and therefore involves interactions between receptors and numerous potential elicitors of resistance (Bushnell and Rowell, 1981).

Speculation abounds on the molecules involved in recognition. Attempts to isolate race- and cultivar-specific elicitors of phytoalexin accumulation and/or the HR from fungi have proved remarkably unsuccessful. Molecules which are potent but non-specific elicitors of phytoalexin accumulation have been isolated from culture filtrates, mycelium and cell walls of several fungi (Darvill and Albersheim, 1984). The list of elicitors of phytoalexin accumulation includes peptides, glycoproteins and oligosaccharides derived from fungal wall polymers such as ß-glucan, chitin and chitosan. Lipid fractions (notably including arachidonic acid) also have activity.

Alkali insoluble ß-1,3-linked glucans are common to the cell walls of many fungi. Partial acid hydrolysis of the glucans recovered from *Phytophthora megasperma* f. sp. *glycinea* released a heptaglucoside (Fig. 5, Sharp *et al.,* 1984) which is a potent fragment, active in the nanomolar range in the elicitation of phytoalexin accumulation in soybean cultivars resistant or susceptible to *P. m.* f. sp. *glycinea.* Closely related variants of the heptaglucoside present in the hydrolysate were found to be inactive, apparently because they lacked the required stereospecificy. High-affinity binding of ß-glucan fragments to soybean microsomal fractions and protoplasts has been reported (Cosio *et al.,* 1988). It would be valuable to use the active heptaglucoside for similar binding studies which may reveal the presence of a receptor which regulates signal transduction in soybean.

Interestingly, the heptaglucoside does not elicit phytoalexin accumulation in parsley cells. Parker *et al.* (1988) examined elicitor activity in fractions of a crude cell wall preparation from *P. m.* f. sp. *glycinea.* They found that proteinaceous constituents were active elicitors in parsley but not in soybean whereas proteinase-treated preparations as well as the heptaglucosides were active in soybean but not in parsley.

Phytoalexin biosynthesis may also be triggered by endogenous elicitors from plant cells (Bailey, 1982). Pathogens have been shown to release pectic fragments from the plant cell wall which are good elicitors of phytoalexin accumulation (Bruce and West, 1982; Davis and Hahlbrock, 1987). Polysaccharide fragments with diverse regulatory activities may be obtained from plant cell walls following enzymic digestion (Albersheim *et al.*, 1986). It is also possible that fungal wall fragments with elicitor activity may be released by enzymes of plant origin, notably chitinase and ß-1,3-glucanases (Mauch and Staehelin, 1989). Work on oligosaccharide elicitors conjures up a fascinating series of interactions involving the release and degradation of active glucans and other fractions from the cell walls of plants and fungi. The role of the non-specific elicitors may be in triggering non-host resistance and also more generally as secondary signals released following a primary *specific* recognition event. There is, however, little evidence to show that cell wall components are released and function during the determinative stages of infection.

Promising results have been obtained by Anderson and co-workers indicating that a galactose and mannose-rich glycoprotein from the α race of *C. lindemuthianum* may elicit phytoalexin accumulation in a resistant-cultivar-specific manner (Tepper and Anderson, 1986; Tepper *et al.*, 1989). However, in only one case, the interaction between the leaf mould fungus *Cladosporium fulvum* and tomato, has an elicitor of necrosis been isolated which shows clear and consistent race and cultivar specificity (De Wit and Toma, 1986; Lamb *et al.*, 1989). De Wit *et al.* (1988) and Lazarovits and Higgins (1979) recovered non-specific elicitors from axenic cultures of the fungus but the molecules with specific activity were recovered from intercellular fluids prepared from leaves heavily infected with virulent isolates of *C. fulvum*. As growth of the leaf mould fungus is completely intercellular they were in effect examining diffusates from *in vivo* mycelium growing biotrophically within a susceptible variety and therefore responding to signals from the plant. Use of a very sensitive tomato variety as a test plant allowed De Wit *et al.* (1988) to identify an elicitor of necrosis as a peptide with a molecular weight of 3049 Da; they suggested that the elicitor is the product of avirulence gene A9 in *C. fulvum*.

The leaf mould fungus is unique in its intercellular development; other fungal pathogens with well defined race structure penetrate into cells in order to establish nutritional relationships. Attempts to detect elicitors in the intercellular fluids recovered from other interactions, for example that between lettuce and the downy mildew fungus *Bremia lactucae* have proved unsuccessful (Crucefix *et al.,* 1984). In view of the somewhat exceptional nature of the tomato leaf mould system it is possible that peptides such as that isolated by De Wit and colleagues may not be the determinants of specificity in other diseases.

The failure to recover specific elicitors from other interactions may be due to the highly localized nature of the molecules involved in recognition; for example they may be components of the fungal cell wall found only on intracellular infection structures such as haustoria (Mansfield *et al.,* 1988). Also, the molecules may be induced by the presence of certain plant products, produced only at certain stages during the development of infection, or they may be unstable.

Why do cells die during the hypersensitive response?

It is becoming increasingly apparent that damage to the plant's plasma membrane is an early event during the HR to fungi and bacteria (Bailey, 1982; Pavlovkin *et al.,* 1986; Koga *et al.,* 1988; Woods *et al.,* 1988b; Ádám *et al.,* 1989). When membrane damage becomes irreversible, the responding cell should be considered to be dead because tonoplast dysfunction and loss of compartmentation soon follow. The onset of irreversible membrane damage (IMD) can be detected at the cellular level by the failure of cells to plasmolyse (Mansfield, 1986; Woods *et al.,* 1988b). In the few hypersensitive reactions to fungi in which it has been studied in detail, the occurrence of IMD has been found to precede major responses such as phytoalexin accumulation (Sato *et al.,* 1971) and the deposition of silica and autofluorescent phenolics (Koga *et al.,* 1988). Studies of the development of infection *in planta* therefore implicate IMD as a very early response, perhaps chronologically close to critical recognition events which control not only race specific but also many examples of non-host resistance.

These findings contrast with many studies on the responses of isolated protoplasts and cultured cells to non-specific elicitors. In some *in vitro* systems IMD apparently does not occur but various processes are rapidly activated following exposure to elicitors (Dangl *et al.,* 1987; Hamdan and Dixon, 1987; Parker *et al.,* 1988; Dron *et al.,* 1988). Schemes for the relationships between recognition, IMD and other responses occurring during the HR are outlined in Fig. 6. As previously discussed (Mansfield, 1986), the end product of each of these schemes would be a hypersensitive reaction in which restriction

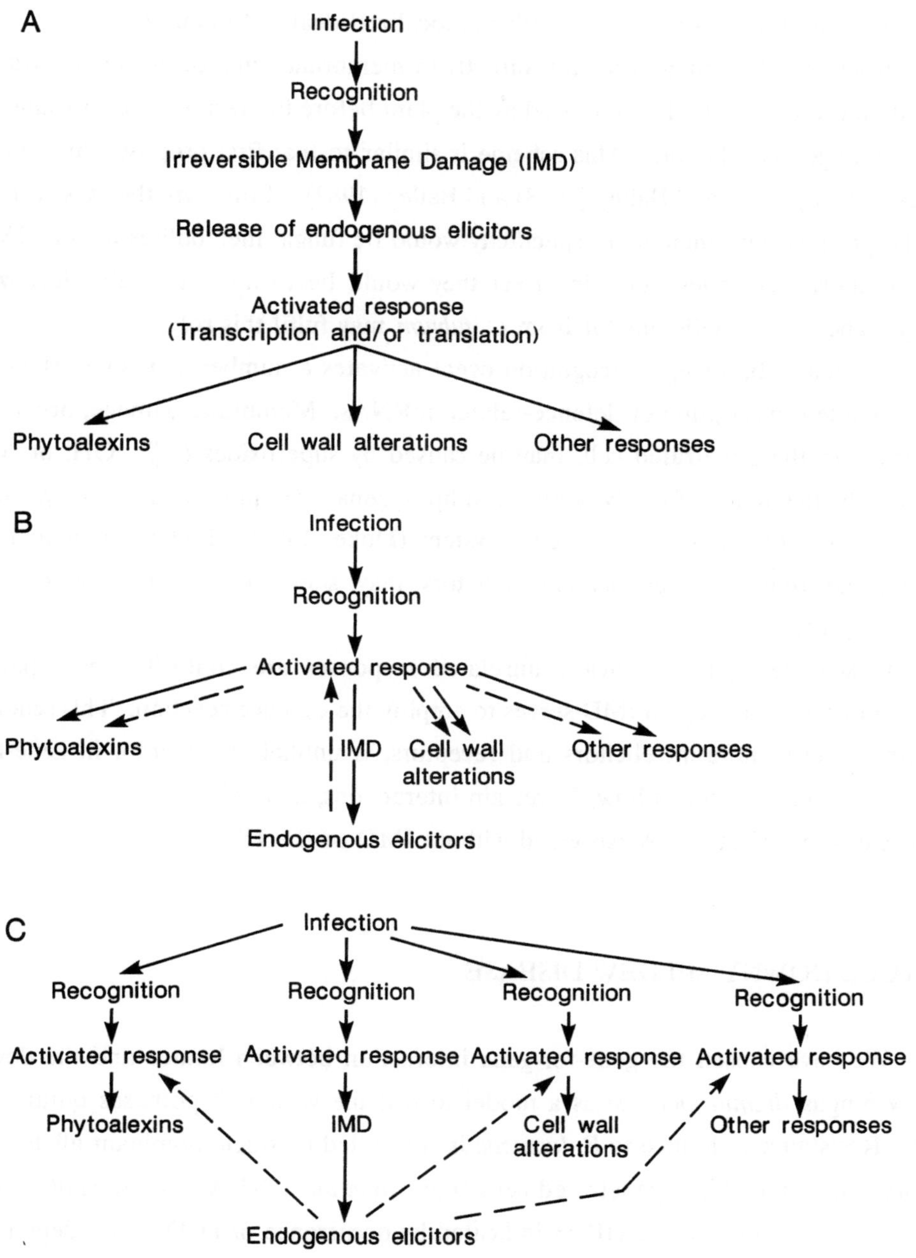

Figure 6. Schemes for the relationships between recognition, IMD and other responses occurring during the HR. The broken line indicates amplification of response by elicitors released as a consequence of IMD.

of the invading fungus is associated with the localized death of plant cells.

In scheme A, recognition leads directly to membrane damage of the penetrated cell and no protein synthesis is required by the plant before the response of surrounding cells to endogenous elicitors. This scheme is similar to that first proposed in seminal articles by Hargreaves and Bailey (1978) and Bailey (1982). If this hypothesis is correct, then the primary determinants of specificity would be fungal metabolites causing IMD within initially challenged cells; in effect they would be toxins with highly localized activity. The race-specific elicitor from *C. fulvum* may fulfil this role.

In scheme B, a single recognition event activates a number of responses which would include transcription of defence-related mRNAs. Membrane damage, occurring specifically in the penetrated cell, may be caused by superoxides ($O_2^{\cdot -}$, $\cdot OH$, or $^1O_2$) generated by the action of newly synthesized lipoxygenase (Keppler and Novacky, 1987) or an $O_2^{\cdot -}$ generating NADPH oxidase system (Doke, 1985). Following membrane damage, the release of endogenous elicitors may serve to amplify responses in surrounding cells.

In scheme C, biochemically unrelated responses are controlled by separate recognition events and again IMD serves to amplify the defence reaction. This scheme, involving several different elicitors and receptors, is unlikely to operate in gene-for-gene systems but illustrates how, in certain interactions, activation of single responses may occur and resistance be expressed without IMD or the HR.

LETTUCE DOWNY MILDEW DISEASE

We have chosen the gene-for-gene interaction between lettuce and the downy mildew fungus *Bremia lactucae* as a model to test the various hypotheses outlined in Fig. 6. Resistance in lettuce to *B. lactucae* is controlled by single dominant alleles and is expressed by the HR in penetrated cells (Farrara *et al.*, 1987; Mansfield *et al.*, 1988). The time of the onset of the HR as indicated by occurrence of IMD varies depending on the resistance (*Dm*) gene being expressed. In cultivars expressing *Dm*5/8, IMD occurs within 1.5h of penetration whereas in expression of *Dm*1 or *Dm*3, IMD is not observed until 2-2.5h later (Woods *et al.*, 1988b).

In view of the speed of IMD, particularly in the expression of *Dm*5/8, it was suggested that the early response was caused by a direct effect of a fungal product on the plant cell membrane as outlined in Fig. 6A. However, Blasticidin S, an inhibitor of protein synthesis, and heat-shock prior to inoculation were found to suppress IMD and allow fungal development to progress within live cells. As heat-shock is known to inhibit the transcription and translation of mRNAs other than those newly synthesized to code for heat-shock proteins, we concluded that IMD was a process requiring protein synthesis. Possible candidates for newly synthesized proteins controlling membrane damage in lettuce are lipoxygenase or $O_2^{\cdot -}$ generating NADPH oxidase as previously discussed.

Heat-shock was also found to have a gene specific effect on the HR. Thus, the HR controlled by *Dm*1 and *Dm*3 was expressed if tissues were incubated overnight (17h) after heat-shock and before inoculation. By contrast, the effect of heat-shock on resistance expressed by *Dm*5/8 did not diminish until 2-3 days after treatment (Woods *et al.,* 1989). Results obtained with *Dm*5/8 indicate that heat-shock is affecting a structural component with a slow turnover rate rather than an inducible protein whose rapid synthesis would be restored within several hours of heat-shock. One explanation for the gene specific nature of the heat-shock phenomenon is that the heat treatment modifies a receptor protein in the host plasma membrane which is specific to the *Dm*5/8 gene and has a low rate of turnover (Morré *et al.,* 1979; Woods *et al.,* 1989).

Lettuce tissues are rich in polyphenols (Sharples, 1964). Histochemical analyses have demonstrated the accumulation of phenolics and lignin in lettuce cells undergoing the HR to *B. lactucae*. Fluorescence microscopy revealed the localized accumulation of wall bound phenolics around penetration sites at the time of IMD during incompatible interactions. This localized reaction may reflect the release of preformed phenolics following membrane dysfunction. Major accumulation of phenols and lignification were not detected within responding tissues until some hours after IMD (J Fagg, unpublished observations). Thus, IMD precedes the major response.

Attempts to isolate elicitors of the HR in lettuce have met with mixed success (Mansfield *et al.,* 1988). However, fractions were recovered from isolated intracellular infection structures (primary and secondary vesicles) which, in some experiments, caused a greater percentage of protoplasts from resistant than from susceptible cultivars to suffer IMD. Some differentiation was also indicated when spores were incubated with protoplasts which had partially regenerated cell walls. Clearly, it will be important to determine the nature of the factor apparently causing cultivar specific IMD in these experiments, but the results obtained suggest that race and cultivar-specific elicitors of

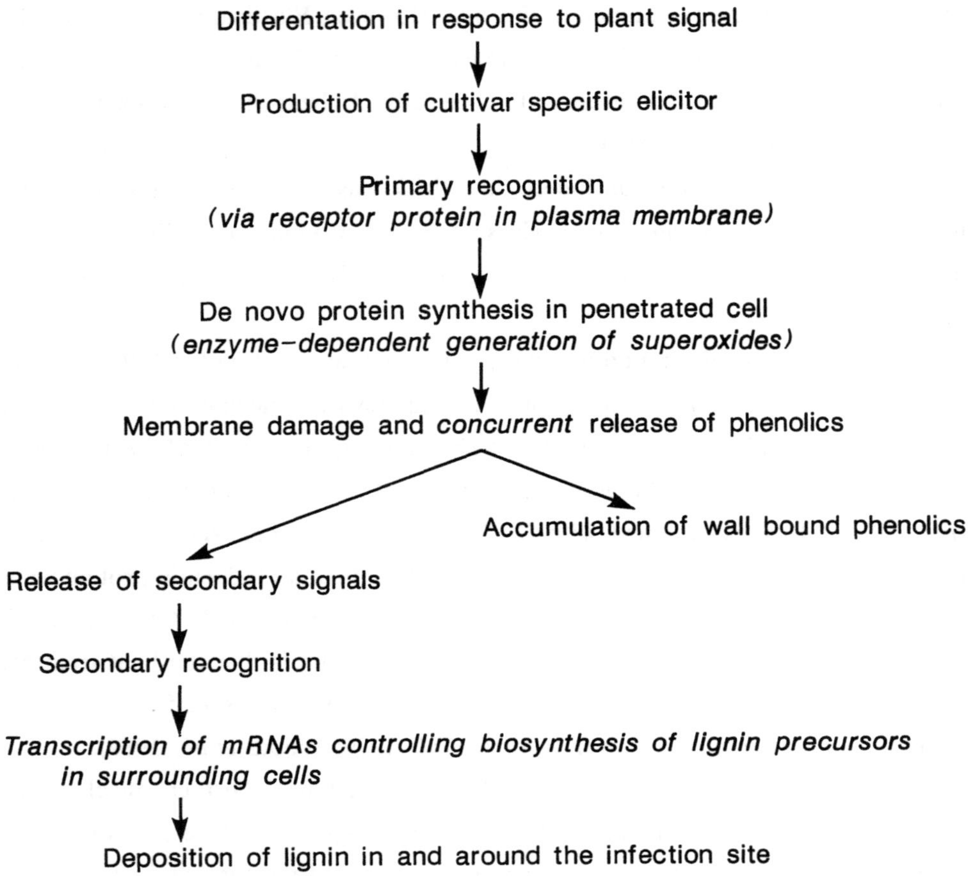

Figure 7. Proposed link between recognition, signal transduction and activated defence during the HR in lettuce to *Bremia lactucae* (the more hypothetical steps are underlined).

IMD and the consequent HR are produced as a result of fungal differentiation in response to features of the plant cell wall.

Having used lettuce downy mildew disease to test certain features of the schemes outlined in Fig. 6, it is possible to propose Fig. 7 as a model for signalling and response more directly applicable to interactions between *B. lactucae* and lettuce. In terms of the plant's response, the model is closely comparable with scheme A in Fig. 6 except that IMD requires protein synthesis by the plant.

CONCLUDING REMARKS

Molecular dissection of plant responses as outlined in Figs 4, 6 and 7 will require examination of mRNA synthesis at the *cellular level* using reporter gene fusions (Dron *et al.*, 1988; Bevan *et al.*, 1989) and *in situ* hybridization (Cuypers *et al.*, 1988; Somssich *et al.*, 1988). Many probes are already available for examination of phenylpropanoid metabolism (Lamb *et al.*, 1989) but attention should also be focussed on enzymes which may be associated with membrane damage in the first challenged cell. If the concept of primary and secondary recognition (in penetrated and surrounding cells) is correct, *in vitro* studies on signal transduction using non-specific elicitors are unlikely to reveal the receptor systems effective during the determinative primary stages of the plant/pathogen interaction. Analysis of recognition and response must be based on a sound knowledge of the biology and genetics of the interaction, as is now available for lettuce downy mildew.

Undoubtedly the major gap in our understanding of plant/fungus interactions concerns the nature of race and cultivar specific elicitors of the HR and their receptors in plant cells. Initial optimism concerning the isolation of avirulence genes from fungi and complementary resistance genes from host plants has been somewhat dented by the discovery of technical and numerical difficulties associated with transposon tagging, in development of efficient transformation systems and in the analysis of eukaryote genomes (Ellingboe, 1985; Day, 1987; Parsons *et al.*, 1987; Pryor, 1987; Bennetzen *et al.*, 1988). Difficulties are compounded when working with an obligate parasite such as *Bremia lactucae*, as discussed by Judelson and Michelmore (1989). There is, therefore, a continuing role for direct biochemical approaches to isolate the signalling molecules which determine specificity in plant/fungus interactions.

ACKNOWLEDGEMENTS

I wish to acknowledge support from the Agricultural and Food Research Council for our work on lettuce downy mildew disease.

REFERENCES

Ádám A, Farkas T, Somlyai G, Hevesi M, Király Z (1989) Consequence of $O_2^{\cdot-}$ generation during bacterially induced hypersensitive reaction in tobacco: deterioration of membrane lipids. Physiol Mol Plant Pathol 34:13-26

Albersheim P, Darvill AG, Sharp J, Davis KR, Doares SH (1986) Studies on the role of carbohydrates in host-microbe interactions. In: Lugtenberg B (ed) Recognition in microbe-plant symbiotic and pathogenic interactions. Springer-Verlag, Berlin, p 297

Bailey JA (1982) Physiological and biochemical events associated with the expression of resistance to disease. In: Wood RKS (ed) Active defense mechanisms in plants. Plenum, New York London, p 39

Beardmore J, Ride JP, Granger JW (1983) Cellular lignification as a factor in the hypersensitive resistance of wheat to stem rust. Physiol Plant Pathol 22:209-220

Beijersbergen JCM, Lemmers CBG (1972) Enzymic and non-enzymic liberation of tulipalin A (α-methylene butyrolactone) in extracts of tulip. Physiol Plant Pathol 2:265-270

Bennetzen JL, Qiu M-M, Ingels S, Ellingboe AH (1988) Allele-specific and *Mutator* associated instability at the *Rp1* disease resistance locus of maize. Nature 332:369-370

Bevan M, Shufflebottom D, Edwards K, Jefferson R, Schuch W (1989) Tissue- and cell-specific activity of a phenylalanine ammonia-lyase promoter in transgenic plants. EMBO J 8:1899-1906

Bishop CD, Cooper RM (1983) An ultrastructural study of vascular colonization in three vascular wilt diseases. I Colonization of susceptible cultivars. Physiol Plant Pathol 23:323-343

Bracker CE, Littlefield LJ (1973) Structural concepts of host-pathogen interfaces. In: Byrde RJW, Cutting CV (eds) Fungal pathogenicity and the plant's response. Academic Press, London p 159

Bruce RJ, West CA (1982) Elicitation of casbene synthetase activity in castor bean. The role of pectic fragments of the plant cell wall in elicitation by a fungal endopolygalacturonase. Plant Physiol 69:1181-1188

Bushnell WR, Rowell JA (1981) Suppressors of defense reactions: a model for roles in specificity. Phytopathology 71:1012-1014

Callow JA, Ray T, Estrada-Garcia TM, Green JR (1988) Molecular signals in plant cell recognition. In: Scannerini S, Smith D, Bonfante-Fasolo P, Gianinazzi-Pearson V (eds) Cell to cell signals in plant, animal and microbial symbiosis. Springer-Verlag, Berlin, p 167

Cosio EG, Popperl H, Schmidt WE, Ebel J (1988) High-affinity binding of fungal ß-glucan fragments to soybean (*Glycine max* L) microsomal fractions and protoplasts. Eur J Biochem 175:309-315

Crucefix DN, Mansfield JW, Wade M (1984) Evidence that determinants of race specificity in lettuce downy mildew disease are highly localized. Physiol Plant Pathol 24:93-106

Crute IR (1985) The genetic bases of relationships between microbial parasites and their hosts. In: Fraser RSS (ed) Mechanisms of resistance to plant diseases. Nijhoff, Dordrecht, p 80

Cuypers B, Schmelzer E, Hahlbrock K (1988) In situ localization of rapidly accumulated phenylalanine ammonia-lyase mRNA around penetration sites of *Phytophthora infestans* in potato leaves. Mol Plant-Microbe Interact 1:157-160

Dangl JL, Hanfle KD, Lipphardt S, Hahlbrock K, Scheel D (1987) Parsley protoplasts retain differential responsiveness to u.v. light and fungal elicitor. EMBO J 6:2551-2556

Darvill AG, Albersheim P (1984) Phytoalexins and their elicitors - a defense against microbial infection in plants. Annu Rev Plant Physiol 35:243-298

Davis KR, Hahlbrock K (1987) Induction of defense responses in cultured parsley cells by plant cell wall fragments. Plant Physiol 85:1286-1290

Day PR (1987) Plant-parasite interaction: a genetical perspective. In: Day PR, Jellis GJ (eds) Genetics and plant pathogenesis. Blackwell, Oxford, p 1

De Wit PJGM (1986) Elicitation of active resistance mechanisms In: Bailey JA (ed) Biology and molecular biology of plant pathogen interactions. Springer-Verlag, Berlin, p 149

De Wit PJGM, Toma IMJ (1986) Occurrence of race-specific elicitors in the host-pathogen interaction tomato-*Cladosporium fulvum*. In: Lugtenberg B (ed) Recognition in microbe-plant symbiotic and pathogenic interactions, Springer-Verlag, Berlin, p 287

De Wit PJGM, Toma IMJ, Joosten MHAJ (1988) Race-specific elicitors and pathogenicity factors in the *Cladosporium fulvum*-tomato interaction. In: Keen NT, Kosuge T, Walling LL (eds) Physiology and biochemistry of plant-microbial interactions. American Society of Plant Physiologists, Rockville, Maryland, p 111

Doke N (1985) NADPH-dependent O_2^- generation in membrane fractions isolated from wounded potato tubers inoculated with *Phytophthora infestans*. Physiol Plant Pathol 27:311-322

Dron M, Clouse SD, Lawton MA, Dixon RA, Lamb CJ (1988) Glutathione and fungal elicitor regulation of a plant defense gene promoter in electroporated protoplasts. Proc Natl Acad Sci USA 85:6738-6742

Ellingboe AH (1985) Prospects for using recombinant DNA technology to study race-specific interactions between host and parasite. In: Groth JV, Bushnell WR (eds) Genetic basis of biochemical mechanisms of plant disease. APS Press, St. Paul Minnesota, p 103

Farrara BF, Ilott TW, Michelmore RW (1987) Genetic analysis of factors for resistance to downy mildew (*Bremia lactucae*) in species of lettuce (*Lactuca sativa* and *L. serriola*). Plant Pathol 36:499-514

Gay JL, Salzberg A, Woods AM (1988) Dynamic experimental evidence for the plasma membrane ATPase domain hypothesis of haustorial transport and for ionic coupling of the haustorium of *Erysiphe graminis* to the host cell (*Hordeum vulgare*). New Phytol 107:541-548

Hamdan MAMS, Dixon RA (1987) Fractionation and properties of elicitors of the phenylpropanoid pathway from culture filtrates of *Colletotrichum lindemuthianum*. Physiol Mol Plant Pathol 31:91-103

Hammerschmidt R, Lamport DTA, Muldoon EP (1984) Cell wall hydroxyproline enhancement and lignin deposition as an early event in the resistance of cucumber to *Cladosporium cucumerinum*. Physiol Plant Pathol 24:43-47

Hargreaves JA, Bailey JA (1978) Phytoalexin production by hypocotyls of *Phaseolus vulgaris* in response to constitutive metabolites released by damaged cells. Physiol Plant Pathol 13:89-100

Heath MC (1974) Light and electron microscope studies of the interactions of host and non-host plants with cowpea rust *Uromyces phaseoli* var. *vignae*. Physiol Plant Pathol 4:403-414

Heath MC (1980) Reactions of nonsuscepts to fungal pathogens. Annu Rev Phytopathol 18:211-236

Heath MC (1986) Fundamental questions related to plant-fungal interactions: can recombinant DNA technology provide the answers? In: Bailey JA (ed) Biology and molecular biology of plant-pathogen interactions. Springer-Verlag, Berlin, p 15

Hutson RA, Mansfield JW (1980) A genetical approach to the analysis of mechanisms of pathogenicity in *Botrytis/Vicia faba* interactions. Physiol Plant Pathol 17:309-317

Judelson HS, Michelmore RW (1989) Strategies for cloning avirulence genes from *Bremia lactucae*. In: Staskawicz B, Ahlquist P, Yoder O (eds) Molecular biology of plant-pathogen interactions. Alan R Liss Inc, New York p 71

Kauss H (1985) Callose biosynthesis as a Ca^{2+} regulated process and possible relations to the induction of other metabolic changes. J Cell Sci Suppl 2:89-103

Keen NT (1986) Pathogenic strategies for fungi. In: Lugtenberg B (ed) Recognition in microbe-plant symbiotic and pathogenic interactions, Springer-Verlag, Berlin, p 171

Keppler LD, Novacky A (1987) The initiation of membrane lipid peroxidation during bacterial-induced hypersensitive reaction. Physiol Mol Plant Pathol 30:233-345

Kistler HC, Van Etten HD (1984) Regulation of pisatin demethylation in *Nectria haematococca* and its influence on pisatin tolerance and virulence. J Gen Microbiol 130:2605-2613

Koga H, Zeyen RJ, Bushnell WR, Ahlstrand GC (1988) Hypersensitive cell death, autofluorescence and insoluble silicon accumulation in barley leaf epidermal cells under attack by *Erysiphe graminis* f. sp. *hordei*. Physiol Mol Plant Pathol 32:395-411

Kolattukudy PE, Podila GK, Roberts E, Dickman MD (1989) Gene expression resulting from early signals in plant-fungus interaction. In: Staskawicz B, Ahlquist P, Yoder O (eds) Molecular biology of plant-pathogen interactions. Alan R Liss Inc, New York, p 87

Lamb CJ, Lawton MA, Dron M, Dixon RA (1989) Signals and transduction mechanisms for activation of plant defenses against microbial attack. Cell 56:215-224

Lazarovits G, Higgins VJ (1979) Biological activity and specificity of a toxin produced by *Cladosporium fulvum*. Phytopathology 69:1056-1061

Mansfield JW (1983) Antimicrobial compounds. In: Callow JA (ed) Biochemical plant pathology. Wiley and Sons, Chichester p 237

Mansfield JW (1984) Plant cell death during infection by fungi. In: Davies I, Sigee DC (eds) Cell ageing and death. Cambridge University Press, Cambridge, p 323

Mansfield JW (1986) Recognition, elicitors and the hypersensitive reaction. In: Lugtenberg B (ed) Recognition in microbe-plant symbiotic and pathogenic interactions. Springer-Verlag, Berlin p 434

Mansfield JW, Porter ACA, Smallman V (1980) Dihydrowyerone derivatives as components of the furanoacetylenic phytoalexin response of tissues of *Vicia faba*. Phytochemistry, 19:1057-1061

Mansfield JW, Richardson A (1981) The ultrastructure of interactions between *Botrytis* species and broad bean leaves. Physiol Plant Pathol 19:41-48

Mansfield JW, Woods AM, Street PFS, Rowell PM (1988) Recognition processes in lettuce downy mildew disease. In: Chapman GP, Ainsworth CC, Chatham CJ (eds) Eukaryote cell recognition. Cambridge University Press, Cambridge, p 241

Mauch F, Mauch-Mani B, Boller T (1988) Antifungal hydrolases in pea tissue II. Inhibition of fungal growth by combinations of chitinase and ß-1,3-glucanase. Plant Physiol 88:936-942

Mauch F, Staehelin LA (1989) Functional implications of the subcellular localization of ethylene-induced chitinase and ß-1,3-glucanase in bean leaves. Plant Cell 1:447-457

Mazau D, Rumeau D, Esquerré-Tugayé MT (1988) Two different families of hydroxyproline-rich glycoproteins in melon callus: biochemical and immunochemical studies. Plant Physiol 86:540-546

Michelmore RW, Ilott TW, Hulbert SH, Farrara BF (1988) The downy mildews. In: Sidhu GS, Williams PH, Ingram DS (eds) Advances in plant pathology Vol. 6. Genetics of pathogenic fungi. Academic Press, London, p 53

Morré DJ, Kartenbeck J, Franke WW (1979) Membrane flow and interconversions among endomembranes. Biochim Biophys Acta (MR) 559:71-152

Newton AC, Crute IR (1989) A consideration of the genetic control of species specificity in fungal plant pathogens and its relevance to a comprehension of the underlying mechanisms. Biol Rev 64:33-50

Niks RE (1988) Non-host plant species as donors for resistance to pathogens with narrow host range. II. Concepts and evidence of non-host resistance. Euphytica 37:89-99

Parker JE, Hahlbrock K, Scheel D (1988) Different cell wall components from *Phytophthora megasperma* f. sp. *glycinea* elicit phytoalexin production in soybean and parsley. Planta 176:75-82

Parsons KA, Chumley FG, Valent B (1987) Genetic transformation of the fungal pathogen responsible for rice blast disease. Proc Natl Acad Sci USA 84:4161-4165

Pavlovkin J, Novacky A, Ullrich-Eberius CI (1986) Membrane potential changes during bacteria-induced hypersensitive reaction. Physiol Plant Pathol 28:125-135

Pryor A (1987) The origin and structure of fungal disease resistance genes in plants. Trends Genet 3:157-161

Ride JP (1986) Induced structural defences in plants. In: Gould GW, Rhodes-Roberts ME, Charnley AK, Cooper RM, Board RG (eds) Natural antimicrobial systems in plants and animals. University Press, Bath, p 159

Sato N, Kitazawa K, Tomiyama K (1971) The role of rishitin in localizing the invading hyphae of *Phytophthora infestans* in infection sites at the cut surface of potato tubers. Physiol Plant Pathol 1:289-295

Sharp JK, Albersheim P, Ossowski P, Pilotti A, Garegg P, Lindberg P (1984) Comparison of the structure and elicitor activities of a synthetic and a mycelial-wall-derived hexa-(ß-D-glucopyranosyl)-D-glucitol isolated from the mycelial walls of *Phytophthora megasperma* f. sp. *glycinea*. J Biol Chem 259:11341-11345

Sharples GC (1964) Polyphenol content of head lettuce. Proc Am Soc Hort Sci 84:356-363

Somssich IE, Schmelzer E, Kawalleck P, Hahlbrock K (1988) Gene structure and *in situ* transcript localization of pathogenesis-related protein 1 in parsley. Mol Gen Genet 213:93-98

Stackman EC (1915) Relations between *Puccinia graminis* and plants highly resistant to its attack. J Agric Res 4:193-200

Staples RC, Yoder OC, Hach HC, Epstein L, Bhairi S (1986) Gene expression during infection structure development by germlings of the rust fungi. In: Bailey JA (ed) Biology and molecular biology of plant pathogen interactions. Springer-Verlag, Berlin, p 331

Staskawicz BJ, Bonas U, Dahlbeck D, Huynh T, Kearney B, Ronald P, Whalen M (1988) Molecular determinants of specificity in plant-bacterial interactions. In: Keen NT, Kosuge T, Walling S (eds) Physiology and biochemistry of plant-microbial interactions. American Society of Plant Physiologists, Rockville,

Maryland p 124

Stewart A, Mansfield JW (1985) The composition of wall alterations and appositions (reaction material) and their role in the resistance of onion bulb scale epidermis to colonization by *Botrytis allii*. Plant Pathol 34:25-37

Tepper CS, Albert FG, Anderson AJ (1989) Differential mRNA accumulation in three cultivars of bean in response to elicitors from *Colletotrichum lindemuthianum* Physiol Mol Plant Pathol 34:85-98

Tepper CS, Anderson AJ (1986) Two cultivars of bean display a differential response to extracellular components from *Colletotrichum lindemuthianum*. Physiol Mol Plant Pathol 29:411-420

Valent B (1986) Genetic studies of fertility and pathogenicity in *Magnaporthe grisea* (*Pyricularia oryzae*). Iowa State J of Res 60:569-594

Vance CP, Sherwood RT (1976) Cycloheximide treatments implicate papilla formation in resistance of reed canarygrass to fungi. Phytopathology 66:498-502

Van Etten HD, Matthews DE, Smith DA (1982) Metabolism of phytoalexins. In: Bailey JA, Mansfield JW (eds) Phytoalexins. Blackie and Sons, Glasgow, p 181

Weltring K-M, Turgeon BG, Yoder OC, Van Etten HD (1988) Isolation of a phytoalexin-detoxification gene from the plant pathogenic fungus *Nectria haematococca* by detecting its expression in *Aspergillus nidulans*. Gene 68:335-344

Woods AM, Didehvar F, Gay JC, Mansfield JW (1988a) Modification of the host plasmalemma in haustorial infections of *Lactuca sativa* by *Bremia lactucae*. Physiol Mol Plant Pathol 33:299-310

Woods AM, Fagg J, Mansfield JW (1988b) Fungal development and irreversible membrane damage in cells of *Lactuca sativa* undergoing the hypersensitive reaction to the downy mildew fungus *Bremia lactucae*. Physiol Mol Plant Pathol 32:483-498

Woods AM, Fagg J, Mansfield JW (1989) Effects of heat-shock and inhibitors of protein synthesis on irreversible membrane damage occurring during the hypersensitive reaction of *Lactuca sativa* L. to *Bremia lactucae* Regel. Physiol Mol Plant Pathol 34:531-544

Wynn WK, Staples RC (1981) Tropisms of fungi in host recognition. In: Staples RC, Toenniessen GH (eds) Plant disease control: resistance and susceptibility. John Wiley and Sons, New York, p 45

SPECIFICITY AND RECOGNITION EVENTS IN THE TRANSMISSION OF PLANT VIRUSES BY VECTORS

A F Murant
Scottish Crop Research Institute
Invergowrie
Dundee
DD2 5DA
UK

ABSTRACT

Most plant viruses need a plant-feeding organism to transmit them between plants. The vector species and the type of relation the virus has with it are highly specific. Vector specificity is determined by recognition events between virus proteins and sites of retention in the vector. Virus particle protein is invariably involved but in some instances recognition is mediated by a non-particle protein (helper component), or by the particle protein of another (helper) virus. Sites of retention in the vector may be external (non-circulative viruses) or internal (circulative viruses).

INTRODUCTION

The barrier presented by the plant cuticle and by the cellulose-containing walls of the epidermal cells confronts plant viruses with problems not shared by viruses that infect animals. Plant viruses have therefore evolved special mechanisms for penetrating this barrier, in most instances depending on some other organism, called a *vector*, to deposit them beyond it. The majority of vectors are insects, especially plant-sucking forms (aphids, leafhoppers, planthoppers and whiteflies) but also leaf-biting insects such as beetles. Other types of vector include leaf-eating eriophyid mites, root-feeding dorylaimid nematodes, and soil-inhabiting fungi with motile zoospores (chytrids, plasmodiophorids). Table 1 gives some examples of types of vector and shows that viruses that fall into the same taxonomic group have vectors of the same (and often only one) type. Furthermore, each virus within a group may be transmitted by only one or a few species of the type of organism that transmits viruses of that group. Moreover,

NATO ASI Series, Vol H 41
Recognition and Response in Plant-Virus Interactions
Edited by R S S Fraser

Table 1. *Some types of association between plant viruses and their vectors*

Type of association	Type of vector	Duration of virus/vector association	Example (virus or virus group and/or typical member
Non-circulative, non-persistent	Aphid	Few hours	Cucumber mosaic cucumovirus
	Aphid	Few hours	Potato Y potyvirus
	Aphid	Few hours	Cauliflower mosaic caulimovirus
	Fungus	Few hours	Tobacco necrosis necrovirus
	Fungus	Few hours	Satellite virus
Non-circulative, semi-persistent	Aphid	Few days	Anthriscus yellows virus
	Aphid	Few days	Beet yellows closterovirus
	Leafhopper	Few days	Maize chlorotic dwarf virus
	Nematode	Weeks/months	Raspberry ringspot nepovirus
Circulative, non-propagative	Aphid	Many days	Barley yellow dwarf luteovirus
	Aphid	Many days	Pea enation mosaic virus
	Whitefly	Many days	African cassava mosaic geminivirus
	Leafhopper	Many days	Maize streak geminivirus
Circulative, propagative	Aphid	Weeks	Lettuce necrotic yellows rhabdovirus
	Leafhopper	Weeks	Potato yellow dwarf rhabdovirus
	Leafhopper	Weeks	Wound tumor phytoreovirus
	Leafhopper	Weeks	Maize rayado fino marafivirus
	Planthopper	Weeks	Rice stripe tenuivirus

different strains of the same virus may have different specific vectors, and different biotypes of the same vector species may differ in their ability to transmit the same virus isolate. Thus a characteristic feature of plant virus/vector relations is their extreme *specificity*. This attracted attention from the earliest days of plant virology, partly because it would seem at first sight to be of little evolutionary advantage to the virus, compared with the alternative of having a very large number of vectors of diverse types.

The nature of the association of the virus with its vector is also very variable (Table 1) and again each virus or virus group exhibits only one of the possible types of association. In the so-called *non-circulative* viruses, the particles are associated with external or exoskeletal surfaces of the vector, and with insect or nematode vectors this means that the viruses are lost on moulting. The duration of this type of association may be only a few minutes or a few hours, as in *non-persistent* transmission by aphids, or may extend up to a few days, as in *semi-persistent* transmission by aphids or leafhoppers. Viruses transmitted by nematodes are retained by them for several weeks or months but are classed as semi-persistent in Table 1 because the nematodes cease to transmit on moulting. Non-persistent viruses are acquired by aphids after feeding on infected plants for periods of only 20-30 seconds up to about 5 minutes, and can then be immediately inoculated to healthy plants; the efficiency of transmission is increased by starving the aphids for 1-2 hours before the acquisition feed but is decreased by prolonging the acquisition feed. With semi-persistent aphid-borne viruses, the efficiency of transmission is increased with increasing acquisition and inoculation feeds up to 1-2 days and is not affected by pre-acquisition starvation. Viruses transmitted in the *circulative* manner by aphids or leafhoppers are able to pass through the gut wall into the haemocoel; they circulate in the haemolymph until they reach the salivary glands and are then secreted with the saliva. This process takes several hours and accounts for the so-called "latent period" - the time between the start of the minimum acquisition feeding time and the onset of ability to transmit. Some circulative viruses multiply in their insect vectors (*propagative*) but most do not (*non-propagative*).

The topic of virus/vector relations and the molecular basis of specificity has been the subject of several reviews in recent years (see for example Harrison and Murant, 1984; Harrison, 1987; Harrison and Robinson, 1988). In this paper therefore no attempt will be made to cover the subject in detail. Instead attention will be concentrated on a few selected groups of plant viruses for which the information available is most extensive and/or illuminating.

NON-CIRCULATIVE VIRUSES

Nepoviruses

The site of retention of virus particles in the vector has been identified directly by electron microscopy for only a few of the non-circulative viruses. Extensive studies with nepoviruses have shown that the 28 nm-diameter particles are retained at specific sites within the alimentary tract, the precise location depending on the particular virus/nematode combination. Particles of the nepoviruses, raspberry ringspot (RRV), tomato black ring (TBRV) and artichoke Italian latent, are found only in the stylet lumen and guiding sheath of *Longidorus* spp. (Taylor and Robertson, 1969; Taylor *et al.*, 1976), whereas particles of arabis mosaic (AMV), grapevine fanleaf, tobacco ringspot and strawberry latent ringspot nepoviruses are found attached only to the cuticular lining of the oesophagus of *Xiphinema* spp. The specific site of retention is often sharply delimited: in *X. diversicaudatum* the particles of AMV were attached to the wall of the odontophore (anterior part of the oesophagus) but not to that of the adjacent odontostyle (stylet) (Taylor and Robertson, 1970).

Presence of the particles of a given virus at a specific retention site in a particular species of nematode is usually well correlated with the ability of that nematode species to transmit the virus. Moreover, Brown and Trudgill (1983) showed that differences in the ability of different populations of *Xiphinema diversicaudatum* to transmit AMV were correlated with differences in the numbers of virus particles attached to the oesophageal linings. These differences between populations are under genetic control, as shown by Brown (1986) who crossed individual nematodes from different populations; his results suggest that frequency of transmission and efficiency of binding of the virus particles to sites of retention may be determined by a single partially dominant gene. In contrast to what was found with *X. diversicaudatum*, Taylor and Robertson (1975) and Trudgill and Brown (1978) observed many particles of a Scottish isolate of RRV attached to the linings of the guiding sheath and odontostyle of the inefficient vector, *L. macrosoma*. Particles of both Scottish and English isolates occurred in more than 50% of individual *L. macrosoma*, but the English isolate was transmitted by fewer than 5% of the nematodes and the Scottish isolate by fewer than 0.3%. These observations suggest that although particles of these isolates are able to bind to a specific site in *L. macrosoma* they are not easily released, especially those of the Scottish isolate. Transmissibility seems therefore to be the resultant of two components, attachment to and release from a specific retention site; efficient transmission depends on achieving an optimum balance between these two processes.

In histochemical studies Robertson and Henry (1986) detected a thin patchy layer of carbohydrate-containing material attached to the lining of the oesophagus of all *X. diversicaudatum* examined, whether carrying AMV or not. This material was probably of nematode origin, but in AMV-carrying individuals the virus particles were associated with these patches and were surrounded by a much thicker cloud-like layer of carbohydrate-containing material whose origin, whether from the nematode or from the plant, was not clear. This material possibly has some role in the binding of the virus particles to, or their release from, the oesophageal lining.

All these observations suggest that the specificity of virus/vector relations is determined by the ability of specific chemical groupings on the virus particle surface, presumably on the particle protein, to interact with other groupings at the specific retention site in the vector. Evidence that the virus particle protein might play an important part in virus/vector specificity was first obtained by Harrison (1964) and Taylor and Murant (1969) in work with RRV and TBRV. The nematodes *Longidorus macrosoma* and *L. attenuatus* transmit isolates of the "English" serotypes of RRV and TBRV respectively, but rarely or never transmit isolates of the "Scottish" serotype. Conversely, *L. elongatus* transmits isolates of the Scottish serotype of both viruses but is a poor vector of the English form of RRV and a very poor vector of the English form of TBRV. This type of correlation between the relative efficiencies of transmission of different virus isolates by a single nematode species and their serological relatedness seems a general phenomenon among nepoviruses (Harrison *et al.*, 1974b). However, Brown *et al.* (1989) recently showed that closely related isolates of the same serotype of TBRV nevertheless differ in transmissibility by a single population of *L. attenuatus*, an unexpected result that needs further investigation.

All nepoviruses have bipartite ssRNA genomes, and "pseudo-recombinant" isolates can be made between some pairs of isolates by taking the larger RNA (RNA-1) from one isolate and the smaller RNA (RNA-2) from another. Experiments of this sort with RRV and TBRV have shown that it is RNA-2 that both determines vector specificity (Harrison *et al.*, 1974a; Harrison and Murant, 1977) and carries the coat protein gene (Harrison *et al.*, 1972; Randles *et al.*, 1977). Although RNA-2 is large enough to code for at least one additional protein, it seems probable that vector specificity is determined by the coat protein.

Potyviruses

Aphid-transmitted potyviruses possess filamentous particles about 750 nm long and have a non-persistent relation with their vectors. They have often been referred to as "stylet-borne", despite the fact that little good evidence has been available until recently to indicate where they are carried in the insect. They display rather less specificity than most other kinds of plant virus in their vector relations, many of them being transmitted by several aphid species.

An intriguing feature of potyviruses is that some can act as helpers for aphid transmission of others. Thus potato virus Y (PVY) can assist the aphid transmission of a closely related strain, potato virus C (PVC), or even of potato aucuba mosaic virus (PAMV), which does not belong to the potyvirus group (Kassanis and Govier, 1971a). On their own, PVC and PAMV are not aphid-transmissible. Kassanis and Govier (1971b) and Govier and Kassanis (1974a,b) showed that aphids could not acquire and transmit PVY by feeding through membranes on purified virus preparations, even though these preparations were highly infective when inoculated to plants manually. However, the aphids could do so when they had previously fed either on PVY-infected leaves that had been u.v.-irradiated to destroy the infectivity of the virus, or through membranes on the non-infective supernatant fluid obtained by centrifuging a fresh extract of PVY-infected leaves. Thus it appeared that PVY-infected plants contain a factor other than the virus particles that is required for the aphid transmission of PVY itself as well as for that of the dependent viruses PVC and PAMV. All potyviruses so far investigated depend for their transmission by aphids on the presence of such a factor, which is a virus-coded non-structural protein, now called "helper component" (HC); non-transmissible potyvirus isolates are unable to induce formation of functional HC but can be aphid-transmitted in the presence of HC induced by transmissible isolates.

Studies by Pirone (1981) and Sako and Ogata (1981) showed that HC from one potyvirus could assist the aphid transmission of several other potyviruses but not all, indicating a degree of specificity. Thornbury and Pirone (1983) showed that the HCs from PVY and tobacco vein mottling potyvirus (TVMV) are serologically distinct, although they can each assist the aphid transmission of the heterologous virus; the helper activity was shown by immunoabsorption experiments to be associated with polypeptides of molecular weights 58,000 and 53,000 respectively. Hellmann *et al.* (1983) showed that antiserum prepared against the TVMV-HC precipitated a specific product of *in vitro* translation of TVMV RNA that was different from the products precipitated by antisera to the particle protein or cylindrical inclusion (pinwheel) proteins of TVMV, or to the two nuclear inclusion proteins of tobacco etch potyvirus (TEV). The HC therefore seems to be the product of a previously undiscovered cistron in TVMV RNA.

The mode of action of HC is still uncertain. Aphids must acquire it either prior to, or along with, the virus; aphids that feed first on virus and then on a source of HC, do not transmit (Kassanis and Govier, 1971a,b; Govier and Kassanis, 1974a,b). Autoradiography of frozen sections of the alimentary tract of aphids fed on ^{125}I-labelled TVMV in the presence of HC (Berger and Pirone, 1986a) revealed accumulations of label along the maxillary stylets, in the alimentary tract between the pharynx and the oesophageal valve, and in the gut. In contrast, label was found only in the gut of aphids fed on labelled virus in the absence of HC, or in the presence of inactivated HC. These observations suggest that HC might act by binding the virus particles to a retention site in the aphid's stylets (although other possibilities, such as that it acts by protecting virus from adverse conditions in the aphid's alimentary tract, cannot be ruled out). Berger and Pirone (1986b) reported evidence that the HC may be glycosylated, and this suggests that carbohydrates may be involved in binding. Stram *et al.* (1987) introduced cloned cDNA representing the HC cistron into *Escherichia coli*; the polypeptides expressed reacted with HC antiserum, but the major product was smaller than authentic HC and lacked biological activity. Berger *et al.* (1988) introduced cloned cDNA representing the HC cistron and the two flanking cistrons into tobacco and this construction "self-processed", resulting in the expression of HC identical in size to that induced in virus infections. Athough the HC in transformed tobacco was expressed at a level about 100-fold less than that in virus-infected plants, it had equal biological activity on a molecule for molecule basis when partially purified and concentrated (PH Berger, AG Hunt, LL Domier, GM Hellmann, Y Stram, DW Thornbury and TP Pirone, unpublished data).

A non-aphid-transmissible isolate of TEV (which can be assisted by PVY-HC) nevertheless produces a HC of its own which is able to assist the aphid transmission of another TEV isolate (Pirone and Thornbury, 1983). This suggests that the defect lies in the inability of the HC of this isolate to recognize its own coat protein. The nucleotide sequence of the particle protein-encoding region of the genome of the non-aphid-transmissible isolate shows six amino acid differences from that of the transmissible isolate and only three of these are on the surface of the virus particle, near the N-terminus (Hiebert *et al.*, 1984; Allison *et al.*, 1985). Two of these three surface replacements destroy DAG amino acid triplets, and Harrison and Robinson (1988) and Laín *et al.* (1988) noted that a DAG triplet occurs near the N-terminus in the amino acid sequence of four other aphid-transmissible potyviruses whose sequences are known. Moreover, a non-aphid transmissible isolate of one of these viruses (plum pox virus) has since been shown also to lack the DAG amino acid triplet (Maiss *et al.*,

1989). Laín *et al.* (1988) suggested that the DAG sequence might be important for the virus/vector interaction. Harrison and Robinson (1988) suggested that it might be involved in recognition of the particle protein by the helper component; they also suggested that the release of the virus particles from their site of retention in the aphid might be brought about by enzymatic cleavage because the exposed N-terminal portion of the coat protein has cleavage sites for trypsin (Hiebert *et al.*, 1984; Shukla *et al.*, 1988).

Caulimoviruses

The caulimoviruses are a group of viruses with 50 nm isometric particles and double-stranded DNA genomes. Studies with aphid-transmissible and non-aphid-transmissible isolates of cauliflower mosaic caulimovirus (CaMV) showed that it, like the potyviruses, depends on a virus-coded HC for transmission by aphids in a non-persistent manner (Lung and Pirone, 1973, 1974). The CaMV HC did not assist aphid-transmission of PVY, nor did the PVY or TEV HCs assist aphid-transmission of CaMV. Woolston *et al.* (1983) showed that aphid-transmissible isolates induce the formation of a protein of molecular weight 18,000 (p18), which co-purifies with the large proteinaceous inclusion bodies in which the virus particles are embedded. By making hybrid virus genomes from the DNA of an aphid-transmissible isolate, Cabb B-JI, and a non-aphid-transmissible isolate, Campbell, they were able to show that both aphid-transmissibility and the production of p18 are determined by open reading frame (ORF) II of the viral genome, indicating that p18 is the HC. However, serological tests showed that a protein that reacts with antiserum to p18 was present in plants infected with the Campbell isolate, although it was absent from plants infected with another non-aphid-transmissible isolate, CM4-184, which has a large deletion in ORF II (Harker *et al.*, 1987). The lack of aphid-transmissibility of the Campbell isolate was narrowed down to a small portion of ORF II that showed only two differences from that of the aphid-transmissible isolate (Woolston *et al.*, 1987); only one of these changes resulted in an amino acid exchange (from glycine to arginine) in the potential protein product, suggesting that it is this single change that causes loss of function.

Semi-persistent aphid-and leafhopper-borne viruses

Anthriscus yellows virus (AYV) is a semi-persistent aphid-borne virus which acts as a helper for the aphid transmission of a second virus, parsnip yellow fleck (PYFV). The 28 nm-diameter isometric particles of AYV were found at a site on the ventral wall of the pharynx of *Cavariella aegopodii* (Murant *et al.*, 1976), attached to a pad of lightly staining material ("M-material") that was also present in virus-free aphids. However, the particles were surrounded immediately by a layer of more densely staining material that resembled in appearance the matrix protein of the inclusion bodies in which they are found embedded in plant cells. Murant (1978) speculated, by analogy with what was known for caulimoviruses and potyviruses, that this densely staining material might represent a virus-coded protein that acts as a helper factor for aphid transmission of AYV as well as that of PYFV. The M-material somewhat resembles the "cloud-like" carbohydrate-containing material found in the nematode *X. diversicaudatum* carrying AMV; its composition is unknown but it seems not to contain carbohydrate (C Fasseas, AF Murant, IM Roberts and WM Robertson, unpublished data).

Recently Childress and Harris (1989) detected the association of the 28 nm-diameter isometric particles of the semi-persistent maize chlorotic dwarf virus (MCDV), and of "M-material", with cuticular linings of the anterior alimentary tract of its leafhopper vector, *Graminella nigrifrons*. As with AYV, the particles were surrounded by more densely staining material, and it is therefore of interest that MCDV is newly reported to have a helper factor (Hunt *et al.*, 1988).

CIRCULATIVE VIRUSES

Pea enation mosaic virus

Pea enation mosaic virus (PEMV) has isometric 28 nm-diameter particles; it is transmitted in the circulative manner by the aphid *Acyrthosiphon pisum* and also by mechanical inoculation. Some isolates (NT) are not aphid-transmissible except from plants that are also infected with transmissible (T) isolates (Adam, 1978). Electrophoretic analysis shows that NT isolates have a single particle protein, of molecular weight about 22,000, which is encoded by the larger of the two species of genomic RNA, RNA-1 (Hull and Lane, 1973). T isolates have an additional protein of molecular weight about 56,000 and an RNA-1 of molecular weight about 1.8×10^6, about 1.2 x 10^5 larger than that of NT isolates (Hull, 1977; Adam *et al.*, 1979). Thus NT isolates appear to be deletion mutants that have lost the part of the genome coding for the 56,000 molecular weight protein. Electron microscopic examination of thin sections of aphids that had had preparations of a T isolate injected into the haemocoel revealed virus-like particles embedded in the basal lamina and in internal cisternae of the aphid's accessory salivary glands, whereas no particles were seen in aphids injected with preparations of a NT isolate (Harris *et al.*, 1975). These results suggest that the 56,000 molecular weight protein plays an important part in aphid transmission and may be responsible for binding of the particles to the basal lamina and for their entry into the accessory salivary gland.

Luteoviruses

Luteoviruses are transmitted by aphids in the circulative manner but there is no hard evidence that they multiply in the vector. They have isometric particles about 25 nm in diameter, are confined to phloem tissue of infected plants and are not transmissible by manual inoculation. There are many viruses in the group, but each seems to be transmitted by only one or at most very few species of aphid. However, in several instances it is known that the particles of a given luteovirus can pass into the haemolymph of an inefficient or non-vector (Harrison, 1958; Rochow, 1969), and it seems that vector specificity is to a large extent determined by the ability of the virus particles to enter the saliva. Rochow (1965) studied this phenomenon with two vector-specific isolates, both considered by him to be strains of barley yellow dwarf luteovirus (BYDV): isolate MAV, transmitted by *Sitobion (Macrosiphum) avenae*, and RPV, transmitted by *Rhopalosiphum padi*. In fact, MAV and RPV have no direct serological relationship to each other and probably only MAV should be regarded as a strain of BYDV, RPV being closer to beet western yellows luteovirus (BWYV). When MAV and RPV were established in mixed infections, both were transmitted by *R. padi*, although only RPV is transmitted by this species from singly infected plants. MAV was not transmitted by *R. padi* fed sequentially, in either order, on plants singly infected with each of the viruses, or by *R. padi* injected into the haemocoel with a mixture of separate purified preparations of the two viruses (Rochow, 1973). However, both viruses were transmitted by *R. padi* injected with extracts of mixedly infected plants (Rochow, 1970). Treatment of such extracts with antiserum to RPV before injection into *R. padi* prevented transmission, but treatment with antiserum to MAV did not. These observations led to the conclusion that in mixed infections MAV RNA can be packaged in RPV coat protein, the particles thus formed being transmissible by *R. padi*. This packaging of the nucleic acid of one (dependent) virus in shells composed entirely of the coat protein of another (helper) virus was termed "transcapsidation" (Rochow, 1972); the term "phenotypic mixing" was used to refer to packaging in shells composed of a mixture of the coat proteins of both viruses. The latter phenomenon has recently (Hu *et al.*, 1988) been shown to occur in mixed infections of MAV with PAV, a serologically related strain of BYDV that is transmitted by both *R. padi* and *S. avenae*; as would be expected from the earlier results, MAV is transmitted by *R. padi* from such mixed infections.

Clearly the coat protein plays a major part in determining vector specificity of luteoviruses, and clues to the way in which this might operate were provided by the work of Gildow and Rochow (1980). They examined thin sections of virus-carrying *S. avenae* and found particles of MAV and RPV in the basal lamina of the accessory salivary glands, and in invaginations of the plasmalemma, but only rarely in the basal lamina of the principal salivary gland, and not at all elsewhere. Only with the *S. avenae*-transmissible MAV were particles found *within* the accessory salivary gland, in cytoplasmic vesicles and secretory canals. Gildow and Rochow (1980) and Gildow (1982) suggested that the virus particles pass from the plasmalemma to the salivary ducts within coated vesicles, which were observed not only in vector aphids carrying MAV and RPV, but also in those carrying BWYV or potato leafroll luteovirus (PLRV). These results suggest that specificity depends on the ability of the particle coat protein to bind to specific receptors, both on the plasmalemma of the accessory salivary gland cells and on the membrane of the coated vesicles.

Viruses dependent on luteoviruses for aphid transmission

In addition to the dependence of some luteoviruses on others for transmission by aphids, several instances are known in which a luteovirus can assist the aphid-transmission, in the persistent manner, of entirely unrelated viruses. For example, carrot red leaf luteovirus (CRLV) assists the transmission of carrot mottle virus (CMotV) by *Cavariella aegopodii* (Watson *et al.*, 1964), and groundnut rosette assistor luteovirus (GRAV) assists the transmission of groundnut rosette virus (GRV) by *Aphis craccivora* (Hull and Adams, 1968). In these systems the dependent virus is transmissible by manual inoculation whereas the luteovirus helper is not. An important way in which the helper dependence of these viruses differs from that of potyviruses and caulimoviruses is that both the helper luteovirus and the dependent virus must be present in the same source plant for aphid transmission to occur; aphids already carrying the helper luteovirus cannot acquire and transmit the dependent virus from a separate plant.

Murant *et al.* (1969) showed that the manually transmissible infectivity in extracts of *Nicotiana clevelandii* infected with CMotV alone is abolished by treatment with ether, chloroform or butan-1-ol and is associated with membrane-bounded structures about 52 nm in diameter. We still do not know whether these structures are vesicles composed of host plant membranes, perhaps associated with viral RNA, or enveloped virus particles of a kind unusual among plant viruses but found among some groups of viruses infecting vertebrates and insects (e.g. alphaviruses and flaviviruses). In contrast, the manually transmissible infectivity in preparations purified from chervil (*Anthriscus cerefolium*) plants infected with both CMotV and CRLV is not destroyed by chloroform, is associated with particles that have the same size and sedimentation behaviour as those of CRLV, and is neutralized by antiserum to CRLV (Waterhouse and Murant, 1983). Moreover, when haemolymph from *C. aegopodii* that had fed on plants infected with both CMotV and CRLV was injected into virus-free aphids, both viruses were transmitted, whereas only CRLV was transmitted when aphids that had fed on plants infected with CRLV alone were injected with extracts from plants infected with CMotV alone (Elnagar and Murant, 1978). Waterhouse and Murant (1983) showed that transmission of both viruses in the first type of test was prevented by treating the aphid extracts with CRLV antiserum. These results suggest that in doubly infected plants the RNA of CMotV becomes packaged in the particle protein of CRLV, in which form CMotV is transmissible by *C. aegopodii*.

Waterhouse and Murant (1983) showed that experimentally CMotV could be associated with two other luteoviruses, BWYV and PLRV, in which event it became transmissible by *Myzus persicae*, the vector of these two viruses. In the light of these experiments, transmission-dependence begins perhaps to seem less of a disadvantage to a virus but possibly a way of extending its range of vectors and thereby possibly even its natural host range.

In its general properties and its dependence on a luteovirus for aphid transmission, GRV closely resembles CMotV (Reddy *et al.*, 1985). Both viruses induce the formation of abundant double-stranded RNA in infected plants but whereas CMotV produces only two major species, dsRNA-1 and dsRNA-2, with molecular weights of about 3.0×10^6 and 0.9×10^6 respectively (Murant *et al.*, 1985), GRV produces not only two dsRNA molecules of about these sizes but in addition a very abundant species of molecular weight about 0.6×10^6, called dsRNA-3, together with a number of minor species of intermediate size (Reddy *et al.*, 1985; Murant *et al.*, 1988). These dsRNA species of CMotV and GRV are not infective unless heat-denatured, and are thought to represent the replicative forms of infective ssRNA counterparts. By establishing a culture of GRV that was free from dsRNA-3 and the numerous minor dsRNA species, Murant *et al.* (1988) obtained evidence which suggested that dsRNA-3 represents the replicative form of a satellite RNA. Moreover, this satellite seems to be largely responsible for the symptoms of groundnut rosette disease, which is the most important virus disease of groundnut in Africa. Here we have a most complicated system of interacting components: a disease caused by a satellite RNA that depends on GRV for its replication, both the satellite RNA and GRV RNA being dependent on GRAV for transmission by aphids, presumably through these molecules becoming packaged in shells composed of GRAV coat protein. A further twist to this story is that preliminary results (AF Murant, unpublished data) suggest that dsRNA-3 may play a key part in mediating the dependence of GRV on GRAV. If that proves to be true we must ask ourselves whether dsRNA-3 is really a satellite: is it perhaps a genome part of GRV?

DISCUSSION

In this short review selected examples have been presented to illustrate the diverse ways in which plant viruses interact with their vectors. Some kinds of plant viruses, such as those transmitted by beetles or fungi, or those transmitted in the circulative, propagative manner by insects, have been omitted entirely. However, it is clear that the transmission of a plant virus by a vector is a culmination of a complex series of recognition events (Table 2). Some of these are the events that occur in the normal processes of RNA packaging and particle assembly, but others are specific to the transmission process.

Few, if any, plant viruses are transmitted as a result of merely superficial contamination of the mouthparts or feeding apparatus of the vector. Most have a specific type of association with a specific vector and, in all the viruses that have been

Table 2. *Recognition events involved in the transmission of plant viruses*

Component A	Component B	Function
Virus nucleic acid	Virus particle protein	Nucleic acid packaging
Virus particle protein	Virus particle protein	Particle assembly
Virus particle protein	Vector retention site	Acquisition by vector
Virus particle protein	Helper factor	Acquisition by vector
Helper factor	Vector retention site	Acquisition by vector
Virus particle protein	Release factor	Inoculation by vector

adequately studied, the virus particles bind to specific sites of retention in the vector. With non-circulative viruses these sites are on the external or exoskeletal surfaces, whereas with circulative insect-borne viruses they have been identified internally, on the basal lamina and plasmalemma of the accessory salivary glands. In these interactions the virus particle protein clearly plays a key part, and this function of the protein is additional to that of packaging and protecting the genomic nucleic acid. With viruses that are circulative in insect vectors the virus particle protein presumably also functions to enable the virus particles to survive in the gut and to pass across the gut wall into the haemolymph. Probably the requirement to maintain recognition between sites on the virus particle surface and at the vector retention site accounts for the high degree of antigenic conservation exhibited by most plant viruses. This contrasts with the majority of vertebrate-infecting viruses which do not need to relate to a vector but depend on antigenic diversity to overcome the immune response.

Some plant viruses induce the formation of a protein additional to the main coat protein, such as the HC of caulimoviruses and potyviruses, or the 56,000 molecular weight protein of pea enation mosaic virus, which seems to act as an intermediary in binding the virus particle to the site of retention in the vector. The reason for the evolution of such helper systems is unclear but Harrison and Murant (1984) suggested that they may allow the virus to develop a broader than usual vector specificity. Certainly this is true of the potyviruses, and is a feature that seems especially important for viruses that persist for very brief periods in the vector and therefore probably spread over only short distances in nature.

Little is known about the nature of the substances to which the virus particles bind in the vector, but the work of Robertson and Henry (1986) indicates that at least in nematodes this material may contain carbohydrate.

Equally as important as the attachment of virus particles to specific retention sites in the vector is their release from these sites for inoculation into a new plant. Although we have no direct evidence about the nature of any such release mechanism, we may suspect that it could happen as a result of the action of enzymes released by the vector, for example in saliva, at specific sites in the virus particle protein. The fact that the exposed N-terminal and C-terminal portions of potyvirus coat proteins are removable by treatment with trypsin seems especially interesting in this regard (Harrison and Robinson, 1988).

REFERENCES

Adam G (1978) Die gemeinsame Übertragung eines vektoriellen und eines avektoriellen Stammes des pea enation mosaic virus durch die Blattlaus *Acyrthosiphon pisum*. Z Pflanzenkr Pflanzenschutz 85:586-593

Adam G, Sander E, Shepherd RJ (1979) Structural differences between pea enation mosaic virus strains affecting transmissibility by *Acyrthosiphon pisum* (Harris). Virology 92:1-14

Allison RF, Dougherty WG, Parks TD, Willis L, Johnston RE, Kelly M, Armstrong FB (1985) Biochemical analysis of the capsid protein gene and capsid protein of tobacco etch virus: N-terminal amino acids are located on the virion's surface. Virology 147:309-316

Berger PH, Pirone TP (1986a) The effect of helper component on the uptake and localization of potyviruses in *Myzus persicae*. Virology 153:256-261

Berger PH, Pirone TP (1986b) Evidence that the potyvirus helper component is a glycoprotein. Phytopathology 76:1063

Berger PH, Hunt AG, Hellmann GM, Pirone TP (1988) Expression of tobacco vein mottling virus helper component protein in transgenic tobacco. Phytopathology 78:1517

Brown DJF (1986) Transmission of virus by the progeny of crosses between *Xiphinema diversicaudatum* (Nematoda: Dorylaimoidea) from Italy and Scotland. Rev Nematol 9:71-74

Brown DJF, Trudgill DL (1983) Differential transmissibility of arabis mosaic and strains of strawberry latent ringspot viruses by three populations of *Xiphinema diversicaudatum* (Nematoda: Dorylaimida) from Scotland, Italy and France. Rev Nematol 6:229-238

Brown DJF, Murant AF, Trudgill DL (1989) Differences between isolates of the English serotype of tomato black ring virus in their transmissibility by an English population of *Longidorus attenuatus* (Nematoda: Dorylaimoidea). Rev Nematol 12:51-56

Childress SA, Harris KF (1989) Localization of virus-like particles in the foreguts of

viruliferous *Graminella nigrifrons* leafhoppers carrying the semi-persistent maize chlorotic dwarf virus. J Gen Virol 70:247-251
Elnagar S, Murant AF (1978) Aphid-injection experiments with carrot mottle virus and its helper virus, carrot red leaf. Ann Appl Biol 89:245-250
Gildow FE (1982) Coated-vesicle transport of luteoviruses through salivary glands of *Myzus persicae*. Phytopathology 72:1289-1296
Gildow FE, Rochow WF (1980) Role of accessory salivary glands in aphid transmission of barley yellow dwarf virus. Virology 104:97-108
Govier DA, Kassanis B (1974a) Evidence that a component other than the virus particle is needed for aphid transmission of potato virus Y. Virology 57:285-286
Govier DA, Kassanis B (1974b) A virus-induced component of plant sap needed when aphids acquire potato virus Y from purified preparations. Virology 61:420-426
Harker CL, Woolston CJ, Markham PG, Maule AJ (1987) Cauliflower mosaic virus aphid transmission factor protein is expressed in cells infected with some aphid nontransmissible isolates. Virology 160:252-254
Harris KF, Bath JE, Thottapilly G, Hooper GR (1975) Fate of pea enation mosaic virus in PEMV-injected pea aphids. Virology 65:148-162
Harrison BD (1958) Studies on the behavior of potato leaf roll and other viruses in the body of their aphid vector *Myzus persicae* (Sulz.). Virology 6:265-277
Harrison BD (1964) Specific nematode vectors for serologically distinctive forms of raspberry ringspot and tomato black ring viruses. Virology 22:544-550
Harrison BD (1987) Plant virus transmission by vectors: mechanisms and consequences. In: Russell WC, Almond JW (eds) Molecular basis of virus disease. Society for General Microbiology Symposium No 40, Cambridge University Press, Cambridge, p 319
Harrison BD, Murant AF (1977) Nematode transmissibility of pseudo-recombinant isolates of tomato black ring virus. Ann Appl Biol 86:209-212
Harrison BD, Murant AF (1984) Involvement of virus-coded proteins in transmission of plant viruses by vectors. In: Mayo MA, Harrap KA (eds) Vectors in virus biology. Academic Press, London, p 1
Harrison BD, Robinson DJ (1988) Molecular variation in vector-borne plant viruses: epidemiological significance. Phil Trans Roy Soc Lond, Ser B 321:447-462
Harrison BD, Murant AF, Mayo MA (1972) Two properties of raspberry ringspot virus determined by its smaller RNA. J Gen Virol 17:137-141
Harrison BD, Murant AF, Mayo MA, Roberts IM (1974a) Distribution of determinants for symptom production, host range and nematode transmissibility between the two RNA components of raspberry ringspot virus. J Gen Virol 22:233-247
Harrison BD, Robertson WM, Taylor CE (1974b) Specificity of retention and transmission of viruses by nematodes. J Nematol 6:155-164
Hellmann GM, Thornbury DW, Hiebert E, Shaw JG, Pirone TP, Rhoads RE (1983) Cell-free translation of tobacco vein mottling virus RNA. II. Immunoprecipitation of products by antisera to cylindrical inclusion, nuclear inclusion, and helper component proteins. Virology 124:434-444
Hiebert E, Thornbury DW, Pirone TP (1984) Immunoprecipitation analysis of potyviral *in vitro* translation products using antisera to helper component of tobacco vein mottling virus and potato virus Y. Virology 135:1-9
Hu JS, Rochow WF, Palukaitis P, Dietert RR (1988) Phenotypic mixing: mechanism of dependent transmission for two related isolates of barley yellow dwarf virus. Phytopathology 78: 1326-1330
Hull R (1977) Particle differences related to aphid-transmissibility of a plant virus. J Gen Virol 34:183-187

Hull R, Adams AN (1968) Groundnut rosette and its assistor virus. Ann Appl Biol 62:139-145

Hull R, Lane LC (1973) The unusual nature of the components of a strain of pea enation mosaic virus. Virology 55:1-13

Hunt RE, Nault LR, Gingery RE (1988) Evidence for infectivity of maize chlorotic dwarf virus and for a helper component in its leafhopper transmission. Phytopathology 78:499-504

Kassanis B, Govier DA (1971a) New evidence on the mechanism of aphid transmission of potato C and potato aucuba mosaic viruses. J Gen Virol 10:99-101

Kassanis B, Govier DA (1971b) The role of the helper virus in aphid transmission of potato aucuba mosaic virus and potato virus C. J Gen Virol 13:221-228

Laín S, Riechmann JL, Méndez E, García JA (1988) Nucleotide sequence of the 3' terminal region of plum pox potyvirus RNA. Virus Res 10:325-342

Lung MCY, Pirone TP (1973) Studies on the reason for differential transmissibility of cauliflower mosaic virus isolates by aphids. Phytopathology 63:910-914

Lung MCY, Pirone TP (1974) Acquisition factor required for aphid transmission of purified cauliflower mosaic virus. Virology 60:260-264

Maiss E, Timpe U, Brisske A, Jelkmann W, Casper R, Himmler G, Mattanovich D, Katinger HWD (1989) The complete nucleotide sequence of plux pox virus RNA. J Gen Virol 70:513-524

Murant AF (1978) Recent studies on association of two plant virus complexes with aphid vectors. In: Scott PR, Bainbridge A (eds) Plant disease epidemiology. Blackwell, Oxford, p 243

Murant AF, Goold RA, Roberts IM, Cathro J (1969) Carrot mottle - a persistent aphid-borne virus with unusual properties and particles. J Gen Virol 4:329-341

Murant AF, Roberts IM, Elnagar S (1976) Association of virus-like particles with the foregut of the aphid *Cavariella aegopodii* transmitting the semi-persistent viruses anthriscus yellows and parsnip yellow fleck. J Gen Virol 31:47-57

Murant AF, Waterhouse PM, Raschké JH, Robinson DJ (1985) Carrot red leaf and carrot mottle viruses: observations on the composition of the particles in single and mixed infections. J Gen Virol 66:1575-1579

Murant AF, Rajeshwari R, Robinson DJ, Raschké JH (1988) A satellite RNA of groundnut rosette virus that is largely responsible for symptoms of groundnut rosette disease. J Gen Virol 69:1479-1486

Pirone TP (1981) Efficiency and selectivity of the helper-component-mediated aphid transmission of purified potyviruses. Phytopathology 71:922-924

Pirone TP, Thornbury DW (1983) Role of virion and helper component in regulating aphid transmission of tobacco etch virus. Phytopathology 73:872-875

Randles JW, Harrison BD, Murant AF, Mayo MA (1977) Packaging and biological activity of the two essential RNA species of tomato black ring virus. J Gen Virol 36:187-193

Reddy DVR, Murant AF, Raschké JH, Mayo MA, Ansa OA (1985) Properties and partial purification of infective material from plants containing groundnut rosette virus. Ann Appl Biol 107:65-78

Robertson WM, Henry CE (1986) An association of carbohydrates with particles of arabis mosaic virus retained within *Xiphinema diversicaudatum*. Ann Appl Biol 109:299-305

Rochow WF (1965) Apparent loss of vector specificity following double infection by two strains of barley yellow dwarf virus. Phytopathology 55:62-68

Rochow WF (1969) Biological properties of four isolates of barley yellow dwarf virus. Phytopathology 59:1580-1589

Rochow WF (1970) Barley yellow dwarf virus: phenotypic mixing and vector specificity. Science 167:875-878

Rochow WF (1972) The role of mixed infections in the transmission of plant viruses by aphids. Annu Rev Phytopathol 10:101-124

Rochow WF (1973) Selective virus transmission by *Rhopalosiphum padi* exposed sequentially to two barley yellow dwarf viruses. Phytopathology 63:1317-1322

Sako N, Ogata K (1981) Different helper factors associated with aphid transmission of some potyviruses. Virology 112:762-765

Shukla DD, Strike PM, Tracy SL, Gough KH, Ward CW (1988) The N and C termini of the coat proteins of potyviruses are surface-located and the N terminus contains the major virus-specific epitopes. J Gen Virol 69:1497-1508

Stram Y, Berger PH, Pirone TP (1987) Cloning and expression of the potyvirus helper component gene. Abstr VII Int Congr Virol, Edmonton, Canada, p 215

Taylor CE, Murant AF (1969) Transmission of strains of raspberry ringspot and tomato black ring viruses by *Longidorus elongatus* (de Man). Ann Appl Biol 64:43-48

Taylor CE, Robertson WM (1969) The location of raspberry ringspot and tomato black ring viruses in the nematode vector, *Longidorus elongatus* (de Man). Ann Appl Biol 64:233-237

Taylor CE, Robertson WM (1970) Sites of virus retention in the alimentary tract of the nematode vectors, *Xiphinema diversicaudatum* (Micol.) and *X. index* (Thorne and Allen). Ann Appl Biol 66:375-380

Taylor CE, Robertson WM (1975) Acquisition, retention and transmission of viruses by nematodes. In: Lamberti F, Taylor CE, Seinhorst JW (eds) Nematode vectors of plant viruses. Plenum, London New York, p 253

Taylor CE, Robertson WM, Roca F (1976) Specific association of artichoke Italian latent virus with the odontostyle of its vector, *Longidorus attenuatus*. Nematol Mediterr 4:23-30

Thornbury DW, Pirone TP (1983) Helper components of two potyviruses are serologically distinct. Virology 125:487-490

Trudgill DL, Brown DJF (1978) Ingestion, retention and transmission of two strains of raspberry ringspot virus by *Longidorus macrosoma*. J Nematol 10:85-89

Waterhouse PM, Murant AF (1983) Further evidence on the nature of the dependence of carrot mottle virus on carrot red leaf virus for transmission by aphids. Ann Appl Biol 103:455-464

Watson M, Serjeant EP, Lennon EA (1964) Carrot motley dwarf and parsnip mottle viruses. Ann Appl Biol 54:153-166

Woolston CJ, Covey SN, Penswick JR, Davies JW (1983) Aphid transmission and a polypeptide are specified by a defined region of the cauliflower mosaic virus genome. Gene 23:15-23

Woolston CJ, Czaplewski LG, Markham PG, Goad AS, Hull R, Davies JW (1987) Location and sequence of a region of cauliflower mosaic virus gene 2 responsible for aphid transmissibility. Virology 160:246-251

THE GENETICS OF PLANT-VIRUS INTERACTIONS: MECHANISMS CONTROLLING HOST RANGE, RESISTANCE AND VIRULENCE

R S S Fraser
AFRC Institute of Horticultural Research
Worthing Road
Littlehampton
West Sussex BN17 6LP
UK

ABSTRACT

This paper reviews the types of genetic interactions between plants and viruses. Some possible models are considered from a mainly theoretical point of view, and are then compared with the experimental data. For non-host immunity, the evidence for positive (inhibitor) and negative (lack of susceptibility) mechanisms is assessed, and the possibility that some cases of apparent non-host immunity operate by preventing cell-to-cell transport is considered. For cultivar resistance, links between different types of genetic control and possible mechanisms are discussed. Virulence (resistance-breaking behaviour by the virus) is considered in terms of frequency, genetic control, and possible association with loss of pathogenic fitness.

INTRODUCTION

Plant species and viruses of plants interact with each other in various ways, producing a variety of outcomes. These range from non-pathogenesis - where there appears to be no interaction - through to symptom formation and virus multiplication. The susceptible (pathogenic) interaction may be modified by host resistance, which in turn may be modified by resistance-breaking behaviour on the part of the virus. All of these types of interactions are ultimately under the control of the plant and viral genomes. In some cases, the outcome is controlled by single or a small number of genes, which may have control of virus-plant interaction as their sole or primary function, or which may have some other primary function in plant growth or viral pathogenesis. These interactions controlled by single genes are likely to involve discrete

NATO ASI Series, Vol H 41
Recognition and Response in Plant-Virus Interactions
Edited by R S S Fraser

recognition events between the products of host and viral genes, and metabolic responses to any signal so generated. In other interactions between plant and virus, the outcome may be influenced by a large number of genes, which may be exerting an indirect effect on pathogenesis and virus multiplication. These interactions may be less likely to involve discrete recognition events with signalling consequences.

The objective of this chapter is to look at some of the types of genetic interaction between plants and viruses, and to consider where recognition and signalling processes may be involved. The approach has to be based on a phenotypic classification of interaction types; this taxonomy is undoubtedly oversimplified. Knowledge of the genetic and biochemical mechanisms underlying an interaction may cause a reappraisal. Having said this, however, it is useful to define certain terms describing types of interaction. These terms fall into sets which apply at different levels of complexity of the plant population.

The first level is the plant species. A species may be either a *host* or a *non-host* for a particular virus; a virus is either *pathogenic* or *non-pathogenic* to a particular species. In the case of a non-host species/non-pathogenic virus, the assumption from visual evidence is that the virus produces no symptoms and does not multiply in the non-host when inoculated. As will be seen below, this assumption may be an oversimplification, but it does provide the starting point for thinking about the types of genetic control that might be involved in non-host resistance.

The second level compares varieties within a species, or closely related species within a genus, where *susceptibility* to a particular virus exists. *Resistant* varieties or related species contain a gene or genes, which inhibit virus multiplication or pathogenicity of *avirulent* virus isolates. Mutants or isolates of the virus which can overcome the resistance gene are referred to as *virulent*.

Finally, the interaction of plant and virus can be modulated at the level of the individual plant: a type of resistance to a virus normally affecting that plant can be conferred by a prior infection, or by chemical or environmental treatment. This *induced* or *acquired* resistance ultimately has a genetic basis. In most cases examined so far, it depends on the presence of a gene for hypersensitive resistance to the virus involved, and the induced resistance may be regarded as a further modulation of the primary activity of that (cultivar) resistance gene. Induced resistance mechanisms are considered elsewhere in this volume, especially in the chapters by Fritig *et al.* and Linthorst *et al.* The topic is not considered further here.

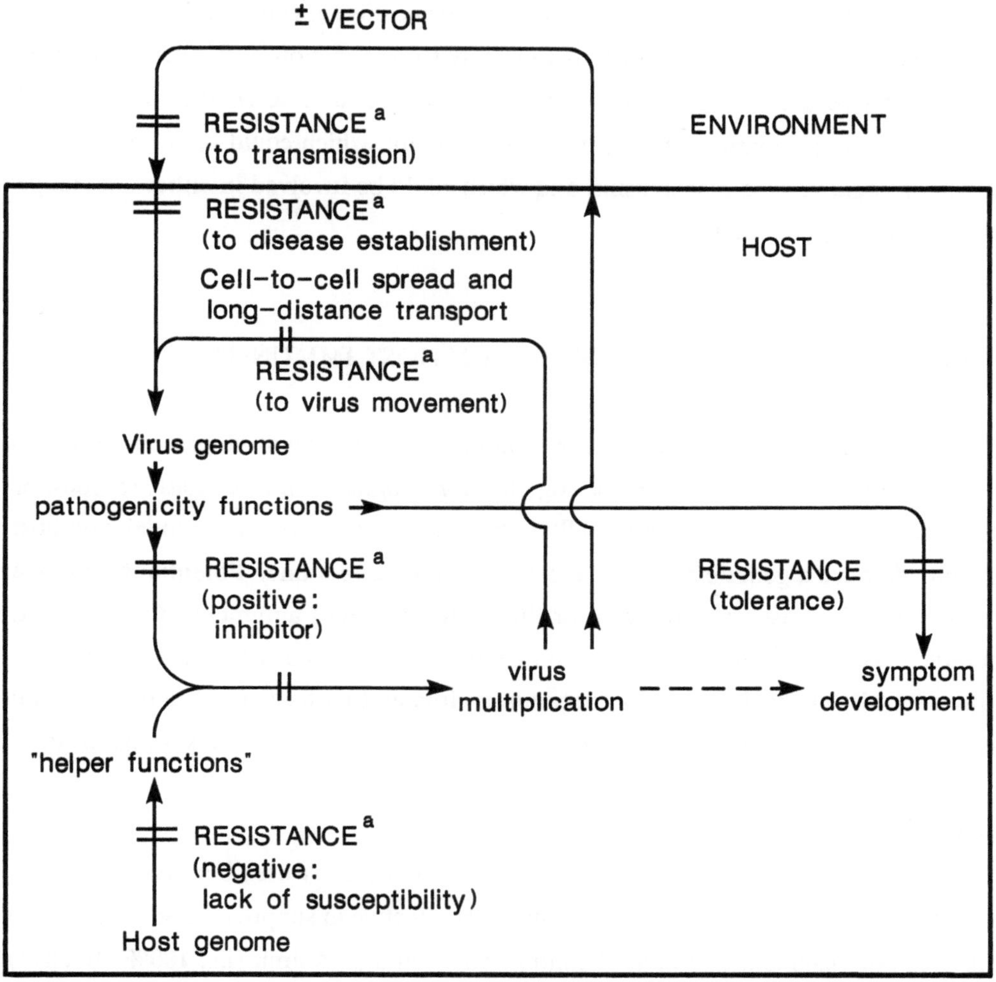

Figure 1. The replicative cycle of a plant virus, showing various possible targets of cultivar resistance mechanisms. [a]Indicates resistance mechanisms which might also be involved in the determination of host range.

Possible models of interaction

Before considering the genetic evidence on non-host resistance and cultivar resistance, it may be useful to consider some possible types of interaction which might apply to both. Fig. 1 attempts to identify points and processes at which determinative interactions or lack of interactions might occur. We can distinguish two basic types of mechanism: *positive*, in which the plant contains, or is induced to produce, some inhibitor of viral multiplication or pathogenesis, and *negative*, in which multiplication and pathogenesis fail because the plant fails to supply some function normally required by the virus. There are also possible determinative events at disease transmission and establishment, and on virus movement within the plant, which could operate by either positive- or negative-type mechanisms, and which could be involved in either host range control, or cultivar resistance.

POSSIBLE GENETIC MECHANISMS IN NON-HOST IMMUNITY

Most plant species are effectively non-hosts to most plant viruses in nature or agriculture, because they do not come together for geographical or climatic reasons, or for the lack of a suitable vector. Ultimately, these types of operational non-host immunity do have a genetic basis, but the effect is an indirect one. Potential host range can also be probed experimentally, by artificial inoculation of a number of species (e.g. Horvath, 1979). Some of the information from this type of study may not be relevant to any naturally-evolved interaction between a virus and its hosts (and non-hosts), but may nonetheless give useful information on potential mechanisms of host range control.

Positive (inhibitor) mechanisms

A plant may be a non-host for a particular virus or viruses because it contains some inhibitor or inhibitors which are completely effective in suppressing some essential feature of establishment and multiplication. This could be a limit (i.e. 100% effective) case of a cultivar-type resistance such as the hypersensitive response, where a restricted amount of multiplication occurs. Holmes (1955) suggested that such non-host immunity might depend on the additive effects of a large number - say 20 to 40 - of individual resistance genes. There is evidence for additive effects of resistance genes at different loci (e.g. Pelham, 1972; Drijfhout, 1978) but not to the scale suggested by Holmes, and only after the genes had been combined by the plant breeder. It is also difficult to envisage the accumulation of a large number of specific resistance genes in nature during evolution of species which would not normally be exposed to a particular virus, but which turn out to be non-hosts when inoculated experimentally.

Positive mechanisms of non-host resistance might also be based on non-specific effects, which generally render the cell unwelcoming to viruses or groups of viruses.

Such mechanisms might include unfavourable cell pH and ionic conditions, or the presence of phenolics and other compounds with deleterious effects on virus particles (Simons and Ross, 1971; Ireland and Pierpoint, 1980). Perhaps one reason for the paucity of reports of virus infections of Gymnosperms is their high content of resins and other compounds with possible activity against virus particles.

Negative (lack of susceptibility) mechanisms

Bald and Tinsley (1967a,b,c) proposed that non-host species were such because they lacked particular *susceptibility genes*. In the terms of Fig. 1, the non-host would fail to provide certain "helper functions" required by the virus.

These susceptibility functions might include host-coded replicase subunits which would interact with the virus-coded component to form a functional replicase (Mouchés *et al.*, 1984; Gargouri and Haenni, this volume). As yet, our knowledge of the biochemistry of these subunits is too limited to say whether they play a part in recognition and determination of host range. While putative virus-coded replicase components have been suggested for numerous virus groups (Goldbach *et al.*, this volume), a host-coded component has been detected for only a few virus infections, and rigorous proof that the host component contributes to the true replicase activity is generally lacking.

Another possible area of interaction is that of specific virus receptors which might be involved in the initial attachment and uncoating. However, the evidence for these in plant virology is poor (Atabekov *et al.*, 1970; Hiebert *et al.*, 1968; reviewed in Fraser, 1987a).

Thus, in sum, the experimental evidence for the Bald and Tinsley "quasi-genetic" model of host range control is still sparse. This seems to be related more to the difficulties of investigating the biochemistry of recognition events in plant-virus interactions, rather than to any discernable inadequacy in the theory. One powerful feature of the model is that it does offer explanations of three aspects of the comparative host ranges of different viruses. Briefly, successful pathogenesis is determined by recognition events between appropriate *susceptibility factors* in the host and matching *pathogenicity factors* in the virus. Host range is determined by the number and nature of the factors in the plant and virus components. Match and mis-match can then explain why two viruses may have non-overlapping host ranges; why the host range of one virus may be entirely contained within the (wider) host range of another virus, and why the host ranges of two viruses may partly overlap.

Subliminal infections: "non-host immunity" reassessed

The operating definition of non-host immunity used above centred on there being no detectable symptoms or virus multiplication in the non-host. Clearly, this depends on how hard one looks. Two lines of evidence suggest that the division between host and true non-host may be rather blurred.

Firstly, several workers have reported "subliminal infections" of ostensibly non-host species, where the virus does multiply, but is restricted to the initially infected cell or to a very few cells around it (Cheo, 1970; Cheo and Garrard, 1971; Hussain *et al.*, 1987). Unfortunately, the genetic basis of this effect is not known, and will be difficult to investigate by normal crossing methods. However, extreme restriction of virus movement also appears to be the mechanism involved in a few cases of cultivar resistance (Nishiguchi and Motoyoshi, 1987; Sulzinski and Zaitlin, 1982), and in at least one case, is known to be controlled by a single dominant gene.

Secondly, experiments with protoplasts have shown, for a small number of plant species and viruses, that some viruses can be inoculated to, and can multiply in protoplasts of ostensibly non-host species (Furusawa and Okuno, 1978; Huber *et al.*, 1981). It would be interesting to know if inoculated leaves of plants from these combinations support subliminal infection. There have also been reports that protoplasts of non-hosts may not support virus multiplication (Motoyoshi *et al.*, 1974), although such negative results should not be over-interpreted, as they might be influenced by trivial factors.

It remains to be seen whether a significant proportion of "non-hosts" can be subliminally infected by a significant proportion of "non-hosted" viruses, or whether subliminal infection is comparatively rare. Studies on infection of protoplasts of non-hosts have also been carried out with too few examples to allow generalizations. However, it does appear from these two types of study that one mechanism of "non-host" immunity at least can operate at the stage of cell-to-cell movement in the viral replicative cycle. It may be that the virus movement protein fails to recognize and interact with the relevant plant components, or the latter may be absent. In the case of single cell infection of tomato lines carrying the *Tm-2* gene for tobacco mosaic virus (TMV) resistance, prior infection with potato virus X (PVX) allowed TMV to spread (Talianski *et al.*, 1982). The implication is that the *Tm-2* gene product inhibited the virus movement function, possibly by reacting with the TMV 30 kDa transport protein, and that the PVX transport protein could complement this defect. Other explanations cannot be excluded, but the results may provide a model for further investigations of this mechanism of host range control.

GENETIC INTERACTIONS IN RESISTANCE AND VIRULENCE

The genetics of resistance

Plant resistance genes have been used as the major strategy for combating virus diseases in numerous crop species (reviewed in Fraser, 1986). Much of our knowledge of the genetics of resistance and virulence is derived from classical plant breeding, and the artificial genetic systems so constructed. These probably tend to be rather simple, given the difficulty of recognizing and breeding for characters controlled by large numbers of genes. The genetics of resistance and virulence may be more complex in the naturally-evolved, wild associations of plants and their viruses, but this topic does not seem to have been investigated.

Certain broad conclusions about the genetics of resistance, and the link with possible biochemical mechanisms, have emerged. I have discussed them in detail in recent reviews (Fraser 1986; 1987a,b), so only the main features will be dealt with briefly in this chapter.

In a survey (Fraser, 1986) of 63 examples of hosts displaying cultivar resistance to particular viruses, taken at random from the literature, resistance was monogenic in at least 50 cases. Of the remainder, a few involved obligate co-operation between resistance genes at a small number of loci, as in resistance to bean common mosaic virus (BCMV) in *Phaseolus vulgaris* (Drijfhout, 1978) (Table 1). Others involved possible epistatic or hypostatic interactions of resistance genes with modifier genes not directly involved in resistance, although in some cases the complex segregation ratios observed could be explained as genotype x environment interactions with a simple (monogenic) genetic basis, rather than as a complex oligogenic system.

For most crop species, resistance to particular viruses seems to be on a very narrow genetic base, often relying on one or only a few genes. There does not seem to be anything comparable to the large numbers of alleles at numerous loci found for some cases of resistance to bacterial or fungal pathogens (Crute, 1985). Perhaps this may reflect a restricted ability of viruses to evolve large numbers of virulence genes; a topic considered in more detail below.

An unpublished survey of the horticultural crops grown in the UK (JA Tomlinson, personal communication) has suggested that 21 crop species or sub-species suffer from at least 25 viruses, for which no sources of resistance are yet known. This may suggest that cultivar resistance is by no means a universal or even frequent phenomenon; again the caveat about comparing the products of the plant breeder with naturally-evolved species applies. It would be interesting to have more information on this question, because it is relevant to the possible evolutionary mechanisms of susceptibility and non-host immunity.

Genetically, the resistance genes in the sample (Fraser, 1986) covered a spectrum from dominant, through incompletely dominant (gene-dosage dependent) to recessive. At least some of the genes in the sample may have been classified as dominant or recessive purely on the basis of their effects on symptom formation: comparison of virus multiplication in homozygous and heterozygous resistant plants clearly suggested gene-dosage dependence (Fraser and Loughlin, 1980; Day, 1984).

These results have implications for whether resistance genes operate by positive or negative mechanisms (Fig. 1). Dominant genes are most likely to involve a specific recognition event and a positive, inhibitor-type mechanism. In the survey sample, dominant genes were strongly associated with mechanisms involving localization, usually with the formation of necrotic lesions. The evidence suggests that this type of mechanism is induced: the very early part of virus multiplication is uninhibited (Fraser, 1988; Fritig *et al.,* this volume).

An extreme case of localization is the restriction of the virus to the directly inoculated cell, as in avirulent TMV inoculated to tomato plants containing the *Tm-2* or *Tm-2*^{2} resistance genes (Nishiguchi and Motoyoshi, 1987). In these cases, it is not yet known whether the resistance - presumably against the transport function involved in cell-to-cell movement of the virus - is constitutive or very rapidly induced after infection. The former explanation seems more likely. In any case, single cell localization mechanisms may be fundamentally different from the slower, local-lesion type mechanisms, and not simply a highly effective form of them.

Gene-dosage dependent mechanisms do not appear to be associated with localization and lesion formation, but may permit some systemic spread of the virus, albeit with more or less limited multiplication and reduced symptom formation. In at least one case, resistance to TMV in tomato controlled by the *Tm-1* gene, the mechanism appears to be constitutive (Fraser and Loughlin, 1980). This is consistent with the possible presence of an inhibitor of the viral replicase in *Tm-1* plants (Watanabe *et al.,* 1988).

There is limited evidence from the survey that fully recessive genes may be associated with negative resistance mechanisms, in that some seem to confer apparent complete "immunity" in the homozygous form. Such mechanisms would, by definition, be constitutive.

The genetics of virulence

The efforts of the plant breeder to create resistant varieties have in several cases been negated by the presence or evolution of virulent, resistance-breaking isolates. There has been considerable argument about the frequency of virulence. Robinson (1976) considered that viral genomes mutated so rapidly (see also Hull, this volume), that virus resistance genes would quickly be overcome. In contrast, Russell (1978) considered that most virus resistance genes had been highly durable in practice. In fact, the answer is probably somewhere between these extremes, and is actually rather difficult to quantify. The chapter by Thresh in this volume assesses the problem in field usage and in epidemiological terms. Here, I am concerned more with the genetical aspects of virulence, and with constraints on its frequency.

In the literature sample used in the survey (Fraser, 1986), virulent isolates had been reported for just over half of the monogenic resistances considered. For the majority of the remainder, the literature is unclear: it is not known whether the genes had been tested by inoculation with a range of virus isolates, from various locations. Only about 10% of the genes in the sample appeared to have been highly durable, with few or no reports of virulent isolates after many years of widespread use in the field and for experimental purposes. Examples of these highly durable genes include R_y for potato virus Y (PVY) resistance in potato (Ross, 1960), and *N* for TMV resistance in tobacco, first described by Holmes in 1938. However, there has recently been a report that a tomato strain of TMV isolated from pepper will cause systemic mosaic on *Nicotiana tabacum* and *N. glutinosa* plants containing the *N* gene (Csillery *et al.,* 1983). Even the most durable genes may eventually succumb.

The implication of these bare statistics is that virulence is common enough to make breeding plants for virus resistance potentially a futile activity. There are certainly cases of resistance genes being introduced in commercial varieties, only to be quickly overcome by matching virulence in the virus. For example, the *Tm-1* gene for TMV resistance in tomato was widely overcome within about one year of the introduction of varieties containing this resistance gene only (Pelham *et al.*, 1970). However, for some "defeated" resistance genes, geographical separation between plant varieties containing the gene, and virulent isolates of the virus, may preserve the utility of the resistance. Furthermore, virulent isolates of a particular virus may be sustainable and definable under experimental conditions, but may lack the pathogenic fitness to allow them to survive or become widespread under agricultural conditions. Some evidence for this conclusion is considered below. The inferences for the plant breeder are that virulence can be a problem, that resistance genes should be tested against numerous isolates of the virus at an early stage in the breeding process for possible breakdown, and that potentially more robust oligogenic resistance systems should be developed if genes are available.

Genetic mapping of virulence determinants

Plant viruses have small genomes, capable of coding for between a few and about a dozen proteins. It is most unlikely that any viral genes could be involved solely in the determination of virulence/avirulence. The determinant is likely to be located in a gene for some other, primary function in viral replication, movement, transmission or structure. Indeed, it remains an open question whether the determinants of virulence/avirulence in microbial pathogens in the wider sense are genes with this as sole function.

The implication of the above discussion is that it is a viral protein which is recognized, or not recognized, by the host resistance mechanism, to determine virulence or avirulence. This borrows heavily from the microbial pathogen model (Crute 1985; and chapters by Vivian and Mansfield in this volume). However, it remains possible that the viral nucleic acid could be the recognition entity in at least some cases. Furthermore, recognition might involve not only sequences coding for functional proteins, but RNA non-coding regions, minus-sense or even double stranded forms.

Before the advent of cDNA techniques, genetic mapping of viruses was hindered by the inability to carry out classical crossing and recombination studies. Limited information about the location of virulence determinants was gained by coarse or indirect mapping methods.

For viruses with a multi-partite genome, the determinant of virulence could be mapped to a particular genomic segment by purifying individual genomic segments from virulent and avirulent isolates, and testing the biological properties of pseudorecombinants. The evidence (reviewed by Fraser, 1986) suggested that in different virus groups, virulence against the resistance genes of various hosts mapped to various genomic segments, probably specifying a number of different viral functions.

An early and ingenious attempt at mapping of a virulence gene on a virus with a monopartite genome was made by Kado and Knight (1966). They studied the mutation by nitrous acid of TMV from the virulent (systemic mosaic) to the avirulent (local lesion) form on *Nicotiana tabacum* plants containing the *N'* gene for TMV resistance. Naked TMV RNA is more susceptible to mutagenesis than encapsidated TMV RNA. Kado and Knight prepared partially-stripped virus particles by treatment with sodium dodecyl sulphate, which under appropriate conditions will strip to varying extents from the 5' end. Particles with different degrees of stripping were separated by centrifugation, then mutagenized. A sharp increase in the number of avirulent mutants became apparent as the region of the TMV RNA encoding the 30 kDa transport protein became exposed, suggesting that this contains the determinant of virulence/avirulence. However, there was also considerable early evidence suggesting that the viral coat protein, which is the gene to the 3' side of the 30 kDa transport protein, is involved in determining the interaction with the *N'* gene (Tsugita, 1962; Fraser, 1983). Modern methods of genetic analysis have also located the determinant of virulence against the *N'* gene in the coat protein sequence (Culver and Dawson, this volume; Saito *et al.*, 1987). Further aspects of virulence against the *N'* gene are considered by Mundry *et al.* in this volume.

A quantum leap in the ability to map viral determinants came about when it became possible to prepare cDNA copies of viral RNA genomes, to prepare artificial recombinants between cDNAs derived from virulent and avirulent isolates, and to derive infectious RNAs from these recombinants using appropriate expression vectors. Recombination can be arranged at various sites in the viral genome, by choice of an appropriate restriction enzyme, and the determinant of virulence/avirulence can then be located to a single base by sequencing of the small part of the total genome so located. This type of genetic mapping can be applied to naturally occurring virulent and avirulent isolates, but with natural isolates, there can be a problem of "background noise" from genetic differences which are not related to the determination of virulence.

A more efficient method is to prepare a virulent or avirulent mutant from a parent with the opposite behaviour, by mild mutagenesis. There is then a higher probability that any difference in sequence detected will be related to virulence.

This type of method is being applied to several viruses. Studies with TMV will serve to illustrate some of the principal conclusions. Virulence against the *N'* gene has been located in the coat protein gene (Knorr and Dawson, 1988; Saito *et al.,* 1987) and is described in more detail in the chapters by Culver and Dawson, and Mundry *et al.* in this volume. Interestingly, TMV coat protein does not seem to be required for induction of the local lesion response controlled by the *N* gene, (Dawson *et al.,* 1988) which is thought to be allelic with *N'* (Valleau, 1943; Dunigan *et al.,* 1987). It would appear that the products of the two alleles may recognize different viral entities.

TMV virulence determinants against two resistance genes in tomato have also been mapped. Virulence against *Tm-1* is located in the 126/183 kDa gene (Meshi *et al.,* 1988a) thought to be a part of the viral replicase (Goldbach, 1987; Goldbach *et al.,* this volume). Interestingly, the *Tm-1* gene allows movement of TMV throughout the plant, but inhibits virus multiplication (Fraser and Loughlin, 1980). Current evidence suggests that *Tm-1* may inhibit replicase activity (Watanabe *et al.,* 1988).

The *Tm-2* gene prevents movement of avirulent strains of TMV from the initially-infected cell (Nishiguchi and Motoyoshi, 1987) and the mutation in the TMV RNA which allows the virus to spread is located in the 30 kDa transport protein (Meshi *et al.,* 1988b). Thus for both *Tm-1* and *Tm-2,* there seems to be a clear link between the mechanism of resistance, and the corresponding change in the virus conferring virulence.

Is there a limit to the number of virulence "genes" a virus isolate can contain?

If, as described above, virulence determinants against different resistance genes can be located in genes specifying different viral functions, and each individual function can be maintained despite the mutation conferring virulence, it seems reasonable to expect that a virus isolate might be able to contain several virulence determinants. However, if the determinants of virulence against two or more resistance genes occur in the same viral gene, there might be constraints on the number of virulence determinants which could be accommodated without loss of pathogenic function. Some pairs of virulence determinants might be mutually exclusive, in that the alterations required in the viral protein might interfere with each other, or might interfere with the primary function of the gene in pathogenesis.

In practice, the number of hosts with a sufficient array of resistance genes against any particular virus to allow testing of these ideas is very small. Consideration must remain largely at the level of speculation until more host resistance genes are discovered, or until plants with large numbers of genetically-engineered resistance elements are constructed. This applies not only to virus-derived resistance genes, such as those based on coat proteins (Beachy *et al.,* this volume), but also to modification of naturally-occurring resistance genes by site-directed mutagenesis. When the naturally-occurring resistance genes have been isolated and are accessible to modification, a viable strategy for creation of durable resistance may be to create a series of related genes which the virus cannot mutate to overcome, for the reasons of mutual exclusivity discussed in this section.

This said, a few examples are known from the present state of viral genetics which may bear on the question of how many virulence genes the viral genome may contain. Thus TMV isolates carrying virulence against both the *Tm-1* and *Tm-2* genes are known (Pelham, 1972), but there do not appear to be any confirmed reports of isolates virulent against *Tm-2* and the allelic *Tm-2*^{2}, which probably operates by a similar mechanism (Nishiguchi and Motoyoshi, 1987). Mutual exclusivity of virulence determinants against these two genes is one possibility, but there may be a more trivial explanation. Virulence against *Tm-2*^{2} is in any case comparatively rare; doubly-virulent mutants might be statistically unlikely rather than biochemically impossible.

The opposite of mutually exclusive virulence determinants may also exist, in that mutation to virulence against one resistance allele or gene might automatically confer virulence against another. In resistance to BCMV in *Phaseolus vulgaris*, all known virus isolates which overcome the *bc-1*^{2} gene also overcome the allelic *bc-1*, although the converse is not true (Drijfhout, 1978). This suggests that *bc-1*^{2} virulence may contain the *bc-1* virulence function.

The number of virulence genes that a single virus can contain has a bearing on whether viruses and their plant hosts can evolve complex gene-for-gene interactions, such as exist between plants and their microbial pathogens (Vanderplank, 1984; and the chapters by Mansfield and Vivian in this volume). Robinson (1976) argued from the supposed ability of viral genomes to mutate rapidly to virulence, that any gene-for-gene interaction, and the relative stability of performance of differential lines of the host, would quickly break down. In fact, where sufficient numbers of resistance genes exist in the host, quite stable gene-for-gene relationships with the respective viruses have

Table 1. *Genetic interactions between* Phaseolus vulgaris *plants with different susceptibility or resistance genes at the* bc-u *and* bc-1 *loci, and strains of bean common mosaic virus with different virulence genes.*

Host resistance genotype[a]	Virus virulence genotype			
	observed			theoretical
	0	1	1.1^2	1^2
bc-u⁺,bc-u⁺/bc-1⁺,bc-1⁺	M	M	M	m
bc-u⁺,bc-u⁺/bc-1,bc-1	M/R	M	M	m/r?
bc-u⁺,bc-u/bc-1,bc-1	M	M	M	m
bc-u,bc-u/bc-1⁺,bc-1	M	M	M	m
bc-u,bc-u/bc-1⁺,bc-1⁺	M	M	M	m
bc-u,bc-u/bc-1,bc-1	R	M	M	r
bc-u,bc-u/bc-1²,bc-1²	R	R	M	m
bc-u,bc-u/bc-1⁺,bc-1²	M	M	M	m
bc-u,bc-u/bc-1,bc-1²	R	R	M	r

[a]Not all possible heterozygotes are shown. R,r = resistant, M,m = mosaic. Based on the results of Drijfhout (1978) and Day (1984). From Fraser (1987).

evolved. Examples include tomato-TMV (Pelham, 1972; Pelham *et al.,* 1970); potato-potato virus X (Jones, 1981); *Capsicum annuum*-TMV (Betti *et al.,* 1988; Tobias *et al.,*1989) and *Phaseolus vulgaris*-BCMV (Drijfhout, 1978; Day, 1984).

The last example is one of the most complex gene-for-gene interactions known for plants and their viruses. The host may contain up to four virus-specific resistance genes: *I* (dominant) and *bc-1, bc-2* and *bc-3* (recessive). Allelic series exist at the *bc-1* and *bc-2* loci. In addition, each *bc* resistance requires for full expression an epistatic effect from an additional recessive allele at a separate locus, *bc-u*. Plants with homozygous *bc-u* alone show no detectable resistance. Virus isolates which overcome

I or alleles at the *bc-1* and *bc-2* loci are known. Resistance gene *bc-3* has not yet been overcome. *bc-u* is, by definition, dependent on the presence of one of the other genes in the *bc* series; it appears to be overcome whenever that other gene is overcome. Clearly, the "checkerboard" (Person, 1959) for the full gene-for-gene interaction between BCMV and *Phaseolus vulgaris* would be an extremely complex one, and would depend on some of the assumptions about "allellism" and mutual exclusivity of virulence determinants discussed here and elsewhere (Day, 1984). A simple version depicting some of the interactions between BCMV and genes at the *bc-1/bc-u* loci only is given in Table 1.

Is virulence accompanied by loss of pathogenic fitness?

As mentioned earlier, it is unlikely that viral genes will have the determination of virulence/avirulence as their sole or even primary function. The examples discussed show that virulence determinants may be located in genes which determine other viral functions. However, it is clear that this primary function can be maintained while allowing mutation to the virulent form. The frequency of successful mutation to virulence does appear to vary for different host resistance genes, and may depend on the sensitivity to mutation of the viral function involved. This can also be examined at a quantitative level, by asking whether the successful mutation to virulence has been achieved at the expense of some impairment of the primary function of the viral gene in pathogenesis (Vanderplank, 1984). Two examples will serve to illustrate the point.

We prepared a number of TMV mutants able to overcome the *Tm-1* resistance gene in tomato, by nitrous acid mutagenesis of the avirulent TMV tomato strain 0. Mutagenesis was under conditions giving a high survival rate, such that there was a high probability that any virulent forms recovered would be single hit mutations (Mundry and Gierer, 1958), and not altered in other viral functions which might affect viability. The mutants were selected by their ability to multiply and cause mosaic symptoms on *Tm-1* plants, and were isolated by single-lesion transfers on Xanthi-nc tobacco. The pathogenic fitness of each mutant was assessed by inoculation to *susceptible* (non-*Tm-1*) tomato plants, and measuring severity of mosaic symptoms produced, and the amount of TMV multiplication. All 19 mutants examined showed a loss of pathogenic fitness, in that they multiplied less well, and caused milder symptoms, than the parent TMV-0 (Fraser and Gerwitz, 1987). This suggests that acquisition of virulence against *Tm-1*, involving a mutation in the putative viral replicase (Watanabe *et al.,* 1988; Meshi *et al.*, 1988a), does lead to a loss of pathogenic fitness. However, a *Tm-1*-breaking isolate from a commercial tomato glasshouse, which was examined in the same study, multiplied as well and caused symptoms as severe as TMV-0. This may suggest that under practical conditions, the virus came under further selection pressure which favoured further mutations restoring pathogenic fitness, while retaining the specific determinant of virulence.

In related work on the *Tm-2*^{2} gene (RSS Fraser, L Betti and A Gerwitz, in preparation), we have examined four naturally-arising isolates which overcome the gene, and which came from widely separated commercial tomato crops. Reports of TMV isolates overcoming the *Tm-2*^{2} gene have been very infrequent, suggesting that this is a difficult evolutionary step for the virus, and that it might be associated with an unacceptable loss of pathogenic fitness. In fact, all four isolates share certain properties suggesting that this is the case. Experimentally, all multiplied to normal concentrations in susceptible tomato plants, and caused symptoms of normal severity. Thus, there is nothing intrinsic to the virus alone which affects pathogenic fitness. However, all four isolates multiplied comparatively poorly in plants containing the *Tm-2*^{2} gene: an example of a "ghost" effect of a defeated resistance gene. Furthermore, all isolates caused extreme inhibition of growth of *Tm-2*^{2} plants, had very low specific infectivity on such plants, and tended to cause very small lesions on tobacco plants containing the *N* gene. Thus in this case, acquisition of virulence does appear to be associated with some loss of pathogenic fitness, partly caused by residual effects of the host genome. The low infectivity of TMV-2^2 on *Tm-2*^{2} plants, especially, may explain why the strain has failed to establish itself as a serious problem in commercial horticulture.

CONCLUSION

More understanding is needed of the mechanisms involved in virus host range control, but the topic is intrinsically difficult to investigate, as many effects are negative or may represent the absence of necessary components. Three basic models have been described in this chapter, but these certainly need amplification and modification. One mistake which should be avoided is to generalize from studies of a very small number of viruses and plant species: we may be dealing with a multiplicity of mechanisms. Studies of cultivar resistance and virulence may provide models which can be tested for relevance to non-host immunity and host range control, but some mechanisms of host range control and non-host immunity might be completely different from those involved in cultivar resistance. Alternatively, as discussed earlier, non-host immunity as defined might be a phenomenon which does not actually exist in nature, or which might exist in a smaller number of interactions than previously thought.

Studies of cultivar resistance have involved some elegant classical genetics, but not generally at the level of complexity of analysis of plant-microbial pathogen interactions. Part of this seems to be because of the comparative simplicity of the genetics of plant-virus interactions. Studies of the genetics of virulence, previously hindered by the lack of analytical methods, have been greatly enhanced by modern approaches using recombinant DNA methods. The prospect for the immediate future is that our understanding of virulence will overtake that of host genes for resistance to viruses, and the mechanisms involved in their action. This situation is likely to persist until host resistance genes can be isolated and studied using methods such as site-directed mutagenesis.

A few host resistance genes may be isolated fairly quickly. Those few with a known product, such as the protease inhibitor which is thought to inhibit cowpea mosaic virus multiplication (Bruening *et al.*, 1987), should be accessible by the usual methods involving expression vectors and genomic libraries of the host. For those genes where the resistance mechanism is not fully understood, and with no known product, the problem of isolation is of a higher order of magnitude. The statistical and methodological problems of screening differential mRNA populations, where the gene may and may not be expressed, and of transposon tagging, are well documented (Hepburn *et al.*, 1985; Bennetzen *et al.*, 1988; Dunigan *et al.*, 1987).

Precise and detailed information on how matching virulence is determined in the virus may help in devising strategies for the isolation of host resistance genes, although experience from the parallel situation in plant-bacterium interactions is not encouraging (see Vivian, this volume).

Despite these difficulties, it is likely that host resistance genes will be isolated and become available for study at some stage in the future. The overall importance of the subject, and the potential for exciting experimentation on isolated genes, means that research effort and resources will continue to be devoted to the topic. Ultimately, this should result in a greater understanding of how resistance and virulence genes function; of the nature of the recognition events between them, and of the signal transduction pathway which invokes other aspects of the host response. From what is already known of the genetics of resistance and virulence, and of resistance mechanisms, it is likely that numerous resistance genes will have to be isolated and studied to give a reasonable overview of the level of diversity involved in different plant-virus interactions.

Will such resistance genes, when isolated, be of use in practical crop protection? Given suitable transformation systems, there is no technical reason why any gene should not be transferred to another species suffering from that virus. Provided that other factors from the host genetic background are not required for operation of the resistance mechanism, the transferred gene may then confer some useful level of resistance. However, two potential problems need to be assessed. Firstly, the virus may, presumably, mutate to overcome the genetically-engineered resistance gene, in the same way as it may overcome those inserted by conventional breeding methods. In practical terms, it would be useful to be able to assess the problem and potential threat of virulence against the scale of investment required to do the gene transfer, by whatever technology. This assessment may be assisted by knowledge of the molecular biology of virulence and resistance, and may be influenced by the ability to modify the interaction by site-directed mutagenesis, as discussed earlier. Secondly, transfer of a host resistance gene to another species may put the virus under completely new types of selection pressure which could have profound effects on its evolution with regard to both host species. These effects would require careful consideration.

REFERENCES

Atabekov JG, Novikov VK, Vishnichenko VK, Javakhia VG (1970) A study of the mechanisms controlling the host range of plant viruses. II. The host range of hybrid viruses reconstituted *in vitro* and of free viral RNA. Virology 41:108-115

Bald JG, Tinsley TW (1967a) A quasi-genetic model for plant virus host ranges. I. Group reactions within taxonomic boundaries. Virology 31:616-624

Bald JG, Tinsley TW (1967b) A quasi-genetic model for plant virus host ranges. II. Differentiation between host ranges. Virology 32:321-327

Bald JG, Tinsley TW (1967c) A quasi-genetic model for plant virus host ranges. III. Congruence and relatedness. Virology 32:328-336

Bennetzen JL, Hulbert SH, Jin Y-K (1988) The use of transposable elements in the study and isolation of plant disease resistance genes. Abstr 5th Intl Congress Plant Pathol, Kyoto p 15

Betti L, Tanzi M, Canova A (1988) Pepper mosaic virus strains and their adaptation to the host I. Biological and serological behaviour. Phytopathol Mediterr 27:7-

17
Bruening G, Ponz F, Glascock C, Russell ML, Rowhani A, Chay C (1987) Resistance of cowpeas to cowpea mosaic virus and to tobacco ringspot virus. In: Evered D, Harnett S (eds). Plant resistance to viruses. Ciba Foundation Symposium 133. Wiley, Chichester, p 23
Cheo PC (1970) Subliminal infection of cotton by tobacco mosaic virus. Phytopathology 60:41-46
Cheo PC, Garard JS (1971) Differences in virus-replicating capacity among species inoculated with tobacco mosaic virus. Phytopathology 61:1010-1012
Crute IR (1985) The genetic bases of relationships between microbial parasites and their hosts. In: RSS Fraser (ed) Mechanisms of resistance to plant diseases. Nijhoff, Dordrecht, p 80
Csillery G, Tobias I, Rusko J (1983) A new pepper strain of tomato mosaic virus. Acta Phytopathol Acad Sci Hung 18:195-200
Day KL (1984) Resistance to bean common mosaic virus in *Phaseolus vulgaris* L. PhD Thesis, University of Birmingham, UK
Dawson WO, Bubrick P, Grantham GL (1988) Modification of the TMV coat protein gene affects replication, movement and symptomatology. Phytopathology 78:783-789
Drijfhout E (1978) Genetic interaction between *Phaseolus vulgaris* and bean common mosaic virus with implications for strain identification and breeding for resistance. Agric Res Rep (Wageningen) 872
Dunigan DD, Golemboski DB, Zaitlin M (1987) Analysis of the N gene of Nicotiana. In: Evered D, Harnett S (eds) Plant resistance to viruses. Ciba Foundation Symposium 133. Wiley, Chichester, p 120
Fraser RSS, Loughlin SAR (1980) Resistance to tobacco mosaic virus in tomato: effects of the *Tm-1* gene on virus multiplication. J Gen Virol 48:87-96
Fraser RSS (1983) Varying effectiveness of the *N'* gene for resistance to tobacco mosaic virus in tobacco infected with virus strains differing in coat protein properties. Physiol Plant Pathol 22:109-119
Fraser RSS (1986) Genes for resistance to plant viruses. CRC Crit Rev Plant Sci 3:257-294
Fraser RSS (1987a) Biochemistry of virus-infected plants. Research Studies Press, Letchworth, Wiley, New York
Fraser RSS (1987b) Resistance to plant viruses. Oxf Surv Plant Molec Cell Biol 4:1-45
Fraser RSS, Gerwitz A (1987) The genetics of resistance and virulence in plant virus disease. In: Day PR, Jellis GJ (eds) Genetics and plant pathogenesis. Blackwell, Oxford, p 33
Fraser RSS (1988) Virus recognition and pathogenicity: implications for resistance mechanisms and breeding. Pestic Sci 23:267-275
Furusawa I, Okuno T (1978) Infection with BMV of mesophyll protoplasts from five plant species. J Gen Virol 40:489-491
Goldbach RW (1987) Genome similarities between plant and animal RNA viruses. Microbiol Rev 4:197-202
Hepburn AG, Wade M, Fraser RSS (1985) Present and future prospects for exploitation of resistance in crop protection by novel means. In: Fraser RSS (ed) Mechanisms of resistance to plant diseases. Nijhoff, Dordrecht, p 425
Hiebert E, Bancroft JB, Bracker CE (1968) The assembly *in vitro* of some small spherical viruses, hybrid viruses and other nucleoproteins. Virology 34:492-508
Holmes FO (1938) Inheritance of resistance to tobacco mosaic disease in tobacco.

Phytopathology 28:553-561
Holmes FO (1955) Additive resistance to specific viral diseases in plants. Ann Appl Biol 42:129-139
Horvath J (1979) New artificial hosts and non-hosts of plant viruses and their rôle in the identification and separation of viruses. X. Cucumovirus group: cucumber mosaic virus. Acta Phytopathol Acad Sci Hung 14:285-295
Huber R, Hontilez J, Van Kammen A (1981) Infection of cowpea protoplasts with both the common strain and the cowpea strain of TMV. J Gen Virol 55:241-245
Hussain MM, Melcher U, Whittle T, Williams A, Brannan CM, Mitchell ED (1987) Replication of cauliflower mosaic virus DNA in leaves and suspension culture protoplasts of cotton. Plant Physiol 83:633-639
Ireland RJ, Pierpoint WS (1980) Reaction of strains of potato virus X with chlorogenoquinone *in vitro*, and the detection of a naturally modified form of the virus. Physiol Plant Pathol 16:81-92
Jones RAC (1981) The ecology of viruses infecting wild and cultivated potatoes in the Andean region of South America. In: Thresh JM (ed) Pests, pathogens and vegetation. Pitman Press, London, p 89
Kado CI, Knight CA (1966) Location of a local lesion gene in tobacco mosaic virus RNA. Proc Natl Acad Sci USA 55:1276-1283
Knorr DA, Dawson WO (1988) A point mutation in the tobacco mosaic virus capsid protein gene induces hypersensitivity in *Nicotiana sylvestris*. Proc Natl Acad Sci USA 85:170-174
Meshi T, Motoyoshi F, Adachi A, Watanabe Y, Takamatsu N, Okada Y (1988a) Two concomitant base substitutions in the putative replicase genes of tobacco mosaic virus confer the ability to overcome the effects of a tomato resistance gene, *Tm-1*. EMBO J 7:1575-1580
Meshi T, Motoyoshi F, Maeda S, Yoshioka H, Watanabe H, Okada Y (1988b) Responsible mutations in the 30K protein gene of TMV for overcoming the *Tm-2* resistance. Abstr 5th Intl Conf Plant Pathol, Kyoto p 38
Motoyoshi F, Bancroft JB, Watts JW (1974) The infection of tobacco protoplasts with a variant of brome mosaic virus. J Gen Virol 25:31-36
Mouchés C, Candresse T, Bové JM (1984) Turnip yellow mosaic virus RNA-replicase contains host and virus-encoded subunits. Virology 134:78-90
Mundry KW, Gierer A (1958) Die Erzeugung von Mutanten des Tabakmosaikvirus durch chemische Behandlung seiner Nukleinsäure *in vitro*. Z Vererbungslehre 89:614-630
Nishiguchi M, Motoyoshi F (1987) Resistance mechanisms of tobacco mosaic virus strains in tomato and tobacco. In: Evered D, Harnett S (eds) Plant resistance to viruses. Ciba Foundation Symposium 133. Wiley, Chichester, p 38
Pelham J, Fletcher JT, Hawkins JH (1970) The establishment of a new strain of tobacco mosaic virus resulting from the use of resistant varieties of tomato. Ann Appl Biol 65:293-297
Pelham J (1972) Strain-genotype interaction of tobacco mosaic virus in tomato. Ann Appl Biol 71:219-228
Person C (1959) Gene-for-gene relationships in host:parasite systems. Can J Bot 37:1101-1130
Robinson RA (1976) Plant pathosystems. Springer-Verlag, Berlin
Ross H (1960) Der Praxis der Züchtung auf Infektionsresistenz und extremer Resistenz (Immunität) gegen das Y-Virus der Kartoffel. Eur Potato J 3:296-306
Russell GE (1978) Plant breeding for pest and disease resistance. Butterworths, London

Saito T, Meshi T, Takamatsu N, Okada Y (1987) Coat protein gene sequence of tobacco mosaic virus encodes host response determinant. Proc Natl Acad Sci USA 84:6074-6077

Simons TJ, Ross AF (1971) Metabolic changes associated with systemic induced resistance to tobacco mosaic virus in Samsun NN tobacco. Phytopathology 61:293-300

Sulzinski MA, Zaitlin M (1982) Tobacco mosaic virus replication in resistant and susceptible plants: in some resistant species virus is confined to a small number of initially infected cells. Virology 121:12-19

Taliansky ME, Malyshenko SI, Pshennikova ES, Atabekov JG (1982) Plant virus-specific transport function. II. A factor controlling virus host range. Virology 122:327-331

Tobias I, Fraser RSS, Gerwitz A (1989) The gene-for-gene relationship between *Capsicum annuum* L. and tobacco mosaic virus: effects on virus multiplication, ethylene synthesis, and accumulation of pathogenesis-related proteins. Physiol Mol Plant Pathol 35:271-286

Tsugita A (1962) The proteins of mutants of TMV: composition and structure of chemically evoked mutants of TMV RNA. J Mol Biol 5:284-292

Valleau WD (1952) The evolution of susceptibility to tobacco mosaic in *Nicotiana* and the origin of the tobacco mosaic virus. Phytopathology 42:40-42

Vanderplank JE (1984) Disease resistance in plants (2nd edition). Academic Press, Orlando

Watanabe Y, Kishibayashi N, Motoyoshi F, Okada Y (1988) Characterization of *Tm-1* gene action on replication of common isolates and a resistance-breaking isolate of TMV. Virology 161:527-532

PLANT VIRUS EPIDEMIOLOGY: THE BATTLE OF THE GENES

J M Thresh
Overseas Development Natural Resources Institute
c/o AFRC Institute of Horticultural Research
East Malling
Maidstone
Kent ME19 6BJ
UK

ABSTRACT

Virus epidemiology involves complex interactions between host and pathogen in which an animal or fungus vector often plays a key role as intermediary. All three components of the pathosystem are subject to genetic variability and there is abundant evidence of the way in which virus strain, host cultivar and vector biotype can influence the onset and course of epidemics. Indeed, virus epidemiology can be considered as a "battle" of virus, host and vector genes. This is the approach adopted in this paper which considers the genetic aspects of the epidemiology of a wide range of viruses of tropical and temperature crops.

INTRODUCTION

Many of the contributions to this volume are concerned with the interactions between viruses and their host plants as assessed in laboratory or glasshouse experiments. These mainly biochemical studies are providing important information on mechanisms of pathogenicity, resistance and symptom expression and there are good prospects of exploiting the results obtained in developing novel methods of disease control. However, it is important to appreciate that other more biological and less biochemical approaches are possible in studying plant-virus interactions. This explains the extensive literature on the epidemiology and population dynamics of plant virus diseases and the emphasis given to the role of animal and fungus vectors in achieving the initial encounters between many viruses and their host plants. Considerable

NATO ASI Series, Vol H 41
Recognition and Response in Plant-Virus Interactions
Edited by R S S Fraser

attention has also been given to the effects of weather and other environmental factors so as to explain changes in the incidence of viruses and vectors and in attempts to develop effective methods of forecasting.

An overall biological approach is adopted in this paper in which the emphasis is on the inherent genetic features of viruses, hosts and vectors that influence their co-evolutionary inter-relationships. Gibbs (1983) adopted a similar attitude in presenting virus ecology as the "struggle of the genes" and others have discussed the relationships between hosts and their pests or pathogens in terms of "the evolutionary dance" (Whittaker, 1969) or "biological war games" (Dawkins and Krebs, 1979). These various approaches are entirely valid in emphasizing the dynamic continually-changing situation and they provide new insights into the processes involved.

HOST VARIATION

Harper (1982) commented that taxonomists seek stable, readily recognizable features to delineate species, whereas ecologists are more concerned with the intra-specific variation that leads to important differences within and between populations. He also stressed that lists and maps of species distribution can be misleading in ignoring polymorphisms within populations and ecotypic differences between them.

These general comments are apposite in considering the behaviour of vascular plants in relation to viruses. Almost invariably there is considerable intra-specific variation in susceptibility or response to infection. This feature is undoubtedly of great evolutionary importance. Moreover, it is one that has been exploited intuitively for millennia by farmers who tend to select vigorous healthy-looking plants for further propagation and reject those that appear weak and unhealthy. More recently, genotypes that are particularly vulnerable to infection have been used by virologists as test or indicator plants and in bio-assays. Conversely, breeders have utilized accessions that withstand infection as sources of resistance in breeding programmes.

Damaging outbreaks of virus disease can occur only if many plants are infected and if they are seriously affected. Thus the overall susceptibility and sensitivity of host populations have a crucial influence on the losses sustained. These can be enhanced or decreased by growers according to their choice of cultivar. This emphasizes the current importance of plant breeders in determining the range of cultivars available (Buddenhagen, 1983).

There are many reports in the literature of serious virus disease outbreaks being largely confined to certain cultivars or occurring soon after new ones were released. Several representative examples are discussed in the following sections to indicate the importance of host genotype in epidemiology and to support the comment of Simmonds (1962) that "disease patterns are to a great extent a product of plant breeding and agricultural practices".

Cocoa swollen shoot virus

This virus would not have caused so much destruction in West Africa if trees of the Amelonado type had not been so widely grown (Thresh *et al.*, 1988). Almost all the plantings made before 1949 were derived from a few initial introductions that originated from South or Central America, where swollen shoot virus does not occur. They proved to be highly susceptible and sensitive to infection and were killed by the most prevalent strains of virus that occur in Ghana. These strains have less drastic effects on some of the Upper Amazon types introduced later in attempts to obtain sources of resistance (Posnette and Todd, 1951).

Red clover necrotic mosaic dianthovirus

This virus was first reported in England from trial plantings incorporating some of the newly-introduced tetraploid cultivars of early-flowering broad red clover (Bowen and Plumb, 1979). Hungaropoly and three other cultivars of similar genetic background were highly susceptible compared with many of the diploid types grown previously. The tetraploids were also vulnerable because they persisted for relatively long periods and so provided an extended opportunity for infection to occur.

Turnip mosaic potyvirus

This virus is prevalent in cruciferous crops in many countries but it was not reported in lettuce until 1966, when damaging outbreaks occurred in California (Zink and Duffus, 1969). They were restricted to the recently-introduced crisp-head cultivar "Calmar" which was being grown widely because of its resistance to downy mildew (*Bremia lactucae*). In comprehensive tests on a wide range of genotypes all seven downy mildew-resistant cultivars of crisp-headed type proved to be susceptible to turnip mosaic virus whereas all others were immune.

There was no segregation within populations of each cultivar and an examination of the available pedigrees suggested that mosaic-susceptibility was introduced to lettuce from two mildew-resistant lines of a closely related wild species (*Lactuca serriola*). Different sources of resistance had been used in developing the mildew-resistant cultivars of butterhead and other types that were immune to turnip mosaic virus. The apparent linkage between the genes for mildew resistance and mosaic susceptibility in crisp-headed cultivars was confirmed in subsequent studies (Zink and Duffus, 1970).

Maize rough dwarf fijivirus

Maize was introduced from the New World to the Old more than 400 years ago, and the crop has long been cultivated extensively in Italy and other Mediterranean countries where rough dwarf disease occurs. However, the disease has a very recent history and its first appearance was closely associated with the introduction of high-yielding American hybrid cultivars (Harpaz, 1972).

Trials of the new hybrids started in Italy in 1946 and the first commercial plantings totalling 1500 ha were in 1948, when rough dwarf disease was first recorded. It attracted little attention until 1949, when outbreaks were so serious as to threaten the whole future of hybrid cultivars. These were much more susceptible than the material grown previously and 90% infection was reported in one area of northern Italy where local types were virtually unaffected (Trebbi, 1950; Grancini, 1962).

There was a similar sequence of events in Israel, where hybrids were first released in 1952. Plantings totalled 7000 ha by 1957, when rough dwarf disease was first recorded. In 1958, major outbreaks occurred in all regions where hybrids were grown. Up to 75% infection occurred in plantings along the coastal plains, with severe dwarfing and much premature death. Infection was only 5-7% in local, open-pollinated, dent types grown under similar conditions (Harpaz, 1972).

Maize rough dwarf provides a striking example of the way in which new and unexpected problems can occur when long-established cultivars are replaced. Rough dwarf does not occur in North America and the hybrids introduced to Mediterranean countries had not previously been exposed to infection. Consequently, it is hardly surprising that they were severely affected by comparison with established local cultivars. These were so resistant or tolerant that infection was entirely overlooked until susceptible genotypes were grown.

Sugarcane fijivirus

Additional ways in which disease problems are closely related to the vulnerability of the cultivars being grown are illustrated by experience with sugarcane Fiji disease in Bundaberg and other parts of Queensland, Australia. This disease is caused by a planthopper-borne virus that first attracted attention in Australia in the 1920s, when existing cultivars were being replaced by P.O.J. 2878, which was less susceptible to a prevalent bacterial gumming disease. The new "wonder" cane proved to be highly susceptible to Fiji disease, which caused continuing difficulties despite a well-organized control campaign involving inspection and eradication and provision of healthy planting material (Egan and Toohey, 1977).

The official Queensland figures for numbers of plants eradicated under-estimate the magnitude of the losses as they exclude data for the worst-affected fields that were abandoned before counts were made. Nevertheless, they reveal a peak of infection during the 1944-1945 season, followed by a decline as P.O.J. 2878 was replaced by a less susceptible cultivar and eventually by resistant ones. Losses became insignificant until there was an upsurge of disease beginning in the late 1960s following an increase in vector populations and the introduction of inherently higher-yielding cultivars that were less resistant to infection (Egan, 1976).

Particular problems were encountered with the widely-grown N.Co.310, which is prone to attack by planthoppers and moderately susceptible to infection with Fiji virus. Infected plants of N.Co.310 are slow to develop obvious symptoms and difficult to detect by the plant health inspectors who are responsible for routine surveys. This decreased the effectiveness of eradication measures and by 1976 it was estimated that there were 10 million infected stocks in the Bundaberg area alone. Virtually all farms were affected and thousands of hectares had to be ploughed out prematurely. In some localities it became difficult to produce adequate supplies of healthy planting material and for any real solution to the problem it was necessary to replace N.Co.310 by less susceptible types (Fig. 1).

N.Co.310 was removed from the list of approved cultivars in much of the Bundaberg area at the end of 1978 and from the remaining localities in 1980. This was a crucial development in bringing the Bundaberg epidemic under control. A similar policy was also effective in preventing the losses that are otherwise likely to have occurred in the Plane Creek area of Queensland where Fiji disease was first reported in 1982 and immediate steps were taken to replace N.Co.310. From these experiences

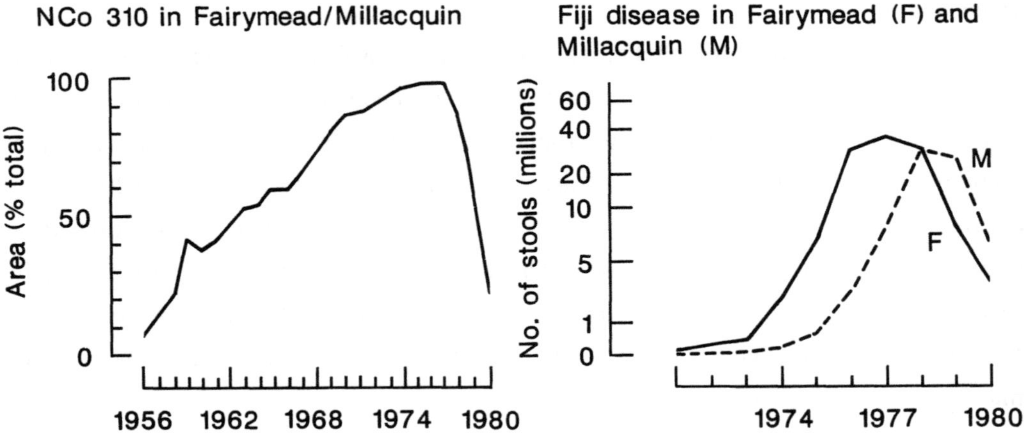

Figure 1. The cultivation of sugarcane cv. N.Co.310 in the Fairymead and Millacquin localities of Bundaberg, Queensland: 1959-1980 (left) and the incidence of sugar cane fijivirus in the Fairymead (F) and Millacquin (M) localities: 1972-1981 (right). Data from Egan (1976) and unpublished.

it is apparent why Fiji disease has been referred to as "N.Co.310 disease" (Ryan, 1988).

Rice waika virus

The history of rice waika disease in Kyushu, Japan is also closely associated with the use of one particular cultivar (Kiritani, 1983). The disease was first reported in 1967 and shown to be caused by a hopper-borne virus that is now considered to be rice tungro spherical virus. The affected area in Kyushu increased to a maximum in 1973 and then declined to zero by 1977. The prevalence of infection was closely linked with the use of cv. Reiho which was introduced in 1968 and by 1972 accounted for more than half the total area of rice grown. The area then decreased when it became apparent that Reiho was particularly vulnerable to waika disease and the cultivar was withdrawn in 1977.

Fruit tree viruses

Citrus tristeza closterovirus would not have caused such devastating losses or attracted so much attention in South America and elsewhere but for the extreme sensitivity of sweet orange scions when grown on sour orange rootstocks (Bar-Joseph *et al.*, 1981). There is a somewhat similar situation with plum pox potyvirus which causes serious diseases of cvs Pozegaca and Victoria. The virus is far less damaging in plum rootstocks and in many other cultivars that have at least some degree of tolerance.

Many other examples of tolerance have been encountered in work on fruit crops (Posnette, 1977). This is largely inevitable because suitable cultivars are propagated vegetatively and retained for many years. Any that are sensitive to commonly occurring viruses are likely to grow or crop badly. They are soon discarded even though the reasons for the unsatisfactory growth are not always apparent. For instance the use of sour orange rootstocks was abandoned in South Africa and some other countries long before the nature of tristeza was elucidated. Similarly, several types of quince were found to be unsuitable for use as pear rootstocks in England because of "incompatibility" effects (Hatton, 1928). These were shown later to be due to the sensitivity of the quinces to viruses that were prevalent at the time in pear cultivars (Cropley, 1967).

The existence or prevalence of several other important viruses of fruit crops was not appreciated until serious disease problems were encountered after attempts to introduce new rootstock or scion cultivars. This is illustrated by experience in North America where the use of Virginia Crab and Spy 227 as apple rootstocks or inter-stocks has been restricted by their extreme sensitivity to viruses that are widespread but latent in many commercial apple cultivars (Gardener *et al.*, 1946; Tukey and Brase, 1943). Similarly, apple chat fruit and rubbery wood diseases were not known until a sensitive cultivar (Lord Lambourne) was introduced and revealed the prevalence of infection in rootstocks and other scion cultivars (Luckwill and Crowdy, 1950).

Hop mosaic potyvirus

The many different hop cultivars grown in England can be divided into two distinct groups according to their sensitivity to hop mosaic virus. When Golding cultivars are infected they develop conspicuous symptoms and usually die or grow so badly that they have to be replaced. All other English cultivars are extremely tolerant and fail to produce symptoms, even though all established clones are infected.

Hop mosaic virus is aphid-borne and serious outbreaks are likely to occur in Goldings whenever they are planted close to tolerant cultivars, or when tolerant male plants are used as pollinators for Goldings. This became apparent at an early stage of the investigations which began soon after mosaic was first recorded in England between 1907 and 1910 (Salmon, 1923; Mackenzie *et al.*, 1929; Keyworth, 1947). It was soon established that losses could be avoided or kept to acceptable levels by growing Goldings with sensitive males at isolated sites or on entirely separate farms. Such measures are effective but not always convenient. It is also expensive for nurserymen to grow Goldings away from other cultivars so as to avoid any risk of mosaic and to qualify for the official certification scheme. Moreover, changes in market demand induced some growers of Goldings to plant tolerant cultivars for the first time and heavy losses then occurred in Goldings (Thresh, 1979a). There is a similar problem in west Germany, where mosaic appeared in Goldings at isolated sites in the Hersbruck area of Bavaria, soon after tolerant cultivars were introduced from the main hop-growing areas of the Hallertau. Mosaic also occurred in the traditional mosaic-sensitive Hersbruck cultivar when it was introduced to the Hallertau (Thresh, 1979b).

VIRUS STRAINS

Many different biological and physio-chemical criteria have been used to distinguish virus strains and there is an extensive literature on strain variation, interrelationships and interactions. Much of the variation reported in physio-chemical or serological characteristics is of uncertain epidemiological significance. However, many differences have also been reported in such biological features as host-range, virulence and transmissibility by vectors. The epidemiological significance of such differences has long been recognized as in studies on sugarcane mosaic and sugar beet curly top viruses in the United States, described later in this chapter. Seasonal and other changes in the prevalence of strains have also been recorded in studying the incidence of cucumber mosaic cucumovirus in vegetable crops and their weeds in south-east France (Quiot *et al.*, 1979). The strain that predominated in spring-sown crops of tomato, celery and pepper was much more sensitive to heat than the one isolated from equivalent summer plantings.

Other complex interactions between the virulence of strains and host response occur with barley stripe mosaic hordeivirus, which is transmitted mainly through seed and by direct contact between plants. Sensitive barley cultivars infected with virulent strains produce a large proportion of infected seed, although few seeds develop and they tend to be small with poor viability. By contrast, tolerant cultivars produce abundant viable seed but few are infected. Thus the virus strains that persist readily for successive generations tend to be of intermediate virulence that infect substantial proportions of seed without drastic effects on fertility or on contact between plants (Timian, 1974). There is a somewhat similar situation with rice stripe tenuivirus, as highly virulent strains that soon kill the host plant are less likely to be acquired and transmitted transovarially by the planthopper vector than less virulent strains (Kisimoto, 1972). Similarly, studies in California have shown the limited epidemiological competence of a highly virulent strain of lettuce mosaic potyvirus (McLean and Kinsey, 1963). This strain occurred in weeds but was seldom isolated from the main lettuce cultivars being grown. These were so severely damaged when infection occurred that they set few viable seed and there was little likelihood of disseminating infected seed lots.

The rapidity with which the predominant strain of virus can change in response to new selection pressure is apparent from studies on tomato mosaic tobamovirus in England. There was a rapid shift in the prevalent virus strain following the introduction of tolerant cultivars which contained the *Tm-1* resistance gene and again when these cultivars were withdrawn (Pelham *et al.,* 1970). In another approach to control, an avirulent strain of tomato mosaic virus was used by some growers in 1972 and more extensively in 1973 and 1974 in attempts to protect plants from the damaging effects of virulent strains. The avirulent mutant was derived from strain "1" and surveys in 1974 revealed virulent forms of this strain in almost all glasshouses where the mutant had been used. Strain "0" predominated in other houses and in a survey done in 1971 immediately before mild strain protection was introduced (Fletcher and Butler, 1975).

Other marked shifts in virulence have occurred following the appearance or introduction of a new virus strain as reported with potato virus Y in Europe during the 1950s. A distinctive strain of the virus that causes a serious necrosis of tobacco was first described in England in 1935 but strains of this type were not known from mainland Europe until reported from Bulgaria in 1950. They then became established in many other European countries including Germany where necrotic strains were first recorded in 1954 and became as prevalent as the original ones by 1956 (Ross, 1959). Major epidemics occurred in potato and tobacco plantings in many regions due to spread by aphids and in potato tubers. Typical necrotic strains decrease potato yields without causing obvious leaf symptoms and they are more difficult to control by long-established certification and roguing procedures than the strains they displaced (Todd, 1960). In the worst-affected parts of mainland Europe there has been no alternative but to replace several popular and widely grown potato cultivars and tobacco growers have also sustained serious losses.

Virulent strains of virus that affect genotypes resistant to other strains are of particular significance in any discussion of "the battle of the genes". This is because the rapidity with which such strains occur and become prevalent is of crucial importance in determining whether resistant cultivars can be utilized successfully. Thus it is hardly surprising that considerable attention has been given to strain virulence by virologists and breeders involved in resistance breeding and also by those concerned with mechanisms of resistance.

Many different examples of resistance due to dominant or recessive genes were considered in a recent compilation which lists virulent strains in 27 of the 54 virus-host combinations assessed (Fraser, 1986). Many other combinations had not been studied intensively and in only six instances had virulent strains not been encountered after what was considered to be a reasonably comprehensive search. An immediate impression from this compilation is that host resistance mechanisms are easily overcome and therefore of limited value as a virus disease control measure. However, before any such conclusion can be drawn it is necessary to consider each virus-host combination in greater detail and in particular the extent to which resistant cultivars have been grown, the circumstances in which virulent strains have been encountered and the time taken for them to become established.

Few comprehensive assessments of this type have been undertaken and yet they could be valuable in indicating whether certain types of resistance are particularly effective and durable and also whether viruses of some taxonomic groups are inherently more variable than others. Meanwhile, it is known that some virulent strains have only been encountered after resistant cultivars have been grown widely and successfully for several years. In other instances virulent strains have been reported, but only in laboratory studies or in areas other than those where the resistant cultivar was first selected and used.

There are also examples of virulent strains that appear to lack epidemiological competence and so have remained limited in distribution. This is the situation with the strain of raspberry ringspot nepovirus which was isolated in Scotland from the raspberry cultivar Lloyd George that is immune to the usual strain of the virus occurring there (Murant *et al.*, 1968). The resistance-breaking strain has a restricted distribution and Harrison (1978) attributed this to a lack of competitive ability that was associated with limited invasiveness and low rates of seed transmission in weed hosts.

Harrison (1981) later drew attention to the limited distribution in the U.K. of resistance-breaking strains of potato virus X that cause a systemic mottle of "King Edward" potato and other cultivars that are immune to the usual strains of the virus. These and other examples led him to conclude that resistance to viruses is more durable than against fungal pathogens. He also suggested two possible reasons. One is that the range of biologically effective virus variants that can be produced is probably restricted not only by the small coding capacity of virus genomes but also by the likelihood that many individual gene products have more than one function, and the need for a mutant gene product to perform all such roles. The second suggestion was that the dissemination of resistance-breaking strains of virus by vectors may be restricted by the relatively small number of vectors that disperse compared with the vast number of air-borne fungal spores. However, this comparison is not altogether valid and does not take account of the many virus particles that may be present in an individual vector and the ability of some vectors to infect many plants and in some instances sustain virus replication throughout their life.

Russell (1978) also concluded that a breakdown of resistance is less likely with viruses than with fungal pathogens. Nevertheless, such generalizations seem premature and various examples of breakdown have occurred, as described later with sugarcane mosaic disease in Louisiana and rice grassy stunt disease in South-East Asia. A third example occurred when mosaic-resistant soy-bean cultivars were introduced to Korea in the early 1970s (Cho *et al.,* 1977). Cultivars that were resistant to the usual strain of soybean mosaic potyvirus developed a severe necrotic disease. This was caused by a hitherto unreported virulent strain that caused typical mosaic symptoms in non-resistant varieties.

These and other examples of breakdown are due to virulent strains of virus that invade, or damage more severely, cultivars that were previously resistant to or tolerant of infection. It is seldom clear whether failure is due to the emergence or introduction of an apparently novel strain or to the increased prevalence of an existing one. Moreover, the type of host resistance utilised and the way it was deployed are not always clear from publications, and yet these factors can greatly influence the quantity and type of inoculum available and the opportunity for virulent, resistance-breaking strains to appear and become established. The opportunity and the overall selection pressures operating can be very different as tolerant cultivars may support considerable multiplication of the usual strain whereas cultivars that are resistant, hypersensitive or immune to infection do not. The performance and durability of resistant cultivars are also influenced by their proximity to susceptible or tolerant ones and the rapidity with which these are displaced.

There is little evidence with virus diseases of the relative merits of single and multiple resistances, or of major and minor genes and the ease with which these are overcome. Moreover, strains of only three viruses have been considered in terms of the gene-for-gene hypothesis (Drijfhout, 1978; Jones, 1981; Pelham, 1972) and there is no information on the effectiveness of intraspecies or varietal mixtures of the type used against some fungal diseases (Wolfe, 1985).

Breeding for resistance has such obvious advantages that it is likely to become increasingly important as a means of avoiding or decreasing the losses due to virus diseases. This is already apparent from much of the current work at the Rice Research and other International Institutes. These serve regions where sophisticated control measures are inappropriate and where there are limited opportunities for controlling vectors with pesticides. Thus, virus strain variation merits increased attention in relation to strategies and methods used in breeding for resistance and in developing appropriate ways of deploying host resistance genes in attempts to achieve durable control. Similar reappraisal will also be required if there is to be increased use of "mild" strains to protect plants from the damaging effects of at least some virulent ones. Moreover, it will eventually be necessary to consider the possible emergence of resistant or tolerant virus strains in deciding how chemotherapeutants or genetically engineered forms of resistance should be used most effectively.

VECTOR BIOTYPES

The variation that occurs in morphological and other features of vector populations has received much attention from biologists as it is of obvious taxonomic importance in relation to the correct identification and precise delineation of vector species (Claridge and Den Hollander, 1983). Much of the variation reported relates to minor differences in morphology or pigmentation that are of little apparent ecological importance. However, differences have also been reported in physiological attributes including rates of growth or reproduction, flight behaviour, host range, response to crowding and ability to withstand pesticides or periods of high or low temperature. These differences are of obvious epidemiological significance, as are those reported between different vector populations in their ability to acquire and transmit viruses. For example, Storey (1932) distinguished "active" and "inactive" races of the leafhopper *Cicadulina mbila* by their ability to transmit maize streak geminivirus. There are many other less extreme examples of intraspecific differences in vector efficiency (Kisimoto, 1967). Such differences can have profound effects on the overall "fitness" of vector populations as demonstrated in studies on rice hoja blanca and other viruses that multiply in insect vectors and have detrimental effects on their fecundity and longevity (Jennings and Pineda, 1971). These effects may account for the generally low proportion of *Sogatodes oryzicola* that transmit hoja blanca virus in the field, for the cyclical periodicity of epidemics and for the difficulty encountered in selecting colonies of highly infective individuals for use in resistance breeding.

A detailed consideration of the whole range of arthropod, nematode and fungus vectors is inappropriate in a volume devoted mainly to virus-host interactions. However, it is important to appreciate that many viruses can only infect their hosts by way of vectors and that the role of vector biotypes can be crucial in determining the outcome of the battle between viral and host genes. This is apparent from experience with hopper-borne viruses of tropical rice that only become prevalent in plantings of vector-susceptible cultivars, or following a shift in vector biotype and the colonization of cultivars that were previously resistant (see below).

Additional evidence of the importance of vector biotypes has been obtained from long-term studies in North America on the green bug aphid *Schizaphis graminum* which transmits some strains of barley yellow dwarf luteovirus and maize dwarf mosaic potyvirus. *S. graminum* was originally a pest of wheat and seldom infested sorghum. The first evidence of a major shift in biotype was obtained in the early 1960s when Biotype A was displaced by Biotype B which caused damaging infestations of several wheat cultivars that were highly resistant to Biotype A (Wood, 1961).

A further shift in population occurred in the late 1960s when sorghum was severely infested in Texas and neighbouring states by a third biotype (Harvey and Hackerott, 1969) and there was a marked change in the amount and season of flight activity (Daniels, 1981). Three additional biotypes of *S. graminum* have since been reported including type "D" that is resistant to organophosphorus and other insecticides. This biotype is now widespread and effective control of aphid infestations and virus spread is much more difficult to achieve than when insecticides were first introduced (Peters *et al.*, 1975).

Insecticides have been used widely to restrict the direct damage caused by *S. graminum* and many other virus vectors and also in attempts to restrict virus spread. Considerable success has been achieved but insecticide-resistant biotypes have been reported of many virus vectors, including whiteflies, aphids, planthoppers and leafhoppers and such important and widely distributed species as *Bemisia tabaci, Myzus persicae* and *Aphis gossypii*. Indeed, resistant populations of *M. persicae* almost completely displaced susceptible populations on sugar beet in eastern England during the 1970s when there were severe epidemics of virus yellows. *M. persicae* and the viruses it transmits have since become increasingly difficult to control in sugar beet, potato, brassicas and other crops and attempts are being made to develop new approaches to control and to devise more effective ways of deploying the insecticides available (Rice *et al.,* 1985; Dewar *et al.,* 1988).

Even greater problems due to the appearance of insecticide-resistant biotypes have been encountered in attempts to control the leafhopper and planthopper vectors of rice viruses in Japan where insecticides have been used very intensively for many years. For example, malathion was adopted against *Nephotettix cincticeps* in the 1950s and resistant populations were first reported in 1960-1961. Resistance was reported later to BHC (1962) and carbamates (1969) and there have been similar developments with *Laodelphax striatellus* and *Nilaparvata lugens* (Miyata and Saito, 1984).

These examples indicate the adaptability of vector species and the way in which problems can arise as populations become able to colonize new hosts or change in response to the introduction of a novel insecticide or resistant cultivar. Changes can also occur in other characteristics and it is clear that the genetic features of vector populations are of great epidemiological importance.

FOUR CASE HISTORIES

The dynamic and complex inter-relationships that can occur between host cultivars, virus strains and vector biotypes are exemplified in the following "case histories".

Rice grassy stunt tenuivirus and its planthopper vector Nilaparvata lugens

The rice brown planthopper (*Ni. lugens*) was a little-known and unimportant pest of tropical rice until very damaging outbreaks were reported in Fiji in 1959 and in many parts of South-East Asia during the 1960s or 1970s. The Asian outbreaks occurred soon after the introduction of the first high-yielding cultivars from the International Rice Research Institute (IRRI) in the Philippines. IR 8 and the four other IRRI cultivars that were released between 1966 and 1971 were all susceptible to *Ni. lugens* which transmits rice grassy stunt virus (GSV). Routine screening for resistance to *Ni. lugens* began in 1968 and against GSV in 1969. Effective sources of resistance were soon identified in rice accessions against *Ni. lugens* and in a wild rice (*Oryza nivara*) against GSV (Khush, 1977). Both sources of resistance are due to dominant major genes that were designated *Bph 1* and *Gs*. Gene *Bph 1* was incorporated in IR 26 (1973) and in four other IRRI cultivars that were released in 1974 or 1975 and these also contained *Gs*.

The resistant cultivars or their derivatives were at first grown widely and successfully. However, damaging infestations of *Ni. lugens* were reported within two or three years of the introduction of the new cultivars in the Philippines, Indonesia and Vietnam. This was due to the emergence of a resistance-breaking biotype (BPH 2) which colonized cultivars that were resistant to the original populations of BPH 1 (Khush, 1984). There was no evidence at the time of any breakdown in the resistance to GSV and *Gs* was incorporated into all the new cultivars and breeding lines being developed at IRRI. This continued routinely until a new resistance-breaking strain of the virus, GSV-2, was identified in parts of the Philippines in 1982-1983 (Hibino *et al.*, 1985). Similar strains have since been found in Indonesia and elsewhere and *Gs* is now regarded as a "failed" or "defeated" resistance gene. It is no longer used in the IRRI breeding programme and no alternative sources of resistance to the virus have been identified (G Khush, personal communication).

New sources of resistance were exploited to overcome the threat posed by BPH 2 and a recessive gene (*bph 2*) was identified in rice accession Ptb-18. The gene was incorporated into IR 32 and twelve other IRRI cultivars that were released between 1975 and 1983. As with the previous cultivars, they were grown successfully for some years until a new biotype (BPH 3) was reported in parts of Indonesia and the Philippines in 1981 (Khush, 1984). BPH 3 was able to damage several cultivars containing *bph 2* but not the earlier ones with *Bph 1* or IR 36 which contains *bph 2* and another gene for resistance. This led to the release of IR 56 and other cultivars with the dominant resistance gene *Bph 3* or recessive gene *bph 4*. These had not previously been utilized and they are now being deployed separately or in combination with *Bph 1* or *bph 2* in attempts to develop more durable forms of resistance (Heinrichs, 1988).

Rice tungro virus and its leafhopper vector Nephotettix virescens

There are reports of what appears to have been tungro disease in Indonesia in 1859 and later in Indonesia, Malaysia and the Philippines (Ou, 1984). However, the disease attracted only limited attention until the 1960s, when more intensive cropping practices and improved high-yielding cultivars were introduced to the tropical rice-growing areas of South-East Asia. Tungro then became prevalent in many areas and it is now regarded as one of the archetypal diseases of "the green revolution" (Thresh, 1989).

The need for cultivars resistant to tungro and its main leafhopper vector *Nephotettix virescens* became apparent at an early stage of the breeding programme at IRRI and routine screening tests have been used since the late 1960s (Khush, 1977). Several different sources of resistance to *Ne. virescens* have been utilized and nine major genes have been distinguished (Khush, 1984 and unpublished data). Apart from IR22, all the IRRI cultivars released since 1966 were rated as having at least some degree of vector resistance in the Philippines *at the time they were introduced.*

Vectors are less likely to settle and feed on resistant varieties than on susceptible ones (Heinrichs and Rapusas, 1983). The reproduction and longevity of vectors are also impaired on resistant varieties and this has been associated with an increased tendency to feed in the xylem rather than in the phloem (Auclair *et al.*, 1982). These features restrict vector populations and also the acquisition and transmission of tungro viruses. Thus, vector-resistant varieties tend to escape tungro disease (Heinrichs and Rapusas, 1983) and they have played a major role in decreasing the losses that would otherwise have occurred.

Problems due to the emergence and build-up of vector populations that are able to thrive on hitherto resistant cultivars have been less acute than those encountered with *Ni. lugens*. One reason for this may be that *Ni. lugens* is restricted to rice whereas *Ne. virescens* has a wide host range. Populations of *Ni. lugens* also tend to be much greater than those of *Ne. virescens* and the higher levels of resistance to *Ni. lugens* impose a greater selection pressure (Khush, 1977). Nevertheless, there have been various reports of tungro epidemics associated with marked shifts in the behaviour of cultivars that were originally regarded as virus- and vector-resistant (Dahal, 1988; Dahal *et al.*, 1988).

One such shift seems to have occurred in the Philippines during the early 1970s when the first two of the IRRI cultivars (IR 5, IR 8) were seriously affected. They had previously tended to escape infection because of their vector resistance due to the dominant gene *Glh 3*. During the epidemics of the 1970s later cultivars IR 36 and IR 42 which contained a different resistance gene (*Glh* 6) were relatively undamaged. However, they succumbed in Southern Cotabato in 1979 and elsewhere in the Philippines in 1983-1984. During this period IR 29 and IR 34 which have a third type of resistance derived from Gam Pai were little damaged and this led to more extensive plantings of other Gam Pai cultivars. They included IR 50, IR 54 and IR 64 which were extensively infected in southern Cotabato and Leyte in the 1986 and 1987 epidemics.

There have been similar experiences in Thailand, Indonesia and other countries where tungro is prevalent. Thus previous views on the apparent durability of resistance to *Ne. virescens* (Khush, 1984) have been changed and it is now accepted that populations are likely to adapt to resistant cultivars within a few years of their general release (Ruangsook and Khush, 1987). This has led to the current interest in mechanisms of resistance, gene deployment and breeding strategies in attempts to provide cultivars with more effective and durable forms of resistance than those used previously (Hibino *et al.*, 1985).

Of particular interest is the promotion of an official variety rotation scheme in Sulewesi and other parts of Indonesia to decrease the likelihood of *Ne. virescens* adapting to particular rice genotypes (Manwan *et al.*, 1985). Three groups of cultivars are distinguished according to the genes for vector resistance that are known or assumed to be present. Each growing season cultivars of only one type are used throughout entire districts and then replaced by cultivars of a different group for the following crop.

Sugarcane mosaic potyvirus

The history of sugarcane production in Louisiana, U.S.A. provides a striking example of the way in which a virus disease can have a drastic and continuing influence on the type of cultivar grown and on the breeding practices adopted.

The disease now known to be caused by sugarcane mosaic virus was first reported in 1892. It has since been found in virtually all countries where sugarcane is grown. The first reports from Louisiana were in 1919 when the main cultivars were traditional "noble" canes (*Saccharum officinarum*) of tropical origin that had been grown successfully for many years. The noble canes proved to be extremely vulnerable to mosaic and infection soon became widespread due to the dissemination of infected planting material and also to spread by aphid vectors (Summers *et al.*, 1948). Serious losses were sustained and by the early 1920s it seemed that sugarcane production might have to be abandoned in many areas. However, the situation changed when "P.O.J." cultivars derived from crosses between *S. officianarum* and *S. barberi* were introduced from Indonesia. They were much more tolerant of infection than the noble canes, which were almost entirely replaced between 1924 and 1929. The "P.O.J." cultivars were then largely superceded by Co.281 and Co.290. These were mosaic-tolerant cultivars from India derived from *S. officianarum* x *S. spontaneum* x *S. barberi*.

The next important development was the release in 1930 of the first of a sequence of locally-produced "C.P." cultivars that were selected at Canal Point, Florida for their resistance to infection. By 1955 C.P. selections accounted for 95% of the total area and led to a big decrease in the incidence of mosaic disease. It was no longer regarded as a serious problem and in 1954 a decision was taken to release an introduced cultivar (N.Co.310) that was high-yielding and cold-resistant but suceptible to infection with mosaic virus and somewhat tolerant. This led to a marked deterioration in the situation. Moreover, virus strain H was first reported in 1956 and shown to infect C.P.33-101 and other widely grown cultivars that had previously been highly resistant (Abbott, 1958).

By 1960 only 12% of the Florida area was planted with cultivars that were still rated as resistant. Infection again became prevalent and several widely grown cultivars had to be totally or partially replaced. It was also necessary to reintroduce roguing as a control measure in attempts to ensure the availability of healthy stocks of planting material (Abbott, 1962).

Roguing is inconvenient, expensive and not always effective, especially in areas where infection is prevalent or where virus-tolerant cultivars are grown. Several of these are still widely planted because it has been difficult to breed virus-resistant cultivars of acceptable yield and quality. A particular problem has been that *S. spontaneum* which was the main source of resistance used in the original C.P. cultivars is susceptible to strain H (Abbott and Todd, 1962). These factors explain why mosaic continues to cause losses, albeit much less than those reported in the 1920s before the first of the resistant cultivars was introduced (Breaux, 1985).

Beet curly top geminivirus

Beet curly top virus has had a profound impact on the history and profitability of sugar beet production in large areas of the United States. Indeed, the losses caused by the virus were at first so great that many beet processing factories had to be abandoned or operated intermittently and inefficiently until resistant cultivars were introduced. They completely transformed the situation and the use of resistant cultivars soon become an indispensible feature of beet production in California and adjoining states (Coons, 1949).

Repeated attempts were made to establish beet production in different parts of the United States during the 19th century. Various difficulties were encountered including the losses caused by beet curly top disease. This was first reported in 1888 and there were many later reports of serious epidemics that led to complete crop failures and serious financial hardship. Production had to be abandoned in some of the most seriously affected areas and there were strong incentives to develop resistant cultivars that would withstand infection.

The types of beet that were grown originally were from mixed stocks of European origin. They were very seriously affected by curly top disease and plants infected early usually died. However, occasional plants were only slightly affected and the first resistant cultivar (U.S. No.1) was derived from selections of this type. They were collected in various areas, grown together at a test site in Utah and released in 1933-1934. Additional selections were made later from within the mixed population and this led to substantial improvements in curly top resistance, yield and resistance to bolting.

By 1938 almost all plantings were of resistant cultivars and their success is indicated by the marked upward trend in productivity recorded in California and in Utah between 1937 and 1963 (Fig. 2). During this period selection and breeding procedures changed as first in-bred populations and then monogerm hybrids were introduced. This involved producing curly top-resistant multigerm inbreds and cytoplasmic male sterile lines for use as parents (McFarlane, 1969). There was also a need to combine curly top resistance with resistance to fungus diseases (*Perenospora schachtii* and *Cercospora beticola*) and to "virus yellows" due to beet yellows closterovirus or beet western yellows luteovirus.

It has long been known that many different strains of beet curly top virus can be distinguished by the severity of their effects on beet. An early trend to become apparent

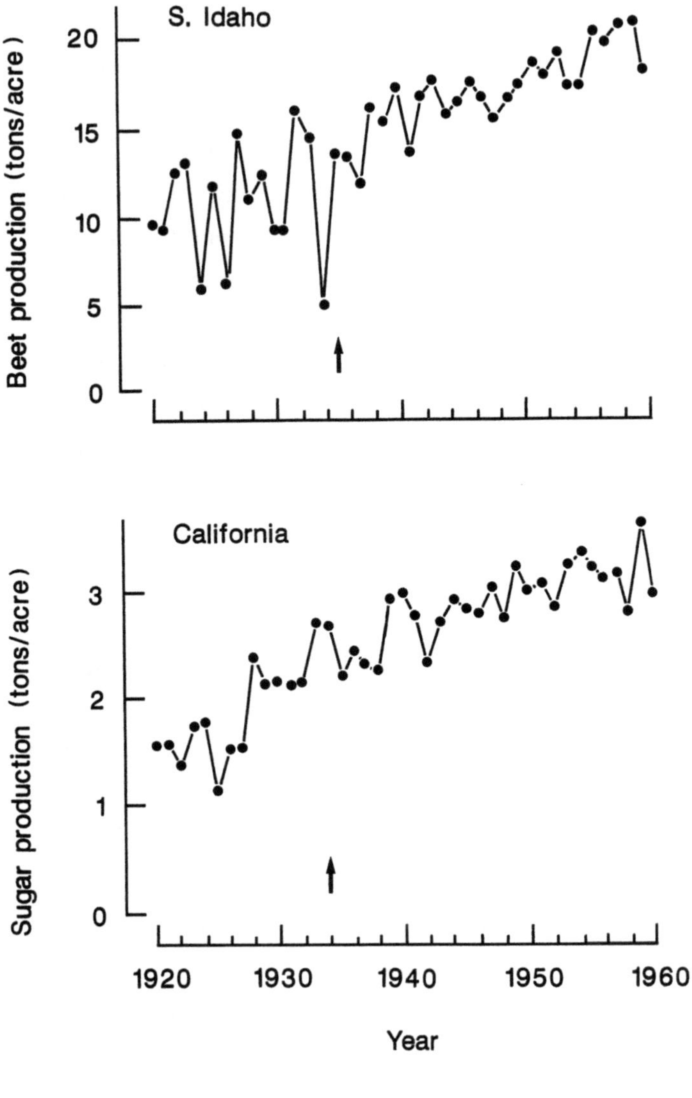

Figure 2. Productivity of sugar beet in Southern Idaho and California: 1920-1960. Arrows indicate the years in which the first curly top-resistant cultivars were released. Data from McFarlane (1969) and Duffus (1977).

was for the strains of virus that were isolated from beet at the beginning of each growing season to be much less damaging than those reported later (Giddings, 1948). This was attributed to the tendency for virulent strains to be selected in beet because the leafhopper vector thrives better in sparse open stands than in closed vigorous ones (Carter, 1930). It was also suggested that virulent strains tended to die out during the winter months because they have such drastic effects on their weed hosts.

Bennett (1963) noted an apparent trend towards increased virulence of curly top strains in different parts of the U.S. This trend seems to have continued in California in recent decades and the virulent strains isolated in the 1970s were much more damaging than those reported in the 1950s or 1960s (Magyarosy and Duffus, 1977). This was associated with marked ecological changes in the region and the year-round production of beet which favours the selection of virulent strains.

These results emphasize the importance of continued attempts to develop beet cultivars more resistant than those now available. Meanwhile, it is necessary to deploy existing cultivars in association with insecticides and other control measures that are intended to decrease the overall "infection pressure" on plantings (Duffus, 1981). It is particularly important to achieve a reduction during the early stages of crop growth when even the most resistant of the cultivars being grown is vulnerable to infection and can be severely damaged.

DISCUSSION

In recent decades there have been substantial increases in the productivity of many crops due to the introduction of new cultivars and changed techniques, including the increased use of irrigation, fertilizers and pesticides. The improved productivity has been achieved despite continuing and sometimes increased losses due to pests and diseases, some of which were previously unknown or unimportant. These developments have led to a vigorous and continuing debate on the underlying causes of pest and disease outbreaks and on the various ways in which cropping practices have influenced the losses sustained (Thresh, 1982).

There has been considerable emphasis on the impact of new cultivars and on the displacement of old "landraces". Much of the debate has been concerned with the most appropriate selection and resistance breeding strategies to adopt, on the relative merits of major and minor genes and on the need for stable, durable forms of resistance. Moreover, attention has also been given to the great differences between crop stands and natural vegetation and to the apparent vulnerability of many crops to pests and diseases compared with their wild progenitors, especially when crops are introduced to entirely new areas. From such considerations it has been suggested that the prolonged

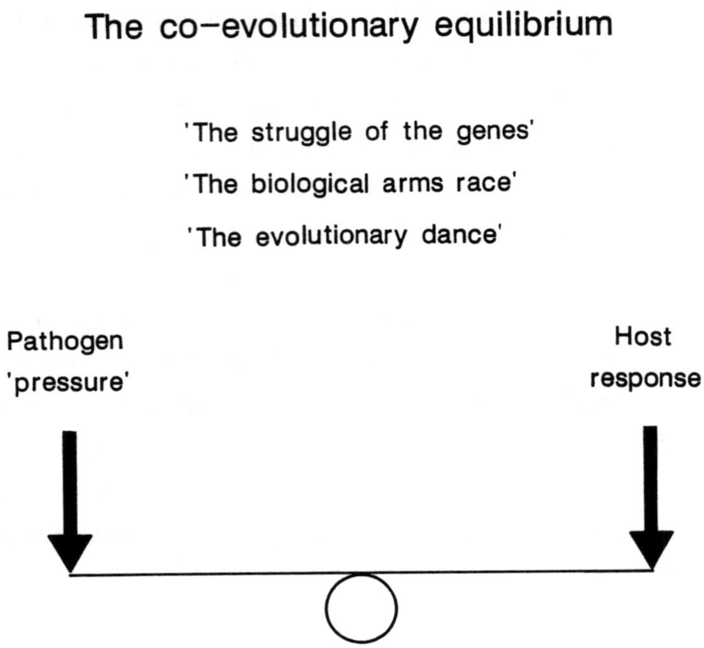

Figure 3. Diagramatic representation of the dynamic equilibrium established between pathogen "pressure" and host response.

exposure of host plants to their pests and pathogens over many generations leads to co-evolutionary adaptation and ultimately to some form of equilibrium. This can be expressed diagramatically as a balance between pest or pathogen "pressure" and host response (Fig. 3). Pathogen "pressure" is greatly influenced by weather and other environmental variables but there are also strong underlying genetic components that determine overall virulence and epidemiological competence. An inherent ability to resist or tolerate infection is an equally important feature of host response. However, host abundance and phenology are also involved because plants that are inherently vulnerable to infection can evade damage if they are few or widely scattered or if they develop at times of low pathogen pressure. This indicates the dynamic nature of the equilibrium which is to be expected as first the pathogen and then the host gains supremacy due to changes in weather conditions or relative abundance, or to the occurrence of new genes for pathogen virulence or host resistance.

Such co-evolutionary sequences of adaptation and counter-adaptation between hosts and their pests or pathogens justify the title of this paper and the similar approach adopted previously (Dawkins and Krebs, 1979; Gibbs, 1983). Indeed, the "war games" analogy can be developed further by distinguishing between occasional "skirmishes" or minor "battles" that occur sporadically and the prolonged, severe "wars of attrition". Sustained encounters necessitate a permanent and substantial diversion of energy and resources to avoid total annihilation of attacker or defender, whereas a relatively minor confrontation is quite different and justifies far less expenditure. There are also parallels between the adaptations and counter-adaptations of biological systems and the measures and counter-measures of a genuine "arms race".

There is little direct evidence concerning viruses that can be used to support the co-evolutionary interpretation of host-pathogen interactions as presented here and elsewhere (Gibbs, 1983; Harlan, 1976; Thresh, 1982). This is partly because of the paucity of data on the incidence and effects of viruses in natural plant communities. Moreover, virology is a relatively young science and there is a dearth of information on the changes in disease prevalence that have been associated with even the most recent phases of agricultural development.

Despite these limitations there are many examples of serious epidemics of virus disease that can be interpreted as instances of biological instability following the disruption of long-standing equilibria (Buddenhagen, 1977; Thresh, 1980, 1982). These include epidemics that have occurred:

- as a direct consequence of a "new encounter" between virus and host, or of a "re-encounter" after a prolonged period of separation;
- following the introduction of a virus or vector to an entirely new area;
  associated with the introduction of new cultivars and the displacement of traditional land races;
- due to the emergence of completely new virus strains or vector biotypes;
  after an intensification of cropping practices or an extension of the growing season following the introduction of new cultivars, irrigation or some form of protected cropping;
- following the widespread use of pesticides that have enhanced vector populations by destroying natural enemies, and
- in association with exceptional weather conditions that have led to increased vector populations or to enhanced host vulnerability.

From this list it is apparent that the crucial overriding factor has in some instances been a change in weather patterns or a chance event over which there was little or no human control. However, cropping practices, breeding policies and other features of agricultural development that are under at least some degree of control are of even greater importance. They have had seriously disruptive effects and led to increased losses that undermine the benefits sought.

This realization is not new as Bawden (1955) commented "if we are to reap the full rewards of high farming, we must also learn to control virus diseases". His statement has even greater validity now than it had when written. Crop protection specialists still face the challenge of devising effective pest and disease control measures that make it possible to exploit fully the benefits of improved high-yielding cultivars and the continuing advances being made in agricultural technology. The aim is to devise control measures that are more effective and more durable than those now available so that crops acquire some of the stability and resistance of natural vegetation. This objective is unlikely to be achieved by reverting to outmoded cropping systems or by attempting to simulate the great genetic diversity found in nature and in primitive agriculture. Nevertheless, there is considerable scope for continuing to develop and utilize resistant cultivars. They have obvious advantages compared with other approaches to control and there have been notable successes. There have also been failures and resistant cultivars can undoubtedly be deployed more effectively and with greater discrimination than previously.

The tendency in the past has been to adopt a simplistic approach in selecting resistant cultivars and in using them without sufficient regard for other possible control measures. Attitudes are now changing and in at least some instances resistant cultivars are being used as key components of integrated management strategies aimed at achieving a reduction in overall infection pressure. Attention is also being given to strategies of gene deployment to decrease the chances of "breakdown" due to the emergence of virus strains and vector biotypes able to overcome the resistance being utilized.

Another recent development in the "battle of the genes" has been the revived interest in mild strain protection as a virus disease control measure following the successful use of this approach to decrease the losses due to tomato mosaic, citrus tristeza and papaya ringspot viruses (Fulton, 1986). The production of genetically-engineered plants with an ability to withstand virus infection due to the incorporation of a viral coat protein gene or a benign satellite is likely to be of even greater significance and of wide application. However, the first field trials have only recently been completed and the full implications of these discoveries are not yet apparent. It will be of particular interest to determine the circumstances in which the different forms of induced resistance are most effective and whether resistance-breaking strains of virus will be encountered.

These various developments are important and long overdue because all the indications are that the "battle of the genes" will intensify as high-yielding cultivars and modern cropping practices are introduced to additional regions of the world and crops are subject to even greater levels of infection pressure than those now encountered. There is an urgent need for changed tactics and also for completely new strategies to restore the balance between pathogen pressure and host response and so avoid the serious epidemics now encountered.

REFERENCES

Abbott EV (1958) Strains of sugarcane mosaic virus in Louisiana. Sugar Bulletin, New Orleans 37:49-51

Abbott EV (1962) Problems in sugar cane disease control in Louisiana. Proc Int Soc Sugar Cane Technol 11:739-742

Abbott EV, Todd EH (1962) Mosaic in clones of *Saccharum spontaneum* and in Kassoer. Proc Int Soc Sugar Cane Technol 11:753-755

Auclair JL, Baldos E, Heinrichs EA (1982) Biochemical evidence for the feeding sites of the leafhopper *Nephotettix virescens* within susceptible and resistant rice plants. Insec Sci Appl 3:29-34

Bar-Joseph M, Roistacher CN, Garnsey SM, Gumpf DJ (1981) A review on Tristeza; an ongoing threat to citriculture. Proc Int Soc Citriculture 1:419-423

Bawden FC (1955) The spread and control of plant virus diseases. Ann Appl Biol 42:140-147

Bennett CW (1963) Highly virulent strains of curly top virus in sugar beet in Western United States. J Am Soc Sugar Beet Technol 12:515-520

Bowen R, Plumb RT (1979) The occurrence and effects of red clover necrotic mosaic virus in red clover (*Trifolium pratense*). Ann Appl Biol 91:227-236

Breaux RD (1985) Controlling diseases of sugarcane in Louisiana by breeding: potentialities and realities. J Am Soc Sugar Cane Technol 4:58-61

Buddenhagen IW (1977) Resistance and vulnerability of tropical crops in relation to their evolution and breeding. Ann NY Acad Sci 287:309-326

Buddenhagen IW (1983) Agroecosystems, disease resistance and crop improvement. In: Kommedahl T and Williams PH (eds) Challenging problems in plant health. American Phytopathological Society, St Paul, Minnesota, p 450

Carter W (1930) Ecological studies of the beet leaf hopper. USDA Tech Bull 206

Cho EK, Chung BJ, Lee SH (1977) Studies on identification and classification of soybean virus diseases in Korea. II Etiology of a necrotic disease of *Glycine max*. Plant Dis Rep 61:313-317

Claridge MF, Den Hollander J (1983) The biotype concept and its application to insect

pests of agriculture. Crop Protection 2:85-95
Coons GH (1949) The sugar beet: product of science. Sci Monthly 68:149-164
Cropley R (1967) Decline and death of pear on quince rootstocks caused by virus infection. J Hort Sci 42:113-115
Dahal G (1988) Transmission of tungro-associated viruses by field and selected colonies of *Nephotettix virescens* Distant and their mode of feeding on selected rice cultivars. PhD Thesis, University of Philippines, Los Baños, The Philippines
Dahal G, Aguiero VM, Cabunagan RC, Hibino H (1988) Varietal reaction to tungro (RTV) with change in leafhopper "virulence". International Rice Research Newsletter 13:12-13
Daniels NE (1981) Migration of greenbugs in the Texas Panhandle in relation to their biotypes. Misc Pub 1487. Texas Agricultural Experiment Station
Dawkins R, Krebs JR (1979) Arms races between and within species. Proc Roy Soc Lond B 205:489-511
Dewar A, Devonshire A, Ffrench-Constant R (1988) The rise and rise of the resistant aphid. Br Sugar Beet Rev 58:40-43
Drijfhout E (1978) Genetic interaction between *Phaseolus vulgaris* and bean common mosaic virus with implications for strain identification and breeding for resistance. Agric Res Rep (Neth) 872
Duffus JE (1977) Beet-free periods - the key to higher sugar beet yields. Calif Agric 31:18-19
Duffus JE (1981) Curly top virus control. Proc IX Int Congr Plant Protect Washington 2:508-510
Egan BT (1976) The fall and rise of Fiji disease in southern Queensland. Proc Queensl Soc Sugar Cane Technol 43:73-77
Egan BT, Toohey CL (1977) The Bundaberg approved plant sources scheme. Proc Queensl Soc Sugar Cane Technol 44:55-59
Fletcher JT, Butler D (1975) Strain changes in populations of tobacco mosaic virus from tomato crops. Ann Appl Biol 81:409-412
Fraser RSS (1986) Genes for resistance to plant viruses. CRC Crit Rev Plant Sci 3:257-294
Fulton RW (1986) Practices and precautions in the use of cross protection for plant virus disease control. Annu Rev Phytopathol 24:67-81
Gardener FE, Marth PC, Magness JR (1946) Lethal effects of certain apple scions on Spy 227 stock. Proc Am Soc Hortic Sci 48:195-199
Gibbs AJ (1983) Virus ecology - "struggle" of the genes. In: Lange OL, Nobel PS, Osmond CB, Ziegler H (eds) Physiological plant ecology. III Response to the chemical and biological environment. Springer Verlag, Berlin, p 537
Giddings NJ (1948) Some studies of curly-top virus in the field. Proc Am Soc Sugar Beet Technol 5:531-538
Grancini P (1962) Ulteriori notizie sul nanismo ruvido del mais. Maydica 17:17-25
Harlan JR (1976) Diseases as a factor in plant evolution. Annu Rev Phytopathol 14:31-51
Harpaz I (1972) Maize Rough Dwarf. Israel Universities Press, Jerusalem
Harper JL (1982) After description. In: Newman EI (ed) The plant community as a working mechanism. British Ecological Society; Blackwell Scientific Publications, Oxford, p 11
Harrison BD (1978) The groups of nematode-transmitted plant viruses and molecular aspects of their variation and ecology. In: Scott PR, Bainbridge A (eds) Plant disease epidemiology, Blackwell, Oxford, p 255
Harrison BD (1981) Plant virus ecology: ingredients, interactions and environmental

influences. Ann Appl Biol 99:195-209
Harvey TL, Hackerott HL (1969) Recognition of a greenbug biotype injurious to sorghum. J Econ Entomol 62:776-779
Hatton RG (1928) The behaviour of certain pears on various quince rootstocks. J Pomol 7:216-233
Heinrichs EA (1988) Variable resistance to homopterans in rice cultivars. ISI Atlas of science: plants and animals 1:213-220
Heinrichs EA, Rapusas H (1983) Correlation of resistance to the green leafhopper, *Nephotettix virescens* (Homoptera: Cicadellidae) with tungro virus infection in rice varieties having different genes for resistance. Environ Entomol 12:201-205
Hibino H, Cabauatan PQ, Omura T, Tsuchizaki T (1985) Rice grassy stunt virus strain causing tungrolike symptoms in the Philippines. Plant Dis 69:538-541
Jennings PR, Pineda AT (1971) The effect of the hoja blanca virus on its insect vector. Phytopathology 61:142-143
Jones RAC (1981) The ecology of viruses infecting wild and cultivated potatoes in the Andean region of South America. In: Thresh JM (ed) Pests, pathogens and vegetation. Pitman, London, p 89
Keyworth WG (1947) Mosaic disease of the hop: a study of tolerant and sensitive varieties. Rep E Malling Res Stn 1946 p 142
Khush GS (1977) Disease and insect resistance in rice. Adv Agron 29:265-341
Khush GS (1984) Breeding rice for resistance to insects. Prot Ecol 7:147-165
Kiritani K (1983) Changes in cropping practices and the incidence of hopper-borne diseases of rice in Japan. In: Plumb RT, Thresh JM (eds) Plant virus epidemiology. Blackwell Scientific, Oxford, p 239
Kisimoto R (1967) Genetic variation in the ability of a planthopper vector; *Laodelphax striatellus* (Fallén) to acquire the rice stripe virus. Virology 32:144-152
Kisimoto R (1972) Biology of rice stripe virus disease (in Japanese). Iden 26:34-40
Luckwill LC, Crowdy SH (1950) Virus diseases of fruit trees. II. Observations on rubbery wood, chat fruit and mosaic in apples. Progress report. Rep Agric Hort Res Stn Univ Bristol 1949, p 68
Mackenzie D, Salmon ES, Ware WM, Williams R (1929) The mosaic disease of the hop; grafting experiments. II. Ann Appl Biol 16:359-381
McLean DL, Kinsey MG (1963) Transmission studies of a highly virulent variant of lettuce mosaic virus. Plant Dis Rep 47:474-476
McFarlane JS (1969) Breeding for resistance to curly top. Int Inst Res Beet 4:73-83
Magyarosy AC, Duffus JE (1977) The occurrence of highly virulent strains of the beet curly top virus in California. Plant Dis Rep 61:248-251
Manwan I, Sama S, Rizvi SA (1985) Use of varietal rotation in the management of tungro disease in Indonesia. Indones Agric Res Dev J 7:43-48
Miyata T, Saito T (1984) Development of insecticide resistance and measures to overcome resistance in rice pests. Prot Ecol 7:183-199
Murant AF, Taylor CE, Chambers J (1968) Properties, relationships and transmission of a strain of raspberry ringspot virus infecting raspberry cultivars immune to the common Scottish strain. Ann Appl Biol 61:175-186
Ou SH (198) Exploring tropical rice diseases: a reminiscence. Annu Rev Phytopathol 22:1-10
Pelham J (1972) Strain-genotype interaction of tobacco mosaic virus in tomato. Ann Appl Biol 71:219-228
Pelham J, Fletcher JT, Hawkins JH (1970) The establishment of a new strain of tobacco mosaic virus resulting from the use of resistant varieties of tomato. Ann Appl Biol 65:293-297

Peters DC, Wood EA, Starks KJ (1975) Insecticide resistance in selections of the greenbug. J Econ Entomol 68:339-340

Posnette AF (1977) Tolerance of virus infection in crop plants. Rev Appl Mycol 48:113-118

Posnette AF, Todd JM (1951) Virus diseases of cacao in West Africa. VIII. The search for virus-resistant cacao. Ann Appl Biol 38:785-800

Quiot JB, Devergne JC, Cardin L, Verbrugghe M, Marchoux G, Labonne G (1979) Écologie et épidémiologie du virus de la mosaïque du concombre dans le sud-est de la France. VII. Répartition de deux types de populations virales dans des cultures sensibles. Ann Phytopathol 11:359-373

Rice AD, Devonshire AL, Gibson RW, Gooding AR, Moores GD, Stribley MF (1985) The problem of aphid resistance to aphicides and alternative chemical methods of preventing virus transmission. Proc 48th Winter Conf IIRB Brussels p 209-228

Ross H (1959) Über die Verbreitung der Tabakrippenbräunestämme des Y Virus der Kartoffel in Deutschland und anderen Ländern. Phytopathol Z 35:97-102

Ruangsook B, Khush GS (1987) Genetic analysis of resistance to green leafhopper *Nephotettix virescens* (Distant) in some selected rice varieties. Crop Prot 6:244-249

Russell GE (1978) Plant breeding for pest and disease resistance. Butterworths, London

Ryan CC (1988) Epidemiology and control of Fiji disease virus of sugarcane. Adv Dis Vector Res 5:163-176

Salmon ES (1923) The "mosaic" disease of the hop. J Minist Agric Fish 29:927-934

Simmonds NW (1962) Variability in crop plants, its use and conservation. Biol Rev 37:422-465

Storey HH (1932) The inheritance by an insect vector of the ability to transmit a plant virus. Proc R Soc Lond B 112:46-60

Summers EM, Brandes EW, Rands RD (1948) Mosaic of sugarcane in the United States, with special reference to strains of the virus. USDA Tech Bull 955 Washington DC

Thresh JM (1979a) Recent changes in the health status of English hop plantings. Proc Br Crop Prot Conf 1979 p 169

Thresh JM (1979b) Hop-growing in Germany. Newsl Fed Br Plant Pathologists 2:47-48

Thresh JM (1980) The origins and epidemiology of some important plant virus diseases. Appl Biol 5:1-65

Thresh JM (1982) Cropping practices and virus spread. Annu Rev Phytopathol 20:193-218

Thresh JM (1989) Insect-borne viruses of rice and "the green revolution". Trop Pest Manage 35:

Thresh JM, Owusu GK, Boamah A, Lockwood G (1988) Ghanaian cocoa varieties and swollen shoot virus. Crop Prot 7:219-231

Timian RG (1974) The range of symbiosis of barley stripe mosaic virus. Phytopathology 64:342-345

Todd JMcA (1960) Tobacco veinal necrosis on potato in Scotland: control of the outbreak and some characters of the virus. Proc 4th Conf Potato Virus Diseases, Braunschweig 82-92

Trebbi T (1950) Il nanismo del mais in Provincia di Brescia nel 1949. Notiz Mal Piante 8:13-16

Tukey HB, Brase KD (1943) An uncongeniality of the McIntosh apple when top-worked onto Virginia Crab. Proc Am Soc Hort Sci 43:139-142

Whittaker RH (1969) Evolution of diversity in plant communities. In: Diversity and stability in ecological systems. GM Goodwell and HH Smith (eds) Brookhaven Symp Biol, New York, p 178
Wolfe MS (1985) The current status and prospects of multiline cultivars and variety mixtures for disease resistance. Annu Rev Phytopathol 23:251-274
Wood EA (1961) Biological studies of a new greenbug biotype. J Econ Entomol 54:1171-1173
Zink FW, Duffus JE (1969) Relationship of turnip mosaic virus susceptibility and downy mildew (*Bremia lactucae*) resistance in lettuce. J Am Soc Hort Sci 94:403-407
Zink FW, Duffus JE (1970) Linkage of turnip mosaic virus susceptibility and downy mildew, *Bremia lactucae*, resistance in lettuce. J Am Soc Hort Sci 95:420-422

SIGNALS AND STRUCTURES INVOLVED IN EARLY INTERACTIONS BETWEEN PLANTS AND VIRUSES OR PSEUDOVIRUSES

T Michael A Wilson[1], Kitty A Plaskitt, John W Watts, Jane K Osbourn and Peter AC Watkins
John Innes Institute
AFRC Institute of Plant Science Research
Colney Lane
Norwich
Norfolk NR4 7UH
UK

ABSTRACT

One mechanism by which stable plant RNA virus particles might be disassembled *in vivo* involves recruitment of cell factors and ribosomes by a short, easily exposed portion of RNA. No cell specificity exists during virus uptake or early cotranslational disassembly. Virus-like reporter ribonucleocapsids were synthesized *in vitro* to study the mechanism, sequence-dependence and site of disassembly and transient gene expression without concomitant replication. "Pseudoviruses" and transcapsidated viral RNAs can elucidate the mechanism(s) of genetically engineered, coat protein-mediated cross-protection in electroporated tobacco protoplasts or plants.

INTRODUCTION

The majority of known plant viruses (77%; Zaitlin and Hull, 1987) contain genomes of single-stranded RNA. They have evolved more-or-less stable protein coats to sequester and protect the genome-sense RNA as it replicates and accumulates in an infected cell, during local or long-range spread within an infected plant, and between separate plants in the field. In contrast to most animal viruses, plant viruses have not developed ingenious, receptor-mediated, cell-specific infection mechanisms - for example

[1]Present address: Center for Agricultural Molecular Biology, Cook College, Rutgers University, PO Box 231, New Brunswick, New Jersey 08903-0231, USA

NATO ASI Series, Vol H 41
Recognition and Response in Plant-Virus Interactions
Edited by R S S Fraser

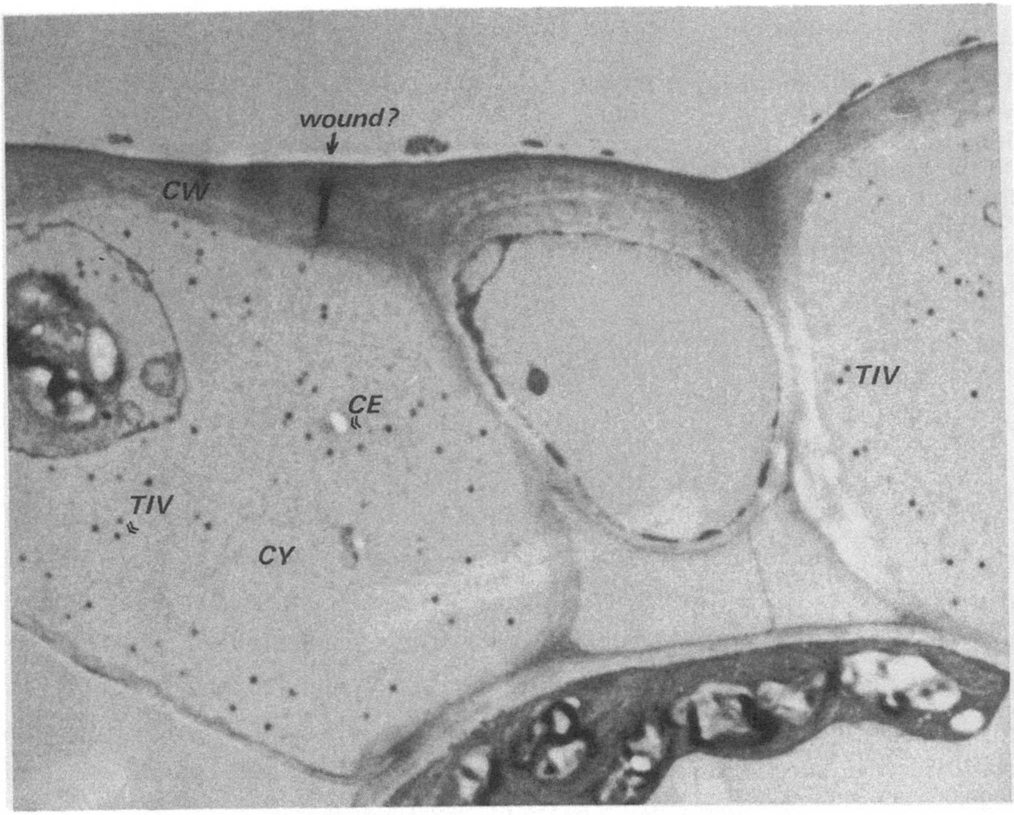

Figure 1. Mechanical "inoculation" of tobacco leaf epidermal cells with *Tipula* irridescent virus (TIV), a large polyhedral dsDNA-containing virus of crane-fly larvae. TIV particles are visible throughout the epidermal cell cytoplasm indicating that uptake is not virus-specific. A possible wound site is indicated. CE = grain of Celite abrasive; CW = cell wall; CY = cytoplasm.

via clathrin-coated, acidified endocytic vesicles (Willingmann *et al.*, 1989). Doubtless this reflects the very different architecture of plant cells. Once a single cell has been infected, access to other cells in the syncytium is possible *via* cytoplasmic connections at plasmodesmata, where virus-coded protein(s) may modify and extend the normally-restricted gating limit.

With the exception of the purely seed- or pollen-transmitted cryptic viruses (Antoniw *et al.*, this volume), all plant viruses are obligate wound pathogens, requiring a third party to assist them in gaining access to one or more infectable cells in a potential host plant. Insects, nematodes, fungal zoospores or Man are the preferred agents or vectors for virus spread and delivery into mechanically damaged cells. To some extent, virus/

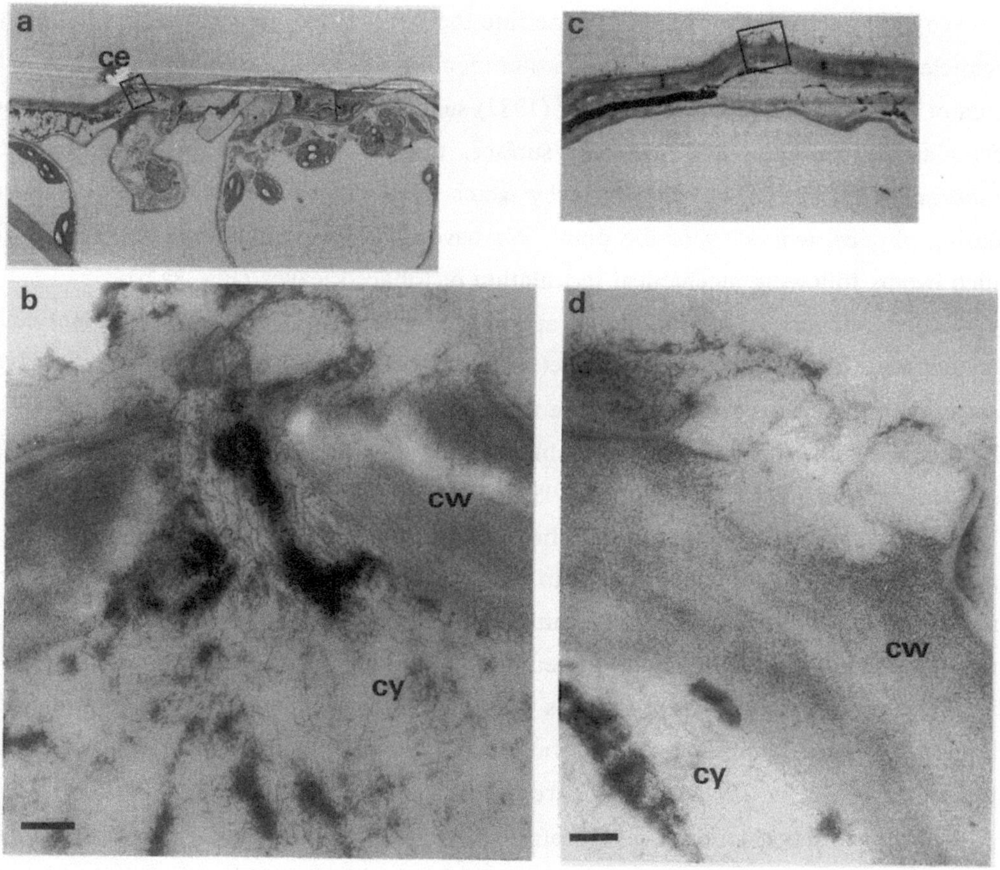

Figure 2. TMV-like pseudovirus particles 60nm in length containing mRNA for chloramphenicol acetyltransferase (CAT) enter cells through wounds in the thick outer cell wall (cw) caused by abrasive. The lower epidermis of a Xanthi tobacco leaf was "inoculated" with CAT rods, left for 2 h then fixed, embedded in resin and sectioned as described (Plaskitt *et al.*, 1988). At very low magnifications (panels a, c) several epidermal and spongy mesophyll cells show glass-knife scoring damage caused by particles of Celite (ce). Panels (b) and (d) show enlarged boxed areas from (a) and (c), respectively and filamentous pseudovirus particles entering or within the cytoplasm (cy). In (b), the extruded "bead" of cytosol (Laidlaw, 1987) is clearly visible outside the cell. Panel (d) may be a repairing wound or a serial section adjacent to a lesion as in (b). Scale bars = 200nm.

plant host-range therefore depends upon plant choice by the vector. Mechanical wounding does not lend itself to selectivity at the stage of virus entry into a cell (Fig. 1).

Wounding must be sufficient to penetrate the outer structural defences of the plant (cuticle, cell wall and probably the plasma membrane) yet not so severe that the cell cannot repair itself. Recently, Laidlaw (1987) suggested that extrusion of a bead of the cytoplasmic contents onto the leaf surface, due to hydrostatic pressure, and its subsequent retraction was the route by which virus nucleocapsids were internalized during physical wounding of the plant. We have observed structures compatible with such events following mechanical inoculation of tobacco leaves (Fig. 2a,b).

In general, very large numbers of apparently intact virus particles enter plant cells in this way. Identifying which particle(s) then proceed to establish the infection has always been a serious obstacle and one which continues to restrict progress in this area. In part, we have side-stepped this problem by creating virus-like reporter structures *in vitro* (Sleat *et al.*, 1986; Gallie *et al.*, 1987a; Jupin *et al.*, 1989). To date, these so-called "pseudovirus" particles have not had the ability to replicate, even if co-inoculated with a true "helper" virus. Thus they can provide a direct measure of the extent of nucleocapsid disassembly and concomitant expression of the encapsidated reporter mRNA. We shall return to these increasingly useful model structures later.

Another approach we have taken is to prepare radiolabelled virions of high specific activity so that we can follow the distribution and fate of only those particles which enter the inoculated cells (the so-called "parental" virus particles; Shaw *et al.*, 1986).

The fundamental paradox is, therefore, how are these necessarily robust structures (viral nucleocapsids) disassembled in the cellular *milieu*? Could one perhaps identify some host-specific factor(s) or site(s) to account for the process of virus destabilization which, in many cases (e.g. turnip yellow mosaic virus (TYMV) or tobacco mosaic virus (TMV); Keeling and Matthews, 1982; Wilson and Shaw, 1985), requires extremely harsh, non-physiological conditions to achieve *in vitro*? Alternatively, the mechanism may be independent of cell type and completely generic, in which case, what are the common features and how is virus-host range later manifest? Put bluntly, what in the cell can mimic the effects of phenol, urea, detergent, alkali or other chemicals on virus stability?

Most of what follows pertains to observations *in vitro* which are providing new clues or directions for future experiments *in vivo* to help resolve the paradox highlighted above. Modest progress has been made with experiments *in vivo* over the past three years (Shaw *et al.*, 1986; Plaskitt *et al.*, 1988), but further efforts are now a priority despite the limitations imposed by the unfavourable ratio of infectious to physical virus particles.

THE EXPERIMENTAL EVIDENCE

Cotranslational disassembly

Five years ago, the first reports of this phenomenon appeared (Wilson, 1984a,b) from experiments using conventional plant- or animal-derived cell-free translation systems. Uncoating of the viral RNA was effected in a 5' to 3' direction by the first ribosome to translate the viral (messenger-sense) RNA, although the uncoating did not go to completion. With hindsight, cotranslational disassembly was appealingly simple and logical. Anthropomorphically, it fulfilled the presumed desire of the virus to maximize protection for its RNA genome, while ensuring complete primary translation of the gene(s) required for subsequent stages of infection (e.g. replication). How distal portions of the genome are uncoated still remains unclear. For several years we have been investigating the possibility of a 3' to 5', co-replicational virus disassembly mechanism using (+) and (-) strand-specific probes (Goelet and Karn, 1984; JG Shaw and TMA Wilson, unpublished results). Unfortunately this work is hampered by the absence of an easily-manipulated *in vitro* replication system for TMV, and so must be pursued in inoculated tobacco leaves or protoplasts.

Probably the most extreme illustration of the strategic benefit of continued protection for the viral genome was seen when TMV particles or unencapsidated TMV RNA were added to a prokaryotic (70S ribosome) cell-free translation system (Wilson, 1986). RNA from particles was translated more efficiently and produced much higher molecular weight polypeptides (e.g. the 126 kDa protein) than the unencapsidated messenger.

In almost all our studies, and in several others reported independently with TMV (Register and Beachy, 1988; Roenhorst *et al.*, 1989; RSS Fraser, personal communication), with soil-borne wheat mosaic virus, a furovirus (Shirako and Ehara, 1986), or with several icosahedral plant RNA viruses (Brisco *et al.*, 1985, 1986; Roenhorst *et al.*, 1989), some pre-treatment of virions was found necessary to initiate cotranslational uncoating. Often this involved exposure to mildly alkaline conditions and/or chelation of divalent metal ions, to swell the protein shells. Of course, the question arises again: how relevant are these treatments to possible events *in vivo*?

It is unlikely that viruses encounter alkaline compartments inside plant cells. However, there is good evidence for a low, calmodulin-regulated level of intracellular Ca^{2+} ions (Allan and Hepler, 1988) and, in concert with membrane phospholipids, this has been claimed to be responsible for disassembly of both isometric and rod-shaped nucleocapsids (Durham *et al.*, 1977; Durham, 1978).

Several independent observations from our own laboratory serve to emphasize the artificiality of pre-treatment with alkaline buffers to render virions susceptible to cotranslational disassembly, and to confirm that events which destabilize particles *in vivo* apply universally:

1. Particles of several members of the potexvirus group were completely inactive as templates for *in vitro* translation, despite prolonged exposure to buffers between pH 8.0-9.0 (Wilson, 1984a; Bendena *et al.*, 1985; JB Bancroft, personal communication). However, inoculation of tobacco leaves with potato virus X and L-[^{35}S]-methionine resulted, within 30-60 minutes, in Cs_2SO_4 gradient-fractionated material indicative of translation complexes containing ribosome(s) and partially uncoated virions (Wilson and Shaw, 1987). Analogous contrasting results have been obtained *in vitro* (DW Thornbury and TMA Wilson, unpublished data; MA Mayo, personal communication) and *in vivo* (Wilson and Shaw, 1987) with members of the potyvirus and tobravirus groups. Whether the *in vitro* data reflect a 5'-terminal encapsidation initiation site on potyviral and tobraviral RNAs (as with potexviruses), or simply the quality of the virus preparations tested remains open to question, certainly for tobacco rattle virus (AbouHaidar and Hirth, 1977; Bergh and Siegel, 1989; JF Bol, personal communication).

2. Similarly, we have encapsidated chimaeric RNA molecules bearing a 5'-terminal "cap" structure and a 5'-proximal sequence (Ω') representing all but the first G of the untranslated TMV (U1 strain) leader (Gallie *et al.*, 1987a; Jupin *et al.*, 1989). Also included were additional nucleotides (most significantly, G-residues) derived from the SP6 transcription vector or binary vector (pROK2) and the cloning strategy used (Gallie *et al.*, 1987b; Sleat *et al.*, 1988a). In all cases to date, we have been unsuccessful in destabilizing these "pseudovirus" particles with mild alkali (up to pH 9.5) to initiate cotranslational disassembly *in vitro* (DE Sleat and JK Osbourn, unpublished results). However, some of these virus-like structures must uncoat *in vivo,* following mechanical "inoculation" of tobacco or pea leaf epidermis (Plaskitt *et* al., 1988; Gallie *et al.*, 1987a), PEG-inoculation or electroporation of tobacco mesophyll protoplasts (Gallie *et* al., 1987a; Osbourn *et al.*, 1989), and even after microinjection into amphibian oocytes (Gallie *et al.*, 1987a), because translation products can be detected. We are presently engaged in experiments to construct nucleocapsids with a systematic range of RNA-leader

sequences, including the true TMV leader correctly positioned, to study their effects on particle stability and downstream gene expression *in vitro* and *in vivo*.
3. When TMV is treated at approximately pH 12 to produce RNA-free coat protein for re-assembly experiments *in vitro* (Durham, 1972), a small number of particles resist uncoating and can be recovered by high-speed centrifugation. These represent an extreme example of the so-called alkali-stable fraction of TMV (Perham, 1969). In the electron microscope these pH 12-resistant particles resembled normal, helical TMV (300nm in length) together with many short (50nm) rodlets (Wilson *et al*., 1978) and an almost equivalent number of what appeared to be "stacked-disk" rods, usually of protein alone (Durham *et al*., 1971). These structures were confirmed by optical diffraction analysis (PJ Shaw, personal communication). However, although CsCl gradient fractionation produced two bands, it failed to separate the "stacked-disk" rods and helical particles. Further work is in progress on this material. *In vitro* translation of the pH 12-resistant particles revealed little or no intrinsic template activity (Wilson, 1984a; J McHale and PAC Watkins, unpublished results). However, after dilution and centrifugation in buffer at pH 8.2 or phenol-extraction of the RNA, a significant amount of 126 kDa polypeptide was synthesized. In local lesion assays on *Nicotiana tabacum* cv. Samsun *NN*, the pH 12-resistant particles had approximately 60% of the infectivity of control TMV. These pH 12-stable particles also produced systemic infections on *N. tabacum* cv. Xanthi (TMA Wilson and J McHale, unpublished results) and generally appeared more viable than previously reported (Wilson *et al*., 1978). Whatever protects this sub-population of virions against uncoating at pH 12 is clearly no defence against the disassembly mechanism operating on this same material in inoculated tobacco cells!

Thus we have yet to identify what triggers the exposure of viral RNA *in vivo*. Once started, limited evidence indicates that ribosome-mediated uncoating, at least up to the end of the first open reading frame, is possible (Shaw *et al*., 1986; Plaskitt *et al*., 1988). Using rabbit polyclonal antisera to a synthetic, NH_2-proximal peptide of the non-structural polypeptide of 126 kDa (amino acids 34-47), we hope to confirm the composition of putative *in vivo* translation complexes ("striposomes"; Shaw *et al*., 1986; X Wu and JG Shaw, unpublished results). Preliminary results with cycloheximide-arrested complexes *in vitro* are shown in Fig. 3.

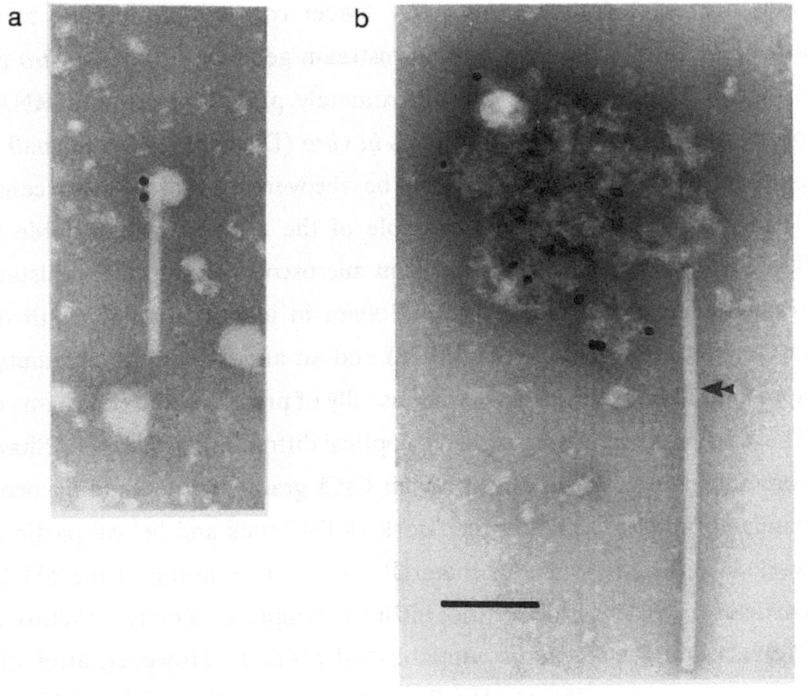

Figure 3. Immunogold labelling of nascent 126 kDa-derived sequences in *in vitro* "striposome" complexes from wheat germ extracts (WG) programmed with pH 8.2-treated TMV particles for 10 min prior to addition of cycloheximide. Diluted samples were applied to carbon-coated rhodium-plated copper grids, blocked with bovine serum albumin in phosphate-buffered saline, then treated sequentially with polyclonal rabbit antiserum to a synthetic peptide (predicted amino acids 34-47 of the 126 kDa polypeptide encoded by U1 TMV RNA; Goelet *et al.*, 1982) and 15nm colloidal gold-conjugated goat anti-rabbit (GAR) IgG. In (b) the double arrowhead marks an end-to-end aggregate of two TMV particles. Scale bar = 100nm.

Rapid, complete exposure of the 5'-leader sequence of TMV RNA

We have studied the *cis*-effect of a derivative (Ω') of the TMV RNA leader sequence on expression of several foreign mRNAs in a wide variety of cells and cell-free systems (Gallie *et al.*, 1987a,b,c; Sleat *et al.*, 1987; 1988b). Several lines of evidence point to the low capacity of this sequence to base-pair with any of the downstream coding regions examined so far (Sleat *et al.*, 1988b).

A computer-predicted secondary structure (Zuker and Stiegler, 1981) for the first 500 nucleotides of TMV RNA also showed no pairing interactions until the G of the first start codon (Fig. 4A). Partial deletions and even an artificial C→G transversion

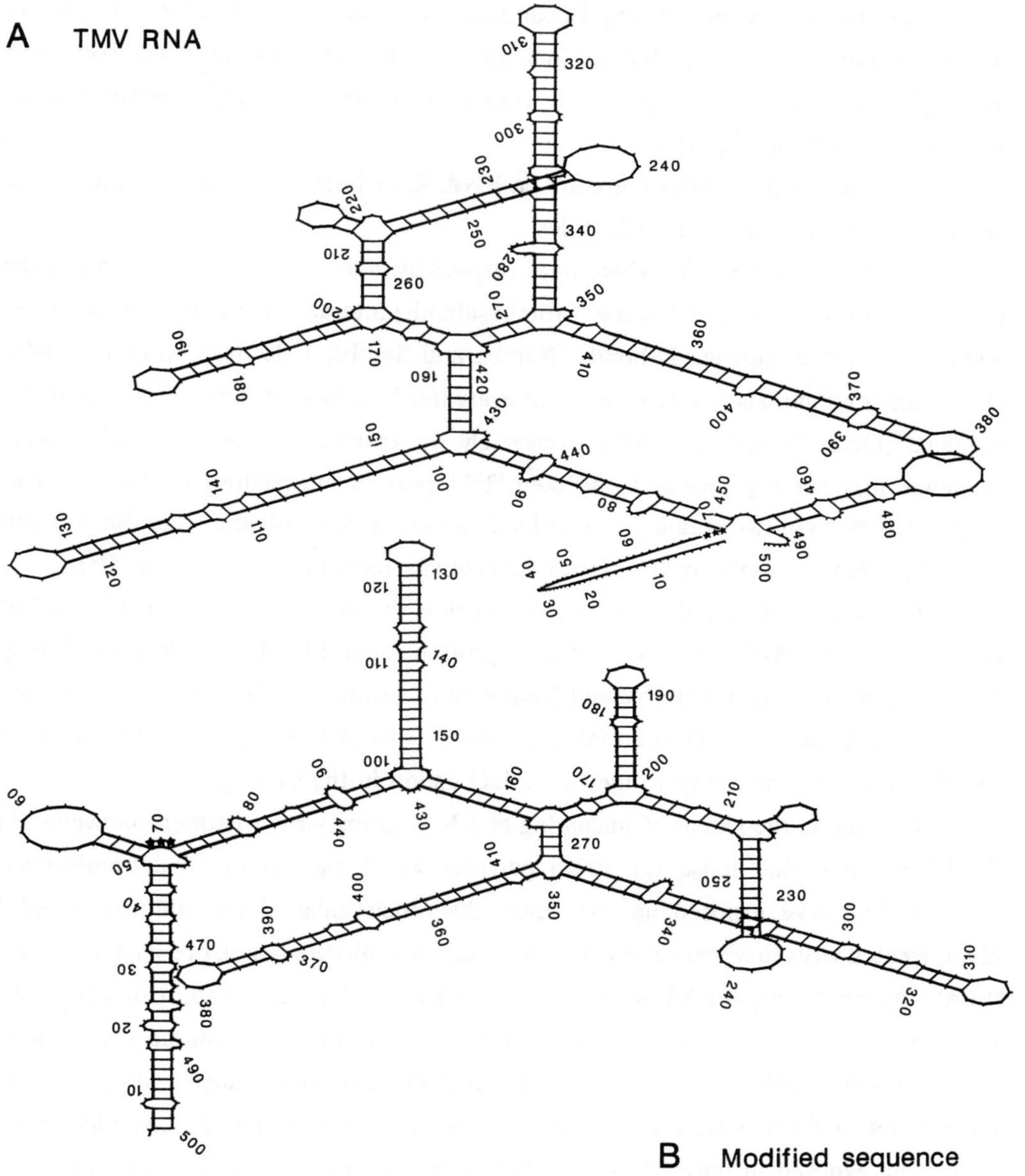

Figure 4. Hypothetical minimum free-energy folding predictions (Zuker and Stiegler, 1981; FOLD and SQUIGGLES on the University of Wisconsin Genetics Computer Group Package) for the first 500 nucleotides of (A) TMV RNA (U1 strain; Goelet *et al.*, 1982); (B) as (A) but with a computer-generated random sequence, $(C,U,A)_{68}$, in place of the untranslated leader. In each case, no allowance can be made for the effect of the 5'-cap in native TMV RNA. The first AUG codon (for the 126 kDa product) is marked by bold stars. In all, 3 different random leader sequences were tried. Two folded as shown in (B), one left an unpaired 5'-tail of 20 residues.

of Ω, behaved similarly (not shown). In contrast, several computer-generated 68-residue random sequences (i.e., up to, but not including, the first AUG codon) which contained only C, U and A residues base-paired to a greater or lesser extent with downstream 126 kDa gene sequences (Fig. 4B).

In helical nucleocapsids there are no RNA-RNA interactions, as the coat protein separates successive turns of nucleotides.

It has been argued that G-residues (especially when in the first position of each trinucleotide binding site on the coat protein subunit) interact most strongly with capsid amino acid side groups *via* H-bonds (Namba and Stubbs, 1986; Namba *et al.*, 1989). Hence the predominance of G-residues at every third position in the origin-of-assembly sequence (OAS; Turner *et al.*, 1988). Conversely, the absence of G-residues in the leader sequence may serve primarily to weaken RNA-protein interactions at the 5'-end of TMV, rather than to create an unfolded structure for efficient recruitment and "scanning" (Kozak, 1989) by 40S ribosomal subunits (compare Figs 4A and 4B).

Related to this, another goal of our work is to resolve the possible "phasing" of encapsidation of TMV RNA. As each coat protein subunit binds 3 nucleotides, it may be significant that the 5'-untranslated leaders of all strains of TMV sequenced to date differ by 3n nucleotides. Thus an integral number of capsid protein subunits can span the distance from the 5'-cap to the first AUG codon in the same phase.

Examining the extent of uncoating of RNA during virus treatment between pH 7.2-9.2 revealed that those U1 strain particles which participate in cotranslational disassembly have exposed approximately 200 nucleotides. Thus addition of mild alkali-treated virus to wheat germ extract or rabbit reticulocyte lysate, pre-incubated with 0.3mM sparsomycin (an inhibitor of peptidyl transferase), revealed approximately 18% of virions in complexes of the type shown in Fig. 5 (9% with one ribosome and 9% with disomes). Since TMV RNA or RNase T1-isolated Ω-fragment (70 nucleotides) can form disomes under these conditions (Tyc *et al.*, 1984), at least the first 71 nucleotides must have been exposed in these virus particles. By measuring the length of the residually packaged portion of RNA in these complexes, it appeared that the majority of rods were 9nm shorter than full-length TMV (300nm), thereby suggesting that 190-200 nucleotides had been uncoated. Primer extension studies will be used to confirm this and define more accurately the 5'-limit of the genome sequence still encapsidated.

Short, synthetic cDNAs (10-19 nucleotides in length) complementary to defined regions of U1 strain TMV RNA between residues 1-117 were hybridized, at cDNA:RNA

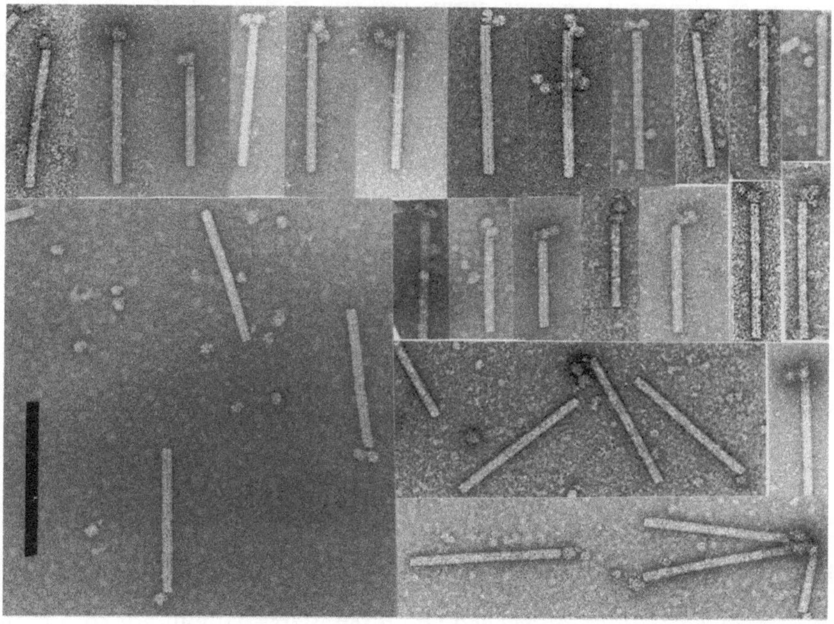

Figure 5. Selected sparsomycin-arrested translation complexes from rabbit reticulocyte lysate and wheat-germ incubations programmed with pH 7.5-treated TMV particles. Of the clearly identifiable, discrete TMV particles scored, approximately 9% had two 80S ribosomes attached, 9% had one 80S, 8% had a complex cluster, possibly of 3 or more, and 74% were devoid of any ribosomes. Scale bar = 300nm (unit length of TMV).

ratios up to 1000:1, with TMV preparations which had been centrifuged in buffers between pH 7.2-9.2 (Wilson, 1984b). After digestion with RNase H, the template activity of each treated virus preparation was extensively or completely inhibited. Again, this suggests that to "activate" the template, virus disassembly had to have proceeded for at least the first 117 nucleotides (KW Mundry, PAC Watkins, T Ashfield, A-G Fernandez, S Eisele-Walter, KA Plaskitt and TMA Wilson, manuscript in preparation). Only a fraction of the virus population, probably 20%+ of rods, behaved in this way. Thus, after cDNA hybridization (to genome co-ordinates 67-77) and RNase H-treatment of alkali-washed virions, intact and functional template mRNA for the 126 kDa polypeptide could be isolated from this virus preparation by phenol-extraction (T Ashfield and TMA Wilson, unpublished results).

Kinetic studies on U-^{32}P-TMV treated with 1% SDS also showed that the RNase T1-resistant oligonucleotide-marker (Ω) could be recovered immediately (in $\leqslant$15secs)

Table 1. *Effect of exogenous proteins during cell-free translation of TMV RNA or pH 8.2-"washed" TMV particles*

Relative incorporation[a] of ^{35}S-methionine into TCA-insoluble polypeptides

Protein[b]	TMV CP		CCMV CP		BSA		Cyto C	
(mg/ml)	0.1	1.0	0.1	1.0	0.1	1.0	0.1	1.0
Template								
TMV RNA	1.81	1.38	1.69	1.54	1.85	1.54	1.23	0.73
Virions	0.77	0.04	1.32	0.68	1.64	1.91	1.32	0.45

[a] Results are expressed as incorporation (cpm) into TCA-insoluble polypeptides for TMV RNA or "washed" virion templates with added proteins, divided by the incorporation with TMV RNA or washed virion templates without added protein. The average absolute incorporation values per μl rabbit reticulocyte lysate were TMV RNA: 0.87 x 10^5 cpm; washed virions: 0.73 x 10^5 cpm. All cpm data are based on averages of pairs of samples from duplicate incubations.

[b] TMV CP: tobacco mosaic virus coat protein; CCMV CP, cowpea chlorotic mottle virus coat protein; BSA, bovine serum albumin; Cyto C, cytochrome C.

and, that during a 15 min incubation in SDS, no further increase occurred in the level of **Ω** exposed (KW Mundry, personal communication). Once more, these data suggest immediate and synchronous virus disassembly for at least 71 nucleotides, and possibly for up to 200-300 residues (as judged by the total ^{32}P radioactivity recovered).

We have shown that only TMV coat protein can selectively inhibit cotranslational disassembly of pH 8.2-"washed" TMV *in vitro* (Wilson and Watkins, 1986; data extended in Table 1). Recently we found that several unrelated proteins (including other viral coat proteins), when added during the centrifugal "washing" step, reduced the template activity of the resulting TMV pellet (Table 2). As these proteins had no deleterious effect on *in vitro* translation of unencapsidated TMV RNA (Table 1), we presume that their presence during centrifugation prevented the partial uncoating of virions rather than degrading any exposed 5'-RNA "tails". Because virions appear to uncoat *in vivo* in the cytosol (Plaskitt *et al.*, 1988), in the presence of many soluble plant proteins, these

Table 2. *Effect of exogenous protein added during the centrifugal "washing" step on cell-free translation of TMV particles*

Relative incorporation[a] of ^{35}S-methionine into TCA-insoluble polypeptides

Protein	TMV CP		BSA		Cyto C
(mg/ml)	0.1	1.0	0.1	1.0	0.1
pH					
7.2	0.09	0.05	0.04	0.03	0.05
8.2	0.25	0.14	0.14	0.02	0.48
9.2	0.44	0.09	0.04	0.02	0.24

[a] Incorporation (cpm) into TCA-insoluble polypeptides divided by the template activity (cpm) of reference virions "washed" at the same pH without added protein. The average absolute incorporation values per μl rabbit reticulocyte lysate were: pH 7.2, 0.83×10^5 cpm; pH 8.2, 1.15×10^5 cpm; pH 9.2, 3.43×10^5 cpm. Other details were as given in the footnote to Table 1.

observations again serve to illustrate the artificiality of our *in vitro* destabilization procedure.

Implications of cotranslational disassembly for coat protein-mediated cross-protection in transgenic tobacco plants

It is generally assumed that the phenomenon of cross-protection between closely related plant viruses depends upon the presence of coat protein - despite the fact that viroids and protein-defective mutant viruses will also cross-protect or interfere with challenge infections in some still-undefined way (Niblett *et al.*, 1978; Zaitlin, 1976; Sarkar and Smitamana, 1981). The presumed role for coat protein involves either RNA-repackaging (de Zoeten and Fulton, 1975), or inhibited disassembly of challenge virus particles (Sherwood and Fulton, 1982; Dodds *et al.*, 1985). Clearly, this area is relevant to studies on, and may even help to elucidate, the early structural events of virus infection.

In collaboration with Professor RN Beachy (Washington University, St. Louis) we have recently been engaged in experiments to address the relationship between coat protein-mediated "protection" or "resistance" in transgenic tobacco plants and the early events of infection by TMV. Further details can be found in the chapter by Nelson *et al.* in this volume.

We have shown that the principal site for parental virus uncoating and early gene expression is the cytoplasm of transiently wounded, epidermal cells and that very large numbers of virions enter *via* lesions in the thick outer cell wall (Plaskitt *et al.*, 1988). In fact, only about 0.4-2.0% of the approximately 1 x 10^6 epidermal cells on a fully-expanded tobacco leaf are involved at this stage (Fannin and Shaw, 1987). Thus the ability of transgenically-expressed coat protein to inhibit, immediately, the cotranslational disassembly of *every* incoming cognate virus particle, if true, is remarkable. Nevertheless, in support of this view, transgenic tobacco lines which express more viral coat protein (up to 0.05-0.1% (w/w) of soluble leaf protein) are generally better protected than those expressing lower levels. Further, by raising the concentration of virus in the inoculum above about 1-2 μg/ml, more plants eventually succumb to virus infection.

Unfortunately, attempts to repeat the experimental approaches of Shaw *et al.* (1986) or Plaskitt *et al.* (1988) with TMV coat protein-transgenic tobacco plants have proved impossible. Both involve following the fate of TMV inoculum which, for experimental reasons, must be applied at high concentration, while "protection" in transgenic plants is only fully effective against very low inoculum concentrations. New approaches, involving Northern blotting for TMV-specific sequences using site-specific M13 cDNA probes (Goelet and Karn, 1984) are presently being examined (JG Shaw, personal communication). At present, research is directed towards finding the haystack; detecting the needle will follow later!

Coat protein-mediated protection occurs in TMV-inoculated mesophyll protoplasts (Register and Beachy, 1988). Protection is overcome by unencapsidated TMV RNA or by pH 8-treated virions, some of which we now know have exposed up to 200 nucleotides at their 5'-end (see above).

Recently we determined that mechanical inoculation of 0.5 μg/ml *unencapsidated* TMV RNA leads to systemic infection of control Xanthi tobacco plants, but *not* the coat protein-transgenic plants of tobacco line 3404 (Powell Abel *et al.*, 1986). Thus, in contrast to earlier reports in which 16 μg/ml TMV RNA was used (Powell Abel *et al.*, 1986), there does seem to be some protection against low (but potentially infectious) levels of naked TMV RNA. Clearly therefore, inhibition of parental challenge virus disassembly is not the sole coat protein-mediated interference mechanism operating.

This view was confirmed using transcapsidated virions made in *vitro* (Osbourn *et al.*, 1989). Infectious RNA from the cowpea strain of TMV (Cc; Sunn-hemp mosaic virus) was encapsidated in coat protein of TMV U1 strain, nuclease-treated to remove any exposed RNA, and then electroporated into mesophyll protoplasts of U1 coat protein-transgenic tobacco line 3404, as described (Register and Beachy, 1988). We too obtained "protection" against native or reconstituted U1 strain virions; but naked U1 or Cc RNA, native Cc virions *and* the transcapsidated hybrid virions (Cc RNA in U1 coat) all gave normal levels of infection (50-55% of protoplasts, screened by fluorescent antibody staining). Thus homologous protein-protein interactions, presumed to inhibit challenge virus uncoating, are not sufficient alone to account for genetically-engineered "resistance". Only free Cc coat protein could protect against the hybrid virions when co-electroporated into control Xanthi mesophyll protoplasts (JK Osbourn, unpublished results).

To address this point further, we constructed TMV-like pseudovirus particles containing mRNA for the reporter enzyme ß-glucuronidase (GUS; Jupin *et al.*, 1989; Fig. 6) and electroporated either naked GUS mRNA or GUS mRNP particles into control Xanthi or 3404-transgenic protoplasts (JK Osbourn *et al.*, 1989). Transient expression of GUS activity then gave a measure of nucleocapsid disassembly. The results are given in Table 3. The data show that 97-99% of GUS rodlets failed to uncoat in the transgenic mesophyll protoplasts, even though these were not "protected" against our transcapsidated Cc/U1 virus. Clearly, the 1-3% or more of replicative RNA molecules which do uncoat can proceed to establish an infection, provided no subsequent stages of the replication cycle are interfered with (as say, for Cc RNA).

In contrast, in parallel experiments where the same GUS rodlets were mechanically inoculated onto tobacco leaves and the epidermal cells were later peeled off and screened for GUS activity, almost identical levels of GUS enzyme activity were found in control Xanthi and coat protein-transgenic tobacco cells (JK Osbourn, unpublished results). Experiments with epidermal cell-derived protoplasts are in progress.

Similarly, we found that mechanical inoculation with 1μg/ml TMV apparently produced equivalent levels of virus replication (measured as coat protein by ELISA) in both types of tobacco epidermal cells up to 20 hours post-infection (JK Osbourn, unpublished results). Nevertheless, this concentration of virus did not lead to systemic

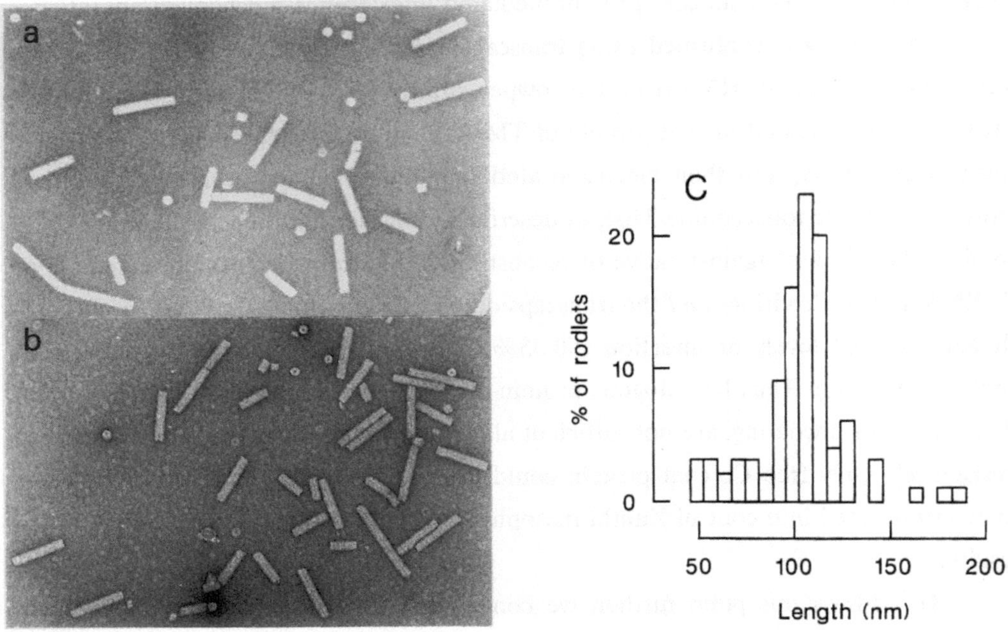

Figure 6. Pseudovirus particles containing mRNA for *Escherichia coli* ß-glucuronidase (GUS; Jupin *et al*., 1989). The 2.4kb transcripts of BglII-cut pJII140 (Sleat *et al*., 1987) predict a rodlet length of 115nm, the mode in the frequency-length histogram (c). Two independent preparations of GUS rodlets are shown (a,b), made by slightly different routes: (a) = directly packaged; (b) = packaging after RNA purification (Jupin *et al*., 1989).

infection of the coat protein-transgenic tobacco.

These recent results imply either (a) that the amount of coat protein in epidermal cells is insufficient to inhibit substantial nucleocapsid disassembly or to interfere with some other step in the replication cycle, and/or (b) that the site or mechanism of virus uncoating is different in mesophyll versus epidermal cells, and/or (c) that virus replication proceeds differently in epidermal versus mesophyll cells (Okamoto *et* al., 1988).

To resolve these options further, we performed immunocytochemical labelling experiments on thin sections of very young (1-2cm long) 3404-coat protein transgenic leaves. Only by cold-fixation and cryosectioning could we detect gold-labelling with a rabbit polyclonal antiserum against TMV (a gift from A Gerwitz, Wellesbourne) or with mouse monoclonal antibodies specific for free coat protein subunits ("cryptotope"; Dore *et al*., 1988). With both primary probes, second antibody gold-labelling was

Table 3. *ß -Glucuronidase (GUS) activities in control or transgenic tobacco mesophyll protoplasts electroporated with naked or encapsidated GUS mRNA*

	GUS activity (pmoles 4-MU[a]/min/mg total soluble protein)		
	Control Xanthi	Coat protein-transgenic	Ratio[b]
Inoculum			
Mock	<6	<6	-
GUS mRNA	169	61	2.8
GUS mRNP	14900	152	98.0

[a] 4-methylumbelliferone (=fluorescent product)

[b] GUS activity in control Xanthi protoplasts divided by GUS activity in U1 coat protein-transgenic protoplasts

associated only with vesicles in the chloroplasts of mesophyll cells (Fig. 7). No labelling was seen in comparable sections of control Xanthi leaves.

Curiously, monoclonal antiserum specific for helically assembled TMV coat protein ("neotope"; Dore *et al.*, 1988) failed to label the 3404 coat protein leaf sections, despite our ability to trap, by ISEM, very low numbers of short-rodlets from leaf sap, mesophyll protoplasts or chloroplast-enriched low-speed pellets of the coat protein-transgenic 3404 tobaccos (JC Register, KA Plaskitt, TMA Wilson, and RN Beachy, manuscript in preparation; KA Plaskitt, unpublished results). It seems most likely that the rare endogenous rodlets extracted from the 3404-transgenic plants contain chloroplast mRNAs, especially the *rbcL* gene transcript (Atreya and Siegel, 1989).

At present we conclude that neither virus (pseudovirus) uncoating nor genome replication is inhibited in mechanically-inoculated epidermal cells of transgenic tobacco plants because the endogenous level of coat protein is either too low or present at the wrong site for interference. This circumvents the intuitive problem of parental virus stoichiometry alluded to above. Only when progeny virus spreads to mesophyll cells is uncoating, but more importantly a later step in replication, efficiently inhibited by the endogenous, closely-related coat protein. If local spread is *via* viral RNA or vRNP

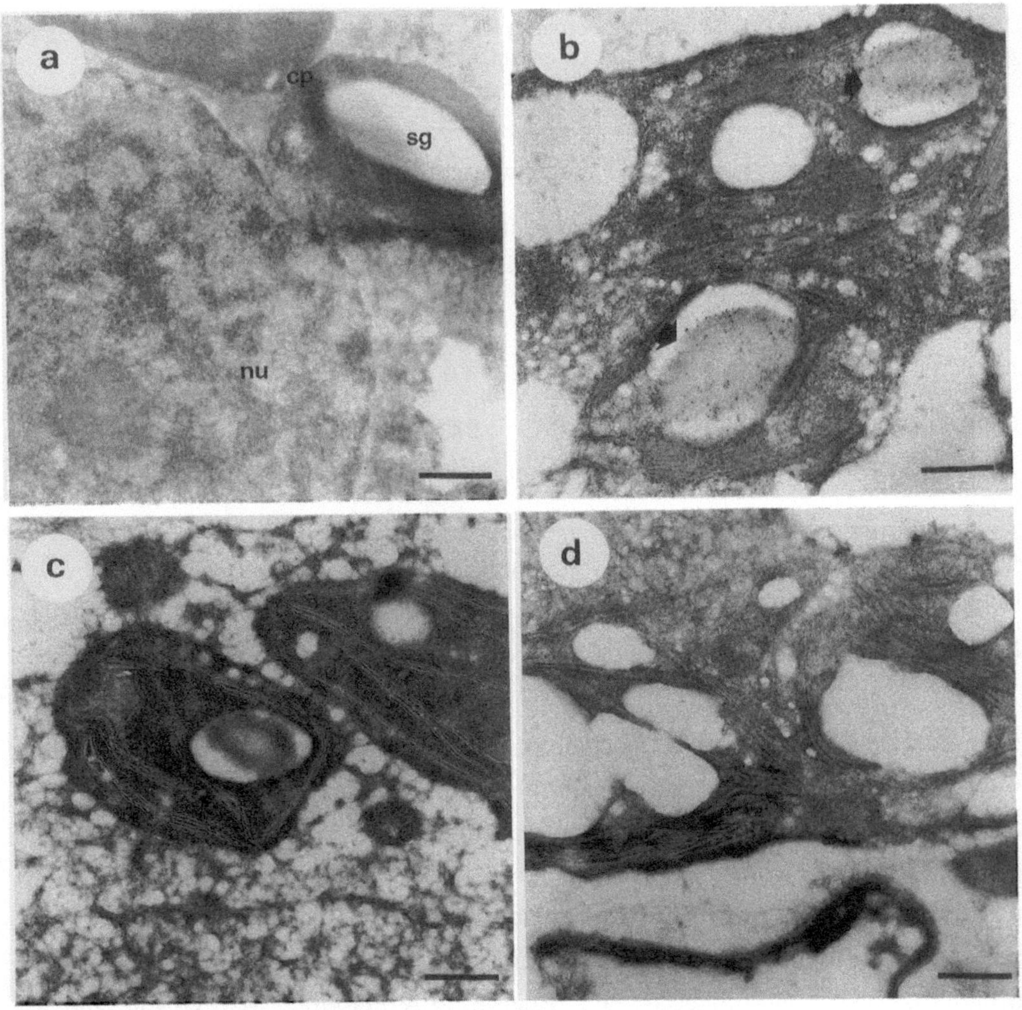

Figure 7; see also facing page. Selected electron micrographs of paraformaldehyde-fixed and cryo-sectioned coat protein-transgenic tobacco leaves (line 3404; Powell Abel *et al.*, 1986) probed with rabbit polyclonal serum against TMV (b), or monoclonal antibodies specific for helically assembled (c,d) or monomeric (e-h) TMV coat protein (Dore *et al.*, 1988). The monoclonal antibody against TMV nucleoprotein helices ("neotope"; panels c,d) also gave little or no labelling of leaf sections (especially LR-White or Lowicryl K4M-embedded material) from systemically TMV-infected control Xanthi plants. Panel (a) is a negative control probed only with a mixture of both gold-conjugated second antibodies (goat anti-rabbit and goat anti-mouse). Panel (a) shows the nucleus (nu), chloroplast (cp) and a starch grain (sg). Colloidal gold labelling is indicated by bold arrows. Scale bars = 500nm.

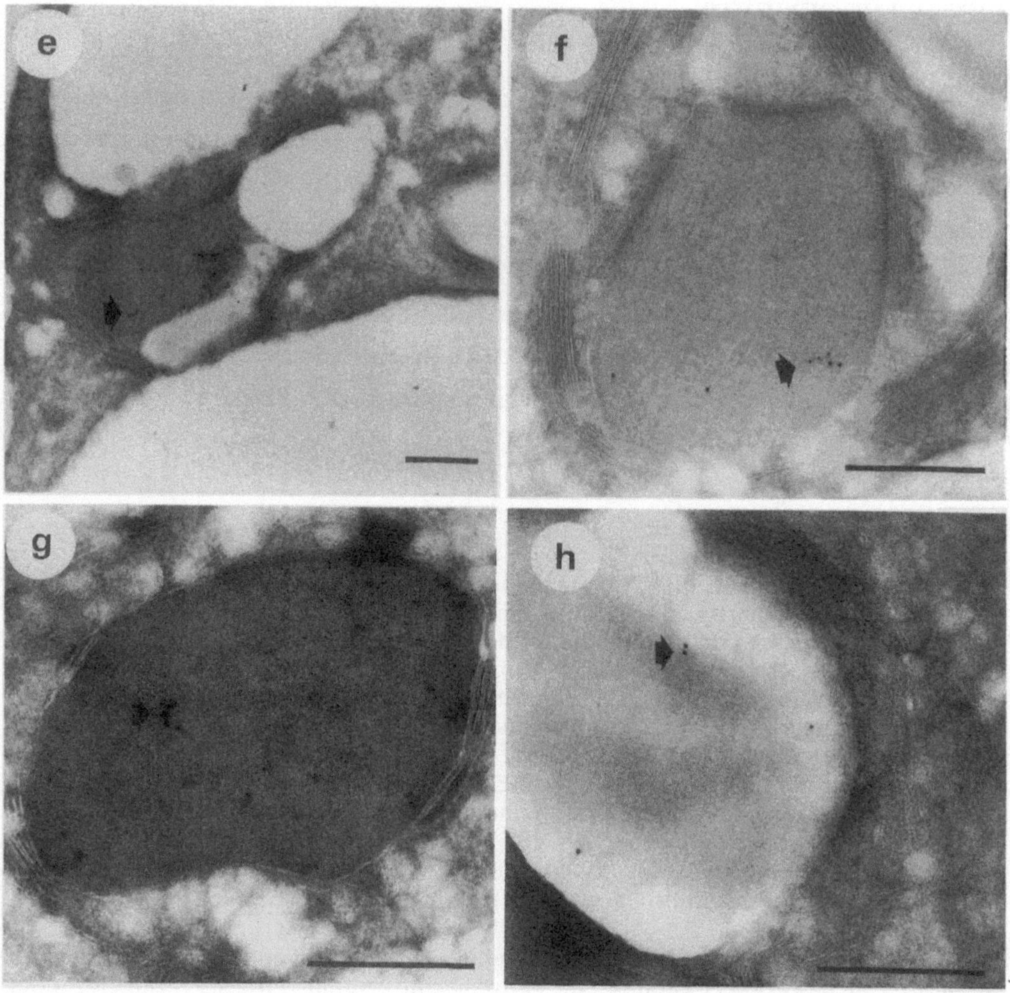

Figure 7. Continued.

(Dorokhov *et al.,* 1984), these must suffer the same interference.

Thus it would seem that the site and mechanism of nucleocapsid disassembly may be different in different cell types - a phenomenon only made apparent by studies with coat protein-transgenic cells and protoplasts. We are continuing to investigate this hypothesis.

ACKNOWLEDGEMENTS

We thank Professor JG Shaw (University of Kentucky, Lexington) for his expertise, advice and many helpful discussions during the execution and interpretation of this work. We are deeply indebted to Professor RN Beachy (Washington University, St. Louis) for generously providing TMV CP-transgenic tobacco plants (line 3404) and for many useful discussions. Professor MHV van Regenmortel (CNRS, IBMC, Strasbourg) kindly supplied mouse monoclonal antibodies to TMV ("cryptotope" and "neotope"). The John Innes Foundation awarded a Research Studentship to JK Osbourn, and Diatech Ltd., London, provided financial support for KA Plaskitt.

REFERENCES

AbouHaidar MG, Hirth L (1977) 5' terminal structure of tobacco rattle virus RNA: evidence for polarity of reconstitution. Virology 76:173-185

Allan EF, Hepler PK (1989) Calmodulin and calcium-binding proteins. In: Marcus A (ed) The biochemistry of plants. Vol. 15. Academic Press, San Diego, p 455

Atreya CD, Siegel A (1989) Localization of multiple TMV encapsidation initiation sites on *rbcL* gene transcripts. Virology 168:388-392

Bendena WG, AbouHaidar MG, Mackie GA (1985) Synthesis *in vitro* of the coat protein of papaya mosaic virus. Virology 140:257-268

Bergh ST, Siegel A (1989) Intraviral homology and subgenomic RNAs of pepper ringspot virus. Virology 168:339-343

Brisco MJ, Hull R, Wilson TMA (1985) Southern bean mosaic virus-specific proteins are synthesized in an *in vitro* system supplemented with intact, treated virions. Virology 143:392-398

Brisco MJ, Hull R, Wilson TMA (1986) Swelling of isometric and of bacilliform plant virus nucleocapsids is required for virus-specific protein synthesis *in vitro*. Virology 148:210-217

De Zoeten GA, Fulton RW (1975) Understanding generates possibilities. Phytopathology 65:221-222

Dodds JA, Lee SQ, Tiffany M (1985) Cross protection between strains of cucumber mosaic virus: effect of host and type of inoculum on accumulation of virions and double-stranded RNA of the challenge strain. Virology 144:301-309

Dore I, Weiss E, Altschuh D, Van Regenmortel MHV (1988) Visualization by electron microscopy of the location of tobacco mosaic virus epitopes reacting with monoclonal antibodies in enzyme immunoassay. Virology 162:279-289

Dorokhov YL, Alexandrova NM, Miroschnichenko NA, Atabekov JG (1984) The informosome-like virus-specific ribonucleoprotein (vRNP) may be involved in the transport of tobacco mosaic virus infection. Virology 137:127-134

Durham ACH (1972) Structures and roles of the polymorphic forms of tobacco mosaic virus protein. J Mol Biol 67:289-305

Durham ACH (1978) The roles of small ions, especially calcium, in virus disassembly, take-over, and transformation. Biomedicine 28:307-314

Durham ACH, Finch JT, Klug A (1971) States of aggregation of tobacco mosaic virus protein. Nature New Biol 229:37-42

Durham ACH, Hendry DA, Von Wechmar MB (1977) Does calcium ion binding control plant virus disassembly? Virology 77:524-533

Fannin FF, Shaw JG (1987) Evidence for concurrent spread of tobacco mosaic virus from infected epidermal cells to neighbouring epidermal and mesophyll cells. Plant Sci 51:305-310

Gallie DR, Sleat DE, Watts JW, Turner PC, Wilson TMA (1987a) *In vivo* uncoating and efficient expression of foreign mRNAs packaged in TMV-like particles. Science 236:1122-1124

Gallie DR, Sleat DE, Watts JW, Turner PC, Wilson TMA (1987b) The 5'-leader sequence of tobacco mosaic virus RNA enhances the expression of foreign gene transcripts *in vitro* and *in vivo*. Nucleic Acids Res 15:3257-3273

Gallie DR, Sleat DE, Watts JW, Turner PC, Wilson TMA (1987c) A comparison of eukaryotic viral 5'-leader sequences as enhancers of mRNA expression *in vivo*. Nucleic Acids Res 15:8693-8711

Goelet P, Karn J (1984) Methods for cDNA cloning and sequencing tobacco mosaic virus RNA. Gene 29:331-342

Goelet P, Lomonossoff GP, Butler PJG, Akam ME, Gait MJ, Karn J (1982) Nucleotide sequence of tobacco mosaic virus RNA. Proc Natl Acad Sci USA 79:5818-5822

Jupin I, Sleat DE, Watkins PAC, Wilson TMA (1989) Direct recovery of *in vitro* transcripts in a protected form suitable for prolonged storage and shipment at ambient temperatures. Nucleic Acids Res 17:815

Keeling J, Matthews REF (1982) Mechanism for release of RNA from turnip yellow mosaic virus at high pH. Virology 119:214-218

Kozak M (1989) The scanning model for translation: an update. J Cell Biol 108:229-241

Laidlaw WMR (1987) A new method for mechanical virus transmission and factors affecting its sensitivity. EPPO Bull 17:81-89

Namba K, Stubbs G (1986) Structure of tobacco mosaic virus at 3.6A resolution: implications for assembly. Science 231:1401-1406

Namba K, Pattanayek R, Stubbs G (1989) Visualization of protein-nucleic acid interactions in a virus. J Mol Biol 208:307-325

Niblett CL, Dickson E, Fernow KH, Horst RK, Zaitlin M (1978) Cross protection among four viroids. Virology 91:198-203

Okamoto S, Machida Y, Takebe I (1988) Subcellular localization of tobacco mosaic virus minus strand RNA in infected protoplasts. Virology 167:194-200

Osbourn JK, Watts JW, Beachy RN, Wilson TMA (1989) Evidence that nucleocapsid disassembly and a later step in virus replication are inhibited in transgenic tobacco protoplasts expressing TMV coat protein. Virology 172: in press

Perham RN (1969) Sucrose density-gradient analysis of the alkaline degradation of tobacco mosaic virus. J Mol Biol 45:439-441

Plaskitt KA, Watkins PAC, Sleat DE, Gallie DR, Shaw JG, Wilson TMA (1988) Immunogold labeling locates the site of disassembly and transient gene expression of tobacco mosaic virus-like pseudovirus particles *in vivo*. Mol Plant-Microbe Interact 1:10-16

Powell Abel P, Nelson RS, De B, Hoffmann N, Rogers SG, Fraley RT, Beachy RN (1986) Delay of disease development in transgenic plants that express the tobacco mosaic virus coat protein gene. Science 232:738-743

Register JC, Beachy RN (1988) Resistance to TMV in transgenic plants results from interference with an early event in infection. Virology 166:524-532

Roenhorst JW, Verduin BJM, Goldbach RW (1989) Virus-ribosome complexes from cell-free translation systems supplemented with cowpea chlorotic mottle virus particles. Virology 168:138-146

Sarkar S, Smitamana P (1981) A proteinless mutant of tobacco mosaic virus: evidence against the role of a viral coat protein for interference. Mol Gen Genet 184:158-159

Shaw JG, Plaskitt KA, Wilson TMA (1986) Evidence that tobacco mosaic virus particles disassemble cotranslationally *in vivo.* Virology 148:326-336

Sherwood JL, Fulton RW (1982) The specific involvement of coat protein in tobacco mosaic virus cross protection. Virology 119:150-158

Shirako Y, Ehara Y (1986) Comparison of the *in vitro* translation products of wild-type and a deletion mutant of soil-borne wheat mosaic virus. J Gen Virol 67:1237-1245

Sleat DE, Turner PC, Finch JT, Butler PJG, Wilson TMA (1986) Packaging of recombinant RNA molecules into pseudovirus particles directed by the origin-of-assembly sequence from tobacco mosaic virus RNA. Virology 155:299-308

Sleat DE, Gallie DR, Jefferson RA, Bevan MW, Turner PC, Wilson TMA (1987) Characterisation of the 5'-leader sequence of tobacco mosaic virus RNA as a general enhancer of translation *in vitro*. Gene 60:217-225

Sleat DE, Gallie DR, Watts JW, Deom CM, Turner PC, Beachy RN, Wilson TMA (1988a) Selective recovery of foreign gene transcripts as virus-like particles in TMV-infected transgenic tobaccos. Nucleic Acids Res 16:3127-3140

Sleat DE, Hull R, Turner PC, Wilson TMA (1988b) Studies on the mechanism of translational enhancement by the 5'-leader sequence of tobacco mosaic virus RNA. Eur J Biochem 175:75-86

Turner DR, Joyce LE, Butler PJG (1988) The tobacco mosaic virus assembly origin RNA: Functional characteristics defined by directed mutagenesis. J Mol Biol 203:531-547

Tyc K, Konarska M, Gross HJ, Filipowicz W (1984) Multiple ribosome binding to the 5'-terminal leader sequence of tobacco mosaic virus RNA. Assembly of an 80S ribosome mRNA complex at the AUU codon. Eur J Biochem 140:503-511

Willingmann P, Barnert H, Zeichhardt, H, Habermehl K-O (1989) Recovery of structurally intact and infectious poliovirus type 1 from HeLa cells during receptor-mediated endocytosis. Virology 168:417-420

Wilson TMA (1984a) Cotranslational disassembly of tobacco mosaic virus *in vitro*. Virology 137:255-265

Wilson TMA (1984b) Cotranslational disassembly increases the efficiency of expression of TMV RNA in wheat germ cell-free extracts. Virology 138:353-356

Wilson TMA (1986) Expression of the large 5'-proximal cistron of tobacco mosaic virus by 70S ribosomes during cotranslational disassembly in a prokaryotic cell-free system. Virology 152: 277-279

Wilson TMA, Shaw JG (1985) Does TMV uncoat cotranslationally in vivo? Trends Biochem Sci 10:57-60.

Wilson TMA, Shaw JG (1987) Cotranslational disassembly of filamentous plant virus nucleocapsids *in vitro* and *in vivo*. In: Brinton MA, Rueckert RR (eds) Positive strand RNA viruses. UCLA Symp Molec Cell Biol Vol.54. Alan Liss, New York p 159

Wilson TMA, Watkins PAC (1986) Influence of exogenous viral coat protein on the cotranslational disassembly of tobacco mosaic virus (TMV) particles *in vitro*. Virology 149:132-135

Wilson TMA, Perham RN, Butler PJG (1978) Intermediates in the disassembly of tobacco mosaic virus at alkaline pH. Virology 89:475-483

Zaitlin M (1976) More understanding is needed. Phytopathology 66:382-383
Zaitlin M, Hull R (1987) Plant virus-host interactions. Annu Rev Plant Physiol 38:291-315
Zuker M, Stiegler P (1981) Optimal computer folding of large RNA sequences using thermodynamics and auxillary information. Nucleic Acids Res 9:133-148

GENETIC ORGANIZATION, EVOLUTION AND EXPRESSION OF PLANT VIRAL RNA GENOMES

Rob Goldbach, Rik Eggen, Cees de Jager, Ab van Kammen,
Jan van Lent, Geertje Rezelman and Joan Wellink
Departments of Virology and Molecular Biology
Agricultural University
Binnenhaven 11
6709 PD Wageningen
The Netherlands

ABSTRACT

Eukaryotic RNA viruses show a wide variation in genomic structure and organization. This is particularly true for RNA viruses infecting plants, which may have either segmented or nonsegmented genomes that are expressed using a variety of translation strategies. Despite this variation, nucleotide sequence data reveal many interviral relationships and indicate that plant RNA viruses can be grouped into a small number of clusters. A review will be given of the genetic relationships that exist between the plant RNA viruses studied so far.

Cowpea mosaic virus is one of the best studied RNA viruses that employ polyprotein processing as a translation strategy. Still, many aspects of the viral infection process have remained obscure so far, e.g. the RNA replication process, translocation through plant tissue, and various aspects of virus-host interaction. A well-defined protoplast system, and the generation of desired virus mutants using cDNA clones will be invaluable in the future unravelling of the viral infection process.

THE ORGANIZATION AND EVOLUTION OF PLANT VIRAL RNA GENOMES

Approximately 75% of the plant viruses studied so far, i.e. at least 470 distinct viruses (see Table 1), have a single-stranded RNA genome of positive polarity. These viruses are classified in distinct taxonomic groups and show a wide variation in capsid morphology. Their genomes may be either segmented or unsegmented, and can have various terminal structures such as genome-linked proteins or cap-structures at the

NATO ASI Series, Vol H 41
Recognition and Response in Plant-Virus Interactions
Edited by R S.S Fraser

Table 1. *Genomes of plant viruses*

Genome Type	Number of viruses	% of total
single-stranded DNA	26	4
double-stranded DNA	13	2
single-stranded RNA (+) sense	470	76
single-stranded RNA (-) sense	85	14
double-stranded RNA	26	4

5'-end, and a poly(A)-tail or tRNA-like structures at the 3'-end. Nevertheless, computer-assisted sequence comparisons of the non-structural proteins they encode have demonstrated that most if not all of them are somehow genetically interrelated. Moreover, most plant RNA viruses appear to have (remote) relatives among the animal RNA viruses. Hence it has been proposed (Goldbach, 1986; 1987) that the como-, nepo- and potyviruses may be arranged in a "super group" of picornavirus-related plant viruses (Fig. 1). What all "picorna-like" plant viruses share, and what indicates that they are all genetically related to the animal picornaviruses, are the following properties:

1. their RNAs have a protein (VPg: Viral Protein genome-linked) covalently linked to the 5'-end and a poly(A) tail at the 3'-end;
2. their RNAs are expressed by the production of so-called "polyproteins" from which the mature functional proteins are derived by proteolytic processing;
3. they encode a number of non-structural proteins exhibiting significant amino acid sequence homology (shaded areas in Fig. 1), and
4. these conserved proteins, which have all been shown or suggested to be involved in the viral RNA replication process, are encoded by similarly ordered gene sets.

As a consequence, it may be anticipated that all viruses of this supergroup employ the same RNA replication mechanism, using the VPg as a primer molecule during initiation.

Of the picorna-like plant viruses the comoviruses seem most closely related to the picornaviruses. Placing the smaller M-RNA left of the larger B-RNA, the genetic maps of both virus groups are almost completely colinear (Fig. 1). This is certainly evident when we realize that the two capsid proteins of cowpea mosaic virus (CPMV) fold into three ß-barrel domains that mimic in many details the ß-barrels of the three

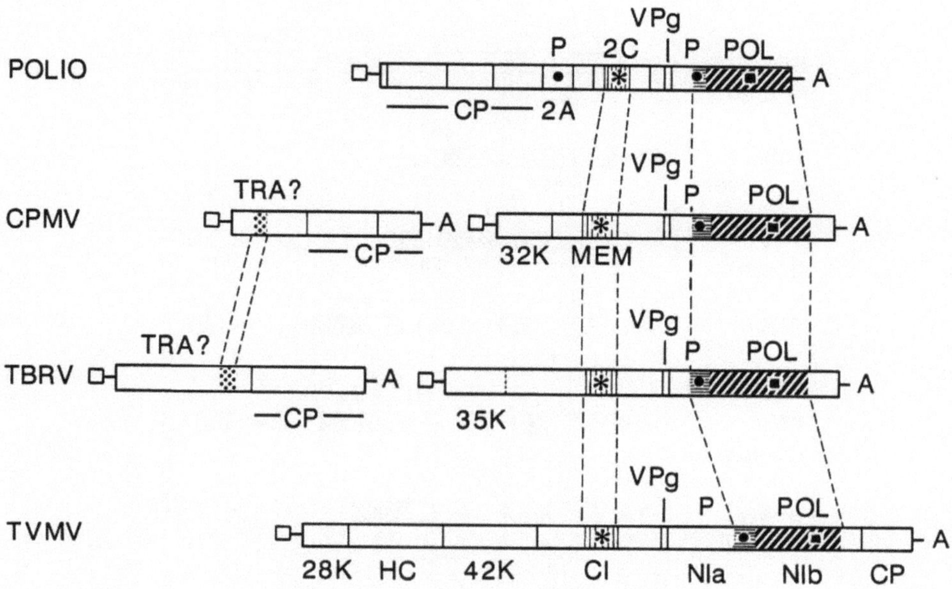

Figure 1. Comparison of the genomes of picornaviruses (polio), comoviruses (CPMV), nepoviruses (TBRV), and potyviruses (TVMV). Coding regions are indicated as open bars, VPg as open squares and polyadenylate sequences as A. Regions of amino acid sequence homology in the gene products are indicated by similar shading. Other symbols: CP, capsid protein(s); TRA, putative transport function; P, protease; MEM, membrane-binding; POL, core RNA-dependent RNA polymerase; HC, helper component, *, nucleotide binding domain; •, cysteine protease domain; ■, conserved polymerase domain.

major capsid proteins of picornaviruses (J Johnson, personal communication). Hence, comoviruses may be regarded as "split" picornaviruses. In view of all the similarities, common ancestry seems to be the major evolutionary mechanism which would explain the genetic relationship with the true picornaviruses.

The second supergroup of plant RNA viruses includes viruses that are - to a lesser or greater extent - related to sindbis virus *(Alphaviridae* - see Fig. 2). This supergroup of "sindbis-like" plant viruses is larger and contains viruses that show more variation with respect to genome size, terminal genomic structures and translation strategy than the picornavirus supergroup. The genetic relationship to the alphaviruses is most evident for the tricorna-, tobamo-, tobra-, furo- and hordeiviruses, since they all specify proteins exhibiting significant sequence homology to three non-structural proteins of sindbis virus, encoded by genes which are similarly arranged. Two of these proteins contain a nucleotide-binding domain sequence and an RNA polymerase domain

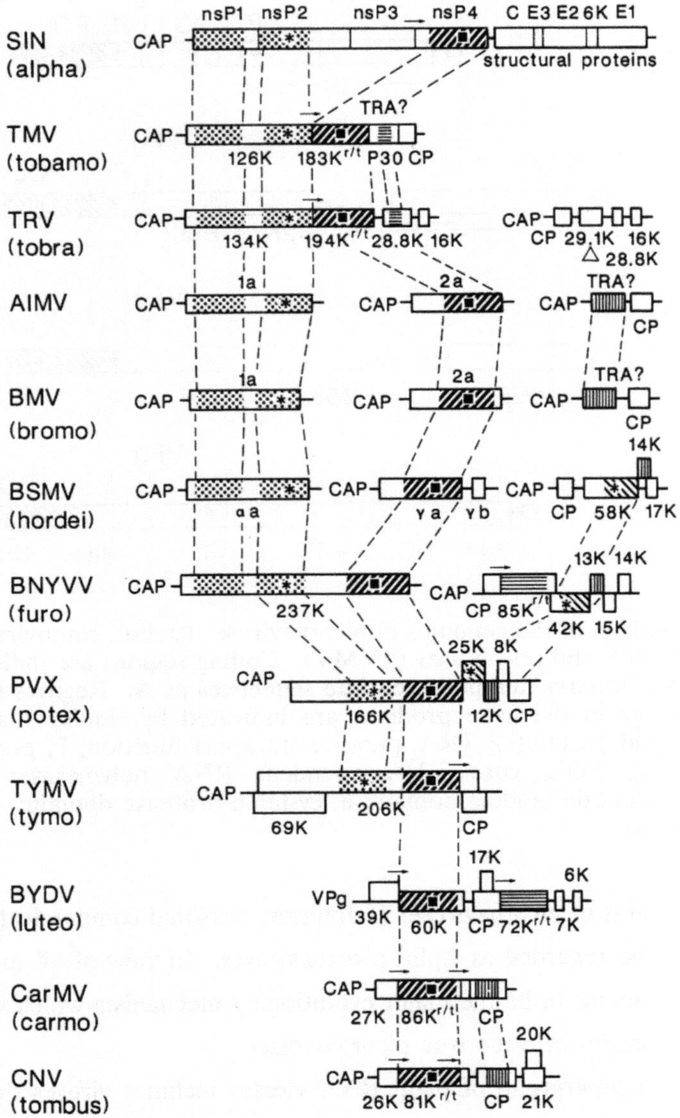

Figure 2. Comparison of the genomes of sindbis virus (Alphaviridae) and the sindbis-like plant viruses. The "supergroup" of sindbis-like viruses includes the tobamovirus (TMV), tobraviruses (TRV), alfalfa mosaic virus group (A1MV), bromoviruses (BMV), hordeiviruses (BSMV), furoviruses (BNYVV), potex-viruses (PVX), tymoviruses (TYMV), luteoviruses (BYDV), carmoviruses (CarMV), tombusviruses (CNV) and sobemoviruses (SBMV, shown in Fig. 3). Coding regions in the genomes are indicated as open bars, and regions of amino acid sequence homology in the gene products indicated by similar shading. Closely adjoining or slightly overlapping cistrons in the TMV and CarMV genome are drawn contiguously. —➤ = leaking termination codon; r/t = read-through. For other symbols see Fig. 1.

respectively. This suggests that the proteins conserved among the sindbis-like plant viruses are, as suggested for the picorna-like plant viruses, also engaged in the viral RNA replication process.

As more RNA sequences become available, the supergroup of sindbis-related plant viruses is becoming more and more heterogenous. In particular, the position of the viruses with smaller RNA genomes, i.e. the carmo-, tombus- and luteoviruses, is debatable. They only possess one of the three conserved genes shared among the other viruses of the supergroup, namely that encoding the putative core polymerase polypeptide (Fig. 2).

The situation becomes even more complicated since more and more evidence reveals that the interviral relationships are not only based on common ancestry, but also (perhaps mainly) on recombination, by which sets of conserved and unique genes may be exchanged among viruses belonging to different taxonomic groups. For instance, the potyviral genome has a unique gene order, with a capsid protein gene downstream of the conserved gene set characteristic of all picorna-like viruses, and, moreover, a number of genes upstream of this set for which no counterparts are found in the como- and nepoviruses (Fig. 1). This is hard to explain without the assumption of recombination events. Various virus groups belonging to the sindbis-related plant viruses have also, in addition to their conserved genes, one or more unique genes.

Interviral recombination may indeed explain the obvious differences between genetically related plant viruses and the idea of "modular evolution" of viruses, i.e. that viral genomes have been built up from different combinations of (conserved) gene modules, seems to become more and more valid. Indeed, RNA recombination is a process that has now been well established for plant and animal RNA viruses (reviewed in Goldbach and Wellink, 1988) and recent evidence shows that it even occurs within the conventional taxonomic groups. Thus it has been shown that some tobravirus isolates (I6, N5) are true, natural recombinants of tobacco rattle virus (TRV) and pea early browning virus (PEBV) (Robinson *et al.*, 1987).

Likewise, sequence comparisons between three different luteoviruses; barley yellow dwarf virus (BYDV) (Miller *et al.*, 1988), beet western yellow virus (BWYV) (Veidt *et al.*, 1988), and potato leafroll virus (PLRV) (Van der Wilk *et al.*, 1989), show that members belonging to the same taxonomic group may undergo different recombination events with other viruses. Hence PLRV and BWYV have putative polymerase genes homologous to that of southern bean mosaic virus (SBMV), whereas

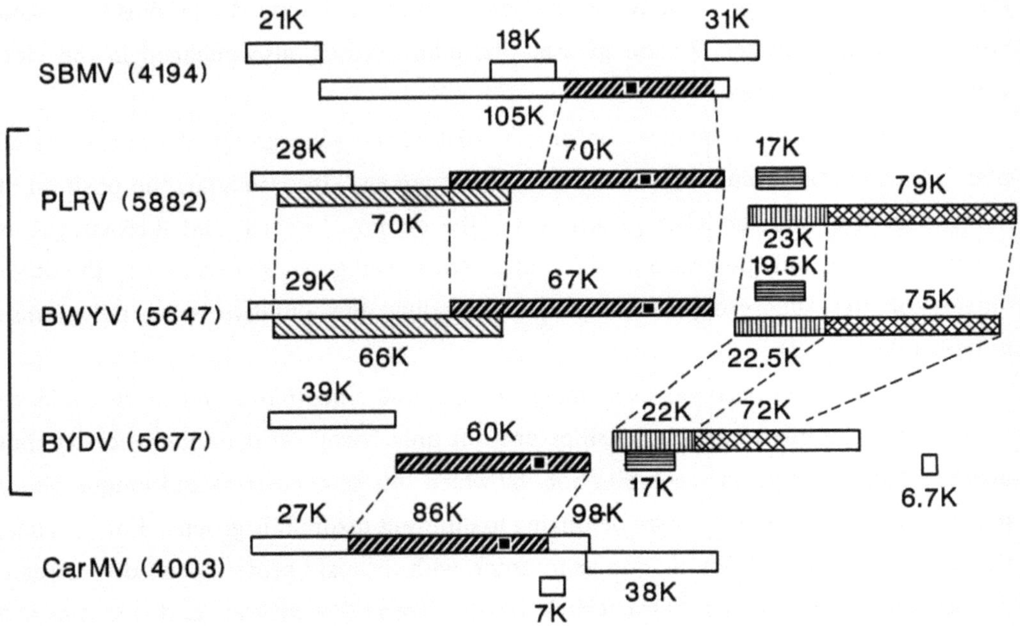

Figure 3. Comparison of the open reading frames in the genomic RNAs of three luteoviruses, PLRV, BYDV and BWYV. The size of the protein encoded by each ORF is shown. Numbers in brackets indicate the length of the RNAs in nucleotides. Similar shading indicates regions of high amino acid sequence homology among the ORFs of the different viruses. Computer-assisted sequence comparisons revealed that the BYDV polymerase is homologous to that of CarMV, while the polymerases of BWYV and PLRV are homologous to that of SBMV (Van der Wilk *et al.*, 1989). The ■ symbol indicates the conserved GDD domain, characteristic of all plant viral RNA polymerases.

BYDV has a polymerase gene homologous to that of carnation mottle virus (CarMV) (see Fig. 3). Such findings illustrate the evolutionary dynamics and flexibility of plant virus RNA genomes. They may disturb attempts by virus taxonomists to set up a true phylogenetic plant virus taxonomy that gives a full account of all interviral relationships.

THE COWPEA MOSAIC VIRUS GENOME

Cowpea mosaic virus (CPMV), type member of the comoviruses, belongs to what is arguably the most studied group of plant RNA viruses. The complete nucleotide sequences of the B- and M-RNA segments have been determined and its RNA

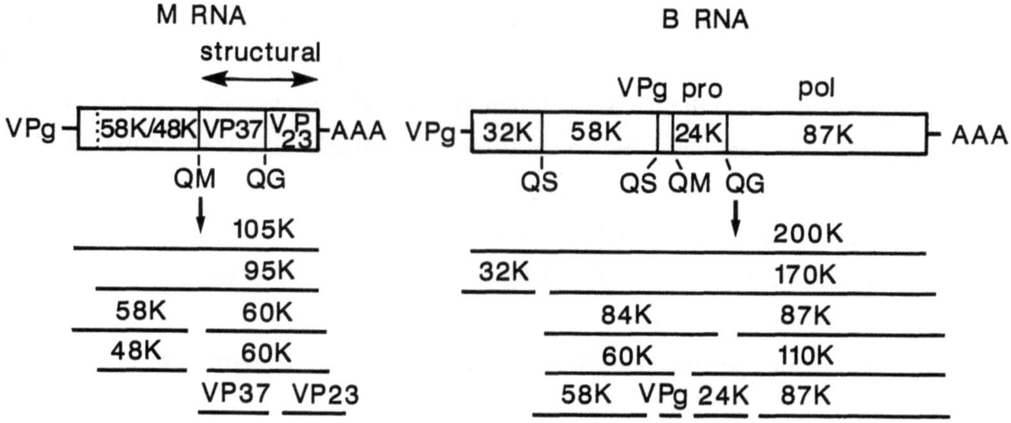

Figure 4. Genetic map of CPMV RNAs. M-RNA encodes two overlapping 105 and 95 kDa polyproteins, that are proteolytically cleaved at Q/M and Q/G cleavage sites to produce the capsid proteins VP37 and VP23 and two proteins of 58 and 48 kDa proposed to be involved in virus movement through plant tissue (see text). B-RNA is translated into a single polyprotein of 200 kDa, from which five functional proteins are derived by proteolytic cleavage at Q/S, Q/M and Q/G sites. All proteolytic cleavages are performed by the viral 24 kDa proteinase.

replication process has been studied in detail. Studies of the translation mechanism were greatly assisted by the availability of a cowpea mesophyll protoplast system (for review see Goldbach and Van Kammen, 1985). As a result, a fairly complete genetic map and expression model has been established (Fig. 4). Since it is known that in cowpea protoplasts B-RNA is able to replicate independently of M-RNA, this RNA is considered to encode all functions required for viral RNA replication (Goldbach *et al.*, 1980).

Identified gene products specified by the B-RNA segment include the 24 kDa viral proteinase (which carries out all proteolytic changes within the M- and B-RNA-coded polyproteins), the 32 kDa proteinase cofactor (essential for the Q/M *in trans* cleavage in the M-polyprotein by the 24 kDa proteinase), the genome-bound protein VPg (which is thought to be involved as primer during initiation of RNA synthesis), and the 87 kDa core polypeptide of the viral replicase. M-RNA encodes the two capsid proteins, VP37 and VP23. Moreover, since B-RNA cannot move from cell-to-cell in the absence of M-RNA, it has been suggested that the non-structural overlapping 58 kDa and 48 kDa proteins encoded by M-RNA are involved in viral transport through plant tissue (Rezelman *et al.,* 1982). Although this conclusion has been strengthened by the observation that all transport mutants (i.e. mutants inducing only small local lesions on *Phaseolus vulgaris* cv. Pinto) map on M-RNA (De Jager, 1976; De Jager *et al.*, 1977) the transport function of CPMV has not yet been identified. Other questions that deal with the organization and genetic functions of the CPMV genome and that still remain to be answered include:

1. what are the template requirements for RNA replication;
2. what are the functions of the B-RNA-coded 58 kDa protein and VPg in the RNA replication process;
3. what are the genetic determinants for host range and symptom expression, and
4. which viral functions are involved in vector (beetle) transmission?

Providing definite answers to these questions will be major tasks for molecular virologists in the next decade, but two important tools already available will certainly be of great help. These are:

1. the availability of full length cDNA clones from which desired mutant RNAs can be generated by *in vitro* transcription, and
2. the well-designed cowpea protoplast system that lends itself particularly to studies of the expression of (mutant) CPMV RNAs, and in which basic cellular processes (RNA replication, translation, polyprotein processing and encapsidation) can be studied in the absence of processes required to support the spread and ecology of the virus.

Expression of CPMV RNAs in the cowpea protoplast system

The expression of the two positive stranded RNAs of CPMV has been studied extensively in cowpea mesophyll protoplasts. Thus for B-RNA a complete translation and proteolytic cleavage map has been established. Upon inoculation of the protoplasts with B components or B-RNA and incubation of the inoculated protoplasts in the presence of ^{35}S methionine, virtually all B-RNA-coded proteins, including those that were theoretically expected on the basis of *in vitro* translation experiments or the B-RNA nucleotide sequence, have been identified and characterized.

In contrast, progress for M-RNA in these respects has been hampered, most probably due to a lower stability of the non-structural proteins coded by M-RNA. In *in vitro* protein-synthesizing systems M-RNA is translated into two overlapping carboxy-coterminal polyproteins of approximately 105 and 95 kDa. These are cleaved by the B-RNA-coded 24 kDa protease into overlapping 58 and 48 kDa proteins and, via a 60 kDa precursor, into the capsid proteins VP37 and VP23 (Fig. 4). *In vitro* translation studies with transcripts from modified cDNA clones of M-RNA have shown that the overlapping polyproteins are the result of initiation at two different AUG start codons, at position 161 and position 512 or 524 (Vos *et al.*, 1984). In cowpea protoplast studies, only the two mature capsid proteins, their common 60 kDa precursor and the 48 kDa protein have been detected so far (Wellink *et al.*, 1987). This leaves open the possibility that *in vivo* M-RNA is translated only into the smaller 95 kDa polyprotein.

Recently, however, Rezelman *et al.* (1989) have established the complete translation and proteolytic processing map for M-RNA *in vivo*. This was achieved by two additional improvements to the protoplast system. Firstly, instead of inoculating the cells with virus or viral RNA using the potassium citrate buffer method (Rottier *et al.*, 1979) or the polyethyleneglycol method (Eggen *et al.*, 1989), viral RNA was efficiently introduced by electroporation, using a chamber with gold-coated glass panel electrodes as described by Hibi *et al.* (1986). Secondly, to detect unstable proteins, the protoplasts were pulse-labelled with ^{35}S-methionine soon (i.e. between 0.5 - 4 hours) after electro-inoculation. This procedure indeed led to the detection of both the 105 and 95 kDa polyproteins, while the amounts of these precursors could be further increased by the addition of 2mM $ZnCl_2$ during the pulse (Rezelman *et al.*, 1989). Zinc ions have previously been shown to inhibit the CMPV proteinase (Wellink *et al.*, 1987). The respective non-capsid 58 and 48 kDa cleavage products of the 105 and 95 kDa polyproteins (see Fig. 4), could also be detected. This was achieved by labelling the protoplasts for one hour at either 18 hours or 43 hours after inoculation and subsequent immunoprecipitation of the protoplast extract using antibodies directed against a synthetic peptide that corresponded to the carboxy-terminus of both 58 and 48 kDa proteins. At both labelling times the 48 kDa protein was present in both the cytoplasmic and membrane fraction of the protoplasts, while the 58 kDa protein was only found in the cytoplasm (Rezelman *et al.*, 1989). The detection *in vivo* of both the 105 and 95 kDa primary translation products as well as their 58 and 48 kDa N-terminal proteolytic cleavage products unequivocally indicates that M-RNA is translated *in vivo* from two different initiation codons. This translation strategy appears to be common to all comoviruses, because it has been observed in *in vitro* protein synthesizing systems for all comoviral M-RNAs tested so far (Goldbach and Krijt, 1982).

Infectious transcripts from cDNA clones

As another approach to studying the genetic organization, replication and expression of the comoviral RNA genomes, we have constructed full-length cDNA clones from which infectious transcripts can be generated by *in vitro* transcription. Using the T7 system and a truncated T7 promoter, *in vitro* transcripts were obtained containing only one non-viral G residue at the 5'-end (instead of a VPg) and, as a result of cleavage of the clones with restriction enzyme ClaI for run-off transcription, 5 extra nucleotides to the 46 residue-long poly(A)-tail. On inoculation of cowpea protoplasts with such transcripts from both B- and M-cDNA clones (denoted pTB1G and pTM1G, respectively), up to 40% of the cells became infected with B-RNA while in 10% of the cells multiplying B- and M-RNA molecules were both detected (Eggen *et al.*, 1989; and Table 2).

Sequence analysis showed that the 5'- and 3'-termini of the replicating transcripts were completely restored, including poly(A)-tails of variable length. The impact of such genetically functional cDNA clones on unravelling the virus multiplication process in detail can already be demonstrated by results from the production and *in vivo* analysis of two types of site-specific mutants.

The first series of mutants was generated (Eggen *et al.*, 1989) to examine whether a stretch of 11 nucleotides, UUUUAUUAAAA, comprising the nucleotides 5883 to 5893 in the 3'-noncoding region of B-RNA, is a crucial regulatory sequence during RNA replication. This stretch is also present in the 3'-terminus of the M-RNA and represents the seven nucleotides preceding the poly(A)-tail and the first 4 A residues of the tail (Fig. 5). By replacing the A residue at position 5887 by a T, an AhaIII restriction site was introduced in the cDNA (mutant pTB1G-Aha), which allowed the introduction of deletions of various length in this region (Table 3). Only those mutant transcripts with minor modifications appeared able to replicate in cowpea protoplasts or cowpea plants, indicating that the modified region indeed has an essential function in viral RNA replication. The results furthermore suggest the importance of a potential small hairpin loop in this region, since deletions causing complete abolition of the hairpin folding gave a drastic reduction in infectivity (Table 3 and Fig. 5).

The second series of mutants was generated (Wellink *et al.,* 1989) to study the role of the M-RNA-coded capsid proteins and overlapping 58 and 48 kDa proteins in virus translocation through plant tissue. In this case a series of insertion and deletion mutants was derived from the infectious wild type M-RNA clone pTMIG (Eggen *et al.,*

Table 2. *Infectivity of CPMV RNA and in vitro transcripts in cowpea mesophyll protoplasts*

Virus/ Clone	RNA or transcript characterization	% protoplasts producing	
		B-RNA encoded proteins[a]	M-RNA encoded proteins[b]
CPMV RNA	vpg-UAUU----A_{50-140}	20	20
pTB35	X_{35}-UAUU----A_{46}UAUCG	0.2	-
PTB2G	GG-UAUU----A_{46}UAUCG	5	-
pTB2G-cap	m^7GpppGG-UAUU----A_{46}UAUCG	10	-
pTB1G	G-UAUU----A_{46}UAUCG	25	-
pTB1G-cap	m^7GpppG-UAUU----A_{46}UAUCG	40	-
pTB1G + pTM2G	GG-UAUU----A_{46}UAUCG	25	0.3
pTB1G + pTM1G	G-UAUU----A_{46}UAUCG	25	10

Detection of viral RNA expression in cowpea protoplasts by the indirect fluorescent antibody technique (Vos *et al.*, 1988). The antisera used were raised against [a] the B-RNA encoded 24 kDa protein or [b] the virus particle.

1989). Modification of the coding regions of the 58 kDa/48 kDa proteins resulted in transcripts that replicated in cowpea protoplasts (as detected by immunofluorescence and Northern analysis) but which remained restricted to the inoculated cells in whole cowpea plants and were thus blocked in inducing systemic infection. Modification of the capsid protein genes had the same transport-inhibiting effect. These results suggest that the M-RNA-coded 58 and 48 kDa proteins are involved in cell-to-cell movement and that for CPMV the viral genome can only be spread through plant tissue in encapsidated form.

Table 3. *Effect of mutations in the 3'-region on the infectivity of B-RNA*

B-cDNA clone	Structure of 3' region of transcript	Infectivity (%)[a]
pTB1G	- GUUUUUAUUA$_{46}$UAUCG	100
pTB1G-Aha	- GUUUUUUUUA$_{46}$UAUCG	60
pTB1G-**Δ**1	- GUUUUUUU A$_{46}$UAUCG	60
pTB1G-**Δ**2	- GUUUUUU A$_{46}$UAUCG	70
pTB1G-**Δ**3	- GUUUUU A$_{46}$UAUCG	70-80
pTB1G-**Δ**4	- GUUUU A$_{46}$UAUCG	20
pTB1G-**Δ**7	- GU A$_{46}$UAUCG	0.5
pTB1G-**Δ**8	- G A$_{46}$UAUCG	0

[a]The infectivity of transcripts from mutant cDNA clones was tested in cowpea protoplasts. 2.5 x 10^6 protoplasts were inoculated with 50μg of transcript and 40 hours later the infectivity was measured by examining the presence of virus-encoded proteins using the indirect immunofluorescent antibody technique (Vos *et al.*, 1988). The infectivity of pTB1G transcripts was arbitrarily set at 100%. This corresponded to 30% successfully infected protoplasts.

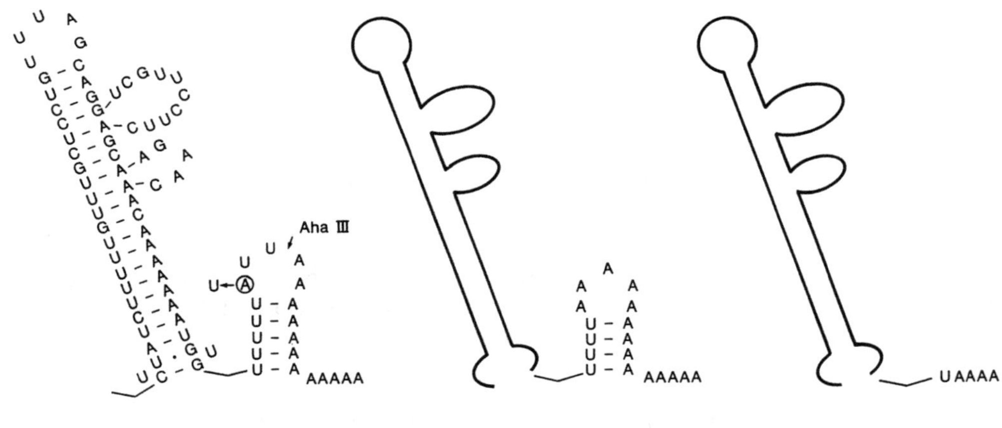

pTB1G (100%) pTB1G Δ 4 (20%) pTB1G Δ 7 (0.5%)

Figure 5. Predicted effect of mutations in the 3'-non-coding region of CPMV B-RNA immediately preceding the poly(A) tail. In transcripts of clones pTB1G (wild type) and pTB1G-Aha (A to U replacement at position 5887) a conserved stretch of 11 nucleotides (see text) is proposed to form a small hairpin. Deletion of 4 nucleotides (mutant pTB1G-**Δ**4; see also Table 3) diminished the stem length and reduced the infectivity of the RNA to 20%. Deletion of 7 nucleotides (mutant pTB1G-**Δ**7) completely abolished the hairpin and led to further reduction of the infectivity to 0.5%.

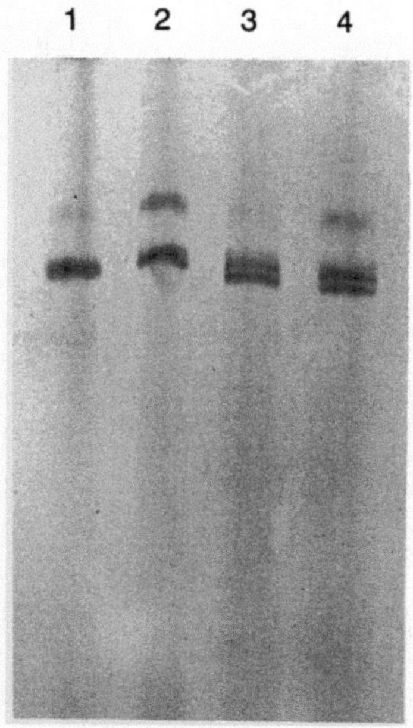

Figure 6. Non-denaturing electrophoresis of wild-type CMPV and mutant M142Bsb in 3% polyacrylamide gel. Lane 1: mutant M142Bsb, lane 2: wild-type CPMV; lanes 3 and 4: a mixture of wild-type CPMV and mutant M142Bsb.

EVIDENCE FOR TWO TYPES OF TRANSPORT MECHANISMS FOR PLANT VIRUSES

Requirements for both 58 and 48 kDa proteins and for one or two capsid proteins for effective local spread of CPMV offers a plausible explanation for the intrastrand complementation described by De Jager and Breekland (1979). Two nitrous-acid induced mutants, N123 and M142Bsb, with mutations in M-RNA, induced only small local lesions in Pinto beans and Blackeye cowpeas when inoculated separately. However, mixtures of the two mutants induced large lesions of wild-type appearance. Since such lesions did not contain wild-type virus, complementation rather than recombination was inferred. Polyacrylamide gel electrophoresis of M142Bsb under non-denaturing conditions showed altered mobility of the fast component, indicating mutation of the coat protein(s) (See Fig. 6).

The involvement of the 58 and 48 kDa proteins in virus transport has been

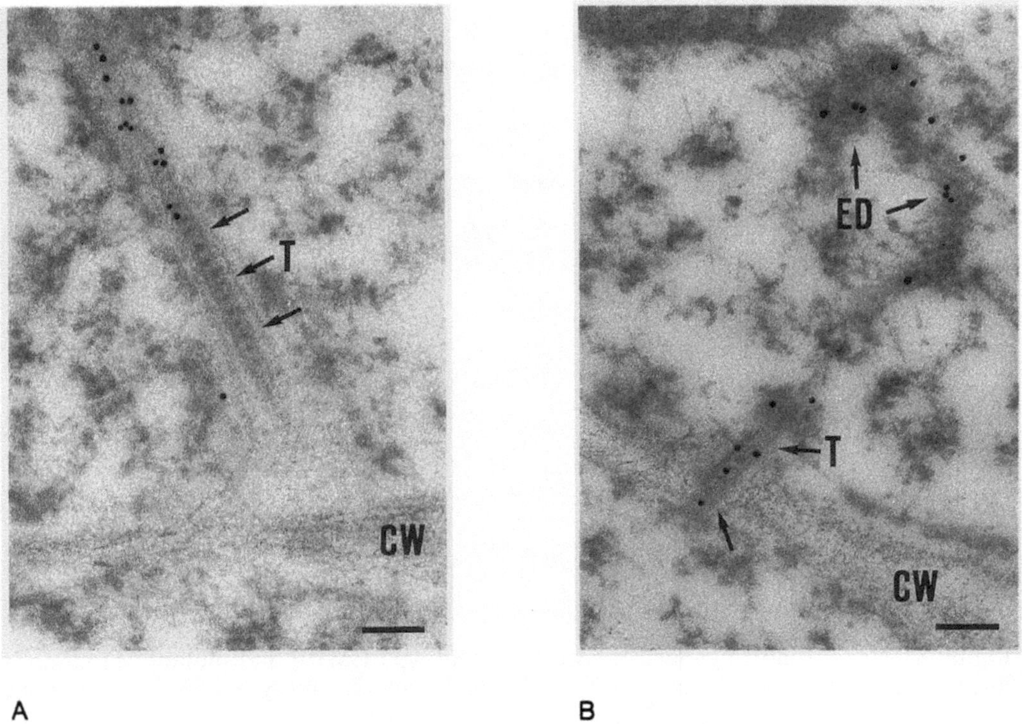

Figure 7. Immunogold localization of the 58 and 48 kDa tentative transport proteins of CPMV. Sections of aldehyde-fixed and LR white-embedded petiolules of infected cowpea plants were treated with antiserum against the 58/48 kDa proteins and protein A-gold complexes (7nm). In infected cells virus-induced tubular structures (T) were observed protruding from (A) or penetrating (B) the cell wall (CW). The tubular structures appear to include virus particles. Gold labelling revealed the location of 58/48 kDa proteins at the tubular structures (A) and in small areas of electron-dense material (ED) associated with these tubular structures (B). The bar represents 100nm.

further substantiated by tracing these proteins in infected cells by immuno-electron microscopy, using antibodies against a synthetic peptide and gold-labelled *Staphylococcus aureus* protein A to visualize immuno-complexes. The 58 kDa and 48 kDa proteins could be specifically detected in tubular structures that protruded from the cell walls and that appeared to be virus-induced extensions or modifications of plasmodesmata (JWM van Lent, J Wellink and R Goldbach, unpublished results, and Fig. 7). Close inspection of a large number of sections demonstrated that the gold-label was exclusively found associated with or in the close neighbourhood of these tubular structures. Also many of these tubular structures, which seemed to be at least partly built up from cell wall components, appeared to contain virus particles (Fig. 7). This finding is in full agreement with our earlier conclusion that encapsidation is a prerequisite for virus movement.

Our findings seem to conflict with the results of Malyshenko *et al.* (1988), who demonstrated for another comovirus, red clover mottle virus, that movement of non-encapsidated B-RNA can occur when host plants are co-infected with a tobamovirus. For tobamoviruses, however, several lines of evidence indicate that these viruses can move for cell-to-cell as naked nucleic acid (Atabekov and Dorokhov, 1984). Hence, it is plausible to assume that in the experiments of Malyshenko *et al.* (1988) the movement of B-RNA was achieved by a transport function that enabled movement of nucleic acid.

We infer from the results described above that for the cell-to-cell movement of plant viruses at least two different types of mechanism exist, one type achieving movement of viral nucleic acid (e.g. the tobamovirus transport mechanism) and another achieving movement of virus particles (e.g. the comovirus transport mechanism).

ACKNOWLEDGEMENT

The authors wish to thank Mary van der Linde for typing the manuscript. This work was supported in part by the Netherlands Foundation for Chemical Research (SON) with financial aid from the Netherlands Organization of Pure Research (NWO).

REFERENCES

Atabekov JG, Dorokhov YL (1984) Plant virus-specific transport function and resistance of plants to viruses. Adv Virus Res 29:313-359

De Jager CP (1976) Genetic analysis of cowpea mosaic virus mutants by supplementation and reassortment tests. Virology 70:151-163

De Jager CP, Zabel P, Van der Beek CP, Van Kammen A (1977) Genetic and physiological characterization of a temperature-sensitive mutant of cowpea mosaic virus. Virology 76:164-172

De Jager CP, Breekland L (1979) Evidence for intrastrand complementation in cowpea mosaic virus infection. Virology 99:312-318

Eggen R, Verver J, Wellink J, Pleij K, Van Kammen A, Goldbach R (1989) Analysis of sequences involved in cowpea mosaic virus RNA replication using site specific mutants. Virology, in press

Goldbach RW (1986) Molecular evolution of plant RNA viruses. Annu Rev Phytopathol 24:289-310

Goldbach RW (1987) Genome similarities between plant and animal RNA viruses. Microbiol Rev 4:197-202

Goldbach RW, Krijt J (1982) Cowpea mosaic virus encoded protease does not recognise primary translation products of M RNAs from other comoviruses. J Virol 43:1151-1154

Goldbach R, Van Kammen A (1985) Structure, replication and expression of the bipartite genome of cowpea mosaic virus. In: Davies JW (ed) Molecular plant virology, Vol. 2. CRC Press, Boca Raton, p 83

Goldbach RW, Wellink J (1988) Evolution of plus-strand RNA virus. Intervirology 29:260-267

Goldbach R, Rezelman G, Van Kammen A (1980) Independent replication and expression of B-component RNA of cowpea mosaic virus. Nature 286:297-300

Hibi T, Kano H, Sugiura M, Kazami T, Kimura S (1986) High efficiency electrotransfection of tobacco mesophyll protoplasts with tobacco mosaic virus RNA. J Gen Virol 67:2037-2042

Malyshenko SI, Lapchic O, Kondakova OA, Kuznetzova LV, Talianski ME, Atabekov JG (1988) Red clover mottle comovirus B-RNA spreads between cells in tobamovirus infected tissues. J Gen Virol 69:407-412

Miller WA, Waterhouse PM, Gerlach WL (1988) Nucleotide sequence and genome organization of barley yellow dwarf virus. Nucleic Acids Res 16:6097-6111

Rezelman G, Franssen HJ, Goldbach RW, Ie TS, Van Kammen A (1982) Limits to the independence of bottom component RNA of cowpea mosaic virus. J Gen Virol 60:335-342

Rezelman G, Van Kammen, Wellink J (1989) Expression of cowpea mosaic virus M RNA in cowpea protoplasts. J Gen Virol, in press

Robinson DJ, Hamilton WDO, Harrison BD, Baulcombe DC (1987) Two anomalous tobra-virus isolates: evidence for RNA recombination in nature. J Gen Virol 68:2551-2561

Rottier PJM, Rezelman G, Van Kammen A (1979) The inhibition of cowpea mosaic virus replication by actinomycin D. Virology 92:299-309

Van der Wilk F, Huisman MJ, Cornelissen BJC, Huttinga H, Goldbach R (1989) Nucleotide sequence and organization of potato leafroll virus genomic RNA. FEBS Lett 245:51-56

Veidt I, Lot H, Leiser M, Scheidecker D, Guilley H, Richards K, Jonard G (1988) Nucleotide sequence of beet western yellows virus RNA. Nucleic Acids Res 16:9917-9932

Vos P, Jaegle M, Wellink J, Van Kammen A, Goldbach R (1988) Infectious RNA transcripts derived from full-length DNA copies of the genomic RNAs of cowpea mosaic virus. Virology 164:33-41

Vos P, Verver J, Van Wezenbeek P, Van Kammen A, Goldbach R (1984) Study of the genetic organisation of a plant viral genome by in vitro expression of a full-length DNA copy. EMBO J 3:3049-3053

Wellink J, Jaegle M, Prinz H, Van Kammen A, Goldbach R (1987) Expression of the middle component RNA of cowpea mosaic virus in vivo. J Gen Virol 68:2577-2585

Wellink J, Van Kammen A (1989) Cell-to-cell transport of cowpea mosaic virus requires both the 58K/48K proteins and the capsid proteins. J Gen Virol, in press

REPLICATION OF PLANT RNA VIRUSES *IN VIVO* IN RELATION TO VIRUS-HOST INTERACTIONS

Radhia Gargouri and Anne-Lise Haenni
Institut Jacques Monod
2 Place Jussieu
75251 Paris Cedex 05
France

ABSTRACT

Various approaches can be used to examine the replication of plant RNA viruses within the host. The modifications incurred by the different cell compartments due to virus infection can indicate the parts of the cell likely to be involved in replication. The sites at which RNA replication occurs can also be ascertained by investigating where virus-specific, newly synthesized or double-stranded RNA species are located, or where the viral nonstructural proteins can be detected. Characterization and analysis of purified double-stranded RNAs has helped to elucidate the mechanism for the synthesis of the coat protein subgenomic RNA. Enzyme complexes (RNA-dependent RNA polymerase or replicase) responsible for RNA replication have been isolated. Although difficult to work with, they provide valuable information regarding the virus-coded nonstructural proteins that participate in RNA replication. The characterization of host-encoded factor(s) possibly involved in this process remains a major problem.

INTRODUCTION

Within the plant kingdom, RNA viruses are far more abundant than DNA viruses, and among RNA viruses, those with a (+) strand RNA genome also greatly outnumber viruses with an RNA genome of (-) polarity or with double-stranded RNA (dsRNA). It is no doubt due to their prevalence that viruses with an RNA genome of (+) polarity have been the most widely studied.

Virus multiplication within the host requires four main steps that overlap chronologically: decapsidation, translation, replication and encapsidation. It should be

NATO ASI Series, Vol H 41
Recognition and Response in Plant-Virus Interactions
Edited by R.S S Fraser

stressed, however, that viral RNA devoid of capsid can be replicated, and that encapsidation derives from the preceding steps and is a non-enzymatic process.

This paper is concerned solely with the replication step, with particular emphasis on host-virus interactions. The review centres on RNA replication *in vivo* of viruses containing a (+) RNA genome with only passing reference to *in vitro* experiments. For further information, the reader is invited to turn to other recent review articles related to viral RNA replication *in vivo* (Martelli and Russo, 1984; Van Kammen, 1985; Francki, 1987; Zaitlin and Hull, 1987).

To examine viral RNA replication, the response of whole plants, of specific plant tissues or of plant cells has been considered. This has led in certain cases to the localization of the site of RNA replication within the cell. Hybridization methods with specific labelled probes have now made it possible to analyze the nature of the newly synthesized nucleic acids that appear during replication. Finally, in a few cases, information is beginning to emerge concerning the composition of the RNA-dependent RNA polymerase (RNA replicase), and the possible involvement of host-derived proteins in replication.

ULTRASTRUCTURAL MODIFICATIONS IN THE HOST AFTER VIRUS INFECTION

Virus infection is generally detected easily by the symptoms produced in the host, and the gross and diverse effects produced can assist in diagnosis of the nature of the pathogen. At the cellular or subcellular level, infection engenders the formation of a large array of cytopathic modifications whose nature depends among other factors on the virus, the host and the growth conditions of the plant. These modifications can serve as the "signature" of the presence of a given virus, as discussed by Martelli and Russo (1984).

Cytopathic structures are of two classes, *crystalline* and *amorphous*. Crystalline structures are basically composed of aggregates of virus particles embedded in a given cellular compartment; they can take the shape of hexagonal crystalline plaques, fibrous or banded arrays, or crystalline needles referred to as paracrystals. Amorphous structures or inclusions are likewise very diverse in location and shape. In addition, they are also disparate in their contents, and may contain not only virus particles, but also ribosomes, membranous material, strands of endoplasmic reticulum, fibrillar material and vacuoles, depending on the type of inclusion. Multiple vesicles also generally appear as part of the inclusions. Since amorphous inclusions are believed also to harbour nonstructural virus-coded proteins, and since the fibrillar material often detected within vesicles has in some cases been shown to be composed of dsRNA, it has been proposed that amorphous inclusions are the site of viral RNA replication (reviewed in Francki, 1987).

Table 1 briefly surveys the virus families for which specific inclusions accompanied by vesiculation have been described, the subcellular compartment in which the inclusions were detected, and the type of inclusion formed. Those viruses for which the inclusions have been shown to contain dsRNA and/or the replicase are indicated. Wherever such investigations have been performed, amorphous cytopathic structures with vesiculation have contained dsRNA and/or the RNA replicase.

Most experiments aimed at defining the site of viral RNA replication have been performed by electron microscopy on thin-sectioned material. Identification of dsRNAs as components of the fibrils residing in certain vesicles has been by enzyme cytochemical tests: the fibrils are resistant to RNase treatment under high salt conditions (as shown for broad bean true mosaic virus (BBTMV), *Cymbidium* ringspot virus (CyRSV), cucumber mosaic virus (CMV) and tomato aspermy virus (TAV)). Other techniques used to detect viral RNA synthesis include specific staining of dsRNAs (e.g. for tobacco rattle virus (TRV), turnip yellow mosaic virus (TYMV)), *in situ* labelling of newly synthesized RNA followed by high resolution autoradiography (TYMV) and the use of molecular probes that specifically recognize the viral RNA or the complementary (-) RNA strands (experiments with cowpea mosaic virus (CPMV), pea enation mosaic virus (PEMV) and tobacco etch virus (TEV)). The recently developed immunocytochemical technique using colloidal gold as marker appears to be the method of choice for localization of the replicase or of specific nonstructural viral proteins presumed to form part of the replicase complex.

A few points warrant additional comments. In some cases, the conclusions reached by examining infected plant tissue have been confirmed by experiments on infected protoplasts (with cucumber green mottle mosaic virus (CGMMV), Sugimura and Ushiyama, 1973; TYMV, Candresse *et al.*, 1987; CPMV, Wellink *et al.*, 1988). For instance, in cowpea protoplasts infected with CPMV B component alone (the RNA of which bears the gene for replication), B-RNA was expressed and replicated, and

Table 1. *Site of viral RNA replication and cytopathological modifications*

Cellular Compartment Affected	Pathogeny	Virus Family
cytoplasm	vesiculate-vacuolate inclusions[a]	comovirus
		tobamovirus
		?[b]
		nepovirus closterovirus potyvirus dianthovirus
nucleus	membranous vesicles fused with nuclear envelope, later released into cytoplasm	pea enation mosaic virus luteovirus bromovirus
chloroplasts	clumping of swollen and peripherally vesiculated chloroplasts	tymovirus
		potyvirus hordeivirus
mitochondria	mitochondrial vesiculation	tobravirus tobamovirus luteovirus
peroxisome-derived vesicles	multivesicular bodies[c]	tombusvirus
vacuoles	vesiculation of tonoplasts	cucumovirus
		tobacco necrosis virus potexvirus tobamovirus

[a] Contain multiple ribosomes, strands of endoplasmic reticulum and membrane vesicles with fine fibrils, as well as crystalline inclusions. In TMV, the cellular structures are also designated "viroplasms", *i.e.* amorphous, nonmembrane-bound structures, electron-dense rod-like structures embedded in ribosome-rich matrix; they give rise when mature to "X-bodies".

Table 1. *Continued*

	Replication	
Virus	Evidence	Reference
CPMV	dsRNA and replicase	De Zoeten *et al*., 1974; Wellink *et al*., 1988
BBTMV	dsRNA in fibrils	Hatta and Francki, 1978
TMV	replicase in viroplasms	Saito *et al*., 1987; Hills *et al.,* 1987
VTMoV	replicase	Rohozinski *et al.,* 1986
PEMV	dsRNA and replicase	De Zoeten *et al*., 1976; Powell and De Zoeten, 1977
TYMV	dsRNA in fibrils, RNA synthesis and replicase in membranes	Laflèche and Bové, 1986; Laflèche *et al*., 1972; Ushiyama and Matthews, 1970; Garnier *et al.,* 1986
TEV	dsRNA in chloroplasts	Gadh and Hari, 1986
TRV	site of RNA synthesis	Harrison *et al*., 1970
CyRSV	dsRNA in fibrils	Russo *et al*., 1983
CMV	dsRNA in fibrils	Hatta and Francki, 1981
TAV	dsRNA in fibrils	Hatta and Francki, 1981

[b] Not assigned to a definite virus family, but VTMoV has similarities with members of the sobemovirus family.

[c] Also observed when plants are infected by fungi.
dsRNA = double-stranded RNA. Virus abbreviations as in text.

vesiculate-vacuolate inclusions developed (Rezelman *et al.*, 1980; Rezelman *et al.*, 1982). This observation confirms other results suggesting that these inclusions are the sites of replication (reviewed in Van Kammen, 1984). In other cases, the activity of isolated organelles has been demonstrated. For instance, the observation that viral dsRNA and replicase are present in the nuclei of plants infected with PEMV has been confirmed by demonstrating that isolated pea nuclei could support the initial stages of PEMV RNA replication (Powell and De Zoeten, 1977).

COMPOSITION OF dsRNAs: SYNTHESIS OF THE COAT PROTEIN SUBGENOMIC RNA

From work on RNA bacteriophages, it has been assumed that in eukaryotic viruses also, the dsRNA species that can theoretically be observed are basically of two types, replicative forms (RF) and replicative intermediates (RI). RFs are full-length stable dsRNAs composed of a (+) and a (-) strand. RIs are branched forms generally made up of a full-length (-) strand and several (+) strands of various lengths that are being elongated. RIs are thus only partially double-stranded structures and probably correspond to the true intermediates of replication. It appears likely that double-stranded structures are largely artifacts produced by phenol deproteinization during RNA extraction; within the cell, proteins would hinder hydrogen-bonding between the two complementary strands (Garnier *et al.*, 1980).

Synthesis of viral progeny RNA of genomic size can, at a first approximation, be regarded as a relatively simple process in virus development. However, an additional mechanism must exist for those viruses that employ the strategy of subgenomic RNA formation for the expression of the internally located genes on their genomic RNA. A large number of viruses resort to this strategy for the synthesis of their coat protein.

In Fig. 1 the two main mechanisms that can be envisaged for the synthesis of subgenomic RNAs are schematized. A (-) RNA of subgenomic size produced by premature termination of the synthesis of genomic (-) RNA could serve as template for the synthesis of (+) stranded subgenomic RNA (Fig. 1A). Alternatively, the subgenomic (+) RNA would be produced by internal initiation of replication on genomic (-) RNA (Fig. 1B). Several attempts have been made to distinguish between these two mechanisms, by isolating the dsRNAs produced *in vivo* and examining their

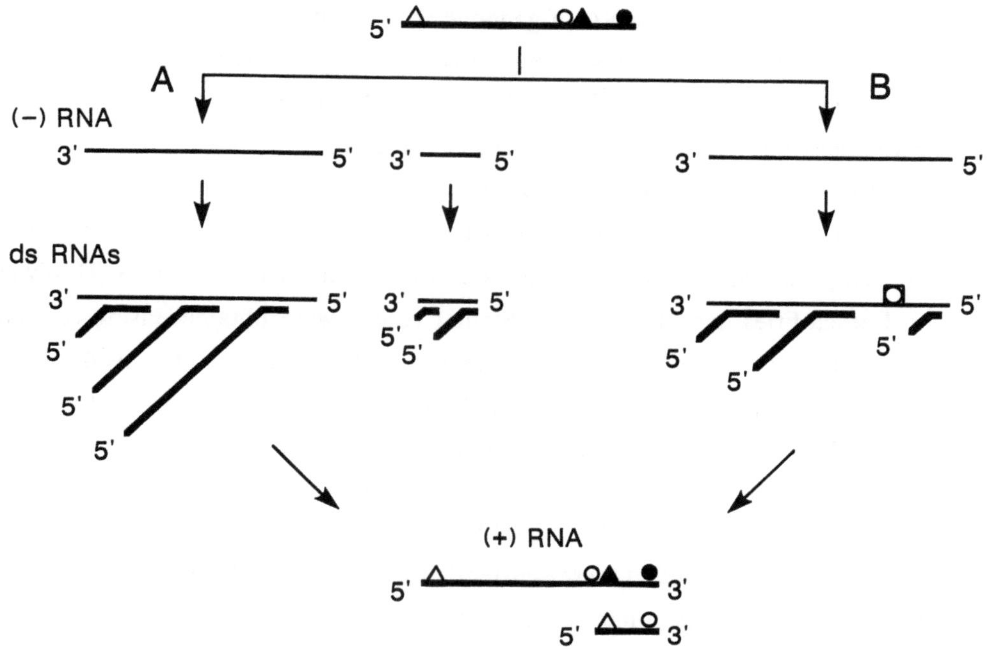

Figure 1. Two possible models (A and B) proposed for the synthesis of the subgenomic RNA *in vivo*. △, accessible and ▲, inaccessible initiation codons; o, accessible and ●, inaccessible termination codons; ▬ (+) RNA; — (-) RNA; ◘, RNA replicase bound to internal promoter.

composition. From the data reported, it is clear that in all cases dsRNA species of genomic length were detected. However, the picture is somewhat confusing with respect to the mechanism of synthesis of subgenomic RNAs.

In the case of tobacco mosaic virus (TMV), several dsRNA species have been detected in infected leaves (Zelcer *et al*., 1981; Bar-Joseph *et al*., 1983; Palukaitis *et al*., 1983; Jordan and Dodds, 1985). The size of one of them that appears in low amounts corresponds to the expected size of the RF and RI of coat protein subgenomic RNA. These results would tend to favour the mechanism shown in Fig. 1A. However, it has been suggested that the presence of dsRNA species of subgenomic size could result from artifacts (Palukaitis *et al*., 1983) since some of these dsRNA species cannot be accounted for by our knowledge of the subgenomic RNA population resulting from TMV infection. On the other hand, a pellet fraction isolated from TMV-infected protoplasts and containing the replicase-template complex was able to synthesize not only genomic RNA, but also subgenomic coat protein mRNA (Watanabe and Okada, 1986). It also synthesized low amounts of (-) RNA of genomic size, but not (-) RNA of subgenomic length. RF molecules of genomic size but no RFs of subgenomic size were detected. These results suggest that the coat protein RNA is produced on a (-) RNA template of genomic size, according to the model presented in Fig. 1B.

For brome mosaic virus (BMV), low amounts of a dsRNA species corresponding to coat protein subgenomic RNA have been detected when total RNA isolated from BMV-infected protoplasts is digested by RNase A under high salt conditions (Hardy *et al.*, 1979; Loesch-Fries and Hall, 1980). Similar results were obtained using cowpea chlorotic mottle virus (CCMV)-infected leaves (Miller and Hall, 1984). In CMV-infected leaves, minor amounts of a dsRNA band corresponding to coat protein subgenomic RNA have been detected (Jaspars *et al.*, 1985); this dsRNA band also appears upon RNase A digestion of dsRNA3 under high salt conditions. Consequently, the reports concerning BMV, CCMV and CMV favour the mechanism proposed in Fig. 1A. However, the possibility that these shorter dsRNA species might result from RNase A digestion of dsRNAs of genomic size has not been unequivocally excluded. On the other hand, Nitta *et al.* (1988) have observed that when tobacco protoplasts are inoculated with CMV RNAs 1+2+4, the double-stranded forms isolated correspond to RNA1 and RNA2, but not to RNA4, suggesting that (+) RNA4 does not seem to serve as a template for the replicase; these results favour the mechanism in Fig. 1B.

Similar analyses failed to detect dsRNA species of subgenomic size in alfalfa mosaic virus (AlMV)-infected protoplasts (Nassuth and Bol, 1983) or in TYMV-infected leaves (Gargouri *et al.*, 1989). In either case, under native and denaturing conditions, analysis of the dsRNAs by hybridization using specific probes detected only dsRNAs of genomic length; this species contains nascent subgenomic as well as genomic (+) RNA molecules. Additionally, in TYMV, direct labelling of the 5' ends of the nascent (+) RNA strands present in the dsRNAs (RF + RI) has revealed that these dsRNAs contain both the genomic and the subgenomic RNA species. These observations are in favour of the mechanism presented in Fig. 1B. However, they are at variance with those of others (Jordan and Dodds, 1985; Valverde *et al.*, 1986) who have provided evidence for the existence of several dsRNA species in TYMV-infected plants, in experiments aimed at using the distribution of dsRNAs to detect and identify viruses.

Several factors may be responsible for the diversity of the results reported. Firstly, host plants cannot be synchronously inoculated, and infected material at various stages of virus development has been used; moreover, different virus strains inoculated on different hosts can produce different patterns of dsRNAs (Dodds *et al.*, 1988). Secondly, various methods have been used to isolate and purify the dsRNAs. Thirdly, there are few reports in which a systematic analysis of the dsRNA species was performed using non-denaturing as well as denaturing gels. Finally, different viruses might use different replication strategies.

Taken together, the results reported in this section do not present a uniform picture. We cannot exclude the possibility that spurious degradation of the dsRNAs of genomic length could be responsible for the presence of shorter dsRNAs (Palukaitis *et al.*, 1983). Experiments along those lines come from work on animal alphaviruses (reviewed in Kääriäinen *et al.*, 1987). Cells infected with such viruses produce a genomic RNA (42S) and a subgenomic RNA (26S) derived from the 3' third of the genome. The subgenomic RNA is synthesized by internal initiation on (-) strand template. It has been proposed that an RNase-sensitive site exists at the position where initiation of the 26S RNA takes place since mild RNase treatment releases the dsRNA of subgenomic size (corresponding to 26S RNA) as well as the other dsRNA species resulting from this cleavage. One cannot exclude the possibility that a similar RNase-sensitive site exists in at least some plant RNA viruses.

The presence of very low levels of dsRNAs of subgenomic length is also not entirely surprising for another reason. In BMV (Ahlquist *et al.*, 1984) and in TYMV (Morch *et al.*, 1987; R Gargouri, C David, MD Morch and AL Haenni, in preparation), the initiation site for (-) RNA strand synthesis is contained within the 3'-terminal 100 nucleotides on the (+) RNA strand. Consequently, since the subgenomic (+) RNA contains this site, it should serve as a template for the synthesis of a complementary (-) RNA strand. Since the resulting (-) RNA in the dsRNA species would lack the internal initiation site for (+) RNA synthesis, it would be unable to serve as template for the production of (+) RNA of subgenomic length, and would therefore fail to amplify the subgenomic RNA.

COMPOSITION OF THE REPLICASE: PARTICIPATION OF VIRUS- AND HOST-CODED FACTORS

Again based on previous work performed on RNA bacteriophages, it is generally assumed that the replicase of all viruses containing an RNA genome of (+) polarity is composed of virus- and most likely also host-encoded proteins.

Among animal viruses, the involvement of a virus-coded subunit has clearly been demonstrated for poliovirus. The virus-encoded protein (designated 3D) contains the amino acid sequence Gly-Asp-Asp (GDD). This constitutes a consensus sequence encountered in one of the nonstructural proteins of all (+) strand plant RNA viruses examined (reviewed in Goldbach and Wellink, 1988 and in Gorbalenya *et al.*, 1989). It is thus generally assumed that the nonstructural protein harbouring the GDD motif is part of the viral replicase. Another consensus sequence commonly found in one of the nonstructural proteins of plant and animal RNA viruses is an NTP-binding motif. This is known to be characteristic of many NTP-utilizing enzymes. It has been proposed that the protein harbouring this motif might perform some function in viral RNA replication, such as NTP-binding and possibly also NTP hydrolysis (Gorbalenya *et al.*, 1989).

Little information is available concerning the putative host-encoded subunit(s) of plant virus RNA replicases. This is largely due to the fact that most enzyme preparations are still contaminated by several proteins that presumably are not all part of the replicase.

Despite considerable efforts to isolate viral replicases, the progress made to date has been relatively slow, for two main reasons. Uninfected plants contain an RNA-dependent RNA polymerase whose activity is strongly enhanced after viral infection. It was therefore felt that this enzyme might be responsible for viral RNA replication (Fraenkel-Conrat, 1983). However, this host-encoded enzyme is not viral RNA-specific. It generates only short RNA strands complementary to the template provided, that have been initiated at multiple sites. The involvement of this host-encoded enzyme in viral RNA synthesis was ruled out when it was established that the CPMV replication complex isolated from CPMV-infected cowpea leaves (Van Der Meer *et al.*, 1984), and the TYMV replication complex isolated from TYMV-infected leaves (Mouchès *et al.*, 1984) were functionally distinct from the host-encoded enzyme. The role of the host-encoded RNA-dependent RNA polymerase in healthy plants is unknown.

The following criteria are now generally considered to be the definition of a *bona fide* viral replicase: it should be dependent on the presence of a virus-specific RNA template, and should be capable of synthesizing full-length complementary (-) RNA strands as well as progeny (+) strands. So far, none of the replicases isolated has been reported to produce progeny (+) RNA molecules.

A second reason for slow progress is that the viral replicases appear to be membrane-bound. Major difficulties have been encountered in releasing the enzyme complex from its insoluble matrix. The use of non-ionic detergents has been a breakthrough in this respect, and should allow rapid progress to be made in the understanding of viral RNA replication.

Four viral replicases have been isolated to date: for TYMV, BMV, CCMV and AlMV. They all comply with the above definition of a viral replicase, to the extent that they are virus-specific, and are capable of synthesizing full-length (-) RNA strands complementary to the (+) RNA template; they not only elongate chains, but are capable of initiating them. In addition, information concerning some of the virus-encoded subunits of the TMV, CMV, CPMV and velvet tobacco mosaic virus (VTMoV) replicases is available, despite the fact that the corresponding enzyme complexes do not depend on added template for their activity. Some characteristics of these replicases are presented below.

TYMV Replicase

TYMV contains a monopartite RNA genome. Also encapsidated in the virion is a subgenomic RNA; it corresponds to the 3' region of the genomic RNA and codes for the viral coat protein (reviewed in Hirth and Givord, 1988). The nucleotide sequence of the genomic RNA (Morch *et al.*, 1988) reveals a large open reading frame capable of coding for a 206 kDa protein, which has also been demonstrated previously by *in vitro* translation of TYMV RNA. This nonstructural protein undergoes co- and/or post-translational cleavage, releasing 150 kDa and 78 kDa polypeptides from the N- and the C-termini respectively (MD Morch, G Drugeon, P Szafranski and AL Haenni, submitted manuscript). The 206 kDa protein, and after cleavage the 78 kDa polypeptide, contain the GDD motif which presumably marks that region of the protein as part of the replicase. The 150 kDa and/or the 206 kDa proteins appear to undergo further cleavage, releasing a 120 kDa polypeptide deriving from the N-terminus of the 206 kDa (or 150 kDa) protein. In the 206 kDa protein, the NTP-binding motif is located just beyond the region corresponding to the 120 kDa polypeptide (Morch *et al.*, 1988).

The replicase isolated from TYMV-infected Chinese cabbage leaves is the one whose composition has been most thoroughly investigated. It preferentially replicated TYMV RNA; replication of the RNA of the closely-related eggplant mosaic virus

(EMV) was 70% that of TYMV RNA, and TMV RNA replication was extremely poor. Replication depended on addition of exogenous RNA. The enzyme preparation contained two major polypeptides of 115 and 45 kDa (Mouchès *et al.*, 1984). With the use of anti-replicase antibodies, it was demonstrated that the 115 kDa subunit is virus-coded (Candresse *et al.*, 1986); its size is comparable to that of the 120 kDa polypeptide observed after *in vitro* translation of TYMV RNA and resulting from maturation of the 206 kDa polyprotein. Thus it is surprising that the GDD motif should reside in the 78 kDa polypeptide rather than in the 115 kDa polypeptide thought to be part of the replicase. In addition, the NTP-binding motif is located neither in the 120 kDa nor in the 78 kDa polypeptide. At present we still cannot exclude the possibility that the 120 and 78 kDa polypeptides, and perhaps a 30 kDa polypeptide harbouring the NTP-binding motif, as well as the uncleaved 206 kDa protein, may participate in replication *in vivo*.

The 45 kDa subunit is of host origin. Because TYMV RNA possesses a tRNA-like structure at its 3' end recognized by several tRNA-specific enzymes (reviewed in Haenni *et al.*, 1982), and since valylated RNA can form a ternary complex with GTP and the wheat germ protein elongation factor EF-1α, experiments were performed to search for the presence of this factor in the replicase preparation. However, neither the elongation factor, the valyl-tRNA synthetase nor the tRNA nucleotidyltransferase could be detected (Joshi *et al.*, 1986). The function of the host-encoded subunit within the host cell remains undefined.

The results reported on the nature of the TYMV replicase are now supported by experiments performed to locate the enzyme within the infected cells. Using immunocytochemical methods with colloidal gold as marker, the replicase and its virus-encoded 115 kDa subunit have been located at the chloroplast periphery, the site where RNA replication occurs. Surprisingly, free replicase molecules were also found in the cytoplasm. Whether these corresponded to recently synthesized replicase molecules or to those that accumulated after viral RNA replication is not known. Unfortunately, using the same technique, no conclusions could be drawn as to the location of the host-encoded 45 kDa protein in healthy leaves (Garner *et al.*, 1986).

BMV Replicase

BMV contains a tripartite genome. In addition to the 3 genomic RNAs (RNA1, 2 and 3) the virion also encapsidates a subgenomic RNA species (RNA4) that derives from the 3' half of RNA3.

Elucidation of the complete nucleotide sequence of the genomic RNAs of BMV has made it possible to search for consensus sequences in the proteins coded. The GDD motif is located in the protein (P2) coded by RNA2, whereas the NTP-binding motif is in protein P1 coded by RNA1. These nonstructural proteins could therefore be involved in replication.

A BMV replicase has been isolated from infected barley leaves (Hardy *et al.*, 1979; Miller and Hall, 1983). The enzyme was almost completely dependent on exogenous BMV RNA. It preferentially replicated BMV RNA, since replication of the RNA of the closely related CCMV was only 20% that of BMV RNA. Initiation of replication occurred on the template supplied (Miller *et al.*, 1986). Electrophoretic analysis of the products revealed the presence of bands resistant to RNase digestion under high salt conditions, that comigrated with BMV dsRNA.

Barley protoplasts are capable of supporting RNA replication when inoculated with RNAs1 and 2 in the absence of RNA3, suggesting that the corresponding translation products P1 and P2 are involved in replication (Kiberstis *et al.*, 1981; French *et al.*, 1986). To investigate this possibility further, antibodies were raised against parts of proteins P1 or P2. These antibodies interacted with the corresponding proteins in the replicase preparation. More importantly, antibodies directed against P1, but not those against P2, completely blocked BMV replication *in vitro*. This demonstrates the involvement of at least part of P1 in replication (Horikoshi *et al.*, 1988; Quadt *et al.*, 1988). The reason why antibodies raised against P2 do not inhibit replication is unclear, although a number of possible explanations have been put forward (Horikoshi *et al*, 1988).

Here again, it is noteworthy that P2, which contains the GDD motif, does not appear to be part of the replicase. Rather, it is protein P1, the protein containing the NTP-binding motif, that is involved in replication.

These results do not prove that P1 is the cataytic subunit of the replicase. Indeed, one cannot exclude the possibility that antibodies raised against P1 could abolish the activity of the replication complex by binding to a component of the complex thereby inducing strong steric modifications within the complex (Quadt *et al.*, 1988).

CCMV Replicase

CCMV differs from the related BMV notably in that the plants infected by the two viruses are not closely related. A template-dependent viral replicase complex has been isolated from CCMV-infected leaves (Miller and Hall, 1984) using conditions comparable to those utilized for isolation of the BMV replicase. Interestingly, the CCMV enzyme complex replicated BMV RNA as efficiently as CCMV RNA, yielding full-length products with either template. The reverse was not true. It has been suggested that this may reflect the situation encountered *in vivo* (Miller and Hall, 1984): when tobacco protoplasts were co-infected with BMV and CCMV, BMV multiplication dominated, as though it could use either replicase for its own benefit (Watts and Dawson, 1980).

The composition of the CCMV replicase has so far not been established.

AlMV Replicase

AlMV contains a tripartite genome, and its overall genome organization is comparable to that of the bromo- and cucumoviruses.

An AlMV RNA-dependent replicase has been isolated from AlMV-infected bean leaves (Howing and Jaspars, 1986). Its activity was dependent on added template and it could synthesize full-length complementary (-) RNA strands. It replicated BMV RNA about half as efficiently as AlMV. This is a rather unexpected observation, since the 3' region of the BMV and AlMV RNAs are very different: the BMV RNAs contain a tRNA-like structure at their 3' end, whereas no comparable structure is present in the AlMV RNAs.

No reports have appeared to date on the subunit composition of the AlMV replicase.

TMV Replicase

TMV contains a monopartite RNA genome. The total nucleotide sequence has revealed a long open reading frame interrupted by an amber (UAG) codon. This open reading frame encodes the nonstructural 126 kDa and 183 kDa proteins. These same proteins are also detected after *in vitro* and *in vivo* translation, the synthesis of the longer protein resulting from recognition of the UAG codon by a suppressor tRNA. Both non-structural proteins are thought to be involved in replication (Kamer and Argos, 1984; Haseloff *et al.*, 1984). The NTP-binding motif is contained in both proteins, and the GDD motif is contained in the 183 kDa readthrough protein.

To date, it has not been possible to isolate from TMV-infected cells an active replicase complex capable of initiating RNA synthesis *in vitro*. However, the intracellular location of the nonstructural proteins has been demonstrated by the immunogold labelling technique. The 126 and 183 kDa proteins are concentrated in granular inclusion bodies also referred to as "viroplasms" (Hills *et al.*, 1987; Saito *et al.*, 1987).

The radioactive photoaffinity analogues 8-azidoadenosine 5'-triphosphate (8-N_3ATP) and 8-azidoguanosine 5'-triphosphate (8-N_3GTP) have been used to determine whether a viral protein might incorporate the label in the presence of UV light (Evans *et al.*, 1985). When such experiments were performed using homogenates from TMV-infected tobacco plants, a 120 kDa protein was specifically labelled. Labelling was prevented by the presence of nucleoside triphosphates. It thus seems likely that the virus-induced 120 kDa protein is part of the replicase, and that it corresponds to the viral nonstructural 126 kDa protein that contains the NTP-binding motif. It would have been particularly interesting to determine whether the 183 kDa readthrough protein could also incorporate the labelled analogues, since the presence of this protein is observed *in vivo*. However, this aspect was not discussed.

Another approach to studying the components of the replicase has made use of mutagenized TMV RNA (Ishikawa *et al.*, 1988). *In vitro*-transcribed TMV RNAs modified near or within the UAG codon ending the 126 kDa protein gene have been used to infect plants. Mutants with an intact 126 kDa and a defective 183 kDa protein were not infectious, whereas an insertion of one amino acid in the 183 kDa protein did not affect infectivity. Removal of the amber codon abolished infectivity, whereas its replacement by the UAA (ochre) or the UAU (tyrosine-recognizing) codon did not. These results favour the idea that the 126 kDa and 183 kDa proteins are both involved in infectivity, and that a balanced level of synthesis of both is required for optimal virus multiplication.

CMV Replicase

CMV contains a tripartite genome. Its overall genome organization is similar to that of the bromoviruses.

A membrane-bound CMV replicase has been isolated from CMV-infected tobacco protoplasts. The replicase exclusively elongated (+) RNA chains initiated *in vivo*, and produced full-length RNA products. RNA replication occurred when protoplasts were inoculated with RNA1 and RNA2. The situation is thus similar to the one encountered with BMV. The composition of the enzyme complex is not known. However, the results obtained suggest that the translation products of RNA1 and RNA2 as well as membrane-bound components of the cell constitute the active enzyme complex (Nitta *et al.*, 1988).

CPMV Replicase

CPMV contains a bipartite RNA genome (B-RNA and M-RNA). Each RNA directs the synthesis of a polyprotein that undergoes post-translational cleavage to yield the finished products. M-RNA codes amongst other things for the viral structural proteins, whereas B-RNA encodes viral nonstructural proteins. The NTP-binding motif and the GDD motif are located in the 58 and 100 kDa polypeptides, respectively: these polypeptides derive from separate regions of the 200 kDa precursor polyprotein encoded by B-RNA.

A replication complex has been isolated from CPMV-infected cowpea leaves (Dorssers *et al.*, 1984). It was capable of elongating viral (+) RNA strands that had been initiated *in vivo*, and complete chains were recovered as dsRNA molecules (RF). Three major protein bands (110, 68 and 57 kDa) were detected in the replication complex. Using antibodies raised against the viral nonstructural proteins, it was demonstrated that the 110 kDa polypeptide was the only known viral polypeptide which was consistently associated with the replicase activity. The other two major protein bands were thus most likely of host origin. It is not known whether they participate in viral replication or were contaminants of the replication complex.

It thus appears that the 110 kDa viral polypeptide constitutes the "core" of the replicase. It is the one containing the GDD motif, and is also the one whose C-terminal part contains 21% sequence homology with the C-terminal part of the poliovirus replicase (protein 3D) that also harbours the GDD motif. The NTP-binding motif resides in the 58 kDa protein, but its possible involvement in replication has not been investigated.

Cytopathic structures appear in infected cowpea cells or protoplasts, composed of amorphous electron-dense material and membranous vesicles. The vesicles are the site of viral RNA replication (De Zoeten *et al.*, 1974). The electron-dense material contains B-RNA-derived nonstructural proteins as detected by immunocytochemical techniques (Wellink *et al.*, 1988). It is not yet known whether the electron-dense material is also involved in viral replication or whether it constitutes a reservoir for the deposit of inactive nonstructural proteins.

VTMoV Replicase

VTMoV contains a monopartite RNA genome designated RNA1. It also contains a viroid-like satellite RNA that is encapsidated in the form of a covalently closed circle (RNA2) or in the linear form (RNA3). An RNA replicase complex has been isolated from the membrane-free cytoplasmic fraction of VTMoV-infected tobacco plants. *In vitro* it appears to elongate (+) strand RNA1 and 3 that have been initiated *in vivo* on their (-) strand template (Rohozinski *et al.*, 1986). These results indicate that synthesis of the viroid-like RNA differs from synthesis of viroids. Indeed, viroids such as potato spindle tuber viroid are located in the nucleus and are replicated by the host DNA-dependent RNA polymerase system (Tabler and Tsagris, this volume).

The composition of the RNA replicase complex of VTMoV is unknown.

CONCLUSIONS

Considerable work remains to be done in the area of viral RNA replication. It is becoming increasingly urgent to characterize the protein subunits that make up the replicase. In particular, one would like to be able to dissect and reconstitute the active replicase as has been possible for the replicase of bacteriophage Qß. Progress along these lines would probably help in determining whether the same factors are required for (-) RNA and for (+) RNA synthesis. It is conceivable that failure to synthesize progeny (+) RNA strands could be due to a defect in the post-transcriptional modification of the (-) strands, possibly because of a defect in the replication complex. Experiments using CMV and its satellite (Collmer and Kaper, 1985) have revealed that in the dsRNAs isolated from infected plants, an unpaired G residue is located at the 3' end of the (-) RNA. The enzyme responsible for this 3'-terminal addition is unknown, and the nature of the 3'-terminal nucleotide of the *in vitro*-synthesized (-) RNA has also not been established.

The ability to obtain protoplasts that can be infected by viruses has been a major step forward in the study of virus development. Protoplasts constitute a simpler model than plant tissues and offer the advantage of studies on synchronized systems. It is with such systems that one can hope to tackle the problem of how the finely-tuned equilibrium is maintained between synthesis of (+) and (-) RNA strands, and between the level of genomic versus subgenomic RNAs that ultimately leads to highly efficient virus multiplication.

ACKNOWLEDGEMENTS

We are indebted to Francois Chapeville and to Fakher Ben Hamida for their constant interest and encouragement. R.G. is grateful to the Ministère des Relations Extérieures, France for a Fellowship. This work was financed in part by the A.I.P. No. 87/4560 - Obtention de Plantes Résistantes aux Virus. The Institute Jacques Monod is an Institut Mixte, CNRS - Université Paris 7.

REFERENCES

Ahlquist P, Bujarski JJ, Kaesberg P, Hall TC (1984) Localization of the replicase recognition site within brome mosaic virus RNA by hybrid-arrested RNA synthesis. Plant Mol Biol 3:37-44

Bar-Joseph M, Rosner A, Moscovitz M, Hull R (1983) A simple procedure for the extraction of double-stranded RNA from virus-infected plants. J Virol Methods 6:1-8

Candresse T, Mouchès C, Bové JM (1986) Characterization of the virus encoded subunit of turnip yellow mosaic virus RNA replicase. Virology 152:322-330

Candresse T, Batisti M, Renaudin J, Mouchès C, Bové JM (1987) Immunodetection of turnip yellow mosaic virus non-structural proteins in infected Chinese cabbage leaves and protoplasts. Ann Inst Pasteur/Virol 138:217-227

Collmer CW, Kaper JM (1985) Double-stranded RNAs of cucumber mosaic virus and its satellite contain an unpaired terminal guanosine: implications for replication. Virology 145:249-259

De Zoeten GA, Assink AM, Van Kammen A (1974) Association of cowpea mosaic virus-induced double-stranded RNA with a cytopathological structure in infected cells. Virology 59:341-355

De Zoeten GA, Powell CA, Gaard G, German TL (1976) *In situ* localization of pea enation mosaic virus double-stranded ribonucleic acid. Virology 70:459-469

Dodds JA, Jarupat T, Lee JG, Roistacher CN (1988) Effects of strain, host, time of harvest, and virus concentration on double-stranded RNA analysis of citrus tristeza virus. Phytopathology 77:442-447

Dorssers L, Van der Krol S, Van der Meer J, Van Kammen A, Zabel P (1984) Purification of cowpea mosaic virus RNA replication complex: identification of a virus-encoded 110,000-dalton polypeptide responsible for RNA chain elongation. Proc Natl Acad Sci USA 81:1951-1955

Evans RK, Haley BE, Roth DA (1985) Photoaffinity labeling of a viral induced protein from tobacco. Characterization of nucleoside-binding properties. J Biol Chem

260:7800-7804
Fraenkel-Conrat H (1983) RNA-dependent RNA polymerases of plants. Proc Natl Acad Sci USA 80:422-424
French R, Janda M, Ahlquist P (1986) Bacterial gene inserted in an engineered RNA virus: efficient expression in monocotyledonous plant cells. Science 231:1294-1297
Francki RIB (1987) Responses of plant cells to virus infection with special reference to the sites of RNA replication. In: Brinton MA and Rueckert RR (eds) Positive strand RNA viruses. Alan R Liss Inc., New York, p 423
Gadh IPS, Hari V (1986) Association of tobacco etch virus related RNA with chloroplasts in extracts of infected plants. Virology 150:304-307
Gargouri R, Joshi RL, Bol JF, Astier-Manifacier S, Haenni AL (1989) Mechanism of synthesis of turnip yellow mosaic virus coat protein subgenomic RNA *in vivo*. Virology, in press
Garnier M, Mamoun R, Bové JM (1980) TYMV RNA replication *in* vivo: replicative intermediate is mainly single stranded. Virology 104:357-374
Garnier M, Candresse T, Bové JM (1986) Immunocytochemical localization of TYMV-coded structural and nonstructural proteins by the protein A-gold technique. Virology 151:100-109
Goldbach R, Wellink J (1988) Evolution of plus-strand RNA viruses. Intervirology 29:260-267
Gorbalenya AE, Blinov VM, Donchenko AP, Koonin EV (1989) An NTP-binding motif is the most conserved sequence in a highly diverged monophyletic group of proteins involved in positive strand RNA viral replication. J Mol Evol 28:256-268
Haenni AL, Joshi S, Chapeville F (1982) tRNA-like structures in the genomes of RNA viruses. Prog Nucleic Acid Res Mol Biol 27:85-104
Hardy SF, German TL, Loesch-Fries LS, Hall TC (1979) Highly active template-specific RNA-dependent RNA polymerase from barley leaves infected with brome mosaic virus. Proc Natl Acad Sci USA 76:4956-4960
Harrison BD, Stefanac Z, Roberts IM (1970) Role of mitochrondria in the formation of X-bodies in cells of *Nicotiana clevelandii* infected by tobacco rattle viruses. J Gen Virol 6:127-140
Haseloff J, Goelet P, Zimmern D, Ahlquist P, Dasgupta R, Kaesberg P (1984) Striking similarities in amino acid sequence among nonstructural proteins encoded by RNA viruses that have dissimilar genomic organization. Proc Natl Acad Sci USA 81:4358-4362
Hatta T, Francki RIB (1978) Enzyme cytochemical identification of single-stranded and double-stranded RNAs in virus-infected plant and insect cells. Virology 88:105-117
Hatta T, Francki RIB (1981) Cytopathic structures associated with tonoplasts of plant cells infected with cucumber mosaic and tomato aspermy viruses. J Gen Virol 53:343-346
Hills GJ, Plaskitt KA, Young ND, Dunigan DD, Watts JW, Wilson TMA, Zaitlin M (1987) Immunogold localization of the intracellular sites of structural and nonstructural tobacco mosaic virus proteins. Virology 161:488-496
Hirth L, Givord L (1988) Tymoviruses. In: Koenig R (ed) The plant viruses. Plenum, New York, p 163
Horikoshi M, Mise K, Furusawa I, Shishiyama J (1988) Immunological analysis of brome mosaic virus replicase. J Gen Virol 69:3081-3087
Howing CJ, Jaspars EMJ (1986) Coat protein blocks the in vitro transcription of the virion RNAs of alfalfa mosaic virus. FEBS Lett 209:284-288

Ishikawa M, Meshi T, Motoyoshi F, Takamatsu N, Okada Y (1988) *In vitro* mutagenesis of the putative replicase genes of tobacco mosaic virus. Nucleic Acids Res 14:8291-8305

Jaspars EMJ, Gill DS, Symons RH (1985) Viral RNA synthesis by a particulate fraction from cucumber seedlings infected with cucumber mosaic virus. Virology 144:410-425

Jordan RL, Dodds JA (1985) Double-stranded RNA in detection of diseases of known and unproven viral etiology. Acta Hortic 164:101-108

Joshi RL, Ravel, JM, Haenni AL (1986) Interaction of turnip yellow mosaic virus Val-RNA with eukaryotic elongation factor EF-1α. Search for a function. EMBO J 5:1143-1148

Kääriäinen L, Takkinen K, Keränen S, Söderlund H (1987) Replication of the genome of alphaviruses. J Cell Sci Suppl 7:231-250

Kamer G, Argos P (1984) Primary structure comparison of RNA-dependent RNA polymerases from plant, animal and bacterial viruses. Nucleic Acids Res 12:7269-7282

Kiberstis PA, Loesch-Fries LS, Hall TC (1981) Viral protein synthesis in barley protoplasts inoculated with native and fractionated brome mosaic virus RNA. Virology 112:804-808

Lafleche D, Bové JM (1986) Sites d'incorporation de l'uridine tritiée dans les cellules du parenchyme foliaire de *Brassica chinensis*, saines ou infectées par le virus de la mosaïque jaune du navet. C R Acad Sci 266:1839-1841

Laflèche D, Bové C, Dupont G, Mouchès C, Astier T, Garnier M, Bové JM (1972) Site of viral RNA replication in the cells of higher plants: TYMV-RNA synthesis on the chloroplast outer membrane system. Proc Fed Eur Biochem Soc 27:43-71

Loesch-Fries LS, Hall TC (1980) Synthesis, accumulation and encapsidation of individual brome mosaic virus RNA components in barley protoplasts. J Gen Virol 47:323-332

Martelli GP, Russo M (1984) Use of thin sectioning for visualization and identification of plant viruses. Methods Virol 8:143-224

Miller WA, Hall TC (1983) Use of micrococcal nuclease in the purification of highly template dependent RNA-dependent RNA polymerase from brome mosaic virus-infected barley. Virology 125:236-241

Miller WA, Hall TC (1984) RNA-dependent RNA polymerase isolated from cowpea chlorotic mottle virus-infected cowpeas is specific for bromoviral RNA. Virology 132:53-60

Miller WA, Bujarski JJ, Dreher TW, Hall TC (1986) Minus-strand initiation by brome mosaic virus replicase within the 3' tRNA-like structure of native and modified RNA templates. J Mol Biol 187:537-546

Morch MD, Boyer JC, Haenni AL (1988) Overlapping open reading frames revealed by complete nucleotide sequencing of turnip yellow mosaic virus genomic RNA. Nucleic Acids Res 16:6157-6173

Mouchès C, Candresse T, Bové JM (1984) Turnip yellow mosaic virus RNA-replicase contains host and virus encoded subunits. Virology 134:78-91

Nassuth A, Bol JF (1983) Altered balance of the synthesis of plus- and minus-strand RNAs induced by RNAs 1 and 2 of alfalfa mosaic virus in the absence of RNA 3. Virology 124:75-85

Nilsson-Tillgren T, Kielland-Brandt MC, Bekke B (1974) On the biosynthesis of tobacco mosaic virus VI. On the subcellular localization of double-stranded viral RNA. Mol Gen Genet 128:157-169

Nitta N, Takanami Y, Kuwata S, Kubo S (1988) Inoculation with RNAs 1 and 2 of

cucumber mosaic virus induces viral RNA replicase activity in tobacco mesophyll protoplasts. J Gen Virol 69:2695-2700
Palukaitis P, Garcia-Arenal F, Sulzinski MA, Zaitlin M (1983) Replication of tobacco mosaic virus VII. Further characterization of single- and double-stranded virus-related RNAs from TMV-infected plants. Virology 131:533-545
Powell CA, De Zoeten GA (1977) Replication of pea enation mosaic virus RNA in isolated pea nuclei. Proc Natl Acad Sci USA 74:2922-2929
Quadt R, Verbeek HJM, Jaspars EMJ (1988) Involvement of a nonstructural protein in the RNA synthesis of brome mosaic virus. Virology 165:256-261
Rezelman G, Goldbach R, Van Kammen A (1980) Expression of bottom component RNA of cowpea mosaic virus in cowpea protoplasts. J Virol 36:366-373
Rezelman G, Franssen HJ, Goldbach RW, Ie TS, Van Kammen A (1982) Limits to the independence of bottom component RNA of cowpea mosaic virus. J Gen Virol 60:335-342
Rohozinski J, Francki RIB, Chu PWG (1986) The *in vitro* synthesis of velvet tobacco mottle virus-specific double-stranded RNA by a soluble fraction in extracts from infected *Nicotiana clevelandii* leaves. Virology 155:27-38
Russo M, Di Franco A, Martelli GP (1983) The fine structure of *Cymbidium* ringspot virus infections in host tissues III. Role of peroxisomes in the genesis of multivesicular bodies. J Ultrastruct Res 82:52-63
Saito T, Hosokawa D, Meshi T, Okada Y (1987) Immunological localization of the 130K and 180K proteins (putative replicase components) of tobacco mosaic virus. Virology 160:477-481
Sugimura Y, Ushiyama R (1975) Cucumber green mottle mosaic virus infection and its bearing on cytological alterations in tobacco mesophyll protoplasts. J Gen Virol 29:93-98
Ushiyama R, Matthews REF (1970) The significance of chloroplast abnormalities associated with infection by turnip yellow mosaic virus. Virology 42:293-303
Valverde RA, Dodds JA, Heick JA (1986) Double-stranded ribonucleic acid from plants infected with viruses having elongated particles and undivided genomes. Phytopathology 76:459-465
Van der Meer J, Dorssers L, Van Kammen A, Zabel P (1984) The RNA-dependent RNA polymerase of cowpea is not involved in cowpea mosaic virus RNA replication: immunological evidence. Virology 132:413-425
Van Kammen A (1985) The replication of plant virus RNA. Microbiol Sci 2:170-174
Watanabe Y, Okada Y (1986) *In vitro* RNA synthesis by a subcellular fraction of TMV-inoculated tobacco protoplasts. Virology 149:64-73
Watts JW, Dawson JRO (1980) Double infection of tobacco protoplasts with brome mosaic virus and cowpea chlorotic mottle virus. Virology 105:501-507
Wellink J, Van Lent J, Goldbach R (1988) Detection of viral proteins in cytopathic structures in cowpea protoplasts infected with cowpea mosaic virus. J Gen Virol 69:751-755
Zaitlin M, Hull R (1987) Plant virus-host interactions. Annu Rev Plant Physiol 38:291-315
Zelcer A, Weaber KF, Balazs E, Zaitlin M (1981) The detection and characterization of viral-related double-stranded RNAs in tobacco mosaic virus-infected plants. Virology 113:417-427

VIROID REPLICATION MECHANISMS

Martin Tabler[1] and Mina Tsagris[1]
Max-Planck-Institut für Biochemie
Abteilung Viroidforschung
D-8033 Planegg-Martinsried
Federal Republic of Germany

ABSTRACT

Viroids are pathogens which are the causal agents of some diseases of higher plants. At the molecular level, viroids consist only of a single-stranded, circular RNA molecule of 246-375 nucleotides. The autonomous replication of viroid RNA in the infected cell requires a close interaction with host factors. At least two replication mechanisms for different viroids can be envisaged. This chapter gives a brief review of possible mechanisms of viroid replication, summarizing our present knowledge of the synthesis of viroid RNA, the processing reactions which form the mature circular RNA molecule, and the relationships between viroid structure and replicability.

INTRODUCTION

Amongst the known autonomously replicating entities, the viroids represent a fascinating and challenging exception in several respects. Firstly, these unconventional plant pathogens are the only replicons with a circular RNA genome. Secondly, the molecular size of the viroid genome is about 10 times smaller than that of the smallest viruses. Thirdly, the infectious and pathogenic RNA is not encapsidated, but the RNA itself, which consists of one molecular species only, is the infectious agent. Finally, for molecular biologists who have become used to the idea that genetic information is primarily expressed in the form of polypeptides, it is striking that the viroid RNA replicon apparently does not code for any peptide at all. Nevertheless, despite their

---

[1] present address: Foundation for Research and Technology, Institute of Molecular Biology and Biotechnology, P.O. Box 1527, GR-71110 Iraklion/Crete, Greece

NATO ASI Series, Vol H 41
Recognition and Response in Plant-Virus Interactions
Edited by R S S Fraser

small molecular size and the lack of viroid-encoded proteins, the viroid genome contains a remarkably high degree of genetic information.

For expression of its genetic content the viroid RNA has to interact closely with the host, and presumably in a more intensive manner than any other parasitic replicon. The circular RNA molecule utilizes host factors for its own replication which are either pre-existing in the cell or induced after infection. The viroid RNA must, for example, contain adequate signals to become transported into the nucleus of the host cell. It must possess also a signal which allows cell-to-cell movement of this "naked" RNA during infection; a process for which tobacco mosaic virus and at least some other plant RNA viruses need to express a specific virus-encoded protein. Likewise, signals which determine host-specificity, pathogenicity and replicability must be encoded in the sequence of the viroid genome, or be provided by the resulting secondary and possibly also tertiary structure. Viroid RNA is, together with tRNA, the best characterized RNA species in terms of molecular structure (for review see Riesner, 1987). But our understanding of its biological function still lags far behind.

THE VIROID REPLICATION CYCLE

Molecular hybridization studies have revealed that the single-stranded circular viroid RNA, which is arbitrarily defined as of (+) polarity, replicates *via* the RNA-RNA route only (reviewed by Sänger, 1987). Cress *et al.* (1983) were first to show that cloned viroid DNA can, nevertheless, induce infections of viroid RNA, but there is no indication that the inoculated DNA can itself replicate.

Various models for viroid replication have been proposed by different laboratories. All involve two RNA-RNA replication steps and at least one processing reaction (reviewed by Sänger, 1987). These models, which are summarized in Fig. 1, take account of certain longer-than-unit-length RNAs of (-) and also of (+) polarity, which can be detected in leaves of viroid-infected plants (Spiesmacher *et al.*, 1983; Mühlbach *et al.*, 1983; Branch and Robertson, 1984; Ishikawa *et al.*, 1984; Hutchins *et al.*, 1985) and also in the cell nuclei where the replication actually proceeds (Spiesmacher *et al.*, 1985). These oligomeric RNAs, which are considered to be replicative intermediates, consist of up to about three monomeric units of (+) RNAs and up to about six units of the corresponding (-) RNAs. There is general agreement

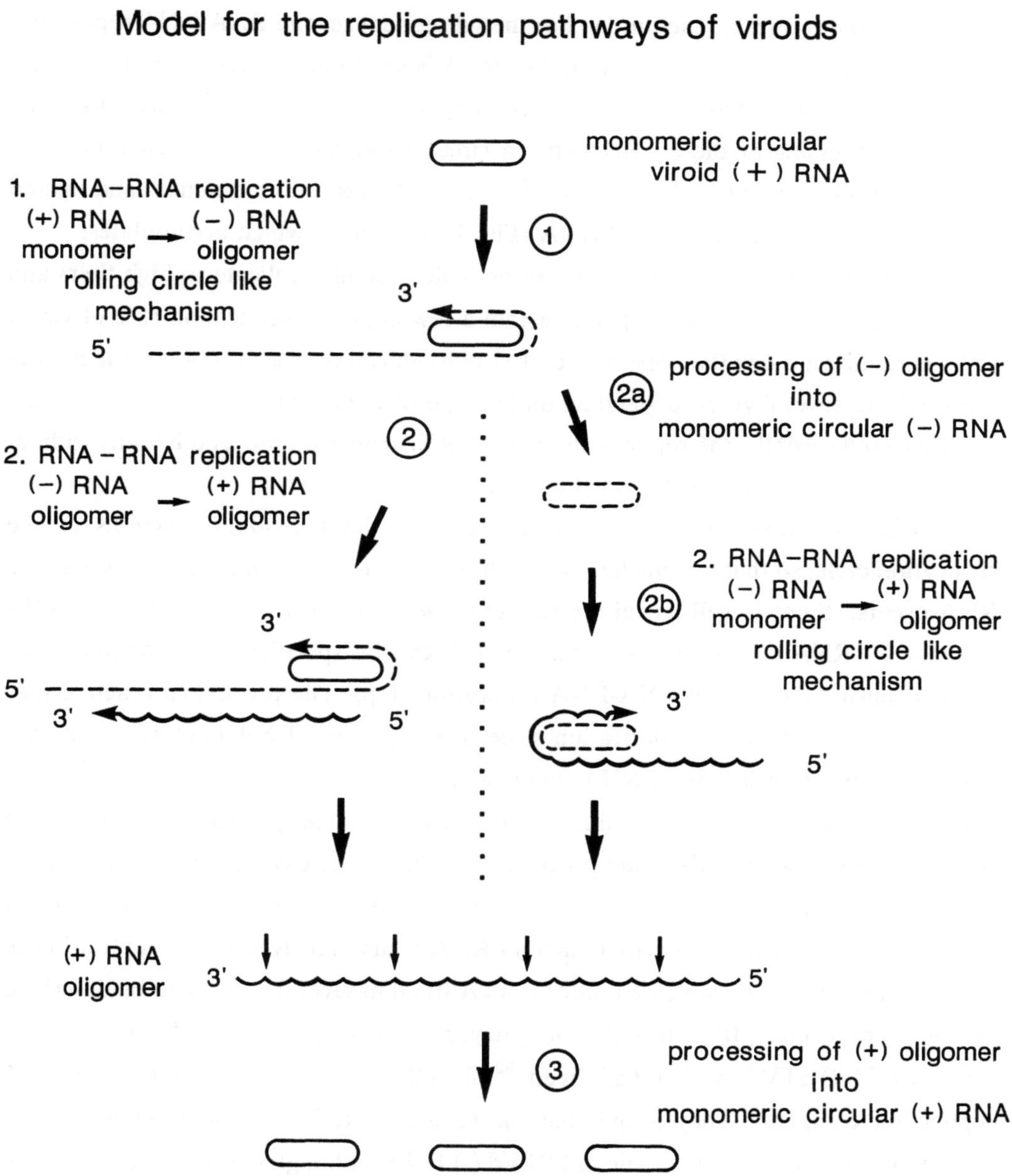

Figure 1. Schematic summary of the replication pathways of circular viroid RNA. For details see text.

that the oligomeric (-) RNA which can be detected by molecular hybridization is synthesized in a rolling-circle-like mechanism using the covalently closed (+) viroid

RNA as template (Fig. 1, reaction 1). For the subsequent second RNA-RNA replication step it is debatable whether oligomeric (+) RNA is synthesized directly on the nascent oligomeric (-) RNA template (Fig. 1, reaction 2) or in a second rolling-circle-like step after the processing of the (-) RNA into a circular monomer (Fig. 1, reaction 2a,b). Finally, the oligomeric (+) RNA is processed, i.e. cleaved into monomers and ligated to form the mature circular viroid RNA (Fig. 1, reaction 3) which accumulates.

Compared to most plant RNA viruses, which usually replicate to high titers and induce symptoms in susceptible plants within a few days, the replication rate of viroid RNA is relatively slow; the appearance of viroid-induced disease symptoms takes from two weeks to several years, depending on the type of viroid and the host plant species. As discovered recently, the replication of viroid RNA can even proceed latently without formation of disease symptoms (Puchta *et al.*, 1988).

Little is known about the stoichiometry of the individual reactions of the replication cycle, so that it is unclear at which step the actual amplification of the viroid RNA occurs. Since in cell nuclei the rate of *de novo* synthesis is higher for (+) RNA than for (-) RNA (Spiesmacher *et al.*, 1985), it can be speculated that amplification happens during the second RNA-RNA replication step. The processing reaction may also be considered as an amplification since one oligomeric RNA precursor molecule delivers several circular viroid RNA monomers.

As mentioned in consideration of Fig. 1, there is some speculation as to whether the oligomeric viroid (-) RNAs are processed into monomeric circles. In the case of the avocado sunblotch viroid (ASBVd), a circular (-) RNA has been detected (Hutchins *et al.* 1985), but it is unclear whether such an RNA is also involved in the replication of the other known viroids, which all differ from ASBVd in several aspects, to be described below. Very recently, Branch *et al.* (1988) fingerprinted circular RNA of potato spindle tuber viroid (PSTVd) which had been ^{32}P-labelled *in vivo* and isolated from the replication complex. They found that the separated RNA fragments contained no oligoribonucleotides characteristic of PSTVd (-) RNA, although these should have been detectable even in the presence of a 100-fold molar excess of the circular (+) RNA. Therefore, the authors concluded that replication of PSTVd proceeds without the formation of a monomeric circular (-) RNA. It has, however, previously been shown that oligomeric PSTVd (-) RNA can be processed into a circular monomeric PSTVd (-) RNA in an *in vitro* processing system derived from the nuclei of healthy potato cells (Tsagris *et al.* 1987b; see also below), so that the participation of a circular (-) RNA in the replication cycle of PSTVd cannot, in principle, be completely ruled out.

VIROID RNA SYNTHESIS

The enzymes catalyzing viroid RNA synthesis

Due to the lack of a viroid-encoded replicase, the viroid RNA must be replicated by a host RNA polymerase. Theoretically, it is conceivable that different RNA polymerases participate in viroid replication since it requires two RNA-RNA replication steps (Fig. 1).

The first indication that the DNA-dependent RNA polymerase II of the host cell is involved in viroid replication was provided by the observation that viroid RNA synthesis in protoplasts is inhibited by the fungal toxin alpha-amanitin (Mühlbach and Sänger, 1979). Later it was also shown that the purified enzyme from plants could accept the viroid RNA as template and "transcribe" it *in vitro* into full-length copies (Rackwitz *et al.*, 1981). Analysis by physical techniques of the complexes formed between viroid RNA and the DNA-dependent RNA polymerase II from wheat germ revealed that of all RNAs tested, the viroid RNA bound most strongly to the enzyme (Goodman *et al.*, 1984). This supports physico-chemical data showing that viroid RNA resembles DNA in some properties (for review see Riesner 1987).

When PSTVd-infected potato plants were used for the establishment of permanent cell suspension cultures capable of continuous viroid RNA synthesis (Mühlbach and Sänger, 1981), it became possible to study viroid replication by monitoring *de novo* synthesis in isolated and highly purified cell nuclei. Spiesmacher *et al.* (1985) showed that the synthesis of PSTVd (-) and (+) RNA proceeds in the nucleus and that alpha-amanitin can inhibit both steps of RNA-RNA replication, further supporting the involvement of the DNA-dependent RNA polymerase II in viroid replication.

From the observation that actinomycin D inhibits viroid (+) RNA synthesis quite effectively, but inhibits the synthesis of (-) RNA to a lesser extent, it was anticipated that DNA-dependent RNA polymerase I, which is less sensitive to this drug, might also be involved in the synthesis of PSTVd (-) RNA. Since the nucleolus is the nuclear compartment where the circular monomeric PSTVd RNA was found to accumulate (Schumacher *et al.*, 1983), and is also the site where the DNA-dependent RNA polymerase I transcribes the ribosomal genes, this concept was quite plausible. Recent and more detailed investigations, however, showed that the alpha-amanitin-sensitive RNA polymerase II is most likely responsible for the entire synthesis of viroid RNA (IM Schindler and HP Mühlbach, personal communication).

Specific initiation of viroid RNA synthesis by DNA-dependent RNA polymerase II

In order to analyze the specific interaction of viroid RNA and DNA-dependent RNA polymerase II further, the enzyme was isolated from tomato leaves and from wheat germ. *In vitro* "transcription" experiments tested whether the purified polymerases can specifically initiate viroid RNA synthesis using circular PSTVd as template, although the viroid cannot provide a promoter in the classical sense, due to its RNA nature. The corresponding RNA-RNA "transcription" was carried out in the presence of the antibiotic cordycepin (2'-deoxy-adenosinetriphosphate). Once incorporated, this stops chain elongation in a manner similar to dideoxy sequencing, to give a distinctive pattern of transcription products (Fig. 2). The length of the fragments was determined on a sequencing gel and could be unambiguously correlated with the known PSTVd sequence. By this experiment we showed that RNA polymerase II, from the host tomato and also from the non-host wheat, initiates the synthesis of PSTVd (-) RNA at nucleotide 168 of the circular PSTVd (+) RNA template (M Tsagris and HL Sänger, unpublished results). We therefore conclude that this is the origin of replication for PSTVd. We also used this approach to determine the polymerase initiation site of citrus exocortis viroid (CEVd) and chrysanthemum stunt viroid (CSVd), at nucleotides 49 and 300 respectively.

These data have confirmed the physical measurements of Goodman *et al.* (1984), who showed that DNA-dependent RNA polymerase II from wheat germ specifically recognizes a viroid RNA sequence and predominantly binds to the ends of the rod-shaped molecule, forming a 1:1 complex. Nevertheless, it was surprising that three viroids which share a great degree of sequence similarity and the same array of specific structural domains (Keese and Symons, 1985) should have the origin of replication at different parts of their rod-shaped molecule.

Since the circular viroid RNA template cannot be compared with the conventional double-stranded DNA promoter region, it is not unexpected that no characteristic upstream sequence elements such as TATA or CAAT boxes can be detected in the primary structures of the three viroids. We have not yet identified any structure common to the three viroids which would indicate why plant RNA polymerase

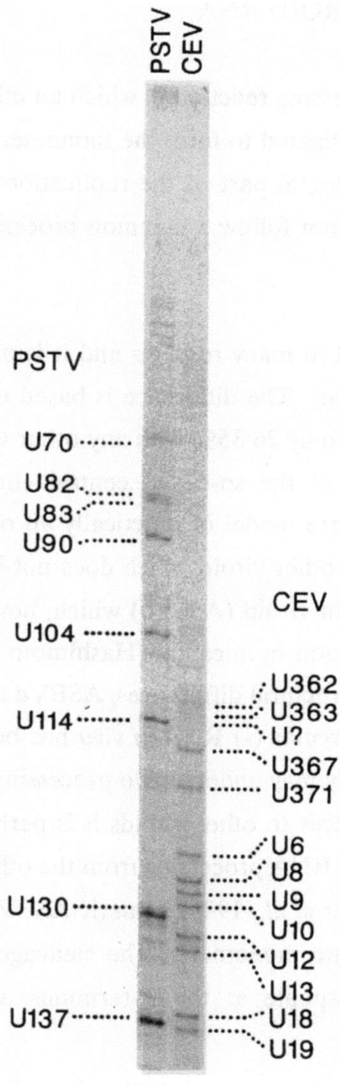

Figure 2. Gel electrophoretic analysis of viroid (-) RNAs synthesized on circular PSTVd and CEVd RNA templates by DNA-dependent RNA polymerase II from wheat germ, visualized by autoradiography. The reaction products were radioactively end-labelled by incorporation of ^{32}P-labelled cordycepin, an antibiotic (3'-deoxy ATP) which terminates chain elongation. The resulting bands could be correlated to specific U residues of the template RNA as indicated for PSTVd (left) and CEVd (right). By determining the length of the reaction products on different polyacrylamide gels, the initiation sites for viroid RNA synthesis could be identified.

II should specifically initiate at the sites determined.

THE PROCESSING OF VIROID RNA

The viroid RNA processing reaction in which an oligomeric (+) RNA is cleaved into a monomeric unit, then ligated to form the monomeric circular viroid RNA which finally accumulates, is an integral part of the replication cycle of each viroid species. Nevertheless, the viroids do not follow a common processing pathway.

The processing of ASBVd

ASBVd is exceptional in many respects and it is practically unrelated to all the other viroids discovered so far. The difference is based not only on the low degree of sequence similarity, which is only 26-35% with any other viroid (Keese *et al.*, 1988), but also on the complete lack of the so-called central conserved region found in the rod-shaped secondary structure model of practically all other viroids which have been sequenced. So far, the only other viroid which does not have this conserved sequence element is the apple scar skin viroid (ASSVd) which, however, contains a comparable symmetric domain at the region in question (Hashimoto and Koganezawa, 1987).

In addition to these structural differences, ASBVd is the only viroid for which the existence of a monomeric circular (-) RNA *in vivo* has been reported (Hutchins *et al.*, 1985) and which consequently may undergo two processing events during its replication.

In view of its differences to other viroids it is perhaps not too surprising that it has a different mechanism of RNA processing from the other viroids. As shown recently (Hutchins *et al.*, 1986; Forster *et al.*, 1988), dimeric ASBVd (+) and (-) RNAs are able to self-cleave autolytically into monomers. The cleavage occurs at a specific site and generates a 2',3'-cyclic phosphate at the 3'-terminus and a hydroxyl group at the 5'-terminus.

This mechanism of RNA cleavage is identical to that originally discovered for the satellite RNA of tobacco ringspot virus (STobRV) (Prody *et al.*, 1986). Forster and Symons (1987) noticed a common sequence element with similar secondary structure within self-cleaving RNAs including STobRV (+) RNA, ASBVd RNAs of both polarities, and also some circular satellite RNAs of plant viruses which are called virusoids (in order to demonstrate their structural relationship to viroids). It was shown that this so-called "hammerhead" structure is the active site for the self-cleavage reaction (Buzayan *et al.*, 1986b; Uhlenbeck, 1987; Haseloff and Gerlach, 1988; Koizumi *et al.*, 1988).

From the "hammerhead" concept, Forster and Symons (1987) also predicted the processing site for the virusoids of *Solanum nodiflorum* mottle virus (vSNMV) and velvet tobacco mottle virus (vVTMoV). They pointed out that the postulated cleavage site correlates with the unique 2'-phosphomonester, 3',5'- phosphodiester bond which had already been found in the RNA progeny of these two virusoids (Kiberstis *et al.*, 1985). The formation of such an untypical linkage was first described for the wheat germ RNA ligase (Konarska *et al.*, 1981) which connects 3'-termini with a 2',3'-cyclic phosphate and 5'-termini with either a hydroxyl or phosphate group (Konarska *et al.*, 1982). The occurrence of an "extra" 2'-phosphate in the progeny of two virusoids indicates, presumably, that a wheat-germ-type RNA ligase finally circularizes the linear unit-length RNA that has been released by self-cleavage *in vivo* from the oligomeric precursor of these satellite RNAs. The self-cleaved monomeric ASBVd RNA can also be ligated *in vitro* with wheat germ RNA ligase (Hutchins *et al.*, 1986), but so far no 2'-phosphate has been discovered in the natural progeny.

In this context it is interesting to note that ASBVd is not only an exception since it is the only viroid species for which self-cleavage has been unambiguously shown, but it is the only example so far of a non-satellite RNA amongst the hammerhead RNAs of plant origin. Since ASBVd seems not to require a helper virus for replication it belongs by definition to the viroids, but in many respects it resembles plant viral satellite RNAs and in particular the virusoids, more closely than it does the other viroids.

The processing of PSTVd and related viroids

In a number of reports and reviews of catalytic RNAs, (e.g. Green, 1988) the viroids are grouped with other examples of self-cleaving RNAs. This statement is not correct, at least in this generalized form.

From sequence analysis, it was suggested that PSTVd RNA could be related to Group I introns, since it contains sequence elements which can be correlated with the boxes characteristic of such introns (Diener, 1986; Dinter-Gottlieb, 1986; Hadidi, 1986). Robertson *et al.* (1985) reported that dimeric PSTVd RNA is able to self-cleave into a monomeric RNA, but not exactly according to the Group I mechanism, and with an efficiency of only 1-5%. In our own studies on the processing of oligomeric PSTVd RNA we originally observed a similar monomerization, and the 5'-end of the monomer could be determined (Tsagris *et al.*, 1987a). In fact, it turned out that the "self-cleavage" was due to a contaminating activity in the batch of ammonium sulphate used for these experiments. When we systematically excluded any possible contamination of our incubation mixture, no self-cleavage of oligomeric PSTVd RNA could be detected under a variety of reaction conditions (Tsagris *et al.*, 1987a). To date we have tested not only the reaction conditions described by Robertson *et al.* (1985), and those for Group I and Group II introns, but also those for hammerhead RNAs, those for the self-cleavage of the sTobRV (-) RNA (Buzayan *et al.*, 1986a) and a number of reaction conditions not described in the literature, involving different combinations of inorganic salts. In none of these cases could any self-cleavage, let alone any circularization, of PSTVd RNA be observed. In view of the surprising discoveries of catalytic activities of different RNAs in the last few years, we emphasize that our results do not exclude that self-cleavage of PSTVd RNA might still occur, but under reaction conditions which have not yet been defined.

However, we anticipated that oligomeric PSTVd (+) RNA may require some host factors for processing into the mature circular viroid RNA. In order to monitor viroid RNA processing we attempted to complement the PSTVd RNA with a nuclear extract, which should contain all the host factors necessary for processing, since the nucleus is the compartment where PSTVd replication proceeds. When such a soluble extract was prepared from potato suspension cells (Tsagris *et al.*, 1987b) or from tomato leaves (Tsagris, 1987), we observed specific monomerization of oligomeric PSTVd (+) RNA, and also the formation of circular PSTVd RNA. By appropriate controls it was shown that PSTVd RNA processing is catalyzed by one or several host proteins. Since the extracts were prepared from uninfected host cells, these enzymes must already pre-exist in the nucleus. The oligomeric (-) forms were also processed to circles when incubated in our *in vitro* processing system, although with a much lower efficiency than for processing of oligomeric PSTVd (+) RNA (Tsagris *et al.*, 1987b).

To examine whether other viroids follow the same mechanism of processing as PSTVd, we tested dimeric forms of CEVd and CSVd (+) RNA which indeed could also be correctly circularized (M Tsagris, M Hillmann and W Linker, unpublished results). For further confirmation of a common processing mechanism, all six possible combinations of artificial chimeric viroids were created by recombination of the *BamHI-AluI* and the *AluI-BamHI* fragments of PSTVd, CEVd and CSVd cDNA (W Linker, unpublished results). The RNAs derived from all dimeric forms of these chimeric constructs were processed *in vitro* with the same efficiency as the corresponding dimeric viroid RNAs, yielding in each case a circle of the expected size (M Tsagris and

W Linker, unpublished results). This finding reveals that the viroids which are related to PSTVd are circularized by the same processing mechanism.

The processing site of PSTVd (+) RNA

In order to identify the site of PSTVd (+) RNA processing, the linear monomeric RNA generated in the nuclear extract was used in primer extension reactions and in RNase H mapping experiments (Tsagris *et al.*, 1987b) for end-determination. These experiments showed that a population of different monomeric RNAs is generated. Amongst these linear unit-length molecules two species dominate, with 5'-ends at around nucleotide 95 and nucleotide 305 of the PSTVd sequence. Since we could not differentiate which of these linear molecules is actually a processing intermediate we had to test several domains of the PSTVd molecule in a more functionally oriented approach. For this, we took advantage of the observation in several infectivity studies with different viroid DNA and RNAs (summarized by Diener, 1986) that only longer-than-unit-length constructs can induce infections. These infectivity experiments show that viroid replication occurs only after the excision of a monomeric unit-length (+) RNA from the mechanically inoculated RNA transcript, and after the subsequent formation of a circular viroid RNA *in vivo*. In the case of inoculation with cloned viroid DNA a transcription reaction is required in addition. It is quite obvious that such an excision of a monomeric RNA *in vitro* and *in vivo* can only be expected if the inoculated precursor RNA itself is greater than unit-length, i.e. if it contains at least some sequence motifs of the circular viroid genome terminally repeated. For a precise monomerization, the two cleavage sites, one 5'-proximal and one 3'-proximal, must be separated by exactly one genome-length from each other, so that cleavage occurs within the identical sequence context of the terminal repeat.

In order to test whether the processing site of PSTVd is located in a particular domain of the molecule, we analyzed PSTVd RNA transcripts which were shorter than a dimer but longer than a monomer. These constructs consist of a unique region which is present just once in the RNA transcript and a region which is present twice, namely once at the 5'-end and once at the 3'-end (Fig. 3). If the processing site were to be located within the unique viroid sequence of a given construct no circular viroid RNA should be formed *in vitro*, and no infections should be observed after mechanical inoculation. If, however, a circular RNA has been formed out of a given RNA construct it should also be infectious, and we conclude that the processing site is located within the terminally extended region.

Various constructs were designed utilizing DNA clones described earlier (Tabler and Sänger, 1985) in order to test different parts of the PSTVd genome for the presence of the processing site (summarized in Fig. 3). The corresponding PSTVd (+) RNAs were synthesized *in vitro* with the help of SP6 or T7 RNA polymerase and subsequently tested for circularization by incubation with the enzymatically active nuclear extract, and for infectivity by mechanical inoculation. In all cases where circles were formed *in vitro*, the corresponding RNA construct proved to be highly infectious, which indicates that the *in vitro* processing system correlates with the bioassays. To our surprise, however, the results suggest that the cleavage/ligation reaction of PSTVd (+) RNA can operate at different parts of the genome. Therefore, we have not been able to identify a single "processing site".

Of particular interest in this context are the results of the processing assays obtained with the constructs duplicated in the so-called "central conserved region". The upper strand of this region ranges in the case of PSTVd from nucleotides 85 to 104 and it contains a number of symmetrical elements which play an important role in the formation of different secondary structures of oligomeric RNAs (Steger *et al.*, 1986). It has been surmized earlier that this viroid domain is involved in viroid RNA processing, since it was observed that a vector-inserted monomeric viroid cDNA unit which does not induce infections becomes infectious when the eleven nucleotides between nucleotides 87-97 are terminally repeated (Tabler and Sänger, 1984). From these data and the results of infectivity tests from other laboratories, Diener (1986) developed a model suggesting that viroid RNA processing occurs in this particular region.

From experiments in which a mutation was introduced into CEVd, Visvader *et al.* (1985) claimed that the *in vivo* processing of longer-than-unit length CEVd (+) RNAs occurs at nucleotides 98-100 which corresponds to nucleotides 95-97 in PSTVd. Surprisingly, the particular construct pSP-Ba(-) with the same eleven nucleotide repeat of nucleotides 87-97 used by Visvader *et al.* (1985), and the related construct pHa97, do not form detectable amounts of circular PSTVd RNA in the *in vitro* processing system (Tsagris *et al.*, 1987b and unpublished results). In line with this observation the specific infectivity of these RNA constructs is relatively low. The RNA transcripts of constructs pHa91 and pHa88 in which the terminally duplicated sequence is much shorter do, however, form circular RNA and are also highly infectious. Although the possible influence of vector-derived sequences flanking the viroid constructs cannot be completely

Construct	5'- terminal region	unique region	3'- terminal region	in vitro processing	in vivo infectivity
pSP-AvI.8(-)	282-359/1-223	224-281	282-359/1-223	+	+++
pA197	282-359/1-97	98-281	282-359/1-97	+	+++
pA191	281-359/1-91	92-281	281-359/1-91	+	+++
pA188	282-359/1-88	89-281	282-359/1-88	+	+++
pA178	282-359/1-78	79-281	282-359/1-78	+	++
pA1342	282-342	343-359/1-281	282-342	+	+++
pBs281	102-281	282-359/1-101	102-281	+	+++
pBa281	87-281	282-359/1-86	87-281	+	+++
pHa106	85-106	107-359/1-84	85-106	+	+++
pHa97	83-97	98-359/1-82	83-97	−	+
pSP-Ba(-)	87-97	98-359/1-86	87-97	−	+
pHa91	83-91	92-359/1-82	83-91	+	+++
pHa88	83-88	89-359/1-82	83-88	+	+++
pBs106	102-106	107-359/1-101	102-106	−	++

Figure 3. Summary of processing tests and bioassays of different PSTVd-specific RNA transcripts with terminally repeated sequence duplications. The partially duplicated PSTVd-RNAs tested can each be divided into a unique region which is represented once in the RNA transcript and the duplicated region which is present twice, namely as a 5'- and 3'-terminal repeat. The numbers given in the figure represent the PSTVd nucleotides and indicate the extension of the corresponding regions of each transcript. The processability of a particular RNA transcript in the *in vitro* processing system is indicated by + (yes) and - (no). The infectivity of the RNAs after mechanical inoculation to tomato is indicated by: + + +, the inoculated RNA caused disease on all test plants within the time necessary to induce infections by inoculation with an oligomeric PSTVd RNA; + +, all plants became diseased but symptom expression was retarded; +, only some of the test plants showed PSTVd symptoms.

ruled out, our combined results nevertheless clearly show that viroid processing can occur at different domains of the molecule, even outside of the central conserved region. This is not completely unexpected, in so far as Hashimoto *et al.* (1986) found that any monomeric linear PSTVd RNA can be infectious, provided it has a 2',3'-cyclic phosphodiester at its 3'-terminus so that it can be ligated to circles by a RNA ligase of the wheat germ type. Therefore, any cleavage reaction at different domains of the PSTVd molecule releasing a monomer with a 2',3'-cyclic phosphodiester at its 3'-terminus will finally lead to the formation of circular RNA, especially since the secondary structure of the linear viroid RNA brings the 3'- and 5'-termini in close proximity to each other. It was shown earlier that linear viroid RNA is an appropriate substrate for such a RNA ligase (Kickuchi *et al.*, 1982; Branch *et al.*, 1982).

Our observation that processing of PSTVd (+) RNA, as tested with various constructs, can operate at different sites of the molecule, should not be interpreted to mean that the processing reaction is more or less random during the replication cycle of viroid RNA *in vivo*. In view of the strict conservation in size, at least in the case of PSTVd, the processing reaction must be rather specific and it presumably does involve the central conserved region. This assumption is based on physico-chemical data which showed that in an oligomeric (+) RNA, this domain is capable of forming 28 stable base pairs, either inter- or intramolecularly, and thereby structurally divides the oligomeric molecule into monomeric units which might be released by the processing enzymes of the host (Steger *et al.*, 1986). Moreover, the involvement of the central conserved region is substantiated by the exceptionally high specific infectivity of the RNA transcript of clone pHa106, which contains the entire domain in terminally duplicated form.

It will be important for the detailed understanding of the viroid RNA processing reactions to identify and characterize the host ribonuclease responsible for the specific cleavage reaction, and the enzyme which catalyzes the circularization. In that context, it is interesting to mention our recent processing experiments with an endoribonuclease, RNase T1, which cleaves the 3'-5'-phosphodiester bond of RNA through an intermediate with a 2',3'-cyclic phosphodiester linkage, which is subsequently hydrolyzed to 3'-phosphate by the enzyme. RNase T1 can not only specifically monomerize oligomeric PSTVd RNA *in vitro*, but the enzyme can also join, in a reversed cleavage reaction, the two ends of the resulting unit-length RNA to form a circular viroid RNA molecule (M Tsagris and M Tabler, unpublished results). The mechanism of this reaction is at present under investigation in our laboratory. In view of these results it is conceivable that a host ribonuclease is catalyzing the entire processing reaction, i.e. cleavage and ligation, of the RNA of PSTVd and related viroids.

VIROID STRUCTURE AND REPLICABILITY

The replication of viroids requires, as detailed in the previous sections, an intimate interaction of the RNA with host factors which are practically unknown, apart from the DNA-dependent RNA polymerase II and a nuclear 41 kDa protein (Wolff *et al.*, 1985), the function of which has yet to be identified. Very recently it was suggested that viroid RNA might interact with the host 7S RNA which is part of the signal recognition particle (SRP). This hypothesis is based on sequence data of tomato 7S RNA which contains a stretch of 36-53 nucleotides exhibiting sequence motifs complementary to five viroids that can replicate in tomato (Haas *et al.*, 1988). As pointed out by these authors, the possible interaction between viroid RNA and 7S RNA should primarily incite disease and therefore affect pathogenicity of viroids rather than be involved in the replication cycle. This is especially likely since the 7S RNA is located in the cytoplasm, in contrast to the viroid's nuclear location.

No matter what the host factors interacting with viroid RNA are, the signals actually controlling replicability are given in the viroid sequence itself and the resulting secondary structure. The complications of unravelling these signals have been illustrated in the case of the determination of the replication origins, which are different for related viroids.

When infectious cDNAs became available, it was expected that the study of the relationship between viroid structure and function could be approached by the introduction of site-specific mutations. To date, the mutagenesis experiments which have been carried out have revealed that the relationship of viroid structure to replicability is remarkably complex. Our understanding of the biological signals encoded in a viroid sequence is therefore still frustratingly poor. Owens *et al.* (1986) found that a single point mutation at various sites on the PSTVd molecule could abolish infectivity. Amongst these mutations the C→U transition at position 284 of the PSTVd sequence was of special interest, since this nucleotide substitution should have no significant effect on the secondary structure, where according to the conventional model, only a G-C base-pair is replaced by a G-U base-pair (formed with the G-residue at position 76). In a subsequent study from the same laboratory (Hammond and Owens, 1987), it was shown that infectivity was restored when the compensatory nucleotide exchange G→A at position 76 was made, which results in a pseudo-recombinant now forming an A-U base-pair. In the same report it was shown that compensating mutations within hairpin I of the PSTVd molecule were, however, not viable.

Our own mutation studies on PSTVd revealed three groups of mutants which are summarized in Fig. 4. As also observed by Owens and co-workers, some point mutations lead to a complete loss of infectivity. Within this group of mutations a single-nucleotide substitution interferes with replicability.

In the second group of mutations we observed infections in several test plants, mostly after a lag phase which varied in length for individual plants. The development of a microscale sequencing technique (Tabler *et al.*, 1989) allowed us to determine the viroid sequences present in an individual test plant. We found that the viroid RNA progeny contained, in these cases, only PSTVd RNA which had reverted to the wild-type sequence. We conclude that in this group of PSTVd mutants the altered viroid RNA is able to replicate at a reduced level unless spontaneous reversion occurs and the infection proceeds at a normal rate.

The third group of mutants involves sequence changes within the so-called virulence modulating (VM) region (Schnölzer *et al.,* 1985). The sequence analysis revealed that the PSTVd progeny which had been formed was neither identical to the inoculated RNA, nor to the wild-type, but that instead a new PSTVd sequence variant had been generated *in vivo*. In view of these rather unexpected results, the primary sequence of the DNA preparation of the construct VM2, which was used as template for the SP6 RNA transcription reaction, was reconfirmed and the infectivity test repeated several times. In all cases tested, the same newly mutated progeny was formed, which itself was stable for at least three further viroid transfers from plant to plant. In addition to the altered sequence, we also observed within this group of PSTVd mutants a different virulence of the newly created PSTVd variant as compared to the wildtype PSTVd.

The results obtained with different mutants of PSTVd indicate that viroids replicate under a strong selection of their sequence. Even minor changes in their primary structure interfere with replicability, which is either completely blocked or largely impaired. Only the VM-region seems to tolerate sequence alterations to some extent, which is also in line with the observation that the naturally occurring sequence variants differ primarily in this domain of the PSTVd molecule. But the "evolution" of the inoculated mutant towards a specific sequence shows that this region too is under selection.

These results suggest that each individual nucleotide of a viroid RNA seems to contribute to the genetic signals which finally lead to replication of the circular molecule. In that respect the viroid RNA can be considered genetically as a much more condensed replicon than viruses which express their genetic information *via* the conventional protein pathway. In order to obtain a better understanding of viroid replication it will be necessary to identify and characterize the host factors which interact

A	Lethal mutations				Reversible mutations				
Name of mutant	LC1	RR5	UC1	UC2	SS7	IP1	IP2	OF1	OF2
Nucleotide number	269	82	91	98	89	168	167	208	153
Wildtype nucleotide	G	U	C	G	A	G	C	C	G
Mutant nucleotide	C	A	G	C	U	A	G	U	A

Effect of point mutations introduced into PSTVd

B	Mutant VM2			
Nucleotide numbers	308			319
Wildtype sequence	UAA		CUUGUUCCG	
Mutant sequence	UAA	U	CUUGUU CG	
Progeny sequence	UAA	(A)	CUUGUU CG	

Mutant VM6					
44		50	308		317
GAC	A	AGA	UA	A	CUUGUUC
GAC	A	AGA	UA	A	CCCGCUC
GAC	(G)	AGA	UA	(U)	CCCGCUC

Effect of mutations introduced into the VM-region of PSTVd

Figure 4. Effect of mutations introduced into PSTVd. Part A indicates the position and the nucleotide changes introduced into PSTVd. The bioassay revealed that four mutations were lethal for PSTVd as indicated, and that the other five mutations lead to PSTVd progeny that reverted to the wildtype sequence during replication. Part B depicts the sequence of the wildtype PSTVd RNA, the mutated form and the progeny obtained after mechanical inoculation. The mutations introduced by mutagenesis are indicated by arrows and the mutations which occurred during replication are marked by circles.

with these unconventional replicating RNAs.

ACKNOWLEDGEMENTS

The experiments described were carried out at the Max-Planck-Institute for Biochemistry, in the Department of Viroid Research headed by Heinz L Sänger. We are very grateful to Prof Sänger for his continuous support and for his stimulating and helpful advice throughout the project. We thank Prof Mühlbach for communicating unpublished data, Miss I Günther and U Eisenrieth for excellent technical assistance, and Miss Cathy Liptrot for proofreading the manuscript. We finally thank Dr Fraser for his invitation to present our data at the NATO Workshop. This work was supported by an Otto Hahn fellowship of the Max-Planck-Society to M Tabler and by an EMBO long term fellowship to M Tsagris.

REFERENCES

Branch AD, Robertson HD (1984) A replication cycle for viroids and other small infectious RNAs. Science 223:450-455

Branch AD, Robertson HD, Greer P, Gegenheimer P, Peebles C, Abelson J (1982) Cell-free circularization of viroid progeny by an RNA ligase from wheat germ. Science 217: 1147-1149

Branch AD, Benenfeld BJ, Robertson HD (1988) Evidence for a single rolling circle in the replication of potato spindle tuber viroid. Proc Natl Acad Sci USA 85:9128-9132

Buzayan JM, Gerlach WL, Bruening G (1986a) Spontaneous ligation of RNA fragments with sequences that are complementary to a plant virus satellite RNA. Nature 323: 349-353

Buzayan JM, Gerlach WL, Bruening G (1986b) Satellite tobacco ringspot virus RNA: A subset of the RNA sequence is sufficient for autolytic processing. Proc Natl Acad Sci USA 83:8859-8862

Cress DE, Kiefer MC, Owens R (1983) Construction of infectious potato spindle tuber viroid cDNA clones. Nucleic Acids Res 11:6821-6835

Diener TO (1986) Viroid processing: A model involving the central conserved region and hairpin I. Proc Natl Acad Sci USA 83:58-62

Dinter-Gottlieb G (1986) Viroids and virusoids are related to group I introns. Proc Natl Acad Sci USA 83:6250-6254

Forster AC, Symons RH (1987) Self-cleaving of plus and minus RNAs of a virusoid and a structural model for the active sites. Cell 49:211-220.

Forster AC, Davies C, Sheldon CC, Jeffries AC, Symons RH (1988) Self-cleaving viroid and newt RNAs may only be active as dimers. Nature 334:265-267

Goodman TC, Nagel L, Rappold W, Klotz G, Riesner D (1984) Viroid replication: equilibrium association constant and comparative activity measurements for the viroid-polymerase interaction. Nucleic Acids Res 12:6231-6246

Green MR (1988) Mobile RNA catalysts. Nature 336:716-718

Haas B, Klanner A, Ramm K, Sänger HL (1988) The 7S RNA from tomato leaf tissue

resembles a signal recognition particle RNA and exhibits a remarkable sequence complementarity to viroids. EMBO J 7:4063-4074
Hadidi A (1986) Relationship of viroids and certain other plant pathogenic nucleic acids to group I and II introns. Plant Mol Biol 7:129-142
Hammond RW, Owens RA (1987) Mutational analysis of potato spindle tuber viroid reveals complex relationships between structure and infectivity. Proc Natl Acad Sci USA 84: 3967-3971
Haseloff J, Gerlach WL (1988) Simple RNA enzymes with new and highly specific endoribonuclease activities. Nature 334:585-591
Hashimoto J, Suzuki K, Uchida T (1986) Infectivity of artificially nicked viroid molecules. J Gen Virol 66:1545-1551
Hashimoto J, Koganezawa H (1987) Nucleotide sequence and secondary structure of apple scar skin viroid. Nucleic Acids Res 15:7045-7052
Hutchins CJ, Keese P, Visvader JE, Rathjen PD, McInnes JL, Symons RH (1985) Comparison of multimeric plus and minus forms of viroids and virusoids. Plant Mol Biol 4:293-304
Hutchins CJ, Rathjen PD, Forster AC, Symons RH (1986) Self-cleavage of plus and minus RNA transcripts of avocado sunblotch viroid. Nucleic Acids Res 14:3627-3640
Ishikawa M, Meshi T, Ohno T, Okada Y, Sano T, Ueda I, Shikata E (1984) A revised replication cycle for viroids: The role of longer than unit length RNA in viroid replication. Mol Gen Genet 196:421-428
Keese P, Symons RH (1985) Domains in viroids: evidence of intermolecular RNA rearrangements and their contribution to viroid evolution. Proc Natl Acad Sci USA 82:4582-4586
Keese P, Visvader JE, Symons RH (1988) Sequence variability in plant viroid RNAs. In: Domingo E, Holland JJ, Ahlquist P (eds) RNA genetics Vol. III. CRC Press, Boca Raton, p 71
Kiberstis PA, Haseloff J, Zimmern D (1985) 2'-phosphomonoester 3',5-phosphodiester bond at a unique site in a circular viral RNA. EMBO J 4:817-827
Kikuchi Y, Tyc K, Filipowicz W, Sänger HL, Gross HJ (1982) Circularization of a linear viroid RNA via 2'-phospho-monoester, 3',5'-phosphodiester bonds by a novel type of RNA ligase from wheat germ and Chlamydomonas. Nucleic Acids Res 10:7521-7529
Koizumi M, Iwai S, Ohtsuka E (1988) Construction of a series of several self-cleaving RNA duplexes using synthetic 21-mers. FEBS Lett 228:228-230
Konarska M, Filipowicz W, Domdey H, Gross HJ (1981) Formation of a 2'-phosphomonoester, 3',5'-phosphodiester linkage by a novel RNA ligase in wheat germ. Nature 293:112-116
Konarska M, Filipowicz W, Gross HJ (1982) RNA ligation via 2'-phosphomonoester, 3'5'-phosphodiester linkage: requirement of 2' 3'-cyclic phosphate termini and involvement of a 5'- hydroxyl polynucleotide kinase. Proc Natl Acad Sci USA 79: 1474-1478
Mühlbach HP, Sänger HL (1979) Viroid replication is inhibited by alpha-amanitin. Nature 278:185-188
Mühlbach HP, Sänger HL (1981) Continuous replication of potato spindle tuber viroid (PSTV) in permanent cell cultures of potato and tomato. Biosci Rep 1:79-87
Mühlbach HP, Faustmann O, Sänger HL (1983) Conditions for optimal growth of a PSTV-infected potato cell suspension and detection of viroid-complementary longer-than-unit-length RNA in these cells. Plant Mol Biol 2:239-247
Owens RA, Hammond RW, Gardner RC, Kiefer MC, Thompson SM, Cress DE (1986)

Site-specific mutagenesis of potato spindle tuber viroid cDNA: alterations within premelting region 2 that abolish infectivity. Plant Mol Biol 6:179-192
Prody GA, Bakos JT, Buzayan JM, Schneider IR, Bruening G (1986) Autolytic processing of dimeric plant virus satellite RNA. Science 231:1577-1580
Puchta H, Ramm K, Sänger HL (1988) The molecular structure of hop latent viroid (HLV), a new viroid occurring worldwide in hops. Nucleic Acids Res 16:4197-4216
Rackwitz HR, Rohde W, Sänger HL (1981) DNA-dependent RNA polymerase II of plant origin transcribes viroid RNA into full-length copies. Nature 291-297-301.
Riesner D (1987) Physical-chemical properties: structure formation. In: Diener TO (ed) The viroids. Plenum Press, New York, p 63
Robertson HD, Rosen DL, Branch AD (1985) Cell-free synthesis and processing of an infectious dimeric transcript of potato spindle tuber viroid RNA. Virology 142: 441-447
Sänger HL (1987) Viroid replication. In: Diener TO (ed) The viroids. Plenum Press, New York, p 117
Schnölzer M, Haas B, Ramm K, Hofmann H, Sänger HL (1985) Correlation between structure and pathogenicity of potato spindle tuber viroid (PSTV). EMBO J 4:2181-290
Schumacher J, Sänger HL, Riesner D (1983) Subcellular localization of viroids in highly purified nuclei from tomato leaf tissue. EMBO J 2:1549-1555
Spiesmacher E, Mühlbach HP, Schnölzer M, Haas B, Sänger HL (1983) Oligomeric forms of potato spindle tuber viroid (PSTV) and its complementary RNA are present in nuclei isolated from viroid-infected potato cells. Biosci Rep 3:767-774
Spiesmacher E, Mühlbach HP, Tabler M, Sänger HL (1985) Synthesis of (+) and (-) RNA molecules of potato spindle tuber viroid (PSTV) in isolated nuclei and its impairment by transcription inhibitors. Biosci Rep 5:251-265
Steger G, Tabler M, Brüggemann W, Colpan, Klotz G, Sänger HL, Riesner D (1986) Structure of viroid replicative intermediates: physico-chemical studies on SP6 transcripts of cloned oligomeric potato spindle tuber viroid. Nucleic Acids Res 14:9613-9630
Tabler M, Sänger HL (1984) Cloned single- and double-stranded DNA copies of potato spindle tuber viroid (PSTV) RNA and co-inoculated subgenomic DNA fragments are infectious. EMBO J 3:3055-3062
Tabler M, Sänger HL (1985) Infectivity studies on different potato spindle tuber viroid (PSTV) RNAs synthesized in vitro with the SP6 transcription system. EMBO J 4:2191-2199
Tabler M, Günther I, Kern R, Sänger HL (1989) A microscale procedure for isolating and sequencing the viroid RNA present in one gram of infected leaf tissue. J Virol Meth 23:111-126
Tsagris EM (1987) Prozessierung der RNA des Potato Spindle Tuber Viroids (PSTV) in vitro. PhD thesis, University of Giessen, Federal Republic of Germany.
Tsagris M, Tabler M, Sänger HL (1987a) Oligomeric potato spindle tuber viroid (PSTV) RNA does not process autocatalytically under conditions where other RNAs do. Virology 157:227-231
Tsagris M, Tabler M, Mühlbach HP, Sänger HL (1987b) Linear oligomeric potato spindle tuber viroid (PSTV) RNAs are accurately processed in vitro to the monomeric circular viroid proper when incubated with a nuclear extract from healthy potato cells. EMBO J 6:2173-2183
Uhlenbeck OC (1987) A small catalytic oligoribonucleotide. Nature 328:596-600
Visvader JE, Forster AC, Symons RH (1985) Infectivity and in vitro mutagenesis of

monomeric cDNA clones of citrus exocortis viroid indicates the site of processing of viroid precursors. Nucleic Acids Res 13:5843-5846
Wolff P, Gilz R, Schumacher J, Riesner D (1985) Complexes of viroids with histones and other proteins. Nucleic Acids Res 13:355-367

THE MOVEMENT PROTEIN OF SOME PLANT VIRUSES

T Godefroy-Colburn, F Schoumacher, C Erny, A Berna, O Moser, M-J Gagey and C Stussi-Garaud
Institut de Biologie Moléculaire des Plantes du CNRS
12 rue du Général Zimmer
67000 Strasbourg
France

ABSTRACT

Plant viral infections spread from cell to cell through plasmodesmata. However, normal plasmodesmata do not allow the transit of the infectious material and viral infection often modifies their structure. Cell-to-cell spread (movement, transport) requires a virus-coded protein which presumably carries out the modification. By Western blot and immunoelectron microscopy, the movement proteins of alfalfa, tobacco and cauliflower mosaic viruses were detected in cell walls and/or plasmodesmata of systemic or hypersensitive hosts, supporting the assumptions made above. To prepare the spread factors of alfalfa mosaic virus (P3, 32 kDa) and tobacco mosaic virus (30 kDa) in the amount required for biochemical studies, we expressed those proteins in *Saccharomyces cerevisiae* using a shuttle expression vector. A 12-amino acid N-terminal deletion of P3 was also produced and purified to near homogeneity.

INTRODUCTION

The most obvious difference between a plant cell and an animal cell is the pectocellulosic cell wall of the former. This has implications in virology. In spite of numerous similarities between plant and animal viruses, e.g. in genomic organization, strategy of expression, mechanism of replication, and protein sequence, plant viruses have a requirement which does not apply to animal viruses: the progeny particles synthesized in the initially-infected cells must be able to cross the cell wall barrier in order to infect the neighbouring healthy cells and, in the case of a systemic infection, invade the whole plant through the vascular system. Thus beside penetration, uncoating, translation and replication of the inoculated virus, and encapsidation of the progeny

NATO ASI Series, Vol H 41
Recognition and Response in Plant-Virus Interactions
Edited by R S S Fraser

particles, the "successful" viral infection of an intact plant must include at least two additional steps: cell-to-cell movement (also called cell-to-cell spread or short-distance virus transport) and systemic spread (or long-distance virus transport). Virus movement is obviously a very important parameter of plant pathogenesis. Thus if means were found to inhibit it in crop plants the losses due to viral infections might be drastically reduced.

This paper will be mainly restricted to cell-to-cell movement. It is not intended to be exhaustive and, to simplify the presentation, it will overlook some of the hypotheses made to account for the particulars of virus spread. These have been thoroughly reviewed (Atabekov and Dorokhov, 1984; Zaitlin and Hull, 1987; Hull, 1989; JG Atabekov and ME Taliansky, submitted manuscript).

The physical constraints of cell-to-cell movement

It is generally taken for granted that the infectious particles move from cell to cell through plasmodesmata, which are strands of cytoplasm connecting adjacent cells through the cell walls. Numerous observations of viral particles in plasmodesmata support this point of view (Esau, 1968; Gibbs, 1976; Weintraub *et al.*, 1976; Francki *et al.*, 1985). However, as pointed out as early as 1968 by Esau and again by Gibbs (1976), virus movement does not appear trivial when one compares the relative sizes of plant viruses and of typical plasmodesmata.

Electron microscopic observation shows that plasmodesmata connecting parenchyma cells of healthy tissue have a diameter of 30 to 80 nm along most of their path, i.e. 2 to 4 times the diameter of the smallest plant viruses. However, they often exhibit constricted apertures at either or both ends and in general the plasmodesmatal space is partially occupied by the desmotubule, a membranous structure in continuity with the endoplasmic reticulum of the two cells (Robards, 1976). This leaves an annular space which is the most likely pathway for the movement of solutes. In many tissues, the cytoplasmic annulus appears to have a width of 5 to 20 nm after fixation and it may contain particles which hold the desmotubule in place, leaving irregular channels for the passage of solutes (Overall *et al.*, 1982; Thomson and Platt-Aloia, 1985). As probed by microinjection of fluorescent conjugates, the molecular exclusion limit of symplastic movement (through plasmodesmata) is 700 to 1000 daltons (Terry and Robards, 1987, and references therein), clearly much lower than the size of any plant virus or even of any plant viral genome.

It is thus difficult to avoid the assumption that viral infection changes the structure and/or function of plasmodesmata to allow the movement of the infectious agent between cells, or that it induces the formation of new plasmodesmata with different properties. The latter hypothesis is unlikely, at least in the case of tobacco mosaic virus (TMV) infection (Shalla *et* al., 1982), therefore we are left with the former. Surprisingly, some thoroughly-studied viruses such as tobamoviruses and alfalfa mosaic virus (AlMV) have never been shown to modify plasmodesmata. Many other infections, however, do induce spectacular changes (reviewed by Esau, 1967; Gibbs, 1976 and Francki *et al.*, 1985). For instance, in phloem parenchyma cells infected with closteroviruses or luteoviruses, the plasmodesmata which contain virus particles do not have a desmotubule. With nepoviruses, comoviruses, cucumoviruses, and caulimoviruses, the plasmodesmatal channel becomes lined with a wide (40 to 80 nm) tubule in which viral particles are sometimes observed and which may extend into the cytoplasm between cell-wall outgrowths. In potyvirus-infected tissue, the association of cylindrical inclusions with plasmodesmata suggested to Langenberg (1986) that the inclusions play an active role in the transport of the infectious material.

The problem of virus movement between cells was summarized 21 years ago, soon after the first observations of viral particles in plasmodesmata were made:

> "Both the longitudinal and the transectional views of plasmodesmata containing (beet yellows) virus particles show no cores and no endoplasmic-reticulum associations. When and in what manner these structural features are removed in the presence of virus is an unanswered question." (Esau, 1968).

This question is still waiting for an answer.

The viral function involved in cell-to-cell movement

A large body of evidence (reviewed by Atabekov and Dorokhov, 1984) shows that cell-to-cell movement is dependent on a function specified by the viral genome. An important additional point made by these authors and by Taliansky *et al.* (1982a,b) is that this function is not virus-specific but rather is host-specific. A virus which is unable to infect a given plant because it is restricted in its movement (but replicates in protoplasts made from this plant) can sometimes be forced to spread by preinfecting with an unrelated, systemic virus acting as helper. For example TMV, which does not normally infect tomato lines homozygous for the *Tm-2* gene (but does replicate in their protoplasts) can be made to infect *Tm-2* lines by preinfecting with potato virus X, which is systemic in tomato (Taliansky *et al.*, 1982b).

The particles which move from cell to cell may not always be virions. This follows from the isolation of coat protein mutants of TMV which spread through leaf tissue, although the genomic RNA is not correctly encapsidated (Siegel *et al.*, 1962; Jockusch, 1968; Sarkar and Smitamana, 1981). Incidentally, these mutants cannot spread systemically through the vascular system, presumably because the infectious particles are not protected from degradation (Siegel *et al.*, 1962). The coat protein mutants (and wild-type TMV as well) may spread from cell to cell as "informosome-like" nucleoproteins (Dorokhov *et al.*, 1984).

Another example of the dispensability of the capsid protein for short-distance movement is found in tobacco rattle virus (TRV). This bipartite virus carries the coat protein cistron on RNA-2. However, "unstable" TRV isolates can be recovered from normal TRV infections. They multiply in the inoculated leaves even though they are made up of nonencapsidated RNA-1 (Bruening, 1977), indicating that cell- to-cell spread does occur in the absence of coat. More recently, it was shown that some geminiviruses can spread systemically without capsid, but that up to three other viral functions are needed (Lazarowitz *et al.*, 1989, and references therein). It is likely that one or several of these functions promote(s) cell-to-cell spread. Hence the hypothesis that, for all viruses, the movement of infectious particles between cells is made possible by a specialized protein, different from the coat protein, termed "movement protein", "spreading factor" or "transport factor".

TMV is the only virus for which the movement protein has been identified without ambiguity, on the basis of well-characterized temperature-sensitive mutations. The most famous of the TMV spread mutants is strain Ls1, which is derived from tomato strain L and carries a non-conservative point mutation in the second open reading frame of the genomic RNA (30 kDa cistron) (Leonard and Zaitlin, 1982; Ohno *et al.*, 1983). The essential role of the 30 kDa in virus spread was definitely proven by Deom *et al.* (1987), who showed that transgenic tobacco plants expressing wild-type 30 kDa were able to complement the defective movement function of the Ls1 strain. Moreover, an artificial mutant of TMV, temperature-sensitive for movement, was engineered by introducing the non-conservative point mutation of Ls1 into the parent strain (Meshi *et al.*, 1987).

For other viruses, identification of the movement protein was done mainly by elimination and is in most cases tentative. When the virus has a segmented genome, the movement-protein cistron is assigned to whatever genome segment is not required for

RNA replication in protoplasts but is required for infection of whole plants. This leaves an ambiguity when the segment in question codes for several proteins. RNA-3 of brome mosaic virus (BMV) for example, which is necessary for spread of infection (Kiberstis *et al.*, 1981; Allison *et al.*, 1988), codes for a 32 kDa protein (P3a) and for the coat protein. The movement protein has been identified to P3a because the coat protein is thought to be unnecessary for cell-to-cell spread and because P3a does not have any other known function. The same reasoning was applied to AlMV and to other tricornaviridae. A doubt remains, however, because what is true of a few plant viruses (dispensability of the coat protein for short-distance transport) may not be true of all of them.

Another type of evidence which is sometimes used to assign functions to viral proteins is sequence homology. This approach has not been very productive in the case of the movement function, in the sense that no strong consensus sequence has ever been reported. Presumptive spreading factors appear very heterogeneous and extensive homologies were reported only for closely related viruses. Two groups can be cited. The first involves the tricornaviridae: BMV and cucumber mosaic virus (CMV) (Murthy, 1983), AlMV and tobacco streak virus (TSV) (Cornelissen *et al.*, 1984; our unpublished results), and to a lesser extent AlMV and BMV (Savithri and Murthy, 1983). The second includes TRV and tobamoviruses (Boccara *et al.*, 1986; Cornelissen *et al.*, 1986; Saito *et al.*, 1988). Furthermore, limited sequence homology with the 30 kDa protein of TMV has suggested a movement function for the gene I product of cauliflower mosaic virus (CaMV) and other caulimoviruses (Hull *et al.*, 1986), the other argument being that this protein had not been assigned any other function. Similarly, limited homologies to the 30 kDa protein of TMV were found in non-structural cleavage products of the polyproteins encoded by the M-RNA of cowpea mosaic virus (CPMV, 48 kDa) and by RNA-2 of tomato black ring virus (Meyer *et al.*, 1986). It is nevertheless conceivable, as pointed out by Meshi *et al.* (1987), that there are several families of movement proteins and several mechanisms of action.

VIRAL MOVEMENT PROTEINS IN INFECTED PLANTS

Localization

Given the necessity of plasmodesmatal modification for virus spread on the one hand, and the existence of a virus-coded movement protein on the other, it was natural to assume that the movement protein carried out that modification. It could act directly at the level of the cell wall/plasma membrane or indirectly, at the nuclear level, by changing the expression of some cellular gene(s). It was therefore important to study the localization of this protein in whole plants. This has been done in the last few years for P3 of AlMV (32.4 kDa), for the 30 kDa protein of TMV, for the gene I product of CaMV (P1, 38 kDa according to sequence data), for the 48 kDa non-structural protein of CPMV, and for P3a of CMV (30 kDa). The problem to be overcome was the very small quantity of these proteins which could be recovered from infected plants. In all cases, detection in whole plants, infected under normal conditions, was by immunochemical methods (western blot and/or immunocytochemistry), with antisera raised against either synthetic peptides representing the C-terminus of the protein (Berna *et al.*, 1986; Tomenius *et al.*, 1987; Wellink *et al.*, 1987; Albrecht *et al.*, 1988), *in-vitro* translation products of viral RNA (Berna *et al.*, 1985), or fusion proteins made in *Escherichia coli* (Linstead *et al.*, 1988; MacKenzie *et al.*, 1988).

The putative movement protein of AlMV was the first to be detected unambiguously in the inoculated leaves of a systemic host, *Nicotiana tabacum* cv. Xanthi nc (Berna *et al.*, 1985, 1986). Under the conditions used, most of the coat protein was synthesized between 1 and 5 days post-inoculation (p.i.) (Fig. 1A). Western blotting with the anti-peptide serum (anti-pep3) detected P3 in a 30,000 x g-sedimentable fraction (Pe-30) which had been cleared of large organelles by centrifugation at 1000 x g. The supernatant (S-30) and the 1000 x g pellet did not react specifically with the antiserum. Accumulation of P3 in Pe-30 was transient, with a peak after 3 days of infection (Fig. 1B). This was consistent with an earlier observation of Joshi *et al.* (1984) in radiolabelled leaf slivers. The short lifetime of P3 in the membrane fraction could be explained in two ways: either the protein was very unstable and disappeared as soon as it ceased to be synthesized, or it was exported from the cytoplasm, possibly to the cell wall or to the plasmodesmata. The latter are known to be severed and retained inside the cell wall after cell breakage (Overall *et al.*, 1982), therefore they were not expected to be present in the lysate.

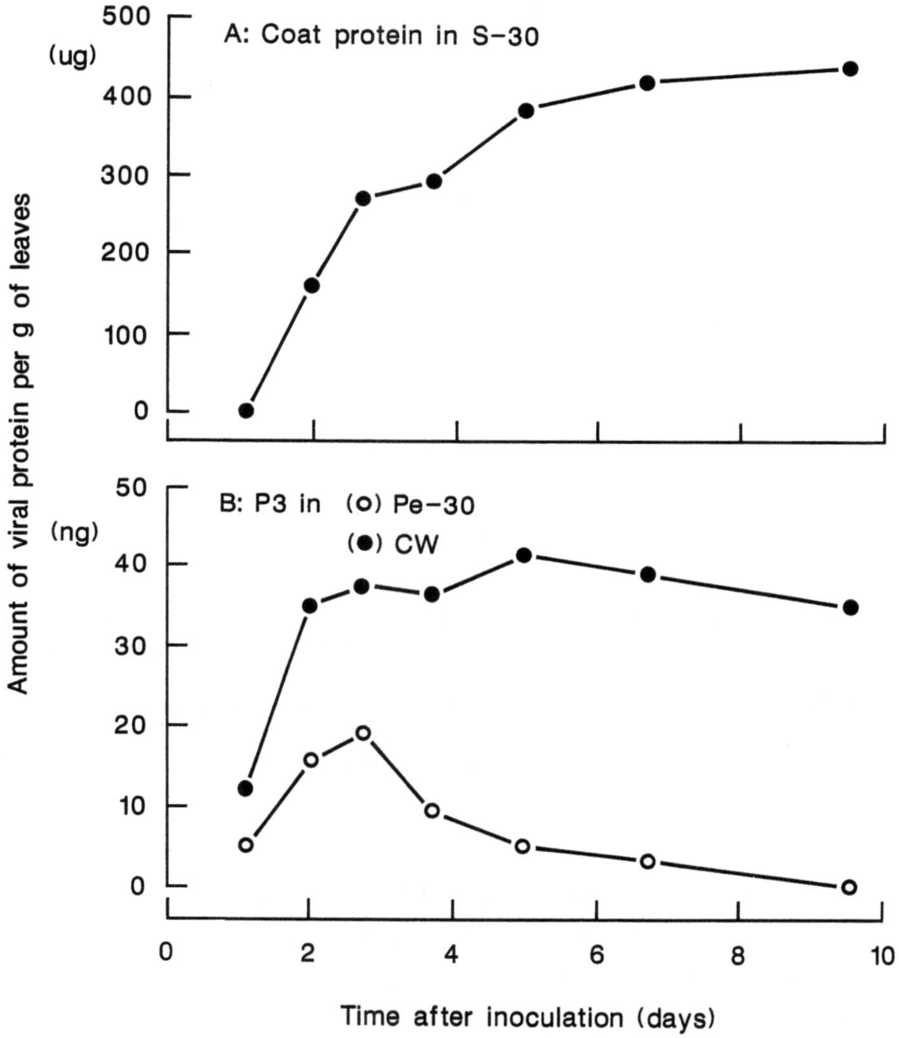

Figure 1. Accumulation of the AlMV movement (P3) and coat proteins as a function of time after inoculation. Subcellular fractions of infected tobacco leaves (1000 x g pellet, 30,000 x g pellet Pe-30, 30,000 x g supernatant S-30, cell-wall fraction CW) were assayed for P3 by Western blotting with an anti-peptide serum and the blots were quantified. The 1000 x g pellet and S-30 did not contain significant amounts of P3. The amount of coat protein in S-30 was similarly quantified by Western blotting with an antiserum to purified coat protein. (From Godefroy-Colburn *et al.*, 1986, by permission)

We were thus prompted to analyse the extraction residue containing cell walls (Godefroy-Colburn *et al.*, 1986). After washing this residue extensively to remove cytoplasmic contaminants, a "cell-wall fraction" (CW) was obtained, in which P3 was

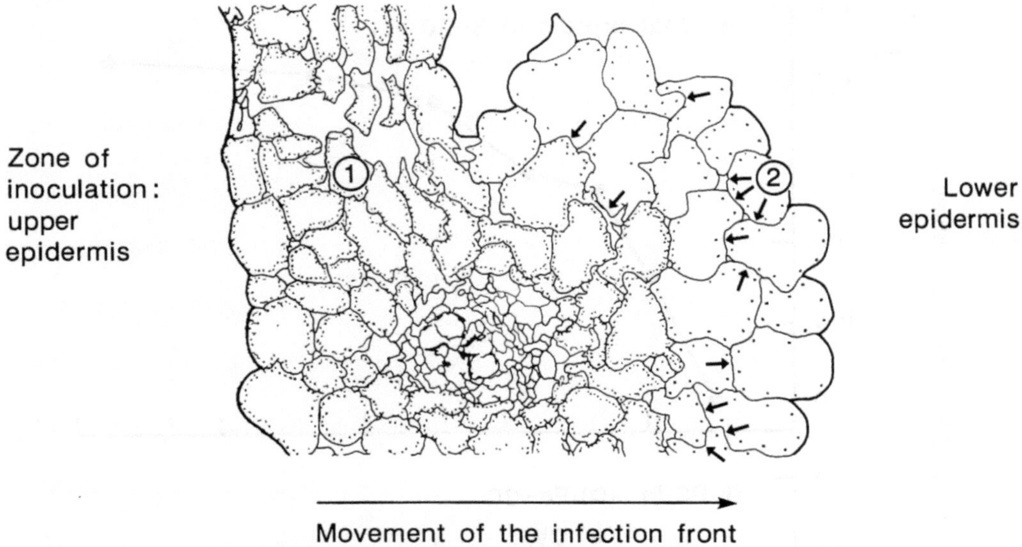

Figure 2. Distribution of the viral movement and coat proteins in an AlMV-inoculated tobacco leaf, 80 h after inoculation. The immuno-electron microscopic observations of P3 in this particular section are indicated by arrows, and the amount of labelling with the anti-AlMV serum (in an adjacent section) is represented by the density of dots along the contours of the cells. The cells of zone 1 were heavily infected and those of zone 2 had just been reached by the front of infection. (From Stussi-Garaud *et al.*, 1987, by permission)

indeed found. Moreover, the amount of P3 in CW was larger than in Pe-30, it was fairly constant from 3 to 5 days p.i. and diminished only slightly thereafter (Fig. 1B). This indicated strongly that the final destination of P3 was the cell wall and/or the plasmodesmata.

The cell-wall location of P3 was confirmed by immunogold labelling of ultrathin sections (Stussi-Garaud *et al.*, 1987). To our surprise, P3-specific label was not observed uniformly in the infected tissue but only in those regions which had just been reached by infection. Fig. 2 is a drawing of the section of inoculated leaf (3.3 days p.i.) on which the first observations of P3 were made. The arrows point to the actual sites where P3 was detected. It appears that those sites lie at the border between heavily-infected tissue (upper epidermis and most of the parenchyma, for example zone 1) which was strongly labelled with an anti-AlMV serum, and cells which had not yet synthesized much virus (near the lower epidermis, for example zone 2). Fig. 3 shows micrographs taken from zone 2, after labelling with the anti-pep3 serum. The negative controls proving the

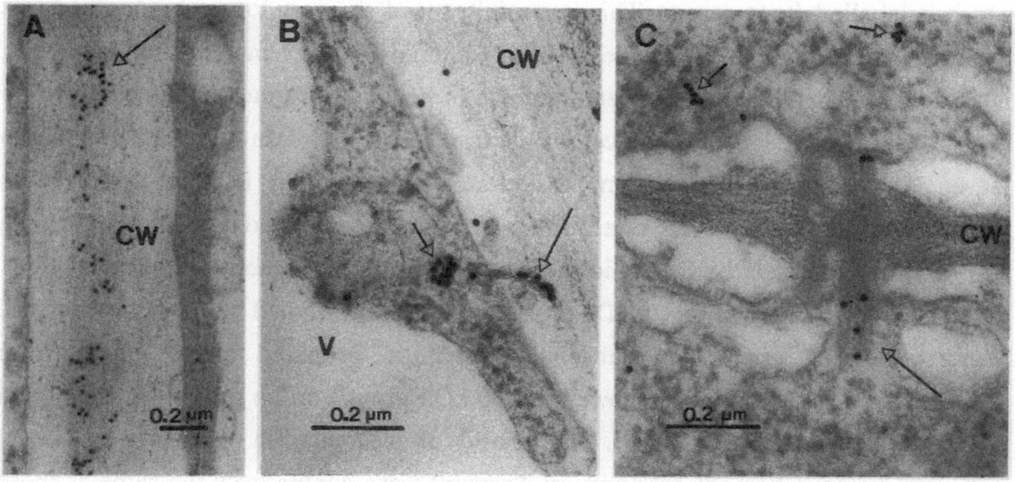

Figure 3. Immunogold-labelling of P3 (A and B) and of the coat protein (C) in leaf tissue at an early stage of infection. The experimental protocol was as described by Stussi-Garaud *et al.*, (1987). Micrographs A and B are from zone 2 of Fig. 2. In A, note the circular aggregate of P3-specific label in the middle-lamellar region (arrow); in B, note the cytoplasmic aggregate of label (short arrow) and the labelled channel penetrating the cell wall (long arrow), which could be either a branch of a plasmodesma, or a secretory vesicle. Micrograph C is from a similar zone of AlMV-infected leaf, taken 24 h after inoculation. The upper cell contains scattered aggregates of AlMV-specific label (short arrows). A tubular structure, extending from a branch of the plasmodesma into the cytoplasm of the lower cell, is labelled (long arrow).

specificity of labelling were done on adjacent sections (Stussi-Garaud *et al.*, 1987). Panel A is typical of most observations made to date on 3-day-infected tissue. The label was mainly localized in the middle lamella of the cell wall, or next to it, in fairly large masses. Circular aggregates (arrow) were fairly common. Panel B is representative of a less frequent distribution of the label. This micrograph shows a cytoplasmic aggregate of label (short arrow) which may be interpreted as a secretory vesicle containing P3, and a labelled channel (long arrow) which may be either the entrance to a plasmodesma or a secretory vesicle integral with the cell wall.

Few plasmodesmata were labelled with the anti-AlMV serum in sections prepared 1 to 3 days p.i and they always connected cells which were not densely labelled, i.e. which were in an early stage of infection (Fig. 3C; the short arrows point to scattered aggregates of label in the cytoplasm of the upper cell whereas no aggregate was visible in the lower cell). An interesting feature of the plasmodesma shown in Fig. 3C is the conspicuous tubular structure which extends into the cytoplasm of the lower cell and which is labelled (long arrow). This is reminiscent of the very long tubular structures extending from modified plasmodesmata of caulimovirus- or comovirus-infected cells, but extensive comparison with plasmodesmata of healthy cells must be done before we can say that AlMV infection induces visible changes in the plasmodesmata of tobacco leaves. The modification is in any case transient because it is never observed in heavily-infected tissue and very rarely observed in recently-infected tissue. We have been unable so far to detect both P3 and the coat protein in the same plasmodesma.

The localization of the 30 kDa protein of TMV was studied by the same methods. Western blotting analysis of subcellular fractions, prepared as in Fig. 1 from TMV-infected tobacco leaves, showed the 30 kDa protein to be present in Pe-30 and CW fractions (Moser *et al.*, 1988). The kinetics of accumulation of 30 kDa protein at 22°C in a systemic host (*N. tabacum*, cv. Samsun) were qualitatively the same as those for AlMV P3 (Fig. 4). Whereas the coat protein accumulated steadily in the inoculated leaves for 5 days, the 30 kDa protein was present transiently in Pe-30 with a maximum after 2-3 days of infection, and accumulated more stably and in larger quantity in the CW fraction. The lifetime of the 30 kDa protein in cell walls seemed to exceed 2 days because the protein was still present in the inoculated leaves 8 days after inoculation, i.e. 2 days after these leaves had stopped synthesizing coat protein. The transient presence of 30 kDa in a membrane fraction of radiolabelled infected leaf slivers was also reported earlier by Joshi *et al.* (1983). Under their conditions (systemic infection of cv. Samsun *NN* at 30°C, in the presence of antibiotics), maximum synthesis of 30 kDa protein was observed after 20 h of infection.

Immunoelectron microscopy gave the best (and earliest) proof that the final destination of the 30 kDa protein was the cell wall, more precisely the plasmodesmata (Tomenius *et al.*, 1987). A very interesting finding was that the peak of 30 kDa protein in plasmodesmata preceded the passage of the virus (or other viral nucleoprotein containing the coat protein) by one hour.

The situation is somewhat different with CaMV because this virus multiplies very slowly in mesophyll cells and moves preferentially through the phloem. However, the results on P1 localization in infected turnip plants are qualitatively the same as for the other two movement proteins. Western blotting analysis of systemically-infected leaves (Albrecht *et al.*, 1988) showed P1 to be present in a particulate fraction containing the replication complex (as well as polysomes etc.) between 10 and 17 days p.i., with a peak at 13 days. In CW on the other hand, P1 accumulated steadily between 13 and 26 days

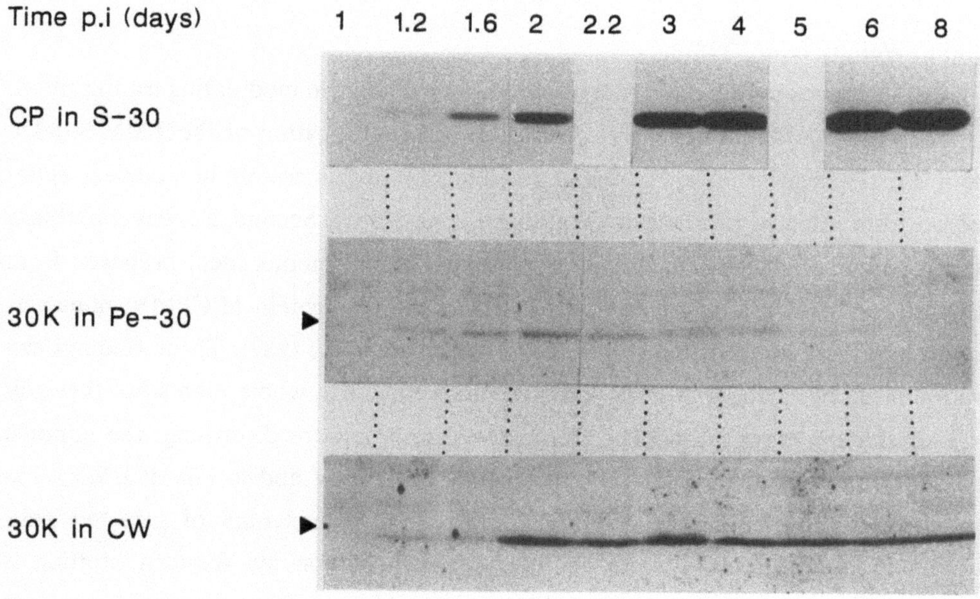

Figure 4. Accumulation of the TMV movement and coat proteins as a function of time after inoculation in a systemic host, *N. tabacum* cv. Samsun. Subcellular fractions were prepared as in Fig. 1 from TMV-inoculated leaves and assayed for 30 kDa and coat protein by Western blotting. (From Moser *et al.*, 1988, by permission)

p.i. Immunoelectron microscopy of systemically-infected leaves (Linstead *et al.*, 1988) visualized P1 at the periphery of some plasmodesmata which had been modified by infection; interestingly P1 was also found in sieve plates of infected turnip, suggesting a role in long-distance transport.

In short, a fairly consistent picture emerges from the localization data presented so far. The three movement proteins (two of them presumptive) are found transiently in the cytoplasm where they are synthesized, and are subsequently exported to the cell wall, more specifically to the plasmodesmata, rather than to the nucleus. In line with these results is the absence of the presumptive movement protein of CPMV (48 kDa) from nuclear fractions of infected cowpea leaves and protoplasts. As with the other three proteins, the 48 kDa protein of CPMV was detected by Western blotting in cytoplasmic membranes of infected leaves and protoplasts. Although no 48 kDa protein was detected in the CW fraction of infected leaves, it is still possible that it was excreted into the extracellular compartment because infected cowpea protoplasts release it into the culture medium (Wellink *et al.*, 1987).

All this tends to rule out a rôle of movement proteins in modulating transcription, but there are some discordant notes. First, the cell wall location of the TMV 30 kDa protein is in apparent contradiction with the detection of this protein in a nuclear pellet prepared from infected protoplasts (Watanabe *et al.,* 1986). Second, P3 was found in a 1000 x g pellet (containing nuclei, chloroplasts and some membranes) prepared from AlMV-infected protoplasts but, in contrast to the 48 kDa protein of CPMV, none was released into the culture medium (van Pelt-Heerschap *et al.*, 1987). These findings can be rationalized by saying that protoplasts do not behave like whole plants but they still lack a satisfactory explanation. The third observation is more disturbing: the putative movement protein of CMV, P3a (equivalent to P3 of AlMV and to P3a of BMV), has been visualized by immunogold cytochemistry in the nucleolus of infected cells (MacKenzie *et al.*, 1988). Unfortunately, no confirmation by Western blotting is available to date. If this observation holds true, our idea of virus movement will have to be revised. In the meantime, however, we will stick to our working hypothesis that the movement proteins act primarily at the cell wall level.

The exact location of the movement protein within the cell wall, however, seems to vary according to the virus. The 30 kDa protein of TMV is visualized inside plasmodesmata (Tomenius *et al.*, 1987) whereas P1 of CaMV is seen at their periphery (Linstead *et al.*, 1988). For AlMV, the evidence goes both ways (Stussi-Garaud *et al.*, 1987; Fig. 3). This variability may be due to intrinsic differences in the properties of the movement proteins but, alternatively, it may have a technical explanation. Ultrathin sections used for immunogold labelling are 50 to 70 nm thick, i.e. they are thicker than the average plasmodesma. Such a structure will be visible as a dark channel after classical staining (osmium, uranyl acetate), if it is sectioned more or less longitudinally, but also if it is entirely included in the section. On the other hand, immunoreagents are large molecules which mainly react with the surface of the section. Therefore the gold label does not correspond exactly to the structure which is visualized. The antigens which are detected by immunogold labelling "inside a plasmodesma" may indeed be incorporated in the structure of the plasmodesma, but they may equally well be part of the cell wall at the periphery of the plasmodesma. Another open question is whether the P3 aggregates which we visualized in the middle lamella were functional or were excretion products.

Timing of action

Electron microscopic observation of whole tissue suggests that the movement proteins act very early in the infection cycle of a given cell. However, no exact timing can be inferred from this type of work, because one cannot tell exactly which stage of the infection cycle each cell has reached.

Fortunately, some data are available about the timing of synthesis of the TMV and AlMV movement proteins in protoplasts. Watanabe *et al.* (1984) showed that the 30 kDa protein of TMV was synthesized between 2 and 9 h p.i. in infected protoplasts at 28°C, with a maximum rate after 6 h, i.e. before any significant amount of virus had been made. Similarly, P3 synthesis was nearly over by 6 h after AlMV infection at 25°C (van Pelt-Heerschap, 1987). The timing in leaf mesophyll should not be very different, although perhaps lower for TMV because of generally lower temperatures. We can safely assume that the movement protein reaches the plasmodesmata within 2 h of its synthesis and should therefore be ready to function at maximum efficiency by 6 to 8 h after infection.

How long does it take for the movement protein to carry out its function which in our hypothesis is modification of existing plasmodesmata? In the case of TMV, Tomenius *et al.* (1987) found the maximum amount of 30 kDa protein-specific label in plasmodesmata one hour before maximum labelling with the anti-TMV serum. Thus it seems that 1 h was sufficient for the movement protein to effect whatever modification was required; 7 to 9 h after infection of a given cell, viral particles (not necessarily virions) should therefore have been passing through plasmodesmata to infect the neighbouring cells. At that time, TMV aggregates were just starting to accumulate in the source cells (Tomenius *et al.*, 1987). This is consistent with our observations on AlMV-infected tissue (Stussi-Garaud *et al.*, 1987; Fig. 3C). As a comparison, 8-hour-infected protoplasts contained about 1% of the maximum amount of TMV (Watanabe *et al.*, 1982).

The timing of the "spreading event" has been determined directly in TMV-infected leaves at 22°C (Fannin and Shaw, 1987) and the results compare favourably with the above figures. Infection spreads from the initially-infected cells to the neighbouring cells between 6 and 8 h p.i. and to the second circle of cells about 5 h later.

What happens to the movement protein after it has done its job? In the case of TMV and AlMV, our interpretation of the immunolabelling patterns is that the protein is eliminated from plasmodesmata very quickly, then retained for some time by a component of the cell wall (e.g. the middle lamella) and eventually degraded or otherwise made unable to react with antibodies. After the movement protein has gone, the modified plasmodesmata seem to be either eliminated or restored to their original state, at least in the case of TMV infection. This follows from two observations:

1. the competence for virus movement of leaves systemically-infected with the TMV-mutant Ls1 was completely lost within 24 h of a shift to the non-permissive temperature (Taliansky *et al.*, 1982a), and
2. the number of plasmodesmata in Ls1-infected leaves was significantly lower at the non-permissive than at the permissive temperature, and also lower than in non-infected leaves or leaves infected with wild-type TMV (Shalla *et al.*, 1982).

In the case of CaMV infection, no precise timing can be given but there are some significant differences to TMV and AlMV infections. Plasmodesmatal modification is quite striking in the case of CaMV whereas it is very subtle and/or very transient in the case of TMV and AlMV. In addition, P1 of CaMV was visualized in all the modified plasmodesmata (Linstead *et al.*, 1988) whereas the 30 kDa protein and P3 were very difficult to detect in plasmodesmata. A naive interpretation would be that P1 is more stable than the other two and functions for a much longer time, thus modifying the plasmodesmata more drastically.

The TMV movement protein in a hypersensitive host

We have seen that the movement protein is the major weapon used by plant viruses to invade their hosts. But some plants fight back. They are able to slow down or even stop viral spread by calling on a battery of defence reactions. The best-known case is the hypersensitive response to TMV in tobacco varieties carrying the *N* gene. As reviewed by van Loon (1983), the hypersensitive response is a very complex series of reactions (e.g. accumulation of callose, hydroxyproline-rich glycoproteins, pathogenesis-related proteins, necrosis of the infected cells, etc.) and it is not clear which one is instrumental in stopping the infection.

Data presented in previous paragraphs tend to show that movement proteins operate at the front of infection in systemic hosts. What happens in hypersensitive hosts? One of the theories of virus spread assumes that the rôle of the movement protein is to inhibit the defence reactions (Atabekov and Dorokhov, 1984). According to this theory, the hypersensitive hosts should be those which can destroy or otherwise inactivate the movement protein before the hypersensitive reaction is triggered, i.e. within a few hours of inoculation.

We tried to test this hypothesis by comparing the kinetics of accumulation of the TMV 30 kDa protein at 22°C in two cultivars of *N. tabacum*: one systemic (Samsun) and one hypersensitive to TMV (Samsun *NN*) (Fig. 5; Moser *et al.*, 1988). As expected (Konate *et al.*, 1984), coat protein synthesis was the same in the two hosts up to 1.7 day (40 h) p.i., that is before the onset of the necrotic response in Samsun *NN*. Accumulation of 30 kDa protein was also nearly the same in the two hosts, whether in the cytoplasm or the cell-wall fraction. This agrees with the more recent finding that the regulation of 30 kDa protein synthesis is the same (and the resulting protein has the same properties) in protoplasts made from systemic and hypersensitive tobacco cultivars (Blum *et al.*, 1989). In whole plants, the difference between Samsun and Samsun *NN* manifested itself after 2 days, the time by which the necrotic lesions became visible and coat protein synthesis stopped in Samsun *NN*. The difference was not so much in the amount of cytoplasmic 30 kDa protein (it decreased in both cases) as in the amount of cell-wall 30 kDa protein. Instead of surviving for at least 2 days as in the systemic host (Fig. 4), the 30 kDa protein of the hypersensitive host disappeared completely within one day.

The disappearance of 30 kDa protein in Samsun *NN* is still unexplained. It could be due to the complete breakdown of the protein, to a modification of the antigen (C-terminus) by one of the enzymes induced by the hypersensitive reaction, or even to a change in the apparent molecular weight of the protein (e.g. by covalent linkage to a cell-wall component). The important point is that the loss of movement protein occurred after, and not before the onset of the necrotic response. This finding therefore does not support the hypothesis that the movement protein inhibits the hypersensitive response but suggests rather that the hypersensitive response affects the integrity of the movement protein.

One could speculate that the apparent loss of the 30 kDa protein corresponds to a functional loss and is related to the fact that TMV stops spreading in Samsun *NN* after 40 h at 22°C. If this hypothesis held true, it would become easy to understand why some strains of TMV keep spreading in hypersensitive tobaccos carrying the *N'* gene even though the infected tissue becomes necrotic, and why the degree of localization varies with the viral strain (van Loon, 1983). The degree of localization would simply reflect the balance between the rate of synthesis (or incorporation into the cell wall) of

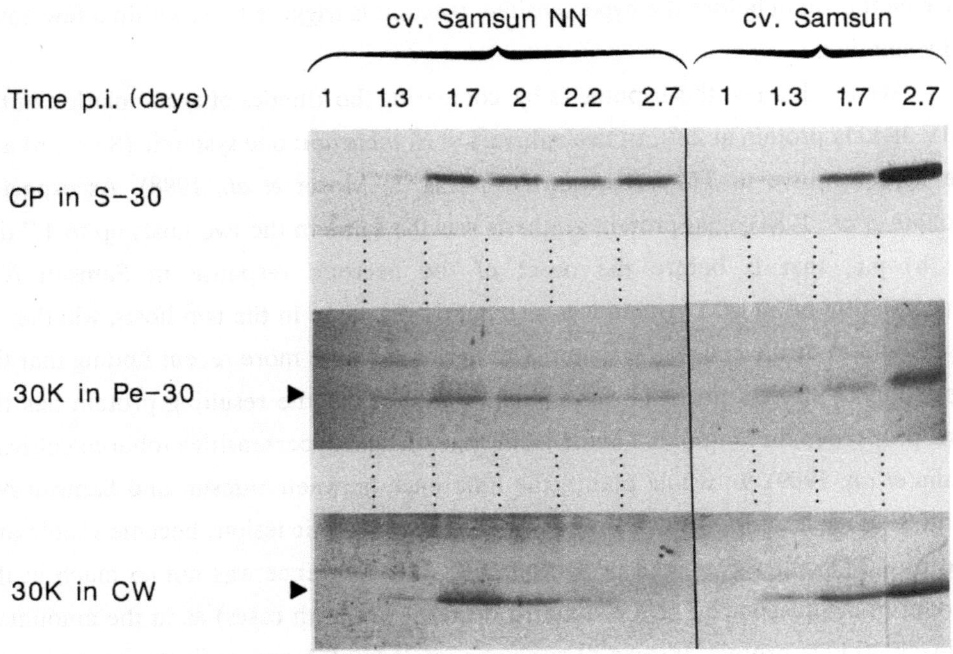

Figure 5. Accumulation of the TMV movement and coat proteins as a function of time after inoculation in a hypersensitive host, *N. tabacum* cv. Samsun *NN*. TMV-infection of Samsun *NN* was done in parallel with infection of cv. Samsun. The 30 kDa and coat proteins were assayed as in Fig. 4. (From Moser *et al.*, 1988, by permission)

the movement protein, and its rate of inactivation by the plant. This is supported by the finding (Watanabe *et al.*, 1987) that the virulence of several strains of TMV is related to the rate of synthesis of the 30 kDa protein in infected protoplasts.

MODE OF ACTION OF THE MOVEMENT PROTEINS

Some speculations

We know next to nothing about this subject and can only make hypotheses. If the movement proteins are to modify plasmodesmata, it is logical to assume that they have an enzymatic activity or that they activate a cellular enzyme. They could also act as carriers for the viral particles and somehow translocate them through plasmodesmata (as the name "transport factor" implies). The data of Tomenius *et al.* (1987) referred to previously, are not in favour of the latter hypothesis. Another possibility, which cannot be ruled out, is that movement proteins become structural components of plasmodesmata in spite of their low abundance in infected plants.

If movement proteins are enzymes or enzyme activators, what are their substrates? Enlargement of plasmodesmata can probably be achieved by loosening the cell wall around them, particularly in the constricted regions at the cytoplasmic ends. This may require digesting some specific polysaccharides or some structural proteins. The possibility that the movement protein might be a protease with the capacity to digest the desmotubule has also been considered by Atabekov and Dorokhov (1984). More complex modifications of the plasmodesmata such as those resulting from CaMV infection would be more difficult to explain by the action of a single enzyme or enzyme activator, except if the enzyme triggered a cascade of reactions.

In order to choose between those hypotheses, we need to test *in vitro* the putative activities of the movement proteins, their binding to different components of the cell wall, to viral RNA, etc. This can only be done with large amounts of pure material. The amount of P3 in AlMV-infected tobacco leaves is, at best, in the neighbourhood of 50 ng/g fresh wt (Fig. 1B). For the 30 kDa protein of TMV, the figure is very similar (Moser *et al.*, 1988). We therefore gave up on the idea of purifying the movement proteins from plants and turned to recombinant DNA methods.

Production of the movement proteins of AlMV and TMV by recombinant yeasts

Saccharomyces cerevisiae, rather than *Escherichia coli*, was chosen as an expression organism for the two movement proteins of interest because yeast cells are expected to translate eukaryotic mRNAs efficiently without the need for splicing a Shine-Dalgarno sequence to the translation initiation site. We used shuttle expression vectors of the pVT series (Vernet *et al.*, 1987) which can replicate both in yeast (with the replication origin of the 2μ plasmid) and in *E. coli* (with the replication origin of pBR322). The selection markers were ampicillin resistance in *E. coli* and complementation of leucine (Leu2) or uracil (Ura3) auxotrophy in yeast (vectors pVT-L and pVT-U respectively). The expression cassette was made of the promoter and 3' end of the yeast ADH1 gene (constitutive alcohol dehydrogenase), separated by a multiple cloning site. The expression cassette did not contain any initiator or terminator codon (except, by chance, in one of the reading frames), therefore the coding sequence to be expressed had to be

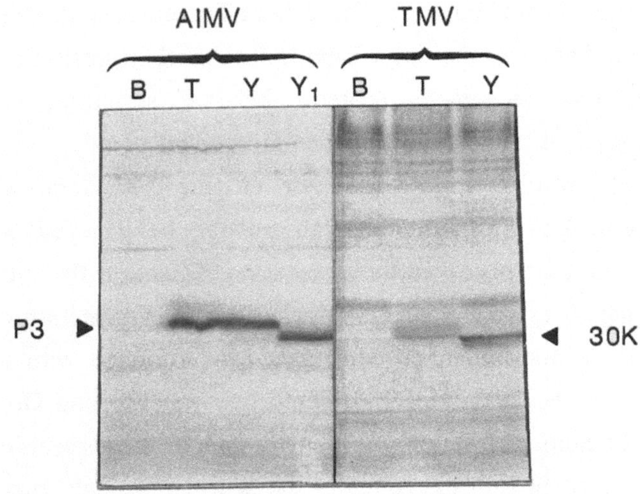

Figure 6. Comparison of the recombinant movement proteins of AlMV and TMV with the corresponding proteins extracted from infected tobacco plants. Analysis was done by SDS-PAGE in 10% gels and Western blotting with antisera to C-terminal peptides. To ensure uniformity of migration of proteins, each lane contained a constant amount of Pe-30 prepared from tobacco plants (equivalent to 180 mg of leaves), either uninfected (lanes B, Y and Y1), or infected by AlMV or TMV as indicated (lanes T). In addition, lanes Y contained extracts of yeasts expressing the complete P3 or 30 kDa genes (amounts corresponding respectively to 30 μg or 300 μg of yeast paste) and lane Y1 an extract of yeast expressing the 12-amino acid N-terminal deletion of P3 (an amount corresponding to 3 μg of yeast paste).

complete.

For practical reasons, the first protein that we expressed in yeast (with vector pVT-L) was a truncated version of AlMV P3 starting at the second AUG of the open reading frame, i.e. lacking 12 amino acids at the N-terminus. More recently, the entire P3 protein was obtained as well as the 30 kDa protein of TMV, with vectors pVT-U and pVT-L respectively. All the recombinant proteins were detected by immunoblotting crude extracts of the expressor yeasts (Fig. 6). The complete P3 made by yeast had the same mobility as bona-fide AlMV protein from the Pe-30 of infected tobacco (compare lanes T and Y). The 12-amino acid deletion of P3 (lane Y1) migrated slightly faster than P3, as expected from its theoretical molecular weight. This result incidentally confirms that the protein translated from AlMV RNA-3 does initiate at the first AUG in infected plants. A comparison of recombinant 30 kDa protein with the protein from TMV-infected Pe-30 shows that, in addition to the full length product, recombinant yeast

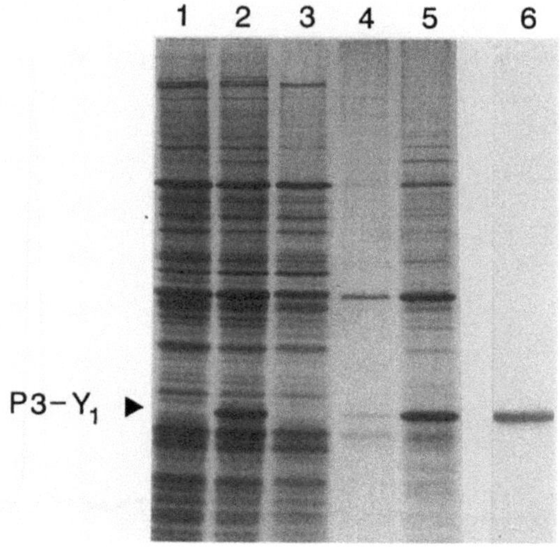

Figure 7. Solubilization of the N-terminally-deleted movement protein of AlMV expressed by yeast. The experiment was done in parallel with expressor and non-expressor yeasts. The yeast paste was ground with glass beads, the extract was clarified of cell debris and centrifuged at 15,000 x g. The pellet was fractionated by equilibrium sedimentation in a sucrose gradient containing EDTA. Material sedimenting between the densities of 1.14 and 1.19 (fraction M) was treated with 1% dodecyl-maltoside (12-M) and centrifuged at 15,000 x g. The resulting pellet was treated with 6M urea and finally recentrifuged at 15,000 x g. The different fractions were analysed by SDS-PAGE in 9% gels followed by Coomassie-blue staining (lanes 1 to 5) or Western blotting with the anti-pep3 serum (lane 6). The samples were from non-expressor (lane 1) or expressor yeasts (lanes 2 to 6) and were loaded in amounts equivalent to 18 mg (lanes 1 to 5) or 0.1 mg of yeast paste (lane 6). Lanes 1 and 2, fractions M; lane 3, 12-M supernatant; lane 4, urea pellet; lanes 5 and 6, urea supernatant.

made a shorter protein. This may have resulted from degradation of the N-terminus or from initiation at the second AUG. The blot also shows that a fraction of the 30 kDa protein made by TMV-infected tobacco migrated more slowly than the full-length translation product, in a diffuse manner reminiscent of glycosylated proteins. As shown in Fig. 7, the three recombinant proteins seemed to have very different efficiencies of expression, TMV 30 kDa protein being the least expressed. These differences may not be significant, however, because the antisera detecting P3 and the 30 kDa protein did not have the same efficiency, and because the metabolic state of the yeast was not rigorously controlled.

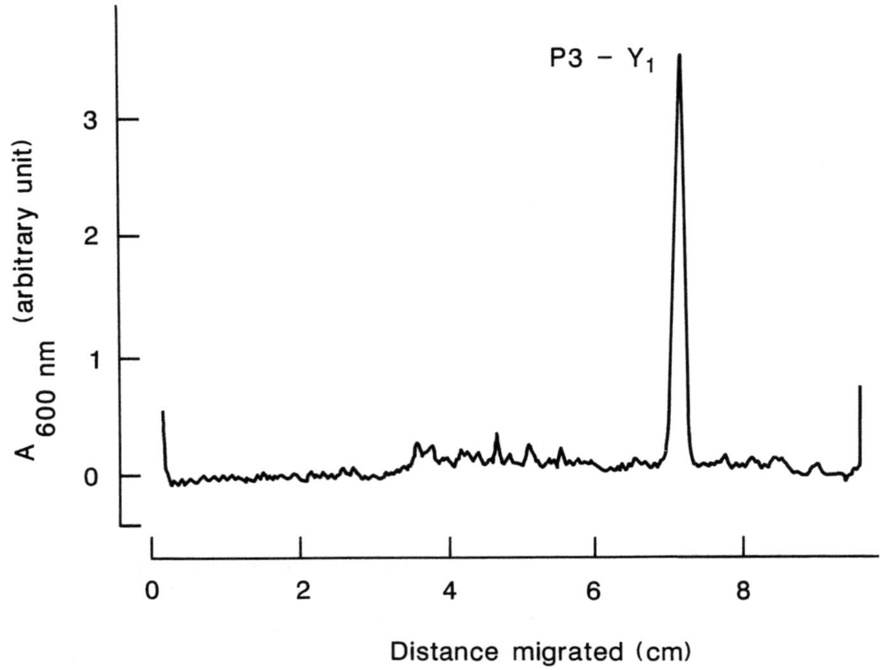

Figure 8. Analysis of the recombinant deleted P3 after purification. The urea supernatant of Fig. 7 was purified by cation-exchange on CM-trisacryl (Industrie Biologique Française) in 6M urea, 30 mM Na-acetate, pH 5.5, and eluted with an NaCl gradient in the same buffer. The fraction eluting at about 80 mM NaCl was analysed by SDS-PAGE in a 9% gel, followed by Coomassie-blue staining and densitometry at 600 nm.

All three recombinant proteins cosedimented with yeast membranes at 15000 x g, and were found at a density of 1.14-1.19 in sucrose-EDTA (fraction M). None of the recombinant proteins was detected in the culture medium but we do not know yet whether any of them were secreted into the periplasmic space.

Further work was done with the 12-amino acid deletion of P3 (P3-Y1). Immunogold cytochemistry of spheroplasts made from expressor yeast confirmed the presence of the P3 antigen. Small aggregates of gold grains were scattered throughout the cytoplasm. Neither the nucleus nor the mitochondria were labelled, and the plasma membrane was not labelled preferentially (not shown).

P3-Y1 was purified from fraction M of expressor yeast. Fig. 7 (lane 2) shows the Coomassie blue-stained SDS-PAGE pattern of this material compared to the equivalent fraction of non-expressor yeast (lane 1). An additional band is visible in lane 2 (arrowhead), which corresponds in position to the band detected by Western blotting (lane 6). Treatment with 1% dodecyl maltoside, a non-ionic detergent, removed most of the cellular proteins from the membrane pellet but did not solubilize any appreciable quantity of the recombinant protein (lane 3). The latter was almost entirely solubilized by 6M urea (lanes 4 and 5). Further purification was done by cation-exchange. Analysis of the eluted material by SDS-PAGE, followed by Coomassie blue staining (Fig. 8) and Western blotting (not presented), showed that P3-Y1 was predominant. The yield was about 50 μg of protein per g of yeast paste. This method of purification is currently being applied to the other recombinant proteins.

CONCLUSION

The number of unanswered questions in this article shows that understanding of how viruses spread is just beginning to develop. There are some fairly hard data (on only three viruses: AlMV, TMV and CaMV) indicating that the target of the movement protein is the cell wall, more specifically some plasmodesmata. The finding that the presumptive movement protein of CPMV is excreted into the culture medium by cowpea protoplasts points in the same direction. However, there are also some conflicting data, especially the immunoelectron microscopic detection of CMV protein 3a in the nucleolus of infected plants. Clearly, more observations need to be made on many more viruses, before we can state any general rules.

As for the mode of action of transport factors, we can only speculate. Biochemical assays need to be carried out with purified material, but the low amount of movement protein in plants infected with viruses such as TMV or AlMV precludes purification by standard methods. Genetic engineering seems to offer a way out and we hope to do some real biochemistry with the recombinant movement proteins which are now available.

An increasing amount of effort, justified by the economic importance of viral spread, is being spent on trying to counteract the action of movement proteins. Some pioneering experiments are currently being done with this goal in mind (Berna and Beachy, 1988). The project involves manipulating the sequence of the 30 kDa gene, introducing the modified gene into tobacco by agroinfection, and testing the activity of the modified protein by complementation of the TMV mutant Ls1. This should give information on the functional domains of the 30 kDa protein. It is also hoped that some of the modified 30 kDa proteins, although inactive, will compete for receptors with the wild-type protein and inhibit the movement function, thus rendering the transformed plants resistant to TMV and, possibly, to some other viruses.

Finally, some important new findings on movement proteins were released during this NATO Workshop. Goldbach and co-workers (this volume) detected the 58 kDa/48 kDa proteins of CPMV in infected leaf tissue, by immunoelectron microscopy. A revealing micrograph showed an accumulation of label at the periphery of a tubular structure crossing the cell wall and containing what appeared to be viral particles. Secondly, Ziegler-Graff and co-workers engineered a frameshift mutant of TRV which did not express the 29 kDa protein. The mutant was deficient in cell-to-cell spread and was complemented by coinoculation of TMV. The 29 kDa of TRV is therefore the second movement protein to be identified on the basis of clear genetic evidence.

ACKNOWLEDGEMENTS

We thank Drs RN Beachy (St Louis) for a clone of the 30 kDa cistron, T Vernet (Montreal) for the yeast expression vectors, K-W Mundry and M Ellwart-Tschürtz (Stuttgart) for an antiserum to the C-terminus of the TMV 30 kDa protein and M-R Chevallier and her colleagues (Strasbourg) for the yeast strains and for helpful advice on yeast transformation.

REFERENCES

Albrecht H, Geldreich A, Menissier-De Murcia J, Kirchherr D, Mesnard JM, Lebeurier G (1988) CaMV gene I product detected in a cell-wall-enriched fraction. Virology 163:503-508

Allison RA, Janda M, Ahlquist P (1988) Infectious *in vitro* transcripts from cowpea chlorotic mottle virus cDNA clones and exchange of individual RNA components with brome mosaic virus. J Virol 62:3581-3588

Atabekov JG, Dorokhov YL (1984) Plant virus-specific transport function and resistance of plants to viruses. Adv Virus Res 29:313-364

Berna A, Beachy RN (1988) The tobacco mosaic virus movement protein: functional domains. Abstracts of Sixteenth International Congress of Genetics, Toronto. Genome 30 (suppl 1):478

Berna A, Briand JP, Stussi-Garaud C, Godefroy-Colburn T (1986) Kinetics of accumulation of the three non-structural proteins of alfalfa mosaic virus in tobacco plants. J Gen Virol 67:1135-1147

Berna A, Godefroy-Colburn, Stussi-Garaud C (1985) Preparation of an antiserum against an *in vitro* translation product of alfalfa mosaic virus RNA3. J Gen Virol 66:1669-1678

Blum H, Gross HJ, Beier H (1989) The expression of the TMV-specific 30-kDa protein in tobacco protoplasts is strongly and selectively enhanced by actinomycin. Virology 169:51-61

Boccara M, Hamilton WDO, Baulcombe DC (1986) The organisation and interviral homologies of genes at the 3' end of tobacco rattle virus RNA1. EMBO J 5:223-229

Bruening G (1977) Plant covirus systems: two-component systems. In: Fraenkel-Conrat H, Wagner RR (eds) Comprehensive virology, Vol. 11. Plenum, New-York, p 55

Cornelissen BJC, Janssen H, Zuidema D, Bol JF (1984) Complete nucleotide sequence of tobacco streak virus RNA 3. Nucleic Acids Res 12:2427-2437

Cornelissen BJC, Linthorst HJM, Brederode FT, Bol JF (1986) Analysis of the genome structure of tobacco rattle virus strain PSG. Nucleic Acids Res 14:2157-2169

Deom CM, Oliver MJ, Beachy RN (1987) The 30-kilodalton gene product of tobacco mosaic virus potentiates virus movement. Science 237:389-394

Dorokhov YL, Alexandrova NM, Miroshnichenko NA, Atabekov JG (1984) The informosome-like virus-specific ribonucleo-protein (vRNP) may be involved in the transport of tobacco mosaic virus infection. Virology 137:127-134

Esau K (1967) Anatomy of plant virus infections. Annu Rev Phytopathol 5:45-76

Esau K (1968) Viruses in plant hosts. Form, distribution, and pathologic effects. University of Wisconsin Press, Madison Milwaukee, London

Fannin FF, Shaw JG (1987) Evidence for concurrent spread of tobacco mosaic virus from infected epidermal cells to neighbouring epidermal and mesophyll cells. Plant Sci 51:305-310

Francki RIB, Milne RG, Hatta T (1985) Atlas of plant viruses, Vol. 1 and 2. CRC Press, Boca Raton

Gibbs A (1976) Viruses and plasmodesmata. In: Gunning BES, Robards AW (eds) Intercellular communication in plants: studies on plasmodesmata. Springer-Verlag, Berlin Heidelberg New-York, p 149

Godefroy-Colburn T, Gagey, MJ, Berna A, Stussi-Garaud C (1986) A non-structural protein of alfalfa mosaic virus in the walls of infected tobacco cells. J Gen Virol 67:2233-2239

Hull R (1989) The movement of viruses in plants. Annu Rev Phytopathol, in press

Hull R, Sadler J, Longstaff M (1986) The sequence of carnation etched ring virus DNA: comparison with cauliflower mosaic virus and retroviruses. EMBO J 5:3083-3090

Jockusch H (1968) Two mutants of tobacco mosaic virus temperature-sensitive in two different functions. Virology 35:94-101

Joshi S, Pleij CWA, Haenni AL, Chapeville F, Bosch L (1983) Properties of the tobacco mosaic virus intermediate length RNA-2 and its translation. Virology 127:100-111

Joshi S, Neeleman L, Pleij CWA, Chapeville F, Bosch L, van Vloten-Doting L (1984) Nonstructural alfalfa mosaic virus RNA-coded proteins present in tobacco leaf tissue. Virology 139:231-242

Kiberstis PA, Loesch-Fries LS, Hall TC (1981) Viral protein synthesis in barley protoplasts inoculated with native and fractionated bromemosaic virus RNA. Virology 112:804-808

Konate G, Kopp M, Fritig B (1984) Multiplication du virus de la mosaique du tabac dans des hôtes à réponse systémique ou nécrotique: approche biochimique à l'étude de la résistance hypersensible. Phytopathol Z 105:214-225

Langenberg WG (1986) Virus protein association with cylindrical inclusions of two

viruses that infect wheat. J Gen Virol 67:1161-1168
Lazarowitz SG, Pinder AL, Damsteeg VD, Rogers SG (1989) Maize streak virus genes essential for systemic spread and symptom development. EMBO J 8:1023-1032
Leonard DA, Zaitlin M (1982) A temperature-sensitive strain of tobacco mosaic virus defective in cell-to-cell movement generates an altered viral-coded protein. Virology 117:416-424
Linstead PJ, Hills GJ, Plaskitt KA, Wilson IG, Harker CL, Maule AJ (1988) The subcellular location of the gene I product of cauliflower mosaic virus is consistent with a function associated with virus spread. J Gen Virol 69:1809-1818
MacKenzie DJ, Tremaine JH (1988) Ultrastructural location of non-structural protein 3A of cucumber mosaic virus in infected tissue using monoclonal antibodies to a cloned chimeric fusion protein. J Gen Virol 69:2387-2395
Meshi T, Watanabe Y, Saito T, Sugimoto A, Maeda T, Okada Y (1987) Function of the 30K protein of tobacco mosaic virus: involvement in cell-to-cell movement and dispensability for replication. EMBO J 6:2557-2563
Meyer M, Hemmer O, Mayo MA, Fritsch C (1986) The nucleotide sequence of tomato black ring virus RNA-2. J Gen Virol 67:1257-1271
Moser O, Gagey MJ, Godefroy-Colburn T, Stussi-Garaud C, Ellwart-Tschürtz M, Nitschko H, Mundry KW (1988) The fate of the transport protein of tobacco mosaic virus in systemic and hypersensitive tobacco hosts. J Gen Virol 69:1367-1373
Murthy MRN (1983) Comparison of the nucleotide sequences of cucumber mosaic virus and brome mosaic virus. J Mol Biol 168:469-475
Ohno T, Takamatsu N, Meshi T, Okada Y, Nishiguchi M, Kiho Y (1983) Single amino acid substitution in 30K protein of TMV defective in virus transport function. Virology 131:255-258
Overall RL, Wolfe J, Gunning BES (1982) Intercellular communication in *Azolla* roots: I. Ultrastructure of plasmodesmata. Protoplasma 111:134-150
Robards AW (1976) Plasmodesmata in higher plants. In: Gunning BES, Robards AW (eds) Intercellular communication in plants: studies on plasmodesmata. Springer-Verlag, Berlin Heidelberg New-York, p 15
Saito T, Imai Y, Meshi T, Okada Y (1988) Interviral homologies of the 30K proteins of tobamoviruses. Virology 167:653-656
Sarkar S, Smitamana P (1981) A proteinless mutant of tobacco mosaic virus: evidence against the role of a viral coat protein for interference. Mol Gen Genet 184:158-159
Savithri HS, Murthy MRN (1983) Evolutionary relationship of alfalfa mosaic virus with cucumber mosaic virus and brome mosaic virus. J Biosci 5:183-187
Shalla TA, Petersen LJ, Zaitlin M (1982) Restricted movement of a temperature-sensitive virus in tobacco leaves is associated with a reduction in numbers of plasmodesmata. J Gen Virol 60:355-358
Siegel A, Zaitlin M, Sehgal OP (1962) The isolation of defective tobacco mosaic virus strains. Proc Natl Acad Sci USA 48:1845-1851
Stussi-Garaud C, Garaud JC, Berna A, Godefroy-Colburn T (1987) *In situ* location of an alfalfa mosaic virus non structural protein in plant cell walls: correlation with virus transport. J Gen Virol 68:1779-1784
Taliansky ME, Malyshenko SI, Pshennikova ES, Kaplan IB, Ulanova EF, Atabekov JG (1982a) Plant virus-specific transport function. I. Virus genetic control required for virus spread. Virology 122, 318-326
Taliansky ME, Malyshenko SI, Pshennikova ES, Atabekov JG (1982b) Plant-virus transport function. II. A factor controlling virus host range. Virology 122:327-331

Terry BR, Robards AW (1987) Hydrodynamic radius alone governs the mobility of molecules through plasmodesmata. Planta 171:145-157

Thomson WW, Platt-Aloia K (1985) The ultrastructure of the plasmodesmata of the salt glands of *Tamarix* as revealed by transmission and freeze-fracture. Protoplasma 125:13-23

Tomenius K, Clapham D, Meshi T (1987) Localization by immunogold cytochemistry of the virus-coded 30K protein in plasmodesmata of leaves infected with tobacco mosaic virus. Virology 160:363-371

Van Loon LC (1983) Mechanisms of resistance in virus-infected plants. In: Bailey JA, Deverall BJ (eds) The dynamics of host defence. Academic Press, New York London, p 123

Van Pelt-Heerschap H (1987) Immunochemical analysis of the alfalfa mosaic virus gene products. PhD Thesis, University of Leiden

Van Pelt-Heerschap H, Verbeek H, Huisman MJ, Loesch-Fries LS, van Vloten-Doting L (1987). Non-structural proteins and RNAs of alfalfa mosaic virus synthesized in tobacco and cowpea protoplasts. Virology 161:190-197

Vernet T, Dignard D, Thomas DY (1987) A family of yeast expression vectors containing the phage f1 intergenic region. Gene 52:225-233

Watanabe Y, Ohno T, Okada Y (1982) Virus multiplication in tobacco protoplasts inoculated with tobacco mosaic virus RNA encapsulated in large unilamellar liposomes. Virology 120:478-480

Watanabe Y, Emori Y, Ooshika I, Meshi T, Ohno T, Okada Y (1984) Synthesis of TMV-specific RNAs and proteins at the early stage of infection in tobacco protoplasts: transient expression of the 30K protein and its mRNA. Virology 133:18-24

Watanabe Y, Ooshika I, Meshi T, Okada Y (1986) Subcellular localization of the 30K protein in TMV-inoculated tobacco protoplasts. Virology 152:414-420

Watanabe Y, Morita N, Nishiguchi M, Okada Y (1987) Attenuated strains of tobacco mosaic virus. Reduced synthesis of a viral protein with a cell-to-cell movement function. J Mol Biol 194:699-704

Weintraub M, Ragetli WJ, Leung E (1976) Elongated virus particles in plasmodesmata. J Ultrastr Res 56:351-364

Wellink J, Jaegle M, Prinz H, van Kammen A, Goldbach R (1987) Expression of the middle component RNA of cowpea mosaic virus *in vivo*. J Gen Virol 68:2577-2585

Zaitlin M, Hull R (1987) Plant virus-host interactions. Annu Rev Plant Physiol 38:291-315

SIGNALLING IN VIROID PATHOGENESIS

V Conejero, JM Bellés, F García-Breijo, R Garro, J Hernández-Yago, I Rodrigo and P Vera
Departamento de Biotechnología
ETSI Agrónomos
Universidad Politécnica de Valencia
Camino de Vera 14
46022 Valencia
Spain

ABSTRACT

Despite the fact that viroids are the infectious agents with the lowest complexity and best known structure, our understanding of their pathogenic interaction with the host plant is still far from complete. Viroid pathogenesis poses many intriguing questions; these have to be answered without any viroid-specified protein to which to attribute a pathogenic rôle. Here we present a view of viroid pathogenesis in which viroid molecules are considered as pure replicating and pathogenic signals. These act as elicitors of host responses which can also be activated by other afflicting agents. A model is outlined to explain how the viroid-induced response at the cellular level becomes a developmental disease and leads to a plant which is more resistant to subsequent infections.

INTRODUCTION

In the early nineteen-eighties, the idea was first proposed that viroids are non-specific elicitors of a host-coded mechanism of response, that can also be triggered by other agents (Conejero 1981; 1982). At that time, the existence of molecular mechanisms for signal perception and transduction in plant cells was barely being considered. Nowadays, signalling has gained recognition and has become a leading area of interest in plant molecular biology.

Unfortunately, the field is at the beginning of its development, and the cumulated knowledge still has numerous gaps, although this disadvantage is also its charm. In this

NATO ASI Series, Vol H 41
Recognition and Response in Plant-Virus Interactions
Edited by R.S S. Fraser

chapter, we are going to attempt to build up a view of viroid pathogenesis within the framework of signalling and signal transduction in plants. We will make suggestions of suitable approaches to increase knowledge of how viroids induce diseases and activate defence reactions in plants. Our approach will be speculative.

VIROID PATHOGENESIS

General considerations

Although viroid replication is a reasonably well understood process (Tabler and Tsagris, this volume), and although there is precise knowledge of their structure and conformation (Gross *et al.*, 1978), the way in which viroid molecules induce diseases and defence reactions is largely unknown.

Viroids have the lowest biological complexity of any pathogen yet known. Their naked RNAs, with a molecular mass of about 10^5 Da, have an informational content of about one tenth that of a minimal virus. Nevertheless, viroids are able to accomplish virus-like functions. They enter the host-plant with the aid of a vector; man with his agricultural practices is probably the main means of transmission. They translocate and spread systemically through the vascular system to reach the target cells. Viroids also replicate autonomously (without the aid of a helper virus), and induce metabolic and developmental alterations reflected in a disease syndrome (plant stunting, leaf malformation and inhibition of root growth). Finally, they are able to induce resistance or protection against subsequent infections.

How do viroids induce these biological processes? Are these biological processes attributable to any specific characteristics of the viroid structure, conformation and mechanism of replication, or are they a response to any specific protein or other kind of molecule induced by infection?

Viroids as non-specific elicitors of responses programmed by the host

Since the evidence indicates that viroids do not code for proteins (Davies *et al.*, 1974; Hall *et al.*, 1974; Semancik *et al.*, 1977; Conejero *et al.*, 1979), it was thought that their pathogenicity had to be exerted through direct interference with some critical cellular targets (Diener, 1979; Semancik and Conejero, 1987). From this, and taking into account the replication and location of viroids in the nucleus (Riesner, 1987; Robertson and Branch, 1987; Sänger, 1987), the obvious strategy was to look for possible interference of viroid RNA molecules (+RNA) or complementary replicative forms, with either host nucleic acids or proteins, as the primary cause of pathogenicity. Thus it has been suggested that viroids express their pathogenicity by altered regulation of gene expression (Rackwitz *et al.*, 1981; Diener, 1981; Dickson, 1981; Solymosy and Kiss, 1985) and altered translocation of proteins, by base pairing with RNA signal recognition particles (Haas *et al.*, 1988). These interpretations are supported by theoretical considerations, but still await full experimental confirmation.

A different type of approach in our laboratory (Conejero and Granell, 1986) led to evidence indicating that viroids do not incite the developmental syndrome and defence reactions by direct and specific interference with the normal flux of genetic information in the host cell. Rather, the evidence suggested that viroids seem to be elicitors of a general response of the host plant, and that this is mediated by ethylene. This response can also be triggered by agents with no genetic information. It could be said that viroids as pathogens are devoid of "personality".

The idea that at least certain physiological alterations induced by viroid infection were not specific arose when it was found that natural senescence of *Gynura aurantiaca* DC plants was accompanied by the accumulation of the same pathogenesis-related (PR) proteins (P1, 15 kDa and P2, 18 kDa) originally described as associated with citrus exocortis viroid (CEVd) infection (Conejero *et al.*, 1979). Later, (Conejero and Granell, 1986), it was shown that high doses of Ag^+ ions elicited a reaction almost indistinguishable from that induced by CEVd in *Gynura*. It was concluded that the pathological syndrome, defence reactions and the physiological condition of enhanced resistance were components of a general mechanism of response triggered by different types of elicitors.

SIGNALLING IN VIROID PATHOGENESIS

Although a reductionist view would consider that all types of information in biological systems are related to the genetic language, there are informational fluxes that have their own structure and dynamics. This is irrespective of the fact that the genetic machinery can be implicated as a recurrent step, at any time when synthesis of new protein is needed.

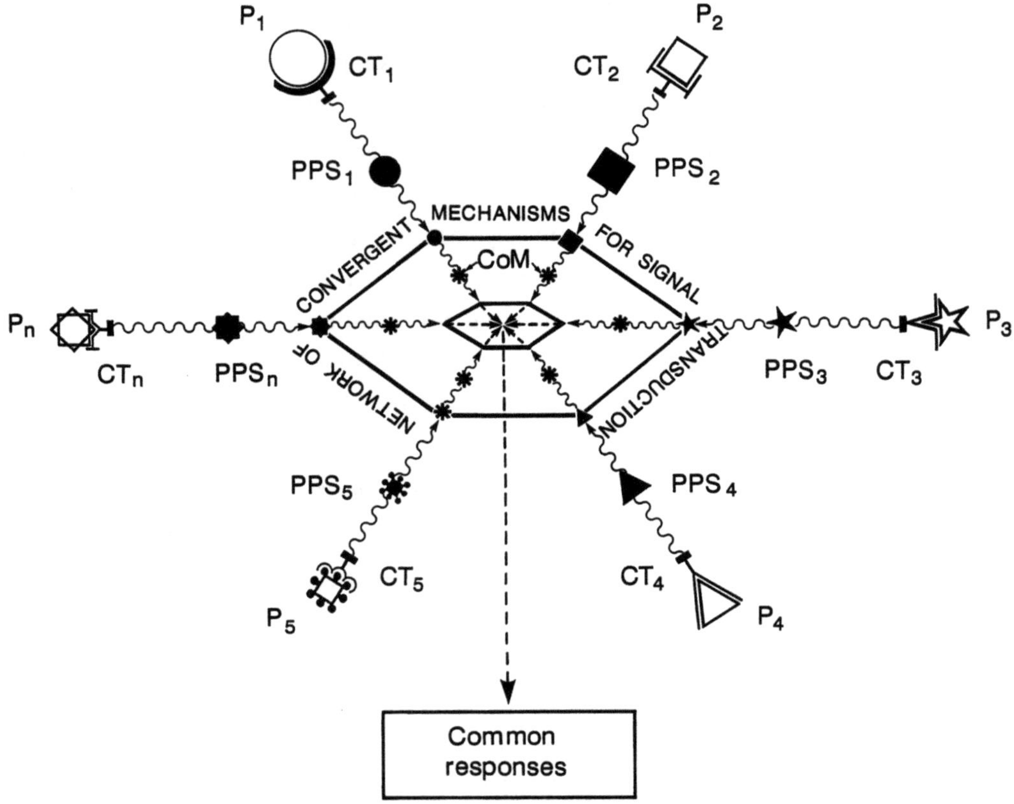

Figure 1. Convergence in transduction and response to pathogenic signals. Originally different pathogenic signals are conveyed, through integrating mechanisms, into a common message leading to elicitation of common responses. P_1-P_n - different pathogens; CT_1-CT_n - cellular targets or sensors; $P_1.CT_1$ etc - primary pathogenic events; PPS_1 etc - primary pathogen signalling; CoM - common messenger.

These fluxes begin with the "perception" of a change in the environment ("primary signal" or "stimulus"). The signal is "transduced" through intermediary steps ("secondary" or higher order "signals" or "messengers") into a message which is understandable by cellular targets with the capability of giving a "short-term" response, or by the genetic machinery when a more permanent response is required.

The signals can be very distinct; examples include radiation, temperature, pressure, non-biotic chemicals, certain regulatory molecules, cell-to-cell structural contacts, structural components of pathogens and breakdown products of cell structures. The capacity of response to certain of these signals is on the basis of the morphogenic processes associated with ontogeny. This capacity would also explain the homeostasis and adaptability that characterize living systems. In this respect plants are no exception. Because of their lack of mobility, they have to accommodate to the environment and their capacity to respond to external signals needs to be highly developed.

Convergence in the mechanisms of signal transduction and response to signals

Have all different signals specific perception and transduction systems? Although there is not enough experimental evidence to answer this question, it is probably not the case. In general, the evolution of these systems appears to have been guided by a principle of mechanistic economy when diversification has not been essential. This tendency has led to development of common mechanisms of information processing, and the idea that plants are endowed with a common mechanism of response to different afflicting agents is in accordance to this principle. But economy is not the only advantage of having a number of interlinked response reactions to form a general system of response (Fig. 1). Such a system would have at least two additional advantages compared with independent transduction chains: control would be easier, and attack by certain pathogens would elicit defence against subsequent infection by other pathogens.

Viroids as signals

Viroids are signals because they induce host responses. Since the molecular message of viroids is not translated into specific proteins that could help in their replication or pathogenicity, we can say that viroids are "pure signals". This is an essential difference from viruses. The "signalling language" of viroids, however, is like that of viruses in that it is recognizable by the host enzymatic machinery as a replicative order or request: "please replicate me" (Fraser, 1987). In susceptible hosts the viroid is also recognized as a "pathogenic message".

Do these signals have any special structural traits enabling them to exert biological functions without the aid of any viroid-specified protein? It is an increasingly accepted idea that viroids probably associate *in vivo* with host proteins in order to accomplish some aspects of their infectivity and pathogenicity, but full experimental confirmation is still awaited.

From the point of view of their nucleotide composition viroids have no special characteristics, such as strange or modified purine or pyrimidine bases. Viroid RNAs, however, have very singular structural features: they are covalently closed circular RNA molecules which under non-denaturing conditions adopt a very compact conformation that confers a rod-like appearance. This conformation is maintained by means of a high proportion of intra-molecular base-pairing, interrupted only in certain regions where the lack of association gives rise to the so called "loops". A number of theoretical models have been developed in attempts to correlate these structural features with the replicability and pathogenicity of viroids (Sänger, 1982; Flores, 1984; Schnölzer *et al.*, 1985; Visvader and Symons, 1985).

The approach has not been restricted to studies of the differential behaviour of naturally existing viroids and viroid strains in combination with different hosts. New technologies for the construction of viroid cDNA clones and infectious RNA transcripts, site-specific mutagenesis and construction of chimeric viroid cDNAs combining fragments from different viroids or strains of a viroid have permitted new approaches (for a review see Owens and Hammond, 1987).

However, a number of difficulties have to be overcome. There is uncertainty about the native conformation of viroids. Viroids might adopt very different conformations when interacting with host components during replication or pathogenesis. Secondly, it is very difficult to relate precisely localized point mutations with loosely defined and evaluated biological properties such as replicability and pathogenicity. A more accurate knowledge of the basic component steps of viroid-host interaction will be necessary to establish meaningful structure-function relationships.

Are the signals between viroid and plant extracellular or intracellular?

To answer this question we need to distinguish between initiation of the infection in the "primary target cells" and the subsequent progress of the infection by cell division. Viroids are extracellular signals whether they arrive in a cell from outside the host plant, or from previously infected cells by systemic spread. However, when spread in association with cell division (progeny viroids are split between daughter cells), viroids must be considered as intracellular signals. In this regard, it is important to note the following: for viroids to be perceived as pathogenic signals they may have to be present at a certain level; this level seems not to be reached before replication; the bulk of the viroid population in the developing leaf tissue (in which the symptoms arise) is probably spread through cell division, so viroids are probably maintained inside the cells most of the time; and there are data indicating that viroids are pathogenic intracellularly. It has been shown that CEVd induces enhanced biosynthesis of ethylene and PR proteins in permanently infected tomato cell cultures (Bellés *et al.*, 1989b).

Are viroids primary signals or secondary messengers?

Again, it is necessary to dissect the question and consider the replicative and pathogenic messages separately. As far as the replicative message is concerned viroids (+RNA) are at the same time the primary signals, recognized by host RNA polymerases, and the response (product) of the replicating machinery. Minus-RNA copies can be regarded as the amplifying intermediary step of the transduction chain. This intermediary form could be considered as a "second messenger". This transducing circuit is different from the more common form where a signal is perceived at the plasma membrane. In this case the second messengers convey information from plasma membrane to cytosol and nuclei (for a review see Boss and Morré, 1989).

As pointed out above, viroids need to be amplified through replication not only to be infectious but also to switch on a pathogenic response. This might imply that viroids as pathogens also have the character of secondary messengers; amplification is a distinctive feature of conventional second messengers (Boss and Morré, 1989). Nevertheless, this role cannot be attributed to viroids since the product of the amplification of the pathogenic message (progeny viroid molecules) and the primary infecting signals are identical. No pathogenic signalling appears to derive from replication as a process (Conejero and Granell, 1986; Diener, 1987; Semancik and Conejero, 1987).

Recognition as infectious signals and perception as pathogenic signals: the question of specificity

Viroids are not infectious to all plants; only certain hosts are susceptible. One may conclude that viroids are specific in both their establishment as infectious entities (recognition, replication and spread) and in their capacity to incite a disease in a given host.

Specificity in viroid-host interactions, as in other systems, cannot be discussed without considering the biological meaning. We need to distinguish a differential affinity for cellular targets leading to damage (fortuitous specificity), from a selective recognition generated by coevolution of viroids (or any pathogen) with the host plant as a consequence of a history of interactions. The fact that a plant is a host or non-host for a given viroid does not imply *a priori* any adaptive or integrating process responsible for the compatibility or non-compatibility. In the case of an ancient viroid-host relationship, the possibility of specific resistance, tolerance or susceptibility can be entertained. Finally, it is interesting to point out that the mechanisms of replication and pathogenesis seem to be non-specific.

It is possible, however, that viroids may have been native components in certain plants. They may have become infectious through evolution, by loss of control of their synthesis, thus leading to a kind of "auto" or "self-infection". In these plants newborn viroids ("protoviroids") would still be native components. These replicating RNAs might be transferred in some unknown way to related plants in the wild. More recently, with the intensification of agricultural practices, viroids would probably have been transmitted to phylogenetically-related cultivated plants with the conserved capability to "recognize" viroids as "self". Viroids would not be normal constituents of these plants but they would have a certain degree of structural homology with host components enabling them to reach and penetrate competent cells and to be replicated by host enzymes. One can speculate that there could also be a specific viroid-host interaction during the establishment of the infection.

How the viroid-cell compatibility is determined is a challenging question. The specificity determinants could be at the level of viroid entrance to the cell, the accessibility of or recognition by the replicative machinery, or specificity could be determined by a combination of both types. In any case, the determinants or compatibility factors would only be present in a combination suitable for initiation of infection in certain meristematic cells at a given phase of the cell cycle (Semancik and Conejero, 1987).

Since the pathogenic encounter of plants and viroids is thought to be relatively recent (Diener, 1987), it is conceivable that the pathogenic components of viroid-host interaction may lack the specificity traits that one would expect from coevolution of very old partners, as seems to be the case for viruses, bacteria and fungi and their host plants.

In conclusion, it is possible to explain a specific recognition of viroids as "self" by the host during the establishment of the infection, leading to compatibility. Disease, however, would be a reflection of the sensing or perception of viroid molecules as "non-self" by the host, and the subsequent pathogenic reaction. If a given host has not the capacity, either to perceive viroids as "non-self" or to react after the perception, then it would be a non-susceptible (tolerant) host.

At this point the question arises: which is the structural or dynamic cellular target that interacts with viroid molecules to trigger the pathogenic response? At present there is no answer to this question; little experimental effort has been devoted to it. Nevertheless, it is worthwhile to comment on work which has looked for complexes of viroids with cell components.

From these studies the clearest conclusion is that the bulk of the viroid population seems to be present in the nucleolus (Schumacher *et al.,* 1983). Thus the possibility that the primary pathogenic encounter takes place in the nucleolus has been proposed (Riesner 1987). Experiments *in vitro* have demonstrated an affinity of potato spindle tuber viroid (PSTVd) for histones and for a non-histone protein (Wolf *et al.,* 1985; Klaff *et al.,* 1989). However, the physiological significance of this finding is unknown.

Recently, Hass *et al.* (1988) have found a sequence complementarity between the 5' terminus of tomato signal recognition particle (SRP) RNA and the "lower strand" of the rod-like RNA secondary structure model of PSTVd and four other viroids (CEVd, chrysanthemum stunt viroid (CSVd), tomato apical stunt viroid (TASVd) and tomato planta macho viroid (TPMVd)), which also replicate and incite disease in tomato plants. Interestingly, this region includes the virulence-modulating (VM) region (Schnölzer *et al.,* 1985) or "P domain" (Visvader and Symons, 1985; 1986). On this basis Haas *et al.* (1988) suggest that SRP RNA would be a possible primary cellular target with which viroid molecules could interact, thus inciting disease. Two of the potential mechanisms proposed involve direct interaction of viroid RNA either with SRP RNA or with SRP protein. In both cases the pathogenesis is explained as an alteration of translocation and integration of membrane proteins into endoplasmic reticulum in differentiating tissue at the growing apex. This would result in abnormal formation of cellular membranes in meristematic cells and initiate the development of symptoms like leaf malformations and retardation of growth.

It is interesting to note that the authors integrate in their model the fact that the mature PSTVd molecules accumulate in the nucleolus with a specific stage of cell division in which the interaction would be possible. In this stage the nucleolus and the nuclear membrane disintegrate and mature viroid molecules could be transiently released into the cytoplasm (Haas *et al.,* 1988). The problem which remains is how to reconcile the exquisite and precise nature of the viroid-SRP RNA interaction that would imply 30 bp complementarity, with the reputed lack of specificity of viroids as pathogenic signals or the commonality of the response elicited by viroids.

If the hypothetical viroid-SRP interaction were the real initiator of pathogenesis, one would have to admit a specific primary signalling and a non-specific response in viroid pathogenesis. If that were the case, it is not difficult to imagine that a distortion at the level of the plasma membrane or other cell membrane could trigger a general mechanism of response, which could be activated at the same membrane level by signals from other pathogenic stress factors. In this concept the primary pathogenic event could be less specific than that proposed by Haas *et al.* (1988). Any interaction giving rise directly or indirectly to signals (biochemical changes) capable of being sensed or perceived by the appropriate cell membrane would trigger the same response, through release or production of common second messengers.

Intermediary messengers of the viroid pathogenic signal

There is increasing information in support of the idea that in the signal transduction systems of plants, the plasma membrane has an important role in conveying information from outside the cell to the cytoplasm (Blowers and Trewavas, 1989). There, a series of transduction fluxes lead up to the responding elements. These compounds are called second messengers. The identification of second messengers in plants and their role in signal transduction is only commencing.

Much effort has been devoted to confirming the postulated rôle of cAMP as a second messenger in plants. Neither adenylate cyclase (Yunghaus and Morré, 1977) nor cAMP-dependent protein kinase, the physiological receptor for cAMP, have been found in plants (Brown and Newton, 1981). Furthermore, cAMP has never been found to be required for any physiological response (Hepler and Wayne, 1985). Later, a number of components of plant metabolism have been suggested to play a rôle as second messengers in plant systems: Ca^{2+}, K^+, H^+, NAD^+/NADH, polyamines, diacylglycerol and phosphoinositides (Boss and Morré, 1989). Among these the most clearly accepted as a second messenger is the Ca^{2+}-calmodulin system (Marmé, 1989).

Another good candidate as a second messenger for signal transduction in plant systems is ethylene. This is a main component of the hormonal system which controls plant development and is involved in plant response to different kinds of stress (Lieberman, 1979; Yang and Hoffman, 1984). The location of its biosynthetic machinery either in the plasma membrane or the tonoplast (Guy and Kende, 1984) would be consistent with the idea that its biosynthesis and release could amplify and transduce signals perceived by a sensor located at these membranes.

Until very recently almost all experimental effort on how second messengers might be involved in signal transduction in host-viroid interactions has been carried out in our laboratory. Our concern has been to demonstrate the role of ethylene as a mediator of the viroid-induced response. A number of considerations and data support this idea:

1. Treatment of *Gynura aurantiaca* plants with silver nitrate (500-1000 mg/l) produces a developmental syndrome in the host which is characteristic of CEVd-infected plants, including stunting, leaf malformations, inhibition of rooting and the accumulation of the same PR proteins (P1 and P2) (Conejero and Granell, 1986). In tomato plants, silver nitrate and CEVd both induced the ten cationic PRs (Granell *et al.,* 1987).
2. The viroid-like effects elicited by Ag^+ ions include the induction of a cross-protection-like phenomenon, since Ag^+ at low concentration impairs symptom expression in CEVd-infected plants. The characteristics of this protection suggest that it is produced by interference with the establishment of the viroid infection (Granell *et al.,* 1987).
3. Both CEVd infection and Ag^+ (Conejero and Granell, 1986) stimulate ethylene production.
4. Treatment of plants with 2-chloroethylphosphonic acid, (ethephon, an ethylene releasing compound) at high doses (5000 mg/l) induces the viroid-like syndrome and the same PR proteins in both *Gynura* (Semancik and Conejero, 1987) and tomato plants (Granell *et al.,* 1987).
5. The relative intensity of PR production in three tomato cultivars (Rutgers, Rentita and Hilda 72) as a response to ethephon treatment was the same as that in response to CEVd infection (Granell *et al.,* 1987).
6. Ethephon at a low rate (500 mg/l) is able to impair the establishment of viroid (CEVd) infection (Semancik and Conejero, 1987).
7. Inhibition of either of the last two steps of ethylene biosynthesis: the conversion of S-adenosylmethionine (SAM) into 1-aminocyclopropane-1-carboxylic acid (ACC) and the conversion of ACC to ethylene, inhibits PR production and the expression of symptoms in silver-treated *G. aurantiaca* plants. The same inhibitory effect was obtained with norbornadiene (an antagonist of ethylene action) (Bellés and Conejero, 1989).

Ethylene is not the only phytohormone to have been implicated in viroid pathogenesis. A decrease in gibberellins related to the stunting reaction in CEVd-infected citrus trees (Hanks and Feldman, 1972) and *G. aurantiaca* plants (Rodríguez *et al.*, 1978) is the clearest effect reported. The effects on auxin-like substances and in ABA in both type of hosts were contradictory (Semancik and Conejero, 1987). In infected citrus, auxin-like substances decreased and an inhibitor presumed to be abscisic acid increased. In *Gynura*, however, an IAA-like auxin remained unchanged. Low levels of an additional auxin-like substance appeared and ABA was not affected by the presence of viroid. Also, a reduced root initiation in CEVd-infected *Gynura* was correlated with a reduction in a diffusible auxin-like substances from the apical buds (Flores and Rodrígues, 1981). That these or other hormonal changes could be involved in transducing viroid pathogenic signalling into the cellular response would be in accordance with the notion of plant growth substances as general second messengers (Blowers and Trewavas, 1989).

More recently, attempts to fill the gaps in our knowledge of viroid pathogenesis have been directed to protein phosphorylation. This is reputedly a key element in the signal transduction network of plants (Ranjeva and Boudet, 1987) and was chosen by two research groups working with different viroids infecting tomato plants: PSTVd (Hiddinga *et al.*, 1988) and CEVd (Vera and Conejero, 1989). The work on PSTVd concentrated on the viroid-induced enhancement of phosphorylation of a 68 kDa host protein. According to the authors, this is homologous to human double stranded (ds) RNA-dependent protein kinase (Hiddinga *et al.*, 1988). In addition to this effect, the work on CEVd described increased phosphorylation of other proteins of 40, 30, 27, 24, 23 and 22 kDa. The CEVd-induced effect on protein phosphorylation was Mn^{2+}-dependent (Vera and Conejero, 1989).

Due to their secondary structure, viroids can mimic a ds-RNA. Therefore, the dsRNA-dependence of the enhanced phosphorylation of the 68 kDa homologous to the human kinase was interpreted as a dsRNA-mimicking effect of PSTVd. Thus the activation of the 68 kDa kinase would be the primary event in viroid pathogenesis (Hiddinga *et al.*, 1988). Unfortunately, results obtained with CEVd (Vera and Conejero, 1989) differ from those with PSTV in that the phosphorylation was not dependent on exogenous dsRNA molecules, and the protein did not cross-react with antibodies against human dsRNA-dependent protein kinase 68. The apparent discrepancy between these results needs further study. The identification of the target proteins, the kinases and phosphatases involved, and their connections with other steps of the transduction network also remain open questions.

THE RESPONSE

Macroscopic components of the response

The response of hosts to pathogens consists of active reactions specifically directed to arrest the establishment of the infection or to avoid or counteract the pathological effects; the disease itself is also part of the response. Furthermore, even if it is necessary or possible to distinguish pathological reactions from defence reactions for analytical purposes, this is not always the best strategy. Sometimes the two main aspects of the response are so intimately related as to be impossible to separate clearly. Host physiological and structural changes involved in disease expression may eventually also become components of resistance mechanisms.

In this chapter, the developmental syndrome, the induced systemic resistance and the physiological condition of resistance that the stress associated with disease produces in certain cells or tissues, will be considered as components of the viroid-elicited response (Fig. 2).

Cytopathic changes

Under this heading we will not attempt a complete review of the subject; for this see Semancik and Conejero (1987). Our aim is to emphasize those data most relevant to our main argument:

1. The absence in infected plants of any effect(s) known to be *exclusively* viroid-induced. Many of the observed host responses to the presence of the viroid resemble responses which result from either virus infection or physiological stress.
2. Even though the nucleus has been generally accepted as the site of viroid replication and, potentially, the site where the primary pathogenic event could take place, there seems to be a general consensus that there is no cytopathic structural alteration at the nuclear level. This fact further supports the idea of segregation of viroid replication and pathogenesis (Conejero and Granell, 1986).
3. Alterations have been described in cell walls (Wahn *et al.,* 1980), membranes such as the plasmalemma (Semancik and Vanderwoude, 1976; Wahn *et al.,* 1980;

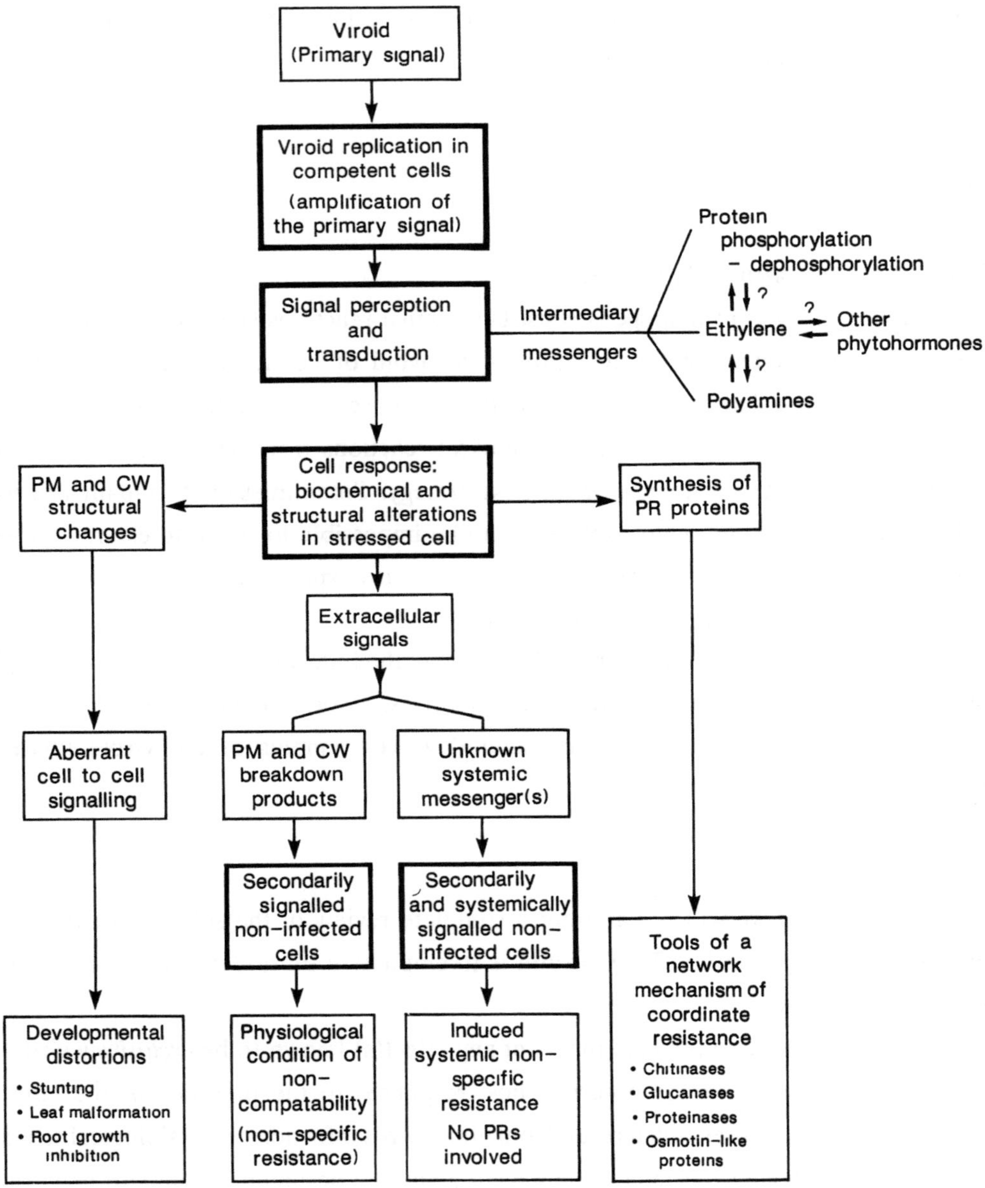

Figure 2. A model to explain how the viroid-induced response at the cellular level becomes a developmental disease and leads to a plant which is more resistant to subsequent infections.

Hari, 1980) the tonoplast (Kojima *et al.,* 1983; Paliwal and Singh, 1981) and the thylakoids (Kojima *et al.,* 1983). Together with the rôle of the cell wall and membranes in recognition reactions for a number of regulatory processes, this has led to speculation that the viroid itself or a viroid-induced signal may be capable of interacting with these reactive interfaces (Semancik and Conejero, 1987).

Molecular components of the response

This is perhaps an over-ambitious heading, considering the lack of knowledge on this subject. For the scanty and scattered data available, terms like "biochemical alterations" or "metabolic changes" might be more appropriate. But the point is that we need to envisage the molecular changes involved in viroid pathogenesis within the framework of signalling, and not only as a more adequate way of organizing the cumulated data. We need this framework as the new paradigm under which to approach future research. We need to know the primary pathogenic event, the intermediate steps, the end products and their mutual regulatory interconnections. Also, a major interesting aspect of these networks is which of their components become regulatory bridges implicating the genetic machinery in signal transduction as well as in the final response.

We will now summarize the data that biochemical studies on viroid pathogenesis have produced. We shall omit the biochemical changes that have already been discussed as intermediary steps of the transduction flux, and restrict the discussion to the end products of the network i.e. the molecular components of the response.

Nucleic acids. Apart from viroid molecules (+RNA) and the intermediary forms involved in their replication (±dsRNA), there is no reported evidence on any qualitative or significant quantitative alteration in the normal nucleic acid constituents of the host cell (Diener, 1987). This restricts the involvement of nucleic acids as components of the pathogenic process to the viroid-related RNA forms, as already discussed.

PR proteins as non-specific components of the response. Attempts to demonstrate the functionality of viroids as mRNAs in *in vitro* (Davies *et al.,* 1974; Hall *et al.,* 1974) as well as in semi *in vivo* (Semancik *et al.,* 1977) protein synthesizing systems gave negative results. *In vivo* studies on viroid infected tissues (Zaitlin and Hariharasubramanian, 1972; Conejero and Semancik, 1977a) did not provide any support for this putative role of viroids. However, the *in vivo* studies revealed that alterations in host protein metabolism are implicated in viroid pathogenesis. Nevertheless, it was not until the idea that viroids do not specify protein synthesis either *in vitro* or *in vivo* became accepted, that the search for host-coded proteins involved in the viroid-incited pathogenic response (PR proteins) commenced in a systematic way.

The first report implicating proteins in viroid pathogenesis was the detection in PSTVd-infected tomato leaves of two proteins with molecular weights 155 and 195 kDa (Zaitlin and Hariharasubramanian, 1972). Five years later, two low molecular weight polypeptides of 15 kDa (CEVd-P1) and 18 kDa (CEVd-P2) were detected in leaves of CEVd-infected *Gynura aurantiaca* (Conejero and Semancik, 1977b).

Enhanced levels of analogous polypeptides were found when the search was extended to additional host plants (Etrog citron, tomato cv. "Rutgers", potato cv. "Marijke") infected with CEVd (Conejero *et al.,* 1979). These polypeptides showed slight variation in size, depending on the host. This fact could be compatible with a viroid-specified nature of these polypeptides and a host-dependent processing reaction. Nevertheless, in one host, *G. aurantiaca,* such polypeptides were also detected in naturally senescing plants, indicating that they were not translation products of the viroid RNA, nor of any mRNA exclusively related to the viroid-host interaction. This finding also excluded the possibility proposed by Matthews (1978) that the polypeptides might be specified by viroid complementary sequences.

Later, the accumulation of several low molecular weight polypeptides was reported for a number of plants infected with several viroids (CEVd, PSTVd, cucumber pale fruit viroid (CPFVd), CSVd) and, in the case of tomato, infected by tobacco mosaic virus (TMV) and cucumber mosaic virus, and the fungus *Cladosporium fulvum* (Camacho-Henriques and Sänger, 1982a,b). More recently, an enhanced level of a high molecular weight protein (140 kDa) was also found in tomato plants infected with TPMVd (Galindo *et al.,* 1984). In addition, decreases in a number of protein species in the range of 14 kDa to 45 kDa were found in different hosts infected by CSVd. Decreased levels of four polypeptides (14.5 kDa to 105 kDa) were associated with pathogenesis in tomato leaves infected either by viroids, viruses or *Cladosporium fulvum* (Camacho-Henriquez and Sänger, 1982b).

All these observations strengthened the idea that the enhanced occurrence of certain proteins which accompanies viroid infection is a host response that may be generalized beyond this particular type of host-pathogen interaction. Thus, these proteins were considered to belong to the class of "pathogenesis-related" or "PR" proteins where the definition includes all proteins which are stimulated in plants as a consequence of pathologically altered metabolism (Van Loon, 1983).

The finding that the polypeptides associated with viroid infection are not specified by the viroid but result from a disease-induced alteration of host metabolism (Conejero *et al.,* 1979; Camacho-Henriquez and Sänger, 1982a,b), prompted the search for other proteins as possible components of the pathological response, as had been described for other systems (Van Loon and Van Kammen, 1970; Gianinazzi *et al.,* 1970). Since the non-denaturing electrophoresis system (Davis, 1964) normally used to detect tobacco PRs excluded basic proteins, a system with the capability of separating cationic proteins (Reisfield *et al.,* 1962) was also tested. This approach led to the detection of ten PR proteins (C1 to C10) in tomato plants infected with CEVd (Granell *et al.,* 1987). Only four cationic PR proteins could be detected in CEVd-infected *G. aurantiaca* (Semancik and Conejero, 1987), again emphasizing the specificity of host control in the biosynthesis of proteins. The basic nature of nine of the tomato proteins contrasts with most PR proteins described so far (Van Loon, 1985). However, the tomato PR proteins do share with tobacco PR proteins the characteristics of being preferentially extracted at low pH and resistant to proteinase digestion (Granell *et al.,* 1987). The tomato PR protein C2, which increases most dramatically, corresponded to the polypeptide previously described as tomato CEVd-P1 (Conejero *et al.,* 1979) or p14 (Camacho-Henriquez and Sänger, 1982b). The basic nature of this protein has been pointed out (Camacho-Henriquez and Sänger 1984) and recently confirmed by sequencing (Lucas *et al.,* 1985). The 69 kDa polypeptide most probably coincides with the 70 kDa polypeptide reported for tomato infected with TPMVd (Diener, 1987) while some of the other SDS-detected proteins matched those reported for tomato infected with PSTVd (Camacho-Henriquez and Sänger, 1982b). These PR proteins were also induced by Ag^+ and ethephon treatments in association with necrotic (localized) lesions or non-necrotic (systemic, viroid-like) reactions (Granell *et al.,* 1987) indicating that their accumulation is not correlated with necrosis. As already discussed, these findings gave support to the idea that PR proteins are components of a non-specific response of the host mediated by ethylene (Granell *et al.,* 1987).

Biological significance of the PR proteins. The discovery of these proteins posed a number of questions, among which their biochemical function and biological role are the most important, not only for understanding of viroid pathogenesis but also as components of a system of response of plants to pathogenic or stress signals.

Even before the concept of PR proteins had been crystallized, attempts to unravel the biochemical activity displayed by viroid-induced proteins had been made with the

P1 protein from CEVd-infected *G. aurantiaca*. In these studies the potential activities of this protein as ribonuclease or hormone-binding protein were tested with no conclusive results (Flores *et al.*, 1978).

Recently, P-69 has been characterized as an alkaline cysteine proteinase (Vera and Conejero, 1988). It has also been found that P-32 and P-34 (C7 and C6 respectively of Granell *et al.*, 1987) are chitinases (García-Breijo *et al.*, 1989). Also, a ß-1,3-glucanase activity has been assigned to a CEVd-induced PR protein in tomato (F García-Breijo, unpublished results).

Critical to the unravelling of the biological role of these proteins is their *in vivo* localization. The electrophoretic detection of PR proteins in extracts obtained by vacuum infiltration of tomato leaves infected with TMV (Parent and Asselin, 1984), treated with chemicals (Hooft van Huijsduijnen *et al.*, 1986), or after infection with *Cladosporium fulvum* (De Wit and Van der Meer, 1986) prompted the idea of the intercellular localization of PR proteins. Recently Carr *et al.* (1987) have reported the localization of PR1 in tobacco leaves by immunofluorescence. A more direct approach, the immuno-gold-EM technique applied to P1 (p14) (Vera *et al.*, 1988) and P69 (Vera *et al.*, 1989b) led to the discovery of two main locations: the vacuole, in association with inclusion bodies, (a newly described location) and the intercellular spaces of CEVd-infected tomato leaves.

We have already commented on the reputed resistance of PR proteins to degradation by endogenous proteases (Van Loon, 1985). Consistently, they have long half-lives (40-70 h) (Matsuoka and Ohashi, 1986), but there must be some turnover. The fact that they accumulate in the intercellular space (Van Loon, 1985) suggests that their degradation might also occur there. Following this idea, evidence has been obtained (Rodrigo *et al.*, 1989) that some tomato PR proteins are degraded upon incubation of intercellular washing fluids (IWF). This degradation occurs at acidic pH and it has been found that the enzyme responsible is a 37-kDa constitutive aspartyl endoproteinase with a pH optimum of 2.5-3.5. The selective hydrolysis of some of the pathogenesis-related proteins by the 37 kDa aspartyl proteinase lends credence to a possible rôle for this proteinase as part of the regulatory mechanism for the biological action of the PR proteins. The biological relevance of this enzyme in the PR metabolism is reinforced by the fact that an analogous enzyme has been isolated from IWF of tobacco plants (I Rodrigo, P Vera and V Conejero, unpublished results).

Although the biological role of most of these tomato PR proteins is as yet unknown the following must be stressed:

1. Chitinases and ß-1,3-glucanases can clearly be implicated in defence against pathogens such as bacteria and fungi, containing chitin or ß-1,3-glucans as a component of their structure (Boller *et al.,* 1983).

2. The rôle of P69 proteinase (Vera and Conejero, 1988) either in symptom production or as a defence tool remains to be elucidated.

3. There is no evidence for any biological activity of P1 (p14), the most intensely accumulated PR protein in leaf tissue as a consequence of the viroid induced response. Nevertheless, P1 has been found in leaves from healthy (non-infected) plants and always in association with cell material undergoing disorganization (Vera *et al.,* 1988). This led these authors to propose the idea that P1 is involved in cell degeneration, either naturally activated as a normal event of the biological cycle of the plant (e.g. in lysogenous development of intercellular spaces (Esau, 1972; Fahn, 1972) or by cell ageing) or provoked by afflicting agents. The possible involvement of P1 in the resistance induced in the systemic reaction of the host or with some other biological role cannot be discarded. The biological significance of P1 is also sustained by the fact that this protein is synthesized as a pre-protein, then targetted to the vacuole and to the apoplast (Vera *et al.,* 1988; 1989a) in association with P69 (Vera *et al.,* 1989b). The vacuolar and apoplastic localization of chitinases have also been reported (Mauch and Staehelin, 1989).

4. The reported evidence to date indicates that ethylene is an intermediary step (second messenger) in the coordinate activation of PR synthesis as part of the response to viroid infection (Conejero and Granell, 1986; Granell *et al.,* 1987; Bellés and Conejero, 1989).

All this suggests that the viroid induced PR proteins, despite their different biochemical activities, might have the common biological role of being part of an adaptive response against potential agressors from the environment.

Cell wall components. Three types of biochemical alterations of the cell wall can be considered for possible involvement in the response of the plant to viroid infection: those involved in limiting cell expansion; those implicated in altering the normal cell-to-cell communication and signalling in developmentally active cells and tissues; and breakdown products with the capability of becoming secondary pathogenic signals for neighbouring cells.

The effects of viroid infection on cell walls were realized in cytological studies.

These cytopathic effects received biochemical confirmation when a differential release of protoplasts from healthy and viroid infected tissues was attributed to abnormal ß-1,3-glucan in viroid infected cells (Marton *et al.,* 1982).

More direct information was obtained with purified cell walls from healthy and CEVd-infected tomato cell suspension cultures. Enhanced levels of the key constituents of extensin, hydroxyproline and specific arabinosyl residues, were detected (Wang *et al.,* 1986), and taken to indicate an increased extensin content. An ethylene-induced accumulation of hydroxyproline rich proteins has also been found in diseased plants, and this accumulation has been correlated with the developmental alterations produced (Wang *et al.,* 1986).

A possible mechanistic relationship between the viroid-enhanced ethylene production and its effects in limiting cell expansion growth involving peroxidases and extensin has been outlined (Semancik and Conejero, 1987).

Finally, prominent protrusions have been detected on the surfaces of CEVd-infected cells (Wang *et al.,* 1986). Although the chemical nature of these structures has not yet been determined, it is possible that they are related to the alterations in cell wall composition. A possible relationship between the specific restriction of cell wall growth and the appearance of endocytic invaginations (plasmalemmasomes) reported in viroid-infected cells has also been suggested.

Individual specific reactions integrated in a general non-specific network system of response. The idea that the whole metabolic structure of a plant consists of pathways of sequential steps which are interconnected in a highly reticulate network seems well accepted. This type of organization, together with cascade systems to amplify signals, permits synchronous control of many biochemical functions (Blowers and Trewavas, 1989). It also explains how very different pathogenic signals can elicit common responses (Fig. 1). Otherwise, it is difficult to explain how a viroid is able to elicit a response including coordinate expression of genes encoding chitinases, glucanases, etc. whose biological role is defence against infections agents or stresses other than viroids.

One could hypothesize that these coordinately interconnected defence tools may have been built up in two steps: specific and individual development of each tool by coevolution with a given pathogen or with aggressive environmental conditions, and then the interlinking of the individual specific responses. Thus, what appears to be a non-specific response may result from evolutionary integration of specific individual components. With this strategy, the battery of defensive weapons would be progressively enriched by new experiences. A prediction of this model would be the possible existence or future development of proteinases and nucleases against viruses and viroids.

The lack of specificity of this response could have biological advantages for the plant. This type of mechanism would broaden the scope of potential pathogens against which the plant would have acquired a certain degree of immunity after having been non-specifically triggered (Fig. 1).

AN INTEGRATIVE MODEL

How a response at the cellular level becomes a developmental disease: intracellular and intercellular stimulus-response processes

The general inhibition of growth in viroid infections is readily understandable. It is proposed that primary infection might occur in the region of differentiation of the procambial initials in the growing apex (Semancik and Conejero, 1987). Then, synthesis and release of ethylene is persistently enhanced during viroid infection resulting in inhibition of growth (Bellés *et al.*, 1989a).

It is more difficult to explain how the pathogenic response at the cellular level has effects at higher levels of organization, thus producing the distorted pattern of leaf development, for example. A precise system of cell-to-cell communication is required for the control of the normal pattern of growth and differentiation of the leaf. Failure or changes of some critical cell-to-cell connections, produced by alterations of the cell wall structure or other signalling events, could distort developmental patterns. A differential susceptibility of cells to becoming infected and to producing a pathogenic response would contribute to the macroscopical anisotropy of development. This hypothesis would also explain how different pathogenic agents such as a viroid (CEVd) and Ag^+ induce the same macroscopic syndrome, if both agents elicited the same general response in the cells. But we still need to explain how two agents with different pathways of access to cells ($AgNO_3$ is applied by spraying, viroids travel through the vascular system) incite indistinguishable developmental aberrations. This difficulty can be overcome if it is assumed that the same juvenile cells with a special reactivity and with the same developmental role are preferentially affected by both agents. We are dealing with an adaptive morphogenetic response to viroids which, following our reductionist ideas, conveys information through common "informational channels". This is true not only at the cellular level (integrating membranes, common second messengers and coordinate activation of certain genes) but also at the supracellular level (commonality of morphogenetic information fluxes which can be distorted in the same way by alteration of certain especially susceptible key cells).

The model involves juvenile compatible cells which have the capacity to become infected and to display a pathogenic response (i.e. to undergo affliction and produce defence reactions) (Semancik and Conejero, 1987). Of these cells, a few would be the primary target cells: those at which the viroid molecules arrive from outside. Also, a small proportion of these cells would be infected by viroid molecules coming from other cells. But the bulk of infection is probably through cell division. The infected cells undergo a series of biochemical changes as a consequence of which they become physiologically and structurally altered. In the language of the signalling model, viroids as "primary signals" are amplified through replication, then "sensed", most probably at the membrane level, thus triggering a network of transductional events leading to cell response. The cell becomes a viroid-diseased cell, then transmits mismessages of aberrant developmental information horizontally, to other cells. Disease mismessages are also transmitted vertically by cell division.

All this conforms the developmental syndrome, which may also involve non-infected cells brought in by secondary signalling. This reflects the existence of a cascade system of signal transduction in the supracellular network, analogous to that at the cellular level (Fig. 2). In this case, one would predict, for example, that ethylene synthesis and PR protein production would occur in "secondarily stressed" but non-infected cells of infected leaves.

How does cellular response lead to a more resistant plant?

Apart from the known phenomena of specific cross-protection among viroids (interference with the establishment of a severe strain of a viroid by prior infection and specific interaction with an homologous milder strain) there are other possible mechanisms of resistance and protection in which signalling and response may be implicated. Chitinases, ß-1,3-glucanases and other PR proteins potentially implicated in defence reactions have already been discussed and may have specific defensive significance (Vera *et al.,* 1989a,b; Mauch and Staehelin, 1989), although none is known against viroids. Non-specific physiological resistance may also occur as a consequence of the stress imposed by infection in either infected or non-infected neighbouring cells which have been signalled. These cells lose the juvenile condition that enables them to be compatible and responsive to viroid infection. This may be the kind of protection produced by Ag^+ and ethephon treatments at low doses. The lack of specificity in the so-called cross-protection between non-homologous viroids (Niblett *et al.,* 1978) could also be an example of non-specific physiological resistance.

In systems reacting hypersensitively (which is not in general the case for viroid infections) there is another type of defensive phenomenon, the so-called induced systemic resistance (as in TMV infection of *N*-gene tobacco). The necrotic localized reaction in fully expanded inoculated leaves induces resistance to virus establishment in younger, upper leaves. Gianinazzi *et al.* (1970) and Van Loon and Van Kammen (1970) suggested that PR proteins could be involved in this phenomenon.

A similar reaction was observed in experiments following the mimicking procedure previously described for tobacco plants (Van Loon, 1977). Fully expanded leaves of *Gynura aurantiaca* plants were pricked with 0.2 M ethephon prior to infection with CEVd. This treatment impaired the establishment of viroid infection. This is taken as additional proof of the general scope of this mechanism of defence. In this case, however, no PR protein was associated with the induced resistance (JM Bellés, unpublished results). This is in accordance with the reported lack of correlation between acquired systemic resistance to viruses and PR proteins (for a review see Fraser, 1987).

CONCLUSIONS AND PROSPECTS

The model outlined in Figure 2 is an attempt to summarize and integrate the data and ideas that have been discussed in the preceeding pages under the concept of host-pathogen interaction as a case of pathogenic signalling and homeostatic response in plants. Although the model has numerous gaps, these are due to lack of experimental data rather than to inconsistency with what is presently known. Deeper knowledge of many mechanistic details is still required, but the model provides a framework within which a number of questions relating to viroid pathogenesis can be considered.

1. How can some host plants actively replicate viroids without any visible pathological syndrome?

2. Why do viroids not "need" to act as messengers to incite disease?
3. How can viroids with very different primary structures produce indistinguishable disease syndromes on a given host?
4. Why is the intensity but not the type of syndrome expressed by the host modulated by the viroid structure?
5. How can the so called cross-protection between rather non-homologous viroids be explained?

Principally, we need to elucidate which is the initial pathogenic alteration or interference produced by viroid molecules; which are the steps leading to the signal that triggers the non-specific network of reactions forming the host response; which are the components of this informational flux; how it is connected with control of gene expression; and what kind of relationship exists between the pathogenic and defence reactions.

Viroids, as the simplest pathogens, offer unique opportunities to study these questions. These opportunities are enhanced by the ease with which viroid molecules can now be engineered. Studies with viroids may give insights into how plants interact with more complex pathogens such as viruses and fungi. However, a major obstacle to progress in studies of plant-viroid interactions is the lack of host systems which are genetically well-defined for variation in response, in terms of resistance/susceptibility and severity of the disease syndrome.

ACKNOWLEDGEMENTS

This chapter is dedicated to Joe Semancik.

REFERENCES

Bellés JM, Granell A, Durán-Vila N, Conejero V (1989a) ACC synthesis as the activated step responsible for the rise of ethylene production accompanying citrus exocortis viroid infection in tomato plants. J Phytopathol 125:198-208

Bellés JM, Vera P, Durán-Vila N, Conejero V (1989b) Ethylene production in tomato cultures infected with citrus exocortis viroid (CEV). Can J Plant Pathol (in press)

Bellés JM, Conejero V (1989) Ethylene mediation of the viroid-like syndrome induced by Ag^+ ions in *Gynura aurantiaca* DC plants. J Phytopathol 124:275-284

Blowers DP, Trewavas AJ (1989) Second messengers: their existence and relationship to protein kinases. In: Boss WF, Morré DJ (eds) Second messengers in plant growth and development. Alan R Liss, New York, in press

Boller T, Gehri A, Mauch F, Vögeli U (1983) Chitinase in bean leaves: induction by ethylene, purification, properties, and possible function. Planta 157:22-31

Boss WF, Morré DJ (1989) Second messengers in plant growth and development. Alan R Liss, New York, in press

Brown EF, Newton RP (1981) Cyclic AMP and higher plants. Phytochemistry 20:2453-2463

Camacho-Henriquez A, Sänger HL (1982a) Gelelectrophoretic analysis of phenol-extractable leaf proteins from different viroid/host combinations. Arch Virol 74:167-180

Camacho-Henriquez A, Sänger HL (1982b) Analysis of acid-extractable tomato leaf proteins after infection with a viroid, two viruses and a fungus, and partial purification of the "pathogenesis-related" protein p14. Arch Virol 74:181-196

Camacho-Henriquez A, Sänger HL (1984) Purification and partial characterization of the major "pathogenesis-related" tomato leaf protein p14 from potato spindle tuber viroid (PSTV)-infected tomato leaves. Arch Virol 81:263-284

Carr JP, Dixon DC, Nikolau BJ, Voelkerding KV, Klessig DF (1987) Synthesis and localization of pathogenesis-related proteins in tobacco. Mol Cell Bol 7:1580-1583

Conejero V (1981) Stimulation of a viroid-like syndrome and inhibition of symptom expression by Ag^+ ion. Abstr 5th Int Congr Virol Strasbourg, France, p 228

Conejero V (1982) Do viroids elicit host pre-existing mechanisms of response? Abstr IV Int Conf Comparative Control of Viral Diseases Banff, Alberta, Canada (Abstr W 12-1)

Conejero V, Granell A (1986) Stimulation of a viroid-like syndrome and the impairment of viroid infection in *Gynura aurantiaca* DC plants by treatment with silver ions. Physiol Mol Plant Pathol 29:317-323

Conejero V, Semancik JS (1977a) Exocortis viroid: alteration in the proteins of *Gynura aurantiaca* accompanying viroid infection. Virology 77, 221-232

Conejero V, Semancik JS (1977b) Analysis of the proteins in crude plant extracts by polyacrylamide slab gel electrophoresis. Phytopathology 67:1424-1426

Conejero V, Picazo I, Segado P (1979) Citrus exocortis viroid (CEV): protein alterations in different hosts following viroid infection. Virology 97:454-456

Davies JW, Kaesberg P, Diener TO (1974) Potato spindle tuber viroid. XII. An investigation of viroid RNA as messenger for protein synthesis. Virology 61:281-286

Davis BJ (1964) Disc electrophoresis. II. Method and application to human serum proteins. Annu NY Acad Sci 121:404-427

De Wit PJGM, Van Der Meer GM (1986) Accumulation of the pathogenesis-related tomato leaf protein P14 as an early indicator of incompatibility in the interaction between *Cladosporium fulvum* (syn *Fulvia fulva*) and tomato. Physiol Plant Pathol 24:17-23

Diener TO (1979) Viroids and viroid diseases. Wiley, New York

Diener TO (1981) Are viroids escaped introns? Proc Natl Acad Sci USA 78:5014-5015

Diener TO (ed) (1987) The viroids. Plenum, New York

Dickson E (1981) A model for the involvement of viroids in RNA splicing. Virology

115:216-221
Esau K (1972) Plant Anatomy, 2nd edn., Wiley, New York
Fahn A (1972) Plant Anatomy, 2nd edn., Pergamon, Oxford
Flores R (1984) Is the conformation of viroids involved in their pathogenicity? J Theor Biol 108:519-527
Flores R, Rodríguez JL (1981) Altered pattern of root formation on cuttings of *Gynura aurantiaca* infected by citrus exocortis viroid. Phytopathology 71:964-966
Flores R, Chroboczek J, Semancik JS (1978) Some properties of the CEV-P1 protein from citrus exocortis viroid-infected *Gynura aurantiaca* DC. Physiol Plant Pathol 13:193-201
Fraser RSS (1987) Biochemistry of virus-infected plants. Research Studies Press, Letchworth; Wiley, New York
Galindo JA, Smith DR, Diener TO (1984) A disease-associated host protein in viroid-infected tomato. Physiol Plant Pathol 24:257-275
García-Breijo F, Garro R, Conejero V (1989) Two pathogenesis-related (PR) proteins C7 (P32) and C6 (P34) from citrus exocortis viroid (CEV)-infected tomato plants are chitinases. Physiol Mol Plant Pathol (in press)
Gianinazzi S, Martin C, Vallée JC (1970) Hypersensibilité aux virus, température et protéines solubles chez le *Nicotiana* Xanthi nc. Apparition de nouvelles macromolécules lors de la répression de la synthèse virale. CR Acad Sci Paris 270D:2383-2386
Granell A, Bellés JM, Conejero V (1987) Induction of pathogenesis-related proteins in tomato by citrus exocortis viroid, silver ion and ethephon. Physiol Mol Plant Pathol 31:83-90
Gross HJ, Domdey H, Losow C, Jank P, Raba M, Alberty H, Sänger HL (1978) Nucleotide sequence and secondary structure of potato spindle tuber viroid. Nature 273:203-208
Guy M, Kende H (1984) Conversion of 1-aminocyclopropane-1-carboxylic acid to ethylene by isolated vacuoles of *Pisum sativum* L. Planta 160:281-287
Haas B, Klanner A, Ramm K, Sänger HL (1988) The 7S RNA from tomato leaf tissue resembles a signal recognition particle RNA and exhibits a remarkable sequence complementarity to viroids. EMBO J 7:4063-4074
Hall TC, Wepprich RK, Davies JW, Weathers LG, Semancik JS (1974) Functional distinctions between the ribonucleic acids from citrus exocortis viroid and plant viruses: Cell-free translation and amino-acylation reactions. Virology 61:486-492
Hanks RW, Feldman AW (1972) Changes in amounts of auxin-like growth promoter, gibberellin, and inhibition in citrus infected with exocortis virus. In: Price WC (ed) Proceedings of the 5th conference of the international organization of citrus virologists, University of Florida, Gainesville, p 244
Hari V (1980) Ultrastructure of potato spindle tuber viroid-infected tomato leaf tissue. Phytopathology 70:385-387
Hepler PK, Wayne RO (1985) Calcium and plant development. Annu Rev Plant Physiol 36:397-439
Hiddinga HJ, Jenssen Crum C, Hu J, Roth DA (1988) Viroid-induced phosphorylation of a host protein related to a dsRNA-dependent protein kinase. Science 241:451-453
Hooft Van Huijsduijnen RAM, Alblas SW, De Rijk RJ, Bol JF (1986) Induction by salicylic acid of pathogenesis-related proteins and resistance to alfalfa mosaic virus infection in various plant species. J Gen Virol 67: 2135-2143
Klaff P, Gruner R, Hecker R, Sättler A, Theissen G, Riesner R (1989) Reconstituted

and cellular viroid-protein complexes. J Gen Virol 70:2257-2270
Kojima M, Murai M, Shikata E (1983) Cytopathic changes in viroid-infected leaf tissues. J Fac Agric Hokkaido Univ 61:219-223
Lieberman M (1979) Biosynthesis and action of ethylene. Annu Rev Plant Physiol 30:533-591
Lucas J, Camacho-Henriquez A, Lottspeich F, Henschen A, Sänger HL (1985) Amino acid sequence of the "pathogenesis-related" leaf protein p14 from viroid-infected tomato reveals a new type of structurally unfamiliar proteins. EMBO J 4:2745-2749
Mauch F, Staehelin LA (1989) Functional implications of the subcellular localization of ethylene-induced chitinase and ß-1,3-glucanase in bean leaves. Plant Cell 1:447-457
Marmé D (1989) The role of calcium and calmodulin in signal transduction. In: Boss WF, Morré DJ (eds) Second messengers in plant growth and development. Alan R Liss Inc, New York, in press
Marton L, Durán-Vila N, Lin J, Semancik JS (1982) Properties of cell cultures containing the citrus exocortis viroid. Virology 122:229-238
Matsuoka M, Ohashi Y (1986) Induction of pathogenesis-related proteins in tobacco leaves. Plant Physiol 80:505-510
Matthews REF (1978) Are viroids negative-strand viruses? Nature 276:850
Niblett CL, Dickson E, Fernow KH, Horst RK, Zaitlin M (1978) Cross protection among four viroids. Virology 61:486-492
Owens RA, Hammond RW (1987) Viroid function: Molecular biology of viroid-host interactions. In: Diener TO (ed) The viroids. Plenum, New York, p 167
Paliwal YC, Singh RP (1981) Cytopathological changes induced by potato spindle tuber viroid in *Scopolia sinensis*. Can J Bot 59:677-682
Parent JG, Asselin A (1984) Detection of pathogenesis-related proteins (PR or b) and other proteins in the intercellular fluid of hypersensitive plants infected with tobacco mosaic virus. Can J Bot 62:564-569
Rackwitz HR, Rohde W, Sänger HL (1981) DNA-dependent RNA polymerase II of plant origin transcribes viroid RNA into full-length copies. Nature 291:297-301
Ranjeva R, Boudet AM (1987) Phosphorylation of proteins in plants: regulatory effect and potential involvement in stimulus/response coupling. Annu Rev Plant Physiol 38:73-93
Reisfeld RA, Lewis VJ, Williams DJ (1962) Disc electrophoresis of basic proteins and peptides on polyacrylamide gels. Nature 195:156-160
Riesner D (1987) Physical-chemical properties. Structure formation. In: Diener TO (ed) The viroids. Plenum, New York, p 63
Robertson HD, Branch AD (1987) The viroid replication process. In: Semanck JS (ed), Viroids and viroid-like pathogens. CRC Press, Boca Raton, p 50
Rodrigo I, Vera P, Conejero V (1989) Degradation of tomato pathogenesis-related proteins by an endogenous 37-kDa aspartyl endoproteinase. Eur J Biochem (in press)
Rodríguez JL, García-Martínez JL, Flores R (1978) The relationship between plant growth substance content and infection of *Gynura aurantiaca* DC by citrus exocortis viroid. Physiol Mol Plant Pathol 13:355-363
Sänger HL (1982) Biology, structure, functions and possible origin of viroids. In: Parthier B, Boulter D (eds) Encyclopedia of plant physiology. New Series Vol. 14B Springer-Verlag, Berlin, p 368
Sänger HL (1987) Viroid function. Viroid replication. In: Diener TO (ed) The viroids. Plenum, New York, p 117

Schnölzer M, Haas B, Ramm K, Hofmann H, Sänger HL (1985) Correlation between structure and pathogenicity of potato spindle tuber viroid (PSTV). EMBO J 4:2181-2190

Schumacher J, Sänger HL, Riesner D (1983) Subcellular localization of viroids in highly purified nuclei from tomato leaf tissue. EMBO J 2:1549-1555

Semancik JS, Conejero V (1987) Viroid pathogenesis and expression of biological activity. In: Semancik JS (ed) Viroids and viroid-like pathogens. CRC Press, Boca Raton, p 71

Semancik JS, Conejero V, Gerhart J (1977) Citrus exocortis viroid: survey of protein synthesis in *Xenopus laevis* oocytes following addition of viroid RNA. Virology 80:218-221

Semancik JS, Vanderwoude WJ (1976) Exocortis viroid: cythopathic effects at the plasma membrane in association with pathogenic RNA. Virology 69:719-726

Solymosy F, Kiss T (1985) Viroids and snRNAs. In: Maramorosch D, McKelvey JJ (eds) Subviral pathogens of plants and animals: viroids and prions. Academic, New York, p 183

Van Loon LC (1977) Induction by 2-chloroethylphosphonic acid of viral-like lesions, associated proteins and systemic resistance in tobacco. Virology 80:417-420

Van Loon LC (ed) (1983) Workshop on pathogenesis-related (b) proteins in plants. Neth J Plant Pathol 89:239-325

Van Loon LC (1985) Pathogenesis-related proteins. Plant Mol Biol 4:111-116

Van Loon LC, Van Kammen (1970) Polyacrylamide disc electrophoresis of the soluble leaf proteins from *Nicotiana tabacum* var. 'Samsun' and 'Samsun NN'. II. Changes in protein constitution after infection with tobacco mosaic virus. Virology 40:199-211

Vera P, Conejero V (1988) Pathogenesis-related proteins of tomato. P69 as an alkaline endoproteinase. Plant Physiol 87:58-63

Vera P, Conejero V (1989) Citrus exocortis viroid (CEV) infection alters the *in vitro* pattern of protein phosphorylation of tomato leaf proteins. Mol Plant-Microbe Interactions, in press

Vera P, Hernández-Yago J, Conejero V (1988) Immunocytological localization of the major "pathogenesis-related" (PR) protein of tomato plants. Plant Sci 55:223-230

Vera P, Hernández-Yago J, Conejero V (1989a) "Pathogenesis-related" P1 (p14) protein. Vacuolar and apoplastic localization in leaf tissue from tomato plants infected with citrus exocortis viroid: *in vitro* synthesis and processing. J Gen Virol 70:1933-1942

Vera P, Hernández-Yago J, Conejero V (1989b) Immunogold localization of the citrus exocortis viroid-induced pathogenesis-related proteinase P69 in tomato leaves. Plant Physiol 91:119-123

Visvader JE, Symons RH (1985) Eleven new sequence variants of citrus exocortis viroid and the correlation of sequence with pathogenicity. Nucleic Acids Res 13:2907-2920

Visvader JE, Symons RH (1986) Replication of *in vitro* constructed viroid mutants: location of the pathogenicity-modulating domain of citrus exocortis viroid. EMBO J 5:2051-2055

Wahn K, Rosenberg-de Gomez F, Sänger HL (1980) Cythopathic changes in leaf tissue of *Gynura aurantiaca* infected with the viroid of citrus exocortis disease. J Gen Virol 49:355-365

Wang MC, Lin JJ, Durán-Vila N, Semancik JS (1986) Alteration in cell-wall composition and structure in viroid-infected cells. Physiol Mol Plant Pathol

28:107-127
Wolf P, Gilz R, Schumacher J, Riesner D (1985) Complexes of viroids with histones and other protein. Nucleic Acids Res 13:355-367
Yang SF, Hoffman N (1984) Ethylene biosynthesis and its regulation in higher plants. Annu Rev Plant Physiol 35:155-159
Yunghaus WN, Morré DJ (1977) Adenylate cyclase activity is not found in soybean hypocotyl and onion meristem. Plant Physiol 60:144-149
Zaitlin M, Hariharasubramanian V (1972) A gel electrophoretic analysis of proteins from plants infected with tobacco mosaic and potato spindle tuber viruses. Virology 47:296-305

THE MOLECULAR BIOLOGY OF SATELLITE RNA FROM CUCUMBER MOSAIC VIRUS

David Baulcombe, Martine Devic and Martine Jaegle[1]
The Sainsbury Laboratory
John Innes Institute
Colney Lane
Norwich NR4 7UH
UK

ABSTRACT

CMV satellite RNA has the capability to modify symptoms when plants are infected with CMV. Benign forms of satellite RNA attenuate symptoms; virulent forms induce new symptoms. The molecular basis of these effects is being investigated as a model for disease processes in plants and so that the attenuation property can be exploited in the genetic engineering of virus resistant plants. Mutation analysis has revealed two separate functional domains in the Y satellite RNA of CMV, which control induction of different types of symptom.

INTRODUCTION

Many viruses are associated with small, extra-genomic RNA molecules which may be considered as molecular parasites. These RNA species, known as satellite RNAs (sat RNA), are parasites in the sense that they are not required by the virus but are themselves dependent on the virus for all functions necessary for multiplication, spread through the infected plant and transmission from plant to plant (Francki, 1985). A previous review suggested that an important feature of sat RNA is a lack of extensive homology with the helper virus (Murant and Mayo, 1982). However, since sat RNAs are recognized by the replicase and other functions of the helper virus, it is likely, even if there is no homology in the primary structure, that features of the secondary or

[1]Department of Molecular Biology, Institute of Plant Science Research, Maris Lane, Trumpington, Cambridge, CB2 2LQ UK

NATO ASI Series, Vol H 41
Recognition and Response in Plant-Virus Interactions
Edited by R S S Fraser

tertiary structure are shared with the helper virus.

Like other parasites, sat RNA may modify its host. In the instance of sat RNA associated with cucumber mosaic virus (CMV), this modification is observed as an effect on the symptoms produced by the infected plant. Depending on the strain of sat RNA, the genotype of the host plant, and the strain of helper virus, the modification is either an amelioration of the viral symptoms or the induction of more severe symptoms. The severe symptoms have sometimes been described as exacerbated viral symptoms. However, as they are quite distinct in type from the viral symptoms, it is more useful to consider the sat RNA-induced symptoms as quite different from the symptoms produced in a plant that is infected with virus alone. This point is illustrated by the effects of CMV infection on tobacco or tomato when the sat RNA of CMV strain Y (Takanami, 1981) is added to the inoculum. In the absence of sat RNA the virus produces mild mosaic symptoms on both plants and a fern leaf effect on tomato. The addition of sat RNA Y to the inoculum results in a bright yellow mosaic symptom on tobacco and a systemic necrosis on tomato. Sat RNA which induces severe symptoms is referred to here as virulent, or benign if it ameliorates symptoms.

It has been demonstrated previously in transgenic plants transformed with a DNA copy of sat RNA from CMV, coupled to a suitable promoter, that a polyadenylated form of the sat RNA was produced (Baulcombe *et al.*, 1986). This transcript had very little biological activity in its transcribed form as the sat RNA sequence was surrounded by non-viral sequence at the 5' and 3' ends (Fig. 1, and our unpublished data). However, when the transgenic plants were infected with CMV, the transcribed sat RNA was replicated and accumulated to high levels. This replicated sat RNA had the full biological activity of the natural sat RNA, including the ability to attenuate symptoms of CMV and also a related virus, tomato aspermy virus (TAV) (Harrison *et al.*, 1987). Gerlach *et al.* (1987) have shown similar resistance against tobacco ringspot (nepo)virus resulting from expression of its sat RNA in transgenic plants. Therefore, this use of satellite RNA is a second example, together with the use of viral coat protein genes (Abel *et al.*, 1986), in which expression of viral RNA in transgenic plants produces resistance to the effects of viral infection. The properties of the sat RNA effect are complementary to those of coat protein-mediated resistance, so that these two types of viral sequence may be a particularly effective combination when expressed in crop plants. For example, the sat RNA effect differs from coat protein-mediated resistance in that no new protein is produced, and in its lack of sensitivity to the level of sat RNA

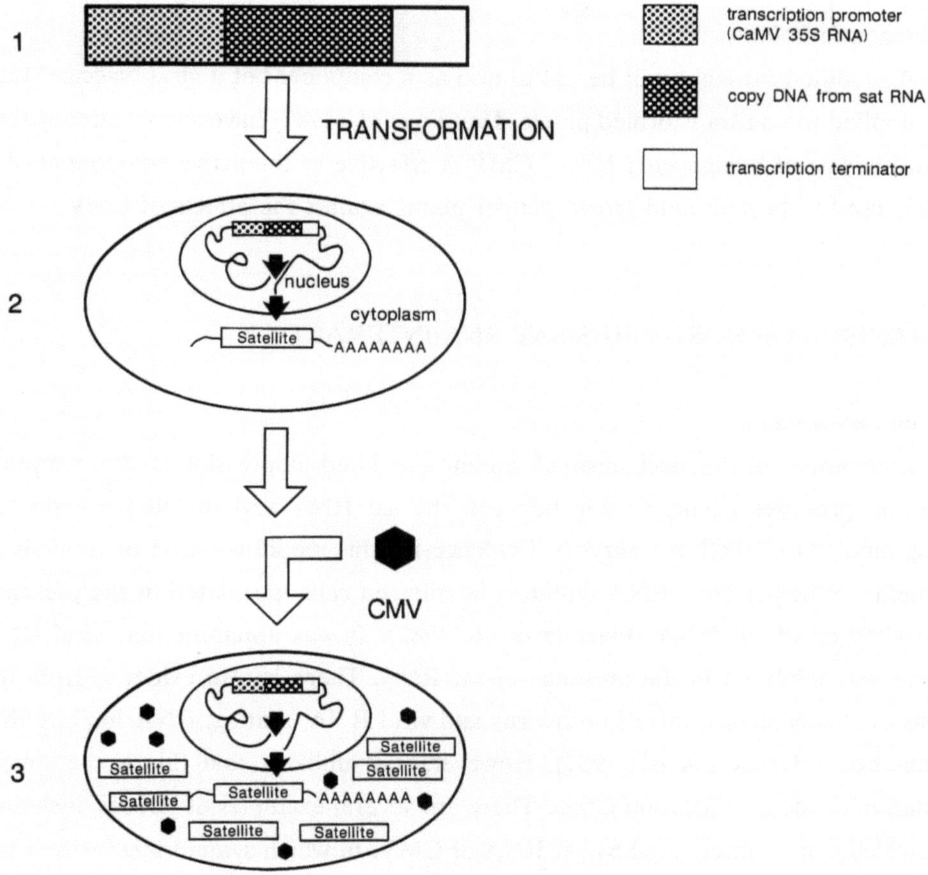

Figure 1. Plant transformation to express biologically active satellite RNA. The figure illustrates steps described in detail by Baulcombe *et al.* (1986). In step 1 a DNA copy of CMV sat RNA was coupled to a promoter and transcriptional terminator. Step 2 was the transfer, by *Agrobacterium*-mediated transformation, of the DNA into the nuclear genome of tobacco plants. The biological activity of the transcripts produced from the integrated DNA was demonstrated when the transformed plants were infected with CMV (step 3). The sat RNA replicated to a high level and was also encapsidated in virus particles.

expression in the plant and the concentration of inoculum (Harrison *et al.*, 1987). However, as the virulent species of sat RNA have a nucleotide sequence which is very similar to the benign sat RNAs (Palukaitis, 1988), it is not yet possible to use this genetically engineered resistance in the field. It is necessary to identify functional domains within the sat RNA so that a modified form can be expressed in plants without the risk of mutation or transmission from the protected plants to nearby non-transformed plants.

A modified sat RNA will be useful also as a component of a viral "vaccine" that can be applied to non-transformed plants. Tien Po *et al.* (1987) have shown already that a natural isolate of benign sat RNA of CMV is effective as the active component of a "vaccine" used to protect field grown pepper plants against the effects of CMV.

MECHANISMS OF HOST-VIRUS-SAT RNA INTERACTION

Symptom amelioration

One model for the mechanism of satellite-mediated suppression of viral symptom production proposes a competition between the sat RNA and the helper virus for limiting amounts of replicase enzyme. Evidence for this model is based on analysis of the kinetics of helper virus RNA synthesis in cultured cells inoculated in the presence or the absence of sat RNA (Piazolla *et al.*, 1982). It was apparent that viral RNA synthesis was inhibited in the presence of sat RNA. There is also evidence from the analysis of steady state levels of both virus and viral RNA, that sat RNA inhibits viral accumulation (Harrison *et al.*, 1987). However, it is unlikely that this is the simple explanation of the amelioration effect. There are several examples of viruses, including TAV, which can be attenuated by sat RNA of CMV, in which symptom severity is not directly related to the amount of virus. Transgenic plants expressing sat RNA were capable of effective inhibition of TAV symptoms despite there being no reduction of the amount of viral RNA compared with control plants which were not producing sat RNA (Harrison *et al.*, 1987). Similarly, there was no effect on production of capsid protein (Harrison *et al.*, 1987) (Fig. 2). It is likely, therefore, that the satellite RNA has the capability to interfere either with the process of symptom induction by the virus, or with the processes of symptom production by the plant.

The sat RNA of tobacco ringspot virus also has the capability to attenuate the effects of viral infection without necessarily inhibiting the helper virus replication (Ponz *et al.*, 1987). In this instance the effect is even more striking than with the sat RNA of CMV and TAV, as the attenuated virus (cherry leafroll nepovirus) cannot replicate the sat RNA of tobacco ringspot virus.

Benign sat RNA of CMV can also protect against the effects of a virulent sat RNA, under certain conditions. This was demonstrated by inoculating plants first with

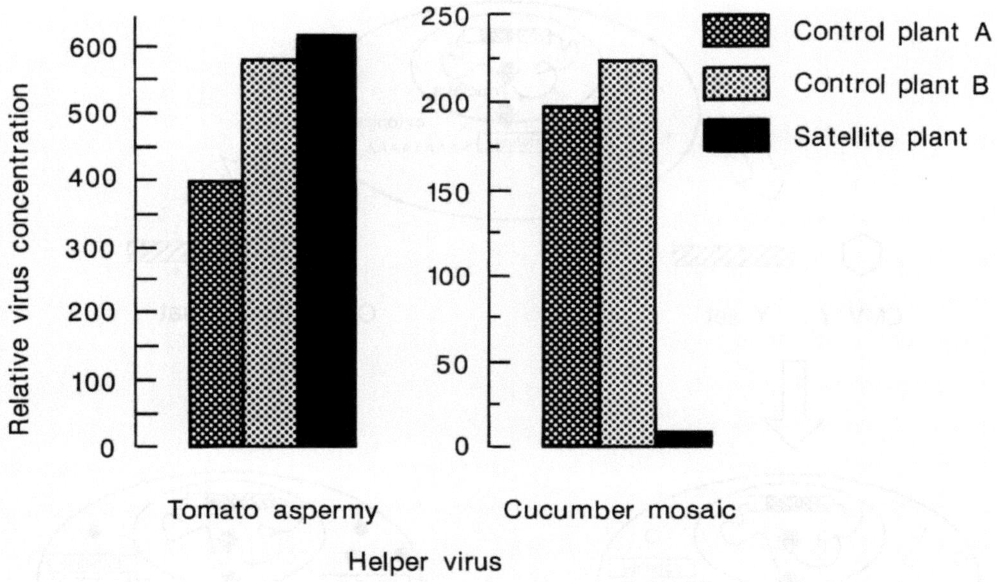

Figure 2. The effect of satellite RNA in transgenic plants on virus accumulation. Transgenic plants either expressing satellite RNA (satellite plant) or not expressing satellite RNA (control plants A and B) were inoculated with satellite-free helper virus. Virus concentration in systemically infected leaves was measured by inoculation of extracts to a local lesion host (*Chenopodium quinoa*).

a viral inoculum containing a benign sat RNA and, after this infection became established, with a viral inoculum supplemented with a virulent sat RNA. An experiment designed to test whether the transformed plants expressing satellite RNA could resist the virulent satellite in the same way is described in Fig. 3. The virulent sat RNA was from the Y strain of CMV (Takanami, 1981) and its effects could be diagnosed by the production of a yellow mosaic symptom on the systemically infected leaves. The two strains of helper virus used in the experiment were either the Y strain of CMV, which is the natural helper of the Y sat RNA, or CMV KIN, which was not associated with a sat RNA when it was isolated (Harrison *et al.*, 1987). The results summarized in Fig. 3 show that the outcome depends on the type of helper strain used. When CMV KIN was used as helper there was inhibition of Y sat, both at the level of symptoms and sat RNA accumulation. This suggests that the transcribed sat RNA was at an advantage over the sat RNA in the challenge inoculum, possibly because it was present in cells before the arrival of the virus. However, this advantage was overridden when the helper virus was the Y strain of CMV (Fig. 3). A likely explanation for this second result is that an

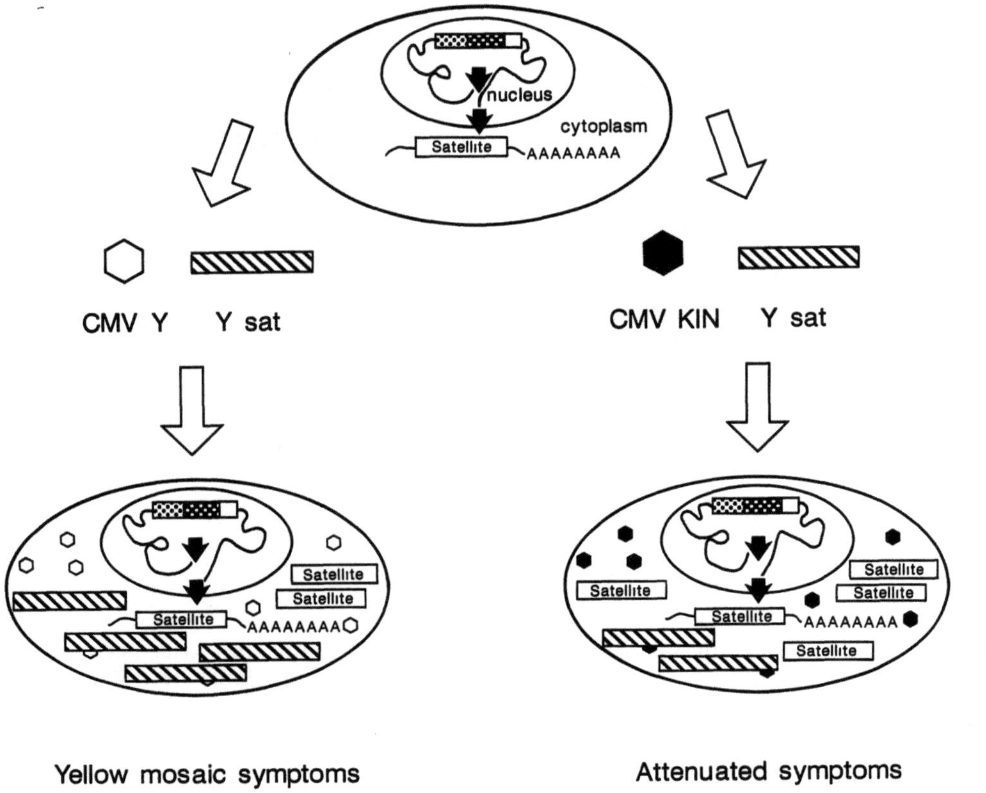

Figure 3. The effect of satellite RNA in transgenic plants on infections with a virulent sat RNA. The diagram illustrates schematically the accumulation of sat RNA and symptom production when the virulent Y sat RNA was introduced with different helper virus strains. Satellite regions are coded as in Fig. 1.

affinity between the CMV Y strain and the Y sat RNA gave the Y sat RNA an advantage in the competition with the sat RNA produced in transgenic plants. At the moment there is no obvious suggestion for the nature of this affinity. Perhaps an anti-sense interaction, such as has been described by Rezaian and Symons (1986) (Fig. 4), is stronger in the homologous combination of helper and sat RNA so that the Y sat RNA is encapsidated and transported from cell to cell more efficiently than with CMV KIN. Alternatively, the viral replicase may operate more efficiently in the homologous combination.

Symptom induction

It is clear that the formation of symptoms in response to the presence of sat

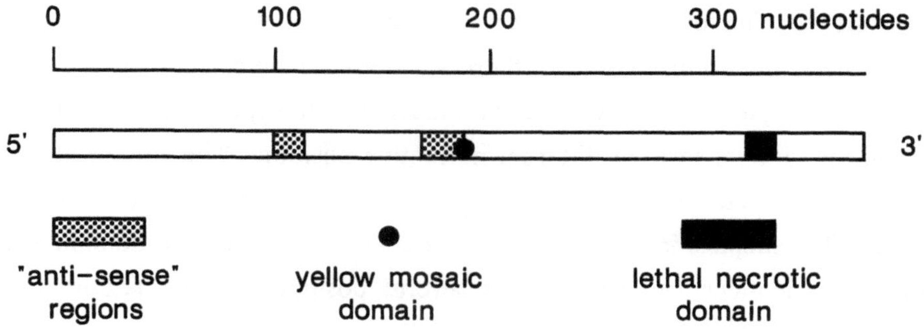

Figure 4. The domain structure of CMV Y sat RNA. The diagram illustrates the location of "anti-sense", yellow mosaic-inducing and lethal-necrotic domains in the nucleotide sequence. The anti-sense domain was defined by Rezaian *et al.* (1986) and the symptom domains by Devic *et al.* (1989) and M Devic, M Jaegle and D Baulcombe (unpublished data).

RNA in an inoculum of CMV involves an interaction, either direct or indirect, between the sat RNA and components of the host plant. As different species of sat RNA induce different responses in the host plant it is likely that this interaction involves variable parts of the sat RNA molecule. These domains have been located in the sat RNA by creating hybrid molecules and correlating the phenotype with the presence of domains from a virulent sat RNA. The virulent sat RNA used in this study was the Y sat RNA described by Takanami (1981) as producing a lethal necrotic disease on tomato and a bright yellow mosaic on tobacco.

A series of hybrid molecules was produced and from the phenotype of these it was concluded that the domain responsible for induction of the yellow mosaic disease was located centrally in the molecule. The lethal necrotic domain was separate and towards the 3' end (Devic *et al.*, 1989)

It was not too surprising that the yellow mosaic domain was located within the central part of the molecule. The Y sat RNA has the unique property amongst known sat RNA species, of inducing the yellow mosaic disease and also has a unique nucleotide sequence in the central region of the molecule. A better resolution of the yellow mosaic domain was obtained by a mutational analysis which identified a site where changing two nucleotides destroyed the yellow mosaic-inducing property.

The localization of the domain responsible for the induction of the lethal necrotic disease was less obvious, as there was no particular area in which a sequence motif was unique to the virulent form of the sat RNA. However, there was a region in which all of the necrosis-inducing forms of sat RNA were identical, and where there was sequence variation in the benign forms. However, there was not a simple correlation between the presence of this sequence and the symptom-inducing phenotype, as each benign form varied from the virulent form at different nucleotide positions. The complexity of the situation was confirmed when various defined mutations were created in that region of the molecule. The simplest interpretation of the data is that the necrosis domain extends over several nucleotides and variation at several positions in that region has the potential to destroy the necrosis-inducing capability of the sat RNA.

CONCLUSIONS

The domain structure of sat RNA is summarized in Fig. 4. This information will be used in the development of sat RNA for use in crop protection and for more fundamental investigation of the mechanism of symptom induction by sat RNA.

In the applied context, the new information suggests regions of the molecule which could be modified to produce a sat RNA that is not capable of inducing symptoms. Ideally, the modified sat RNA would differ from the virulent forms at several positions, so that back-mutation would not occur readily. In the course of this work several mutant forms of the satellite have been created, illustrating that modified forms of the sat RNA are replicable, can move systemically in a plant and can be transmitted by aphids from plant to plant. One mutation which was functional in these respects (M Jaegle, unpublished data) involved a large deletion within the central part of the molecule. Surprisingly, a smaller deletion from within the region of the larger deletion did not accumulate when inoculated on to plants together with helper virus. It appears therefore, that any modifications must leave secondary or higher order structure features of the sat RNA molecule intact. Unfortunately, understanding of higher order structures involving the sat RNA molecule is not very sophisticated. Current models of sat RNA structure do not explain why the smaller deletion described above does not accumulate in plants and the possibility remains that interactions of sat RNA with other viral or host plant factors determine the structure of sat RNA *in vivo.*

In principle, the sat RNA could induce symptoms via the action of small sat RNA-encoded peptides or directly as an RNA molecule. Currently, it appears that the latter is the more likely alternative, as various mutations have been created which interrupt the small open-reading frames and which have no effect on the ability of the sat RNA to induce symptoms. How RNA molecules might induce symptoms is not immediately obvious. A feasible model proposes that sat RNA interferes in the RNA-mediated processes of the healthy cell. It is now clear that there are several potential targets in this area, as RNA is known as a component in ribonucleases and protein transport complexes as well as in the protein synthetic apparatus.

In future work we shall investigate these RNA-mediated processes and look for proteins and other components that bind to sat RNA of CMV. The outcome should be a greater understanding of the molecular biology of sat RNA, so that it can be used safely in crop protection. It is also anticipated that the detailed knowledge of the pathogenic process will provide information about the healthy cell and suggest new methods of engineering disease resistance in plants.

REFERENCES

Abel PP, Nelson RS, De B, Hoffmann N, Rogers SG, Fraley RT, Beachy RN (1986) Delay of disease development in transgenic plants that express the tobacco mosaic virus coat protein gene. Science 232:738-743

Baulcombe DC, Saunders GR, Bevan MW, Mayo MA, Harrison BD (1986) Expression of biologically active viral satellite RNA from the nuclear genome of transformed plants. Nature 321:446-449

Devic M, Jaegle M, Baulcombe DC (1989) Symptom production on tobacco and tomato is determined by two distinct domains of the satellite RNA of cucumber mosaic virus (strain Y). J Gen Virol, in press

Francki, RIB (1985) Plant virus satellites. Annu Rev Microbiol 39:151-174

Gerlach WL, Llewellyn D, Haseloff J (1987) Construction of a plant disease resistance gene from the satellite RNA of tobacco ringspot virus. Nature 328:802-805

Harrison BD, Mayo MA, Baulcombe DC (1987) Virus resistance in transgenic plants that express cucumber mosaic virus satellite RNA. Nature 328:799-802

Murant AF, Mayo MA (1982) Satellites of plant viruses. Annu Rev Phytopathol 20:49-70

Palukaitis P (1988) Pathogenicity regulation by satellite RNAs of cucumber mosaic virus: minor nucleotide sequence changes alter host responses. Mol Plant-Microbe Interact 1:175-181

Piazzolla P, Tousignant ME, Kaper JM (1982) Cucumber mosaic virus-associated RNA 5 - the overtaking of viral RNA synthesis by CARNA 5 and dsCARNA 5 in tobacco. Virology 122:147-157

Ponz F, Rowhani A, Mircetich SM, Bruening G (1987) Cherry leafroll virus infections are affected by a satellite RNA that the virus does not support. Virology 160:183-190

Rezaian MA, Symons RH (1986) Anti-sense regions in satellite RNA of cucumber mosaic virus form stable complexes with the viral coat protein gene. Nucleic

Acids Res 14:3229-3239
Takanami Y (1981) A striking change in symptoms on cucumber mosaic virus-infected tobacco plants induced by a satellite RNA. Virology 109:120-126
Tien P, Zhang X, Qiu B, Qin B, Wu G (1987) Satellite RNA for the control of plant diseases caused by cucumber mosaic virus. Ann Appl Biol 111:143-152

CRYPTIC VIRUSES OF BEET AND OTHER PLANTS

J F Antoniw, R F White and W Xie
Plant Pathology Department
Rothamsted Experimental Station
AFRC Institute of Arable Crops Research
Harpenden
Herts AL5 2JQ
UK

ABSTRACT

The discovery and properties of cryptic viruses are described with particular reference to the cryptic viruses of sugar beet. These unique viruses are characterized by a lack of symptoms in the infected plant, transmission only through seed and pollen and a dsRNA genome. cDNA clones, obtained by molecular cloning of the dsRNA, are being used to further characterize the structure and replication strategies of these viruses and as probes to detect them.

INTRODUCTION

Interactions between plants and viruses are shown most obviously by the production of symptoms. Some viruses are symptomless in certain plants but when transmitted to other species they can produce symptoms; the symptomless infections are often called latent. Cryptic viruses are unusual in that they do not cause obvious symptoms in any host plant. Because of this and their low concentration they remained undiscovered until the late 1960s. In this paper we examine the properties of cryptic viruses, their unusual transmission characteristics and their unique form of association with their host plants. We will concentrate mainly on the cryptic viruses of sugar beet because they were the first to be discovered and have been the most intensively studied. Other cryptic viruses will also be discussed where they provide new insights or more detailed information.

NATO ASI Series, Vol H 41
Recognition and Response in Plant-Virus Interactions
Edited by R S S Fraser

Table 1 *Plants containing cryptic viruses*

Plant	No. of dsRNAs	No. of cryptic viruses	References
alfalfa (*Medicago sativa*)	3	1-2	Boccardo *et al.* (1983; 1987)
amaryllis (*Hippeastrum* sp.)	?	?	Raizada *et al.* (1983)
beet (*Beta vulgaris*)	4	2	Kassanis *et al.* (1977); Accotto and Boccardo (1986); Natsuaki *et al.* (1986); Antoniw *et al.* (1986)
(*Beta maritima*)	2	1	Xie *et al.* (1989)
broad bean (*Phaseolus vulgaris*)	?	?	Kenten *et al.* (1978)
carnation (*Dianthus caryophyllus*)	3	1	Lisa *et al.* (1981a,b)
carrot (*Daucus carota*)	?	?	Natsuaki *et al.* (1983a)
chrysanthemum (*Chrysanthemum coronarium*)	?	?	Natsuaki *et al.* (1983a)
cucumber (*Cucumis sativus*)	3	1	Jelkmann *et al.* (1988)
fescue (*Festuca pratensis*)	5	2?	Boccardo *et al.* (1983,1987)
field bean (*Vicia faba*)	3	1	Kenten *et al.* (1978); Abou-Elnasr *et al.* (1985)
hop trefoil (*Medicago lupulina*)	5	3?	Boccardo *et al.* (1983, 1987)
mibuna (*Brassica rapa* var. *laciniifolia*)	?	?	Natsuaki *et al.* (1983a)
poinsettia (*Euphorbia pulcherrima*)	?	?	Paludan and Begtrup (1986)
radish (*Raphanus sativus*)	3	1	Natsuaki *et al.* (1979, 1983a) Natsuaki (1985)
red clover (*Trifolium pratense*)	2	1	Boccardo *et al.* (1983)
rhubarb (*Rheum rhaponticum*)	?	?	Natsuaki *et al.* (1983a)
ryegrass (*Lolium multiflorum*)	2	1	Plumb (1973); Boccardo *et al.* (1983)
santosai (*Brassica rapa* var. *amplexicaulus*)	?	?	Natsuaki *et al.* (1983a)
spinach (*Spinacia oleracea*)	3	1	Natsuaki *et al.* (1983b)
white clover (*Trifolium repens*)	6	3	Boccardo *et al.* (1985)
white mustard (*Sinapsis alba*)	?	?	Kuhne *et al.* (1987)

DISCOVERY

The first report of a cryptic virus, although not by that name, was the observation of spherical virus-like particles in sugar beet and other *Beta* spp. by Pullen (1968, 1969). From the beginning, two most striking properties of these viruses became apparent; that they do not produce symptoms in infected plants and that they are not transmissible by either aphids or by mechanical inoculation (Pullen, 1968). Pullen (1969) was unable to find plants free of these particles and was unable to produce particle-free plants either by heat treatment or apical meristem culture. Therefore, she could not be certain that they were viruses. Nearly ten years later Kassanis *et al.* (1977) rediscovered these particles, found some plants to be free of them and gave them the name of beet cryptic virus (BCV), first suggested by M Watson, to emphasize the fact that because they cause no symptoms, their presence remains hidden.

Beet cryptic virus was the first cryptic virus to be purified (Kassanis *et al.*, 1977). The low concentration of virus in sugar beet leaf tissue resulted in yields of only about 250 μg virus/kg leaf. The purified virus viewed in the electron microscope consisted of spherical particles 30 nm in diameter.

Similar virus-like particles were found in *Vicia faba* (vicia cryptic virus) (VCV) (Kenten *et al.*, 1978) and in carnation (carnation cryptic virus) (CCV) (Lisa *et al.*, 1981b). A further unusual property of these viruses was discovered when the nucleic acid of CCV was shown to be double-stranded RNA (Lisa *et al.*, 1981a). The only other plant viruses with a dsRNA genome are the plant reoviruses, for example wound tumour virus, but these are insect transmitted and have much larger particles of about 65-75 nm diameter (Boccardo and Milne, 1984). Cryptic virus particles containing dsRNA were then found in a range of plants including alfalfa, hop trefoil, red clover, white clover and meadow fescue (Boccardo *et al.*, 1983). Similar seed-transmitted spherical particles found in Italian ryegrass (ryegrass spherical virus (Plumb, 1973) or ryegrass seed-borne virus (Plumb and Lennon, 1981)) were also found to contain dsRNA and the virus was renamed ryegrass cryptic virus (Boccardo *et al.*, 1983). Cryptic viruses have now been found in a wide range of plants, some of which contain more than one cryptic virus (Table 1), and have been recognized as a new group of plant viruses, the cryptovirus group.

Most of the cryptic viruses discovered so far have a particle size similar to that of BCV (30 nm diameter), but four cryptic viruses found in alfalfa, hop trefoil, red clover and white clover have particles about 38 nm in diameter. Consequently, plant cryptoviruses have been divided into two subgroups dependent upon their size: subgroup A with particles about 30 nm and subgroup B with particles about 38 nm diameter.

TRANSMISSION

Kassanis *et al.* (1977) were unable to transmit BCV by mechanical inoculation to healthy sugar beet plants, to a range of indicator plants, by aphid inoculation using *Myzus persicae*, or by dodder (Kassanis *et al.*, 1977). On one occasion there was apparent transmission of BCV by grafting young root crowns of infected and healthy plants together (Kassanis *et al.*, 1977) but all attempts to repeat this have failed. Attempts to transmit other cryptic viruses by grafting e.g. VCV (Kenten *et al.*, 1979, 1980), CCV (Lisa *et al.,* 1981b) and white clover cryptic viruses 1 and 2 (Boccardo *et al.*, 1985) were all unsuccessful. It seems unlikely therefore that these and other cryptic viruses can be transmitted by mechanical inoculation or the usual virus vectors.

Vegetative propagation of white clover stolons for three years resulted in several stolons that produced leaves containing large concentrations of cryptic virus on one side and virus-free leaves on the other (Boccardo *et al.*, 1987). This observation suggests that cryptic viruses may not be able to move from cell to cell but are limited to propagation by division of the host cell. The low concentration of cryptic virus particles in meristematic cells could occasionally give rise to daughter cells containing no virus thus producing virus-free sectors within the plant. The inability of cryptic viruses to spread from cell to cell could also explain their inability to be transmitted by grafting.

However, BCV is transmitted very efficiently through seed; 90% of plants raised from seed of three commercial cultivars of sugar beet were infected with the virus (Kassanis *et al.*, 1977). White and Woods (1978) found that the frequency of BCV infection was also high in other types of beet, including fodder beet, mangel and red beet. Large proportions of seeds of other plants were also infected with cryptic viruses e.g. *Vicia faba* (Kenten *et al.*, 1979) and white clover (Boccardo *et al.*, 1985).

Cross pollination experiments between healthy beet and BCV-infected plants and analysis of the 138 progeny showed that when both parents were healthy all of the progeny were healthy, when the female parent was healthy and the pollen was from an infected plant 43% of the progeny were infected, when the female parent was infected

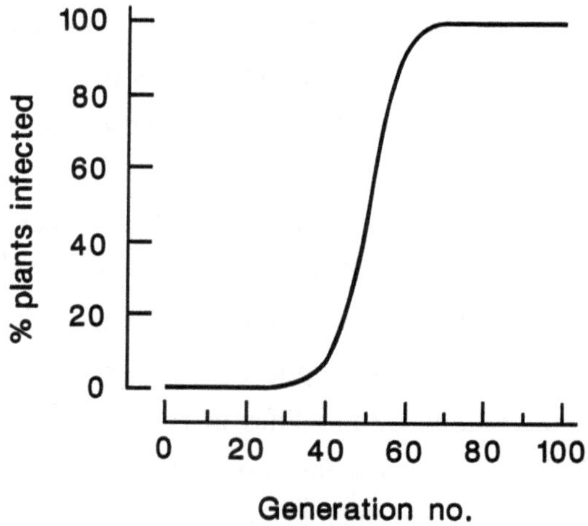

Figure 1. Computer simulation of the spread of beet cryptic virus through a population of 100,000 plants beginning with one infected plant, using the transmission rates given in the text.

and the pollen was from a healthy plant 82% of the progeny were infected, and when both parents were infected then all of the progeny were infected (Kassanis *et al.*, 1978). Similar results were obtained using *Vicia faba* (Kenten *et al.*, 1980). This suggests that the rate of transmission of cryptic viruses by the ovum is nearly twice that of the pollen and may reflect the low concentration of virus particles within cells and the relative sizes of the female and male gametes.

We used the transmission rates for BCV as the basis for a computer model to look at the spread of cryptic viruses through a population of 100,000 plants starting with a single infected plant. The model assumed random segregation of the virus and equal viability of healthy and infected plants. The model predicts that almost the entire population would be infected after about 80 generations (Fig. 1) indicating that the observed transmission rates through ovum and pollen could easily account for the high proportions of infected seed, and therefore that no other means of transmission is necessary.

The unusual transmission characteristics of cryptic viruses led to concern over whether these spherical virus-like particles were really viruses. The main problem was the inability to transmit cryptic viruses to healthy plants, which is essential to distinguish virus diseases from natural variation or physiological effects. Kassanis (1984) argued strongly that they should be considered as viruses because modern analytical techniques showed that these particles were not normal constituents of cells, and that they had a structure reasonably similar to normal plant viruses, consisting of a dsRNA genome within a protein envelope.

Similar spherical virus-like particles with dsRNA genomes were also known to occur in fungi (Buck and Ratti, 1975). This raised the possibility that the cryptic viruses may not be viruses of plants but could be mycoviruses present in associated fungi. It would also account for the low levels of cryptic virus particles normally found in plant sap. This possibility was discounted by two lines of research. Firstly, Boccardo *et al.* (1985) showed that the incidence and concentration of two cryptic viruses in white clover were unaffected by treating plants with high levels of fungicides and by raising seedlings axenically. Secondly, Abou-Elnasr *et al.* (1985) found similar concentrations of VCV dsRNA in intact *Vicia faba* leaves and in isolated washed mesophyll protoplasts, which should be free from most surface and intercellular microflora.

The host range of cryptic viruses is restricted to plants in the same species. Boccardo *et al.* (1987) have shown that the three white clover cryptic viruses are serologically unrelated, and using cDNA hybridization techniques we have shown that there are at least two unrelated cryptic viruses in beet (Antoniw *et al.*, 1986; Xie *et al.*, 1989). Thus it may be common for one host plant to contain more than one independent, unrelated cryptic virus (Table 1).

No serological relationships have been detected between cryptic viruses from different plant families (Boccardo *et al.*, 1987), but relationships have been found between cryptic viruses in related host species e.g. the two cryptic viruses in red clover are serologically related to white clover cryptic viruses 1 and 2 and serological relationships have been shown for some of the cryptic viruses in different species of *Medicago* (alfalfa and hop trefoil) (Boccardo *et al.*, 1987). This suggests that these cryptic viruses may have existed in the ancestors of these plants before they separated during evolution, and that even then the predecessor of these viruses was able to be transmitted only through seed and pollen.

SYMPTOMS

Cryptic viruses do not seem to cause the obvious symptoms normally associated with other viruses, but there are some reports of symptoms apparently associated with with cryptic virus infection.

In most cultivars of Japanese radish (*Raphanus sativus*) no symptoms are produced, but in some cultivars cryptic virus infected seedlings sometimes showed smaller lower leaves and mild yellowing of the leaf edges, hence the name radish yellow edge virus (Natsuaki *et al.*, 1979).

Plumb and Lennon (1981) found that ryegrass mosaic virus (RMV) infection of ryegrass cryptic virus (RCV)-infected ryegrass resulted in increased concentrations of RCV particles. Initial experiments also suggested that ryegrass infected with both RMV and RCV produced more severe symptoms than when infected with RMV alone (Plumb, 1973; Plumb and Misari, 1974) but later work was unable to confirm this (Plumb and Lennon, 1981). Stanarius *et al.* (1984) also reported that higher concentrations of BCV virus particles were produced in sugar beet plants infected with both BCV and beet mild yellowing virus than in plants infected with BCV alone. These results suggest that synergistic interactions can occur between cryptic and other viruses.

Kassanis *et al.* (1977) found that about 15% of seedlings of a Sutton's selection of spinach beet died at the cotyledon stage and about 30% of the remaining plants showed some leaf yellowing. The dying seedlings contained large concentrations of BCV and there was no evidence of infection by other viruses or root rotting pathogens. The apparently normal and yellowing seedlings also contained BCV but at lower concentrations than those that died. As beet cryptic virus was present in all of these plants other factors were probably also involved. Perhaps the wide genetic diversity of beet plants resulted in plants with different levels of sensitivity to cryptic viruses.

Even though cryptic viruses usually appear to cause no symptoms, they may affect the growth and development of the host plant. Pot experiments in a glasshouse showed that healthy sugar beet plants produced significantly larger roots than BCV-infected plants (Kassanis *et al.*, 1978). In small-scale field trials over three years cryptic virus-infected sugar beet plants produced 3-11% less sugar than virus-free plants, but only in the final year was this effect statistically significant (Moir, 1983a,b). However, similar small-scale field trials showed that VCV had little or no effect on the growth and yield of field bean (Kenten *et al.*, 1980, 1981).

In conclusion, the vast majority of cryptic viruses produce no discernible symptoms in their host plants. As cryptic viruses appear to be transmitted only through seed and pollen to the progeny of infected plants, their continued existence depends to a much larger extent than most plant viruses on the successful growth and reproduction of the host plant. Thus there are probably strong selection pressures to ensure that cryptic viruses do not adversely affect the reproductive efficiency of their hosts, and there may well be selection for beneficial effects in the host plants.

MOLECULAR BIOLOGY

Accotto and Boccardo (1986) showed that BCV consists of two coat proteins with molecular weights of 54 kDa and 52 kDa and four dsRNA components with sizes of about 2080, 1760, 1440 and 1330 bps (Fig. 2). Natsuaki *et al.* (1986) showed that beet temperate virus, discovered in Japanese leaf beet cv. Fudanso, was serologically related to BCV but contained only two dsRNA components the same size as the two largest dsRNAs of BCV. Thus it was suggested that BCV was a mixture of two viruses, BCV1 consisting of the two largest and BCV2 of the two smallest dsRNAs. However, other explanations were possible: dsRNAs 3 and 4 could have been subgenomic or defective copies of dsRNAs 1 and 2, or dsRNAs 3 and 4 could have been satellite dsRNAs dependent on dsRNAs 1 and 2 for their replication.

We have cloned all four dsRNAs of BCV, purified from sugar beet cv. Sharpes Klein E, into pUC9 and have isolated cDNA clones made from each dsRNA (Antoniw *et al.*, 1986; Xie *et al.*, 1989). The inserts of the clones used as cDNA probes in Northern blot hybridization experiments hybridized only with the dsRNA from which they were made (Fig. 3, SKE tracks 1-4) showing that the probes are highly specific and suggesting that there is little or no sequence homology between the four dsRNAs (Antoniw *et* al., 1986; Xie *et al.*, 1989). When nucleic acid extracted from 20 individual sugar beet cv. Regina plants was analysed by hybridization of Northern blots with the four cDNA probes, two different patterns of hybridization were observed (Fig. 4): dsRNAs 1 and 2 were always found in the same plants as each other, as were dsRNAs 3 and 4, but the distribution of dsRNAs 1 and 2 was independent of the distribution of dsRNAs 3 and 4 (Xie *et al.*, 1989). These results confirm that there are two cryptic viruses in sugar beet. The analysis of these 20 plants showed that three of the plants were infected with BCV1 alone, seven with BCV2 alone, eight plants contained both viruses and two plants were healthy. Thus BCV1 and BCV2 are two unrelated, independently replicating viruses.

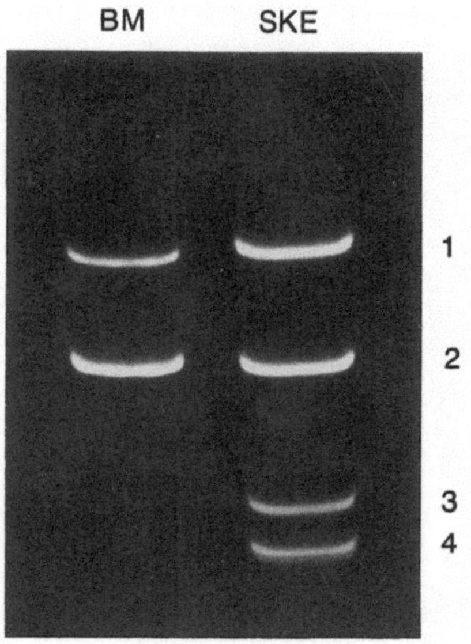

Figure 2. Electrophoresis in 5% polyacrylamide gels of dsRNA extracted from leaves of (BM) *Beta maritima* plants containing BCV1 alone, and (SKE) sugar beet cv. Sharpes Klein E plants containing BCV1 and BCV2.

Using the cDNA clones as probes to assay for both BCV1 and BCV2 we found that these viruses were present in all five sugar beet cultivars tested (Sharpes Klein E, Regina, Salohill, Amethyst and Bravo), and in each cultivar a high proportion (73-100%) of individual plants was infected (Xie *et al.*, 1989). Both viruses were also found in other types of beet including red beet, leaf beet and fodder beet (Xie *et al.*, 1989).

Northern blot analysis also showed that the two dsRNAs of beet temperate virus, isolated from the Japanese leaf beet cv. Fudanso, hybridized with the cDNA probes specific for dsRNAs 1 and 2 but not with the other probes, showing that there are nucleic acid sequence similarities between beet temperate virus and BCV1 (Fig. 3, Fd tracks 1 and 2) (Xie *et al.*, 1989).

Two dsRNA components were also found in *Beta maritima* with sizes similar to dsRNAs 1 and 2 of BCV (Fig. 2). These hybridized with the cDNA probes specific for dsRNAs 1 and 2 showing that they share nucleic acid sequences with the genome of BCV1 (Fig. 3, BM tracks 1 and 2). No evidence was found for BCV2 or a related virus in *Beta maritima*, but only a small number of plants from one geographical location

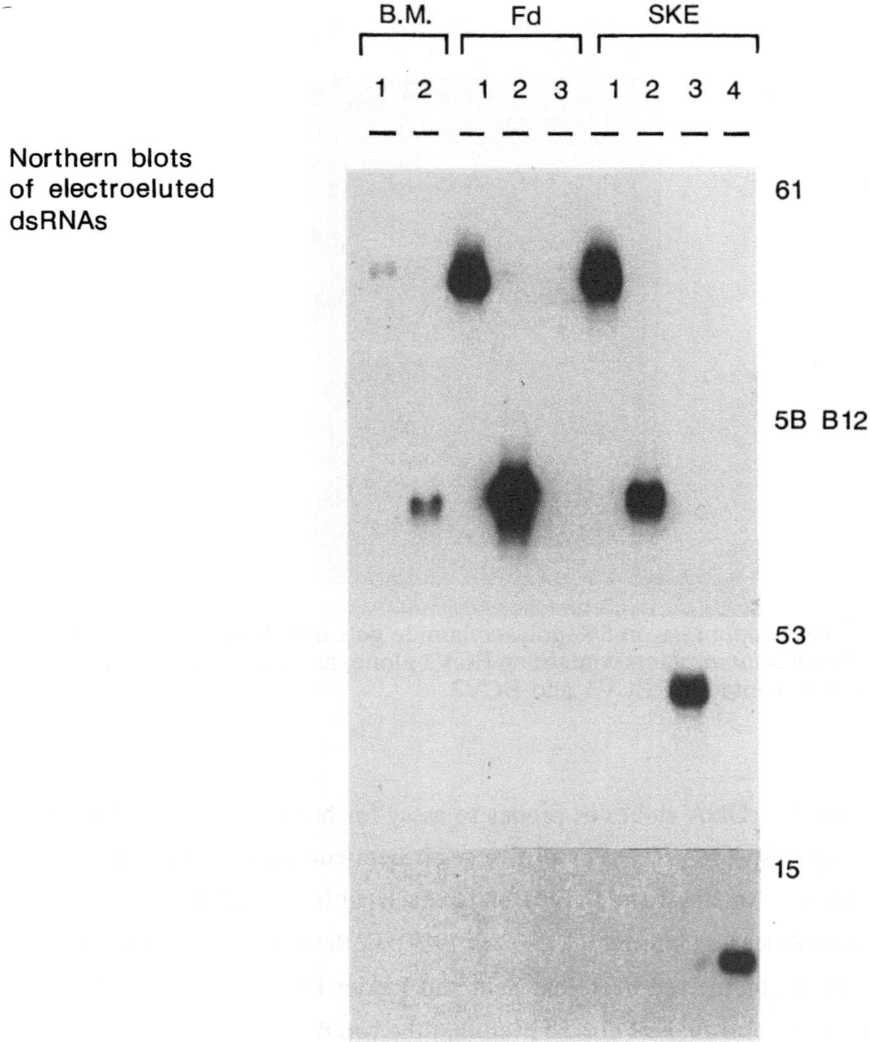

Figure 3. dsRNAs extracted from leaves of *Beta maritima* (BM), leaf beet cv. Fudanso (Fd) and sugar beet cv. Sharpes Klein E (SKE) were separated by electrophoresis in 5% polyacrylamide gels, individual bands were excised from the gel, the dsRNA electroeluted, denatured and re-electrophoresed in 1% agarose denaturing gels (4 replicates), then capillary blotted onto Hybond N. Each Northern blot was hybridized with the ^{32}P-labelled insert of one of the four cDNA clones. Clone 61 is specific for dsRNA1, clone 5B.B12 for dsRNA2, clone 53 for dsRNA3 and clone 15 for dsRNA4. Reproduced from Plant Pathology by permission.

were analysed (Xie *et al.*, 1989). The dsRNAs of purified ryegrass cryptic virus and fescue cryptic virus did not hybridize with cDNA clones of BCV1 or BCV2 (JF Antoniw,

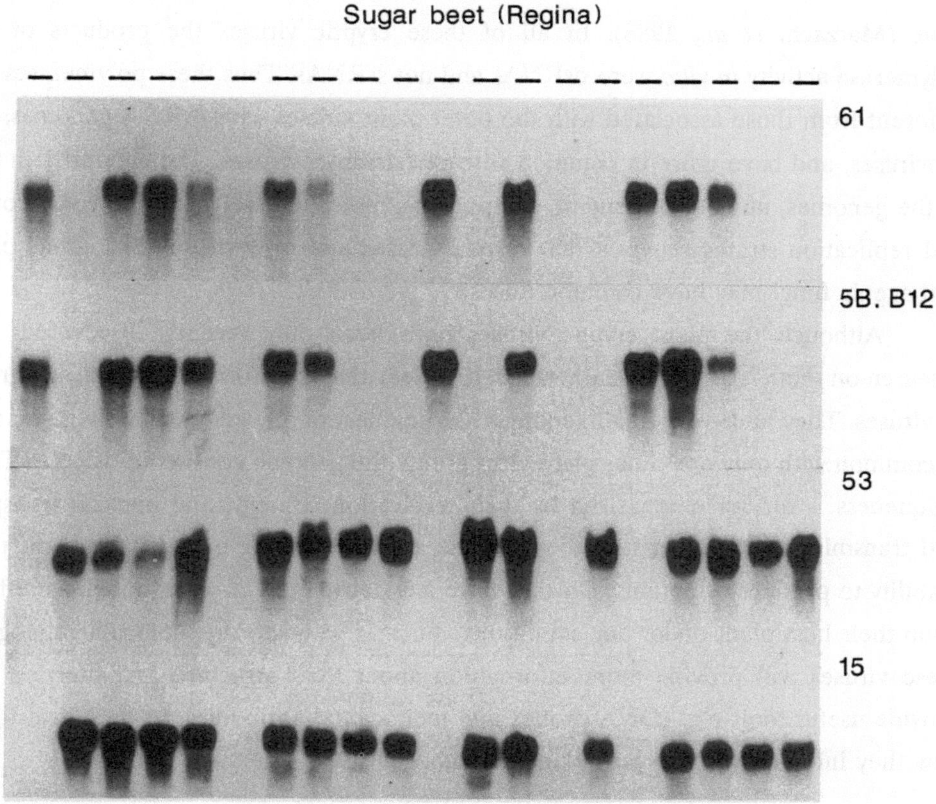

Figure 4. Nucleic acid extracts from leaves of 20 individual plants of sugar beet cv. Regina were denatured and electrophoresed in 1% agarose denaturing gels (4 replicates), then capillary blotted onto Hybond N. Each Northern blot was hybridized with the ^{32}P-labelled insert of one of the four cDNA clones (61, 5B.B12, 53 and 15; specific for dsRNAs 1, 2, 3 and 4 respectively of beet cryptic virus).
Reproduced from Plant Pathology by permission.

RF White, GP Accotto and G Boccardo, unpublished).

Accotto *et al.* (1987) translated the two dsRNA components of BCV1 *in vitro* in a wheat germ cell-free system and found that dsRNA1 encoded a 67 kDa protein, thought to be a viral transcriptase, whereas dsRNA2 encoded a 52 kDa protein, which was immunoprecipitated by an antiserum prepared against the virus particles and is therefore probably the coat protein. These two proteins represent most of the coding capacity of the BCV1 genome (88% for dsRNA1 and 82% for dsRNA2) suggesting that the two dsRNAs are monocistronic genes.

RNA-dependent RNA polymerase activity was detected in purified particles of white clover cryptic viruses 1 and 2 (Boccardo and Accotto, 1988) and carnation cryptic virus (Marzachi *et al.*, 1988). In all of these cryptic viruses the products of the polymerase activity *in vitro* were dsRNAs and not ssRNAs. Thus these polymerases are different from those associated with the other plant viruses with dsRNA genomes, the reoviruses, and have more in common with certain mycoviruses. The similarity in size of the genomes, number of genome components, molecular weight of the coat protein and replication strategy suggest that cryptic viruses and mycoviruses that infect plant pathogenic fungi may have common ancestry.

Although the plant cryptic viruses have been only recently discovered, and research on them is still at an early stage, it is clear that they form a very unusual group of viruses. They have very small genomes, the smallest of any known plant viruses, and, in common with only one other plant virus group, the genome consists of dsRNA. Their uniqueness is further emphasized by their replication strategy, and unusual transport and transmission characteristics. Possibly the most interesting question concerns their inability to produce symptoms. Do they have any detrimental or even beneficial effects upon their host plant under any conditions? Further work on the molecular biology of these viruses will provide more information about their structure and function and provide useful tools e.g. cDNA probes and monoclonal antibodies, to help investigate how they interact with the host plant and other viruses.

ACKNOWLEDGEMENTS

W Xie is a postgradate student registered for a PhD at Imperial College, University of London, and financed by the British Council. We are grateful to Professor K Buck for advice.

REFERENCES

Abou-Elnasr MA, Jones AT, Mayo MA (1985) Detection of dsRNA in particles of vicia cryptic virus and in *Vicia faba* tissues and protoplasts. J Gen Virol 66:2453-2460

Accotto GP, Boccardo G (1986) The coat proteins and nucleic acids of two beet cryptic viruses. J Gen Virol 67:363-366

Accotto GP, Brisco MJ, Hull R (1987) *In vitro* translation of the double-stranded RNA genome from beet cryptic virus 1. J Gen Virol 68:1417-1422

Antoniw JF, Linthorst HJM, White RF, Bol JF (1986) Molecular cloning of the double-stranded RNA of beet cryptic viruses. J Gen Virol 67:2047-2051

Boccardo G, Accotto GP (1988) RNA-dependent RNA polymerase activity in two

morphologically different white clover cryptic viruses. Virology 163:413-419
Boccardo G, Lisa V, Luisoni E, Milne RG (1987) Cryptic plant viruses. Adv Virus Res 32:171-214
Boccardo G, Lisa V, Milne RG (1983) Cryptic viruses in plants. In: Compans RW, Bishop DHL (eds) Double stranded RNA viruses. Elsevier, New York, p 425
Boccardo G, Milne RG (1984) Plant reovirus group. CMI/AAB Descriptions of Plant Viruses No. 294
Boccardo G, Milne RG, Luisoni E, Lisa V, Accotto GP (1985) Three seedborne cryptic viruses containing double-stranded RNA isolated from white clover. Virology 147:29-40
Jelkmann W, Maiss E, Casper R, Lesemann DE (1988) *Cucumis sativus* cryptic virus, a new virus in cucumber. J Phytopathol 121:233-238
Kassanis B (1984) When a virus-like particle is a virus. Phytopathol Z 111:363-366
Kassanis B, Russell GE, White RF (1978) Seed and pollen transmission of beet cryptic virus in sugar beet plants. Phytopathol Z 91:76-79
Kassanis B, White RF, Woods RD (1977) Beet cryptic virus. Phytopathol Z 90:350-360
Kenten RH, Cockbain AJ, Woods RD (1978) Rothamsted Exp Stn Rep 1977 Part 1 p 222
Kenten RH, Cockbain AJ, Woods RD (1979) Rothamsted Exp Stn Rep 1978 Part 1 p 220
Kenten RH, Cockbain AJ, Woods RD (1980) Rothamsted Exp Stn Rep 1979 Part 1 p 176
Kenten RH, Cockbain AJ, Woods RD (1981) Rothamsted Exp Stn Rep 1980 Part 1 p 188
Kuhne T, Stanarius A, Schubert J (1987) Latent virus showing properties of cryptic viruses detected in *Sinapsis alba.* Arch Phytopathol Pflanzenschutz 23:499-500
Lisa V, Boccardo G, Milne RG (1981a) Double-stranded ribonucleic acid from carnation cryptic virus. Virology 115:410-413
Lisa V, Luisoni E, Milne RG (1981b) A possible virus cryptic in carnation. Ann Appl Biol 98:431-437
Marzachi C, Milne RG, Boccardo G (1988) *In vitro* synthesis of double-stranded RNA by carnation cryptic virus-associated RNA-dependent RNA polymerase. Virology 165:115-121
Moir DL (1983a) Rothamsted Exp Stn Rep for 1982 Part 1 p 80
Moir DL (1983b) Some effects of beet cryptic virus in sugar beet. PhD Thesis, University of Newcastle-Upon-Tyne
Natsuaki T (1985) Radish yellow edge virus. CMI/AAB Descriptions of Plant Viruses No. 298
Natsuaki T, Natsuaki KT, Okuda S, Teranaka M, Milne RG, Boccardo G, Luisoni E (1986) Relationship between the cryptic and temperate viruses of alfalfa, beet and white clover. Intervirology 25:69-75
Natsuaki T, Yamashita S, Doi Y, Okuda S, Teranaka M (1983a) Radish yellow edge virus, a seed-borne virus with double-stranded RNA, of a possible new group. Ann Phytopathol Soc Jpn 49:593-599
Natsuaki T, Yamashita S, Doi Y, Okuda S, Teranaka M (1983b) Two seed-borne double-stranded RNA viruses, beet temperate and spinach temperate virus. Ann Phytopathol Soc Jpn 49:709-712
Natsuaki T, Yamashita S, Doi Y, Yora K (1979) Radish yellow edge virus. Ann Phytopathol Soc Jpn 45:313-320
Paludan N, Begtrup J (1986) Inactivation of poinsettia mosaic virus and poinsettia cryptic virus in *Euphorbia pulcherrima* using heat treated mini-cuttings and

meristem-tip culture. Tiddsskr Planteavl 90:283-290
Plumb RT (1973) Rothamsted Exp Stn Rep 1972 Part 1 p 125
Plumb RT, Lennon EA (1981) Rothamsted Exp Stn Rep 1980 Part 1 p 187
Plumb RT, Misari SM (1974) Rothamsted Exp Stn Rep 1973 Part 1 p 137
Pullen ME (1968) Rothamsted Exp Stn Rep 1967 Part 1 p 124
Pullen ME (1969) Rothamsted Exp Stn Rep 1968 Part 1 p 127
Raizada RK, Srivastava KM, Singh BP (1983) Detection of amaryllis cryptic virus by enzyme linked immuno sorbent assay. Indian J Exp Biol 21:299-300
Stanarius A, Kuhne T, Proeseler G, Kleinhempel H (1984) Beet cryptic and beet mild yellowing viruses - morphologically distinguishable by electron microscopy. Arch Phytopathol Pflanzenschutz 20:351-353
White RF, Woods RD (1978) Beet cryptic virus in some varieties of *Beta vulgaris*. Phytopathol Z 91:91-93
Xie W, Antoniw JF, White RF (1989) Detection of beet cryptic viruses 1 and 2 in a wide range of beet plants using cDNA probes. Plant Pathol, in press

PLANT GROWTH REGULATORS, VIRUSES AND PLANT GROWTH

R J Whenham and R S S Fraser[1]
Biochemistry and Molecular Biology Department
AFRC Institute of Horticultural Research
Wellesbourne
Warwick CV35 9EF
UK

ABSTRACT

Virus infection alters plant growth and development. The role of plant growth regulators (PGRs) in control of these processes in diseased plants is reviewed, taking tobacco mosaic virus infections of tobacco and tomato as the main model systems. Growth inhibition was related primarily to severity of visible mosaic symptoms rather than to the extent of virus multiplication. In tobacco, infection increased abscisic acid and decreased active cytokinin concentration, thus inhibiting cell division activity and reducing plant growth. Changes in other growth regulators after infection are briefly reviewed. Auxin and gibberellin activity are generally decreased by infection in diverse combinations of plants and viruses, while ethylene production is stimulated in hypersensitive reactions to infection. The chloroplast may be a major site of interaction between the virus and PGR metabolism.

INTRODUCTION

Virus infection often produces an inhibition of plant growth and causes other visible symptoms of disease such as the light-green/dark-green mosaic or "green islands" seen in systemically-infected leaves. The biochemical mechanisms involved in this viral control over host growth and development remain largely speculative. Since plant

[1] AFRC Institute of Horticultural Research, Worthing Road, Littlehampton, West Sussex BN17 6LP, UK

NATO ASI Series, Vol H 41
Recognition and Response in Plant-Virus Interactions
Edited by R S S Fraser

growth regulators (PGRs) are intimately involved in control of growth and senescence in healthy plants there has been considerable interest in whether they may be involved in control of growth and symptom formation in virus-infected plants. This review is concerned primarily with how virus infection alters growth and how this effect is related to the development of other visible symptoms of infection. Early work on PGR metabolism in virus infected plants has been reviewed previously (Fraser and Whenham, 1982; Fraser, 1987).

COMPETITIVE VERSUS NON-COMPETITIVE INHIBITION OF GROWTH

There are two possible models to explain quantitatively the reduced growth of virus-infected plants. In the competitive model, accumulations of viral nucleic acid and proteins form sufficiently active sinks for nucleoside triphosphates and amino acids, that host nucleic acid and protein synthesis are inhibited and growth of the host plant is reduced. This model requires that large amounts of viral nucleoprotein accumulate within the plant. In tobacco systemically-infected with tobacco mosaic virus (TMV) for example, viral components can become the commonest RNA and protein species in the leaf, representing up to 10% of leaf dry weight. The competitive model predicts that the degree of inhibition of host growth should be proportional to the amount of virus accumulated.

Many viruses, however, do not accumulate to such high concentrations as TMV in tobacco, but nevertheless cause equally severe growth inhibition (Zaitlin, 1979). In these cases a simple model of competitive inhibition is not applicable. Inhibition of host growth must therefore occur by some other, non-competitive means. Such non-competitive mechanisms of growth inhibition might be related to the cytopathic effects of infection and to the severity of visible disease symptoms.

The relative importance of the two models can be tested experimentally. Tobacco (Whenham *et al.*, 1985) and tomato (Fraser *et al.*, 1986) plants were infected with a range of TMV isolates which multiplied to varying extents and caused symptoms of varying degrees of severity. Plant growth was correlated with symptom severity and with TMV multiplication in both experiments. In tobacco the correlations were similar, but in tomato, growth was more strongly related to symptom severity than to virus multiplication.

Table 1. *Multiple regression analysis attributing variation in the growth of TMV-infected plants to various sources*

Host	TMV concentration (mg/g) mean	maximum	Factor accounting for variation in growth	% of variation accounted for
Tomato (+/+)	1.8	3.2	Symptom severity	54
			TMV multiplication	1
			Plant - intrinsic	9
			Unattributed	36
				100
Tomato (*Tm-1/Tm-1*)	1.4	3.0	Symptom severity	53
			TMV multiplication	7
			Plant - intrinsic	15
			Unattributed	25
				100
Tobacco	4.2	6.0	Symptom severity	36
			TMV multiplication	40
			Plant - intrinsic	10
			Unattributed	14
				100

From Fraser *et al.* (1986).

Multiple regression analysis allowed the variation in growth rate to be attributed quantitatively to these two parameters and removed the effects of interactions (Table 1). The intrinsic variation was the variation in growth rate which would have occurred without any virus infection. Of the remaining 85 to 90% of variation, caused by infection, most could be accounted for by variation in symptom severity plus variation in TMV multiplication: only comparatively small percentages of variation in growth could not be accounted for in each of the hosts. In the case of tomato, the multiple regression analysis suggested that only a small proportion of the variation in growth was attributable to variation in TMV multiplication; most was attributable to variation in symptom severity. In tobacco, TMV multiplication and symptom severity contributed about equally to the variation of growth. It is notable that overall, TMV multiplied to significantly higher levels in tobacco than in tomato (Table 1). These results suggest that competitive inhibition of plant growth does occur, but only at the top end of the scale of virus multiplication. In the tomato experiment the concentrations of virus reached, although high in comparison with many other viruses, were not sufficient to cause serious competitive inhibition of plant growth. It is necessary therefore to seek other explanations for the inhibition of host growth, based on non-competitive mechanisms.

What mechanisms might be involved? One possibility is that growth inhibition might involve virus-induced changes in plant growth regulator metabolism which might in turn be related to the cytopathic effects of infection. Development of visible symptoms suggests changes in the chloroplasts, and changes in chloroplast ultrastructure have been reported for many virus infections (reviewed in Fraser, 1987). The central role of the chloroplast in the localization and possible modes of action of PGRs within the cell (Loveys, 1977; Heilmann *et al.*, 1980; Hartung *et al.*, 1982) suggests a mechanism by which a virus might interact directly with the growth regulator metabolism of the host plant.

ABSCISIC ACID

Interest has centered on abscisic acid (ABA) as the best characterized endogenous inhibitor of plant growth (Zeevaart and Creelman, 1988). Early studies of ABA metabolism in virus-infected plants, based on bioassay of semi-purified fractions, yielded very varied results (reviewed in Fraser and Whenham, 1982). Levels of ß-inhibitor, of which ABA is a major component (Milborrow, 1978), were increased in lupins infected with pea mosaic virus (Van Steveninck, 1959) and in rice seedlings infected with rice tungro virus (Mohanty *et al.*, 1979) where growth inhibition was pronounced. In contrast, inhibitor level was apparently unchanged in field-grown tobacco, systemically-infected with TMV, although stunting of growth was also severe (Steadman and Sequeira, 1969). Repeated application of exogenous ABA reduced the growth of healthy plants (Steadman and Sequeira, 1970) and caused further inhibition in infected stems and leaves (Bailiss, 1977; Bailiss *et al.*, 1977). Inhibitor level was decreased in virus-infected *Euphorbia pulcherrima* buds (Mandahar *et al.*, 1979).

Bioassays are insufficient for positive identification of the inhibitor as ABA, as the biological responses employed are unspecific (Milborrow, 1978). Physical methods have been used to measure ABA in a few cases. ABA concentration was decreased during the very early stages of systemic infection of tobacco plants with the tomato streak strain of TMV (Rajagopal, 1977), before any effect on growth was noted. In contrast ABA concentration was increased in cucumber mosaic virus (CMV)-infected cucumber shoots in the early period, up to 4 days after infection, before reduction of hypocotyl elongation was apparent (Aharoni *et al.*, 1977). There was no consistent difference in ABA concentration between healthy and infected shoots at times later than 4 days after infection (Bailiss *et al.*, 1977). Application of low concentrations of exogenous ABA reduced growth of infected but not healthy hypocotyls (Aharoni *et al.*, 1977). It was suggested, but not confirmed (Bailiss *et al.*, 1977), that sensitivity to ABA was altered in infected plants. These experiments used methods which, although based on gas chromatography (GC), lacked adequate standardization to allow correction for the considerable losses in ABA which occur during purification of plant extracts. This makes interpretation of the results difficult.

We have used capillary GC with electron capture detection, coupled to standardization with radiolabelled ABA of high specific activity, to study ABA metabolism in TMV-infected tobacco and tomato plants. In tobacco, ABA concentration was increased soon after infection by both a local-lesion-forming strain of TMV and by systemic infection (Table 2; Whenham and Fraser, 1981; Whenham *et al.*, 1985; Whenham *et al.*, 1986). The greatest increase in ABA concentration, up to 20-fold, occurred in leaves showing necrotic lesions. Systemic infection produced increases of up to 6-fold in inoculated and in systemically-infected leaves which persisted for the duration of viral multiplication.

ABA turnover was also stimulated by infection. ABA is metabolized by several routes in the cell including conjugation with glucose and by an oxidative pathway to yield phaseic (PA) and dihydrophaseic (DPA) acids (reviewed in Zeevaart and Creelman, 1988). The route to PA predominates in tobacco. Systemic infection produced a large increase in PA concentration (Table 2), equivalent in this experiment to an increase of 113 ng/g/day in the rate of ABA synthesis. This is greater than the increase in steady state free ABA concentration. Concentration of ABA glucose ester was little changed

Table 2. *Effects of tobacco mosaic virus infection on the concentrations of abscisic (ABA) and phaseic (PA) acids in tomato and tobacco*

Host	TMV isolate[a]	ABA (ng/g)	PA (ng/g)
Tobacco	healthy	30 ± 4	745 ± 86
	vulgare	82 ± 9	2436 ± 375
	flavum	280 ± 94	nd
Tomato	healthy	266 ± 10	124 ± 32
	MII-16	284 ± 23	62 ± 9
	vulgare	338 ± 11	54 ± 7

[a] *Vulgare* caused severe systemic mosaic and MII-16 caused mild mosaic. *Flavum* caused necrotic local lesions. ABA and PA concentrations were measured 15 days after inoculation. Values are means ± standard errors. nd = not determined. From Fraser, 1987.

by systemic infection (Whenham and Fraser, 1981), suggesting that the increase in free ABA was not a result of release from a "bound" form.

When tobacco was infected with different TMV strains producing systemic symptoms of varying severity, the virus-induced increase in ABA was strongly correlated with the severity of the mosaic produced and much less so with the extent of virus multiplication (Whenham *et al.*, 1985). Inhibition of growth (measured as relative growth rate) was correlated about equally with virus-stimulated increase in ABA concentration, and with differences in TMV multiplication, for the different strains (Fig. 1).

The results suggest that at least some of the effects of TMV on growth of infected tobacco plants may be due to virus-induced ABA accumulation. This was further tested by spraying healthy tobacco plants each day with a low concentration solution of ABA, so as to maintain an increase in leaf ABA concentration similar to that caused by virus infection. ABA treatment of healthy plants, or TMV infection, both caused similar degrees of inhibition of leaf growth, measured as leaf area or leaf fresh weight. The younger the leaf at the time of infection or commencement of ABA sprays,

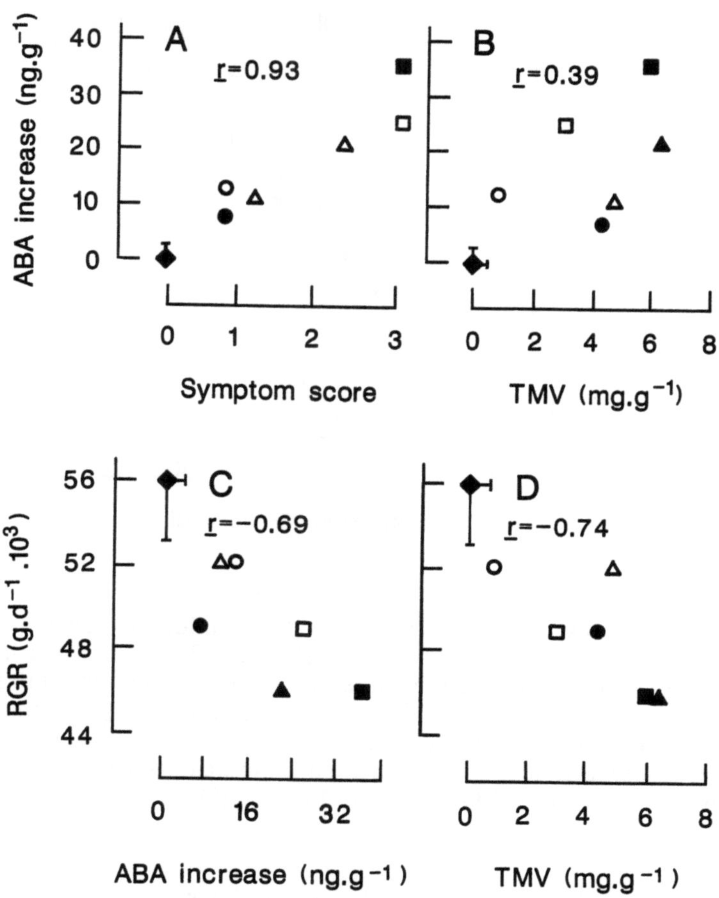

Figure 1. Changes in abscisic acid concentration in TMV-infected tobacco leaves, and relationships between abscisic acid metabolism, virus multiplication and shoot growth rate. Plants were inoculated with TMV isolates causing mosaic symptoms of varying severity. (◆), healthy (sham-inoculated) control. TMV strains in order of increasing symptom severity: (o), M11-16; (●), V4; (△), N2; (▲), *vulgare*, (□), *flavum*; (■), U5. The bars on the healthy plant points indicate least significant differences at P = 0.05. (A), relationship between increase in ABA concentration and mosaic symptom severity. The ABA concentration in healthy plants was 9.8 ng/g. (B), relationship between increase in ABA concentration and virus multiplication. (C), relationship between shoot relative growth rate (RGR) and increase in ABA concentration caused by infection. (D), relationship between shoot relative growth rate and virus multiplication. From Whenham *et al.* (1985).

the greater the growth inhibition recorded. Qualitatively, both systemic TMV infection and applications of ABA to healthy plants were shown to reduce leaf growth by inhibiting cell division rather than cell expansion (Fig. 2). Virus infection (Russell and

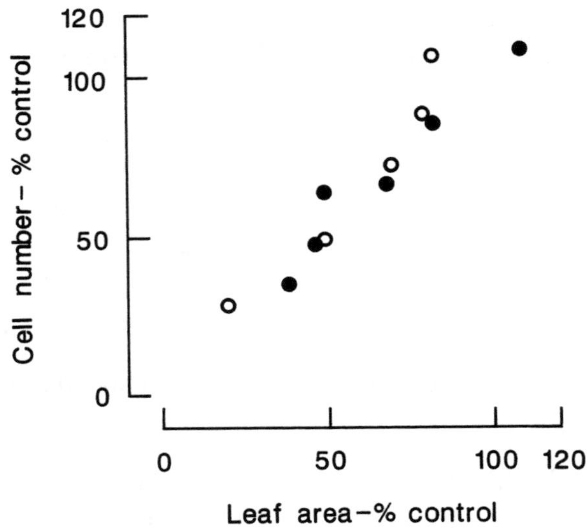

Figure 2. Effect of ABA and TMV *vulgare* (systemic infection) on leaf cell number. Leaf cell number and area in TMV-infected (●) and ABA-treated (o) healthy leaves were expressed as a percentage of the same parameters in healthy, untreated control plants and were plotted for leaves of different ages, at different positions on the stem. Leaf areas were measured with a photoelectric digital scanner. Leaf cell number was determined by macerating leaf discs in 5% chromic acid and counting samples in a haemocytometer slide. From Whenham and Fraser (1981).

Kimmins, 1971) and ABA (Quarrie and Jones, 1977) have been shown to inhibit cell division preferentially in other systems.

The finding that ABA is transported from infected to non-infected parts of plants inoculated with a local-lesion-forming strain of TMV (Whenham and Fraser, 1981) provides a possible explanation of the reduced growth of non-infected leaves in these plants (Fraser, 1979).

The virus-induced increase in leaf ABA concentration may produce a situation favourable for virus multiplication. Inhibition of host growth would reduce competition for biosynthetic precursors required for virus synthesis. Increased ABA concentration itself has a stimulatory effect on virus synthesis and accumulation (Whenham and Fraser, 1980).

ABA is important in control of stomatal closure (reviewed in Zeevaart and Creelman, 1988). However, it is unlikely that enhanced ABA levels in TMV-infected tobacco plants caused reduced growth by closing stomata and hence inhibiting gas exchange and assimilation rate. There was little difference in stomatal aperture, measured by diffusive resistance, between healthy leaves and leaves systemically infected with TMV (Whenham and Fraser, 1981).

The mechanism(s) by which virus infection stimulates ABA accumulation are of interest. Necrotic strains stimulated a greater increase in ABA concentration than systemic strains although virus multiplied to a much lower concentration. The results suggest that virus infection altered ABA metabolism in two separate ways; by causing necrosis, and by some mechanism related to symptom formation in non-necrotizing tissue. Necrosis causes localized water loss and water deficit induces large and rapid increases in ABA concentration (Wright and Hiron, 1969).

The means by which TMV infection increases ABA accumulation in non-necrotizing tissue are less clear. Systemic infection did not cause ABA accumulation by inducing water stress in leaves as healthy and infected leaves did not differ in this respect (RJ Whenham, unpublished data).

An alternative explanation is that the virus may induce ABA accumulation by an effect on chloroplasts. These organelles are the major site of ABA accumulation within the cell (Loveys, 1977; Hartung *et al.*, 1982; Heilmann *et al.*, 1980). ABA is a weak acid (pK_a 4.8) and dissociates at physiological pH. Only the undissociated molecule can penetrate biomembranes. The chloroplast stroma is maintained at a higher pH than the cytoplasm by a light-driven proton pump. The accumulation of ABA in chloroplasts occurs when undissociated molecules pass through the chloroplast membrane, then dissociate in the stroma and are then unable to leave there. Table 3 shows that in healthy, unstressed leaves, almost all of the ABA was sequestered within the chloroplast. TMV infection had little effect on ABA concentration within the chloroplast but increased ABA concentration in the extrachloroplastic fraction. Virus strains which produced the most severe symptoms produced the greatest change in the concentration of extrachloroplastic ABA (up to 80-fold). These results may be of particular relevance to the mechanism of growth control, especially if the receptor for ABA-inhibition of growth is situated outside the chloroplast (Whenham *et al.*, 1985).

How does TMV increase the concentration of ABA outside the chloroplast? Water stress decreases stromal pH and allows leakage of ABA to the cytoplasm. However, there is no evidence that TMV infection causes water stress, nor does TMV alter stromal pH or the ability of the chloroplast to raise pH in the light (Whenham *et al.*, 1986).

Table 3. *Intracellular distribution of ABA in TMV-infected tobacco leaves*[a]

Virus strain	Symptoms	ABA ($ng\ mg^{-1}$ chlorophyll)		% ABA outside chloroplast
		Leaf	Chloroplast	
sham-inoculated control		31.1	30.8	1
MII-16	very mild green mosaic	24.8	17.8	28
V4	Mild green mosaic	50.0	32.9	34
N2	Moderate green mosaic	42.2	27.1	36
vulgare	Severe light green/dark green mosaic	128.6	28.6	78
flavum	Severe yellow-green mosaic	77.9	29.0	63
U5	Very severe green mosaic	86.1	23.8	72

[a]Tobacco plants were inoculated with various isolates of TMV. These multiplied to varying extents, and caused visible symptoms of varying degrees of severity. At 24 days after inoculation, chloroplasts were isolated in non-aqueous solvents to minimize ABA loss during purification. From Whenham *et al.*, 1985.

An alternative explanation of the altered ABA gradient in infected cells is required. There are numerous reports of the effects of virus on chloroplast ultrastructure (reviewed in Fraser, 1987). An early effect of TMV is to degrade chloroplast ribosomal RNA (Fraser, 1969) and to alter the properties of chloroplast proteins (Ladygina *et al.*, 1979). TMV coat protein can accumulate within the chloroplast (Zaitlin *et al.*, this volume). It seems likely, therefore, that TMV alters the permeability properties of the chloroplast envelope in some way, allowing ABA leakage into the cytoplasm. Reduced chloroplastic or increased cytoplasmic ABA concentration then, in some way as yet unknown, switches on further ABA synthesis in a location which is also not yet fully understood.

The conclusions above concerning the role of ABA need not apply to other plants and viruses. In tomato infected with TMV (Table 2), infection did not greatly increase ABA concentration, and the rate of catabolism was not increased, suggesting that ABA plays little role in the growth inhibition in this case. However, basal ABA levels in tomato are much higher than in tobacco plants and exogenous ABA has little effect on growth of tomato, even at high concentrations (Fraser and Whenham, 1989). The results suggest that changes in hormones other than ABA may also be important in other host/virus combinations.

CYTOKININS

The cytokinins are a chemically diverse group of growth regulators which occur in very low concentrations in most plant tissues (McGaw *et al.*, 1984). In virus-infected plants attention has focussed on the senescence-retarding effect of cytokinins and their possible rôle in the formation of virus-resistant "green islands" in systemically-infected leaves. However, the cytokinins were originally discovered by their controlling effects on cell division (reviewed in Letham, 1978), which is reduced in infected plants (Russell and Kimmins, 1971; Whenham and Fraser, 1981). It is therefore also of interest to measure changes in cytokinin concentration in relation to altered growth. Measurement of cytokinin concentration by physical methods is especially difficult. Consequently, many studies of cytokinin metabolism and its relation to growth in virus-infected plants have relied on relatively unspecific and semi-quantitative bioassays and the results have been equivocal.

Bioassay studies have shown that cytokinin activity was reduced in tobacco ringspot virus-infected roots of cowpea and *Nicotiana glutinosa* leaves (Kuriger and Agrios, 1977; Tavantzis *et al.*, 1979). In contrast, cytokinin activity was increased in rice leaves infected with rice tungro virus (Sridhar *et al.*, 1978) and in tobacco leaves and stems systemically-infected with TMV or CMV (Sziraki and Gaborjanyi, 1974; Sziraki and Balazs, 1977; Sziraki *et al.*, 1980). TMV infection of hypersensitively-reacting tobacco increased cytokinin activity both in inoculated leaves that developed lesions (Sziraki and Balazs, 1977) and in non-infected leaves of the same plant (Balazs *et al.*, 1977). Increased cytokinin activity was detected in inoculated leaves before the

Table 4. *Cytokinin concentration in healthy and TMV-infected tobacco leaves*[a]

Leaf tissue	Cytokinin concentration[a] (ng g^{-1} fresh weight)					
	Z[b]	ZOG	ZR	ZROG	Z7G	ZXP[c]
Healthy	16.1	67.6	5.1	6.6	10.3	8.2
Light-green	13.7	65.6	5.7	11.4	15.6	6.4
Dark-green	6.9	120.1	6.1	30.0	62.4	9.8
LSD P = 0.05	6.1	13.1	2.5	5.8	15.2	5.6

[a]Cytokinins were purified from approximately 50g fresh weight of tobacco leaves. Values are the combined results from three separate experiments. Taken from Whenham (1989) and unpublished data.

[b]Abbreviations are as defined in the text.

[c]ZXP is total zeatin nucleotide, measured by a method based on alkaline phosphatase digestion which does not distinguish between mono-, di- or triphosphates.

appearance of visible lesions.

We have measured cytokinin concentrations in highly-purified extracts of TMV-infected tobacco using capillary GC with a nitrogen-specific detector which is highly sensitive and selective for cytokinins and using appropriate radioactive recovery standards to correct for losses during extraction (Whenham, 1983; Whenham, 1989). Tobacco leaves were shown to contain five major cytokinins: zeatin (Z), zeatin riboside (ZR), zeatin-O-glucoside (ZOG), zeatin-O-glucoside riboside (ZROG) and zeatin-7-glucoside (Z7G) as well as zeatin nucleotides (ZXP) which are associated with cytokinin uptake (McGaw, 1987). TMV infection reduced the concentration of Z (Table 4), which is considered to be the active form of cytokinin within the cell (Letham and Palni, 1983). The greatest decrease in Z occurred in the virus-free, dark-green areas of the mosaic which formed in systemically-infected leaves. The concentrations of ZR and ZXP were unchanged by infection.

Free cytokinins are rapidly turned over in leaves to form glucosides, particularly O-glucosides (Palmer *et al.*, 1981; Wang *et al.*, 1977). The effect of TMV infection was to stimulate cytokinin catabolism leading to the accumulation of high concentrations of

Table 5. *Effect of exogenous ABA on cytokinin concentration in healthy and TMV-infected tobacco*[a]

Leaf tissue	Cytokinin concentration ng g^{-1} fresh weight			
	Z	ZOG	ZR	ZROG
Healthy	16.6	49.0	4.3	3.3
Healthy + ABA	11.7	68.8	1.3	16.2
TMV; Light-green	15.7	47.5	2.3	4.3
TMV; Light-green + ABA	10.2	62.3	1.9	9.8
TMV; Dark-green	13.3	72.4	5.1	12.0
TMV; Dark-green + ABA	6.5	94.9	6.8	38.9
LSD, $P = 0.05$	3.2	4.2	3.0	4.0

[a]From Whenham 1989. Plants treated with ABA were sprayed daily with a low concentration solution, so as to maintain an increase in ABA concentration similar to that caused by TMV infection.

cytokinin glucosides (ZOG, ZROG, Z7G) in dark-green but not in light-green tissue. The cellular function of cytokinin-O-glucosides remains unclear; they may be cytokinin storage or transport forms but they are not metabolically inert and are rapidly metabolized to the aglycones when applied exogenously to a variety of tissues (McGaw *et al.*, 1984). Cytokinin-7-glucosides, in contrast, are highly stable and probably represent inactive end-products of cytokinin metabolism.

The large increase in concentration of cytokinin conjugates in infected plants may help to explain early bioassay results which showed increased cytokinin activity, ascribed to Z and ZR, in TMV-infected tobacco (Sziraki and Balazs, 1977; Sziraki and Gaborjanyi, 1974). In these experiments the chromatographic systems employed would not have distinguished cytokinin bases or ribosides from the corresponding cytokinin-O-glucosides present in larger amounts. Z and ZOG are equally active in callus bioassays (Letham *et al.*, 1983), although some of the apparent bioactivity of the O-glucosides may be attributed to breakdown to free zeatin during the assays. It is probable that the enhanced activity observed in these and other earlier experiments may have been due to increased concentrations of O-glucosides present in infected leaves and not to increased concentrations of free zeatin or zeatin riboside.

The interactions of cytokinins and ABA and their effects on growth have also been studied (Table 5; Whenham, 1989). The increase in ABA concentration in leaves systemically-infected with TMV occurs mainly in the uninfected dark green tissue (Whenham *et al.*, 1986). It has been suggested that this elevated ABA concentration might be responsible, directly or indirectly, for the increased chlorophyll concentration of this tissue, and for its resistance to infection. Increasing the ABA concentration of healthy plants by spray treatments stimulated the conversion of free cytokinins to glucosides in a similar manner to that observed in non-ABA-treated, TMV-infected plants (Table 5). ABA treatment of infected plants further enhanced the turnover of free cytokinins to cytokinin glucosides. Zeatin is active in stimulating cell division, so that the reduction of zeatin concentration in infected plants may decrease cell division activity and hence contribute to the reduced growth of infected leaves. Exogenously applied ABA further reduces the growth of infected plants by inhibiting cell division activity. Thus the effects of cytokinins and ABA on growth appear to be closely related. Further study of the sequence of virus-induced changes in ABA and cytokinin metabolism is in progress to establish which is the primary and which the secondary effect.

The results extend the possible model for the molecular interaction of the virus and the host plant growth regulator metabolism via an early effect of virus on chloroplasts. Initially, infection of lower inoculated leaves induces changes in chloroplast membrane permeability. ABA leaks from the chloroplast and ABA synthesis is turned on, increasing cellular ABA concentration. Increased ABA concentration enhances the conversion of cytokinins to inactive glucosides leading to reduced cell division and hence decreased leaf growth. In localized infections, some of the ABA accumulated in inoculated leaves is transported to upper, non-infected leaves where it reduces cytokinin (zeatin) concentration and growth.

Virus-induced changes in ABA and cytokinins may also be involved in the formation and properties of the dark green islands of mosaic. The senescence-retarding effects of cytokinins, in preventing chlorophyll loss, are well known (Noodén and Leopold, 1978). ABA treatment of healthy tobacco leaves increases their chlorophyll concentration (Whenham *et al.*, 1986) although it is not known whether this is a direct effect, or one operating through induced changes in cytokinin metabolism. Concerning the resistance of the dark green areas to virus invasion or direct inoculation, ABA has been shown to reduce the susceptibility of healthy plants to infection (Fraser, 1982), and some compounds with cytokinin activity can inhibit TMV multiplication (Fraser and Whenham, 1978).

ETHYLENE

Ethylene, unlike other plant growth regulators, is a gas and is readily measured by GC without the need for recovery standards (Burg and Burg, 1962).

Early evidence suggested an important role for ethylene in the necrotic response in hypersensitively-reacting plants. Necrogenesis is invariably accompanied by increased ethylene production. Infection of *N* gene tobacco or bean with tobacco necrosis virus (Gaborjanyi *et al*., 1971) or TMV stimulated ethylene production in attached (Pritchard and Ross, 1975; Gaborjanyi *et al*., 1971; De Laat *et al*., 1981) and detached leaves (Nakagaki *et al*., 1970). Ethylene production increased before visible lesion formation as a result of an increase in the activity of the ethylene biosynthetic enzyme 1-amino cyclopropane carboxylic acid (ACC) synthase and an increase in ACC concentration (De Laat and van Loon, 1982).

Conclusive evidence that ethylene is the causal factor in necrogenesis has not been obtained. Treatment with exogenous ethylene after lesion appearance accelerated necrosis but treatment before inoculation actually reduced lesion size (Ross and Pritchard, 1972), although this may have been an indirect effect of ethylene on cell turgor, which affects physical susceptibility to infection. Carbon dioxide and "Alar" (daminozide), inhibitors of ethylene action or synthesis, had little effect on lesion development (Pritchard and Ross, 1975). Treatment with the ethylene-releasing compound ethephon (ethrel), but not with ethylene, induced ethylene formation in healthy leaves (Van Loon, 1977). Overall, the results suggest that ethylene and virus infection produce necrosis by independent mechanisms.

Strong evidence against a causal role for ethylene in virus-induced necrosis is provided by the action of amino vinyl glycine (AVG) which blocks the conversion of S-adenosyl methionine to ACC in ethylene biosynthesis (Adams and Yang, 1981). AVG inhibited ethylene production in TMV-infected tobacco but did not prevent virus-induced lesion formation (De Laat *et al.*, 1981).

Increased ethylene production can occur in the absence of necrosis, for example in *Tetragonia expansa* infected with beet yellow mosaic virus (Gaborjanyi *et al.*, 1971; Bailiss *et al.*, 1977) and in cucumber forming chlorotic lesions after infection with CMV (Marco and Levy, 1979). Necrosis could be induced in *T. expansa* leaves by treatment with exogenous ethylene (Bailiss *et al.*, 1977) whereas carbon dioxide treatment prevented both chlorotic and necrotic lesion formation.

Healthy plants produce ethylene in response to a variety of stress conditions including wounding (Lieberman, 1979). It is probable that some at least of the ethylene production in a hypersensitive response results from non-specific tissue damage caused by the virus.

Increased ethylene production was correlated with reduced growth and epinasty in CMV-infected cucumber hypocotyls and cotyledons (Marco *et al.*, 1976; Marco and Levy, 1979; Levy and Marco, 1976; 1977) and reduced growth of cowpea epicotyls infected with cowpea mosaic virus (Lockhart and Semancik, 1970) or CMV (Kato, 1976). Ethrel treatment mimicked the growth effects of virus infection in healthy plants (Levy and Marco, 1976; Marco *et al.*, 1976). However, removal of endogenous ethylene by potassium permanganate oxidation did not completely restore normal growth in infected cucumber.

Systemic infection of tobacco by TMV (Balazs *et al.*, 1969; Gaborjanyi *et al.*, 1971), in which stunting was severe, did not increase ethylene production, except in the late stages of infection in senescent leaves (Nakagaki *et al.*, 1970). Ethylene synthesis was unaffected in beet systemically infected with beet mosaic and beet yellow mosaic virus (Koch *et al.*, 1979). It seems unlikely therefore that ethylene production has a major role in producing the generalized inhibition of growth seen in some systemically-infected plants and that other PGRs may be more important in this respect.

AUXINS

Auxin metabolism in virus-infected plants has been less extensively studied than ABA, cytokinins and ethylene although altered auxin activity may be important in control of growth and symptom development. There are unfortunately very few data, based on reliable methods of measurement, on auxin relations in virus-infected plants.

This reflects the lack of knowledge of how auxin metabolism is regulated in healthy plants (Reinecke and Bandurski, 1987). The ready availability of proven physical and immunological methods for auxin analysis may improve this situation in the future.

Early work, based on bioassay, showed that auxin activity was generally reduced in infected plants, particularly in the late stage of infection when growth stunting was pronounced (reviewed by Pegg, 1976). The concentrations of indole acetic acid (IAA) and phenoxyacetic acid (PAA), measured by GC, were markedly reduced in tobacco in the very early stages of systemic infection with the tomato streak strain of TMV (Rajagopal, 1977).

Application of exogenous auxin reportedly reduced the severity of some of the visible symptoms produced by systemic TMV infection (Nichols, 1952) but no examples of the reversal of virus-induced stunting have been reported. Recent work (Van Loon *et al.*, this volume) using a physiologically realistic concentration of the natural auxin IAA, has shown that treatment actually worsened symptoms in TMV-infected tobacco.

The mechanism by which virus may reduce auxin concentration in systemically-infected tissues is not known. Decreased auxin activity might result from enhanced IAA oxidase activity (Lockhart and Semancik, 1970; Ghabrial and Pirone, 1967) or reduced auxin transport (Smith *et al.*, 1968).

Auxin activity is not always reduced by virus infection. Auxin activity was unaffected in barley infected with barley yellow dwarf virus (BYDV), although stunting of growth caused by reduced cell division was severe (Russell and Kimmins, 1971). Auxin accumulated in lupins infected with pea mosaic virus (Van Steveninck, 1959) and in hypersensitively-reacting tobacco infected with TMV (Van Loon and Berbee, 1978). The infection stage when auxin activity is measured is probably important. In potato infected with potato virus X, auxin activity was increased during the early stages of infection, but was reduced compared to healthy plants in later stages (Jaros, 1963; cited in Pegg, 1976).

GIBBERELLINS

Vegetative tissues contain very low concentrations of gibberellins (Kuroguchi *et al.*, 1979). Together with the large number of naturally-occurring gibberellins, this makes their measurement by physical methods extremely difficult. As for auxins, studies of gibberellin metabolism in virus-infected plants have relied mainly on bioassay methods. Gibberellin bioassays are, however, particularly unreliable. Differences in bioactivity of less than three- to four-fold are probably not significant and should be regarded as consequences of inaccurate estimation (Crozier *et al.*, 1973). Reports of virus-induced changes in gibberellin metabolism, based on bioassay, should be interpreted with caution.

Apart from a single example of enhanced gibberellin activity in buds of *Euphorbia pulcherrima*, infected with an unidentified virus (Nath *et al.*, 1979), reports show gibberellin activity unchanged or reduced by virus infection. The concentration of gibberellins, tentatively identified as GA_1 and GA_3, was reduced in cucumber infected with CMV (Bailiss, 1974; 1977; Aharoni *et al.*, 1977), in which stem and leaf growth were severly reduced. Treatment with exogenous GA_3 at the same concentration stimulated growth. Infection of barley with BYDV caused stunting and also reduced gibberellin concentration (Russell and Kimmins, 1971).

GA_3 suppressed visible symptoms of tobacco leaf curl virus (Nariami, 1963) but was only partially effective in reversing stunt in a number of other host-virus combinations (Maramorosch, 1957; Stein, 1962; Fernandez and Gaborjanyi, 1976; Yerkes, 1960; Chessin, 1957). Differences in gibberellin catabolism between healthy and infected plants (Ben-Tal and Marco, 1980) may partly explain the lack of effect of exogenous gibberellin. Additionally, altered metabolism of other PGRs may also be involved (Aharoni *et al.*, 1977; Marco *et al.*, 1976; Fraser and Whenham, 1982; Whenham, 1989). Endogenous gibberellin activity was unchanged in tobacco infected with TMV (Budagyan *et al.*, 1964) and in tomato infected with tomato aspermy virus (Bailiss, 1968).

CONCLUSIONS

Two approaches have been used to probe the role of PGRs in viral pathogenesis: the effects of application of exogenous PGRs, and correlative changes in endogenous PGR pools.

Mimicking disease symptoms in healthy plants or altering symptom severity in infected plants by treatment with exogenous growth regulators has often been taken to indicate a role for a particular PGR in disease development. Such experiments have, however, sometimes given misleading results. Synthetic PGRs which do not occur naturally are frequently used. The uptake and metabolism of these may be entirely different from that of endogenous compounds. Clearly, more emphasis should be given to the use of the naturally-occurring compounds, applied in physiological amounts to test virus/PGR interactions. The use of high specific activity radiolabelled PGRs would allow ready measurement of amounts of uptake and, equally important, the inter- or intracellular localization of the regulator in treated plants.

Demonstration of a direct effect of virus infection on metabolism of the endogenous PGR pool provides better evidence for its involvement in pathogenesis. This approach has, in the past, been complicated by the difficulty of measuring PGR concentrations by reliable, physical techniques. This has been particularly true of studies of auxins and gibberellins in virus-infected plants. Now, with the wide availability of powerful new techniques of hormone assay such as HPLC and immunoassay, reliable data should be easier to obtain.

Existing results have emphasized the importance of hormone interactions in provoking the overall response: for example the close relationship between ABA and cytokinins in virus-induced growth inhibition and "green island" formation. A complete picture of the interaction of all PGR classes, or the temporal relationship of observed changes in individual PGRs is not available for any host/virus combination. This now seems to be achievable with currently available assay methods.

Other approaches which seem likely to be fruitful involve the use of new, specific inhibitors of PGR synthesis, action and metabolism, which are becoming available particularly for cytokinins and ABA. This approach has already been of use in probing the rôle of ethylene in pathogenesis. Alternatively, the effect of virus infection of new mutant plant lines, altered in endogenous hormone synthesis, metabolism or sensitivity, might usefully be studied.

Despite the shortcomings of many of the studies of PGR metabolism in virus-infected plants some generalizations are possible. Ethylene seems clearly linked in some way to necrogenesis with only marginal effects on growth and development in systemically-infected plants. ABA and cytokinins have closely interrelated effects on growth and symptom development, which may centre on the chloroplast as the primary site of virus induced changes.

Chloroplasts have also been linked to the intracellular localization and function of PGRs other than ABA and possibly cytokinins. A critical stage may be passage of the

PGRs through the chloroplast membrane. As the plasmalemma is involved in ethylene synthesis, biochemical changes in membrane function after virus infection might usefully be investigated further. Finally, virus-infected plants may well provide a useful model system for the study of the molecular basis of PGR action in healthy plants.

REFERENCES

Adams DO, Yang SF (1981) Ethylene the gaseous plant hormone: mechanism and regulation of biosynthesis. Trends Biochem Sci 6:161-164

Aharoni N, Marco S, Levy D (1977) Involvement of gibberellins and abscisic acid in suppression of hypocotyl elongation in CMV-infected cucumbers. Physiol Plant Pathol 11:189-194

Bailiss KW (1968) Gibberellins and the early disease syndrome of aspermy virus in tomato (*Lycopersicon esculentum* Mill). Ann Bot 32:543-552

Bailiss KW (1974) The relationship of gibberellin content to cucumber mosaic virus infection of cucumber. Physiol Plant Pathol 4:73-79

Bailiss KW (1977) Gibberellins, abscisic acid and virus-induced stunting. In: Kiraly Z (ed) Current topics in plant pathology Akademiai Kiado, Budapest p 361

Bailiss KW, Balazs E, Kiraly Z (1977) The role of ethylene and abscisic acid in TMV-induced symptoms in tobacco. Acta Phytopathol Acad Sci Hung 12:133-140

Balazs E, Gaborjanyi R, Toth A, Kiraly Z (1969) Ethylene production in Xanthi tobacco after systemic and local virus infections. Acta Phytopathol Acad Sci Hung 4:355-358

Balazs E, Sziraki I, Kiraly Z (1977) The role of cytokinins in the systemic acquired resistance of tobacco hypersensitive to tobacco mosaic virus. Physiol Plant Pathol 11:29-37

Ben-Tal Y, Marco S (1980) Qualitative changes in cucumber gibberellins following cucumber mosaic virus infection. Physiol Plant Pathol 16:327-336

Budagyan EG, Lozhnikova VN, Gol'din MI, Chailakhyan MK (1964) On the effect of gibberellin-like substances on tobacco mosaic virus. Proc Acad Sci Armenia S S R 36:111-116

Burg SP, Burg EA (1962) Role of ethylene in fruit ripening. Plant Physiol 37:179-189

Chessin M (1957) Growth substances and stunting in virus-infected plants. Proceedings Third Conference on Potato Virus Diseases, Lisse-Wageningen 80-84

Crozier A, Reid DM, Reeve DR (1973) Effects of AMO-1618 on growth, morphology, and gibberellin content of *Phaseolus coccineus* seedlings. J Exp Bot 24:923-934

De Laat AMM, Van Loon LC, Vonk CR (1981) Regulation of ethylene biosynthesis in virus-infected tobacco leaves. I. Determination of the role of methionine as the precursor of ethylene. Plant Physiol 68:256-260

De Laat AMM, Van Loon LC (1982) Regulation of ethylene biosynthesis in virus-infected tobacco leaves. II. Time course of levels of intermediates and *in vivo* conversion rates. Plant Physiol 69:240-245

Fernandez TF, Gaborjanyi R (1976) Reversion of dwarfing induced by virus infection: effect of polyacrylic and gibberellic acids. Acta Phytopathol Acad Sci Hung 11:271-275

Fraser RSS (1969) Effects of two TMV strains on the synthesis of chloroplast ribosomal RNA in tobacco leaves. Mol Gen Genet 106:73-79

Fraser RSS (1979) Systemic consequences of the local lesion reaction to tobacco mosaic virus in a tobacco variety lacking the *N* gene for hypersensitivity. Physiol Plant

Pathol 14:382-394
Fraser RSS (1982) Are 'pathogenesis-related' proteins involved in acquired systemic resistance of tobacco plants to tobacco mosaic virus? J Gen Virol 58:305-313
Fraser RSS (1987) Biochemistry of virus-infected plants. Research Studies Press, Letchworth; Wiley, New York, Chichester
Fraser RSS, Gerwitz A, Morris GEL (1986) Multiple regression analysis of the relationships between tobacco mosaic virus multiplication, the severity of mosaic symptoms, and the growth of tobacco and tomato. Physiol Mol Plant Pathol 29:239-249
Fraser RSS, Whenham RJ (1978) Inhibition of the multiplication of tobacco mosaic virus by methyl benzimidazol-2-yl carbamate. J Gen Virol 39:191-194
Fraser RSS, Whenham RJ (1982) Plant growth regulators and virus infection: a critical review. Plant Growth Regul 1:37-59
Fraser RSS, Whenham RJ (1989) Abscisic acid metabolism in tomato plants infected with tobacco mosaic virus: relationships with growth, symptoms and the *Tm-1* gene for TMV resistance. Physiol Mol Plant Pathol 34:215-226
Gaborjanyi R, Balazs E, Kiraly, Z (1971) Ethylene production, tissue senescence and local virus infections. Acta Phytopathol Acad Sci Hung 6:57-65
Ghabrial SA, Pirone TP (1967) Physiology of tobacco etch virus-induced wilt of tabasco peppers. Virology 31:154-162
Hartung W, Gimmler H, Heilmann B (1982) The compartmentation of abscisic acid (ABA), of ABA-biosynthesis, ABA-metabolism and ABA-conjugation. In: Wareing PF (ed) Plant growth substances 1982. Academic Press, London, p 325
Heilmann B, Hartung W and Gimmler H (1980) The distribution of abscisic acid between chloroplasts and cytoplasm of leaf cells and the permeability of the chloroplast envelope for abscisic acid. Z Pflanzenphysiol 97:67-78
Kato Y (1976) Ethylene production during lipid peroxidation in cowpea leaves infected with CMV. Ann Phytopathol Soc Jpn 43:587-589
Koch F, Baur M, Burba M, Elstner EF (1980) Ethylene formation by *Beta vulgaris* leaves during systemic (beet mosaic virus and beet mild yellowing virus, BMV + BMYV) or necrotic (*Cerocospora beticola* Sacc.) diseases. Phytopathol Z 98:40-46
Kuriger WE, Agrios GN (1977) Cytokinin levels and kinetin-virus interactions in tobacco ringspot virus infected cowpea plants. Phytopathology 67:604-609
Kuroguchi S, Murofushi N, Ota Y, Takahashi N (1979) Identification of gibberellins in the rice plant and quantitative changes of gibberellin A19 throughout its life cycle. Planta 146:185-191
Ladygina ME, Grishkova VP, Alyoshina NV (1979) Membrane proteins of chloroplasts of intact and TMV-infected tobacco plants. Biokhimiya 44:1635-1642
Letham DS (1978) Cytokinins. In: Letham DS, Goodwin PB, Higgins TJV (eds) Phytohormones and related compounds: a comprehensive treatise. Vol. 1. Elsevier/North Holland, Amsterdam, p 205
Letham DS, Palni LMS (1983) The biosynthesis and metabolism of cytokinins. Annu Rev Plant Physiol 34:163-197
Letham DS, Palni LMS, Tao G-Q, Gollnow, Bates CM (1983) Regulators of cell division in plant tissues XXIX. The activities of cytokinin glucosides and alanine conjugates in cytokinin bioassays. J Plant Growth Regul 2:103-115
Levy D, Marco S (1976) Involvement of ethylene in epinasty of CMV-infected cucumber cotyledons which exhibit increased resistance to gaseous diffusion. Physiol Plant Pathol 9:121-126
Levy D, Marco S (1977) Involvement of ethylene in CMV-induced symptoms in cucumber. Isr J Bot 26:48-49

Lieberman M (1979) Biosynthesis and action of ethylene. Annu Rev Plant Physiol 30:533-591

Lockhart BEL, Semancik JS (1970) Growth inhibition, peroxidase and 3-indole-acetic acid oxidase activity, and ethylene production of cowpea mosaic virus-infected cowpea seedlings. Phytopathology 60:553-556

Loveys BR (1977) The intracellular location of abscisic acid in stressed and non-stressed leaf tissue. Physiol Plant 40:6-10

McGaw BA (1987) Cytokinin biosynthesis and metabolism. In: Davies PJ (ed) Plant hormones and their role in plant growth and development. Martinus Nijhoff, Dordrecht, p 76

McGaw BA, Scott IM, Horgan R (1984) Cytokinin biosynthesis and metabolism. In: Crozier A, Hillman JR (eds) The biosynthesis and metabolism of plant hormones, Cambridge University Press, London, p 105

Mandahar CL, Nath S, Gulati A (1979) Role of growth inhibitors in bud dormancy in virus-infected stem cuttings of *Euphorbia pulcherrima*. Experientia 35:315-316

Maramorosch K (1957) Reversal of virus-caused stunting in plants by gibberellic acid. Science 126:651-652

Marco S, Levy D, Aharoni N (1976) Involvement of ethylene in the suppression of hypocotyl elongation in CMV-infected cucumbers. Physiol Plant Pathol 8:1-7

Marco S, Levy D (1979) Involvement of ethylene in the development of cucumber mosaic virus-induced chlorotic lesions in cucumber cotyledons. Physiol Plant Pathol 14:235-244

Milborrow BV (1978) Abscisic acid. In: Letham DS, Goodwin PB, Higgins TJV (eds) Phytohormones and related compounds: A comprehensive treatise, Vol. 1. Elsevier/North-Holland, Amsterdam, p 295

Mohanty SK, Mohanty SK, Anjaneyulu A, Sridhar R (1979) Physiology of rice tungro disease; involvement of abscisic acid-like substances in susceptible host-virus interactions. Physiol Plant 45:132-136

Nakagaki Y, Hirai T, Stahmann MA (1970) Ethylene production by detached leaves infected with tobacco mosaic virus. Virology 40:1-9

Nariami TK (1963) Effect of gibberellic acid on tobacco leaf-curl affected plants. Indian Phytopathol 16:101-102

Nath S, Mandahar CL, Gulati A (1979) Gibberellins and the break of bud dormancy in virus-infected stem cuttings of *Euphorbia pulcherrima*. Experientia 35:1328-1329

Nichols CW (1952) The retarding effect of certain plant hormones on tobacco mosaic symptoms. Phytopathology 42:579-580

Noodén LD, Leopold AC (1978) Phytohormones and the endogenous regulation of senescence and abscission. In: Letham DS, Goodwin PB, Higgins, TJV (eds) Phytohormones and related compounds: a comprehensive treatise. Vol. 2. Elsevier/North Holland, Amsterdam p 329

Palmer MV, Horgan R, Wareing PF (1981) Cytokinin metabolism in *Phaseolus vulgaris* L. I: variations in cytokinin level in leaves of decapitated plants in relation to lateral bud outgrowth. J Exp Bot 32:1231-1241

Pegg GR (1976) Endogenous auxins in healthy and diseased plants. In: Heitefuss R, Williams PM (eds) Encyclopedia of plant physiology new series, Vol. 4. Springer-Verlag, Berlin, p 560

Pritchard DW, Ross AF (1975) The relationship of ethylene to formation of tobacco mosaic virus lesions in hypersensitive responding tobacco leaves with and without induced resistance. Virology 64:295-307

Quarrie SA, Jones HG (1977) Effects of abscisic acid and water stress on development

and morphology of wheat. J Exp Bot 28:192-203
Rajagopal R (1977) Effect of tobacco mosaic virus infection on the endogenous levels of indoleacetic phenylacetic and abscisic acids of tobacco leaves in various stages of development. Z Pflanzenphysiol 83:403-409
Reinecke DM, Bandurski RS (1987) Auxin biosynthesis and metabolism. In: Davies PJ (ed) Plant hormones and their role in plant growth and development. Martinus Nijhoff, Dordrecht p 24
Ross AF, Pritchard DW (1972) Local and systemic effects of ethylene on tobacco mosaic virus lesions in tobacco. Phytopathology 62:786
Russell SL, Kimmins WC (1971) Growth regulators and the effect of BYDV on barley (*Hordeum vulgare* L.). Ann Bot 35:1037-1043
Smith SH, McCall SR, Harris JH (1968) Auxin transport in curly-top virus-infected tomato. Phytopathology 58:1669-1670
Sridhar R, Mohanty SK, Anjaneyulu A (1978) Physiology of rice tungro virus disease: increased cytokinin activity in tungro-infected rice cultivars. Physiol Plant 43:363-367
Steadman JR, Sequeira L (1969) A growth inhibitor from tobacco and its possible involvement in pathogenesis. Phytopathology 59:499-503
Steadman JR, Sequeira L (1970) Abscisic acid in tobacco plants: tentative identification and its relation to stunting induced by *Pseudomonas solanacearum*. Plant Physiol 45:691-697
Stein DB (1962) The developmental morphology of *Nicotiana tabacum* "White Burley" as influenced by virus infection and gibberellic acid. Am J Bot 49:437-443
Sziraki I, Gaborjanyi R (1974) Effect of systemic infection by TMV on cytokinin level of tobacco leaves and stems. Acta Phytopathol Acad Sci Hung 9:195-199
Sziraki I, Balazs E (1977) The effect of infection by TMV on cytokinin level of tobacco plants and cytokinins in TMV-RNA. In: Kirlay Z (ed) Current topics in plant pathology. Akademiai Kiado, Budapest, p 345
Sziraki L, Balazs E, Kiraly Z (1980) Role of different stresses in inducing systemic acquired resistance to TMV and increasing cytokinin level in tobacco. Physiol Plant Pathol 16:277-284
Tavantzis SM, Smith SH, Witham FA (1979) The influence of kinetin on tobacco ringspot virus infectivity and the effect of virus infection on the cytokinin activity in intact leaves of *Nicotiana glutinosa* L. Physiol Plant Pathol 14:227-233
Van Loon LC (1977) Induction by 2-chloroethylphosphonic acid of viral-like lesions, associated proteins and systemic resistance in tobacco. Virology 80:417-420
Van Loon LC, Berbee AT (1978) Endogenous levels of indole acetic acid in leaves of tobacco reacting hypersensitively to TMV. Z Pflanzenphysiol 89:373-375
Van Steveninck RFM (1959) Factors affecting the abscission of reproductive organs in yellow lupins (*Lupinus luteus* L.). III. Endogenous growth substances in virus-infected and healthy plants and their effect on abscission. J Exp Bot 10:367-376
Wang TL, Thompson AG, Horgan R (1977) A cytokinin glucoside from the leaves of *Phaseolus vulgaris* L. Planta 135:285-288
Whenham RJ (1983) Evaluation of selective detectors for the rapid and sensitive gas chromatographic assay of cytokinins, and application to the analysis of cytokinins in plant extracts. Planta 157:554-560
Whenham RJ (1989) Effect of systemic tobacco mosaic virus infection on endogenous cytokinin concentration in tobacco (*Nicotiana tabacum* L.) leaves: consequences for the control of resistance and symptom development. Physiol Mol Plant Pathol 35:85-95

Whenham RJ, Fraser RSS (1980) Stimulation by abscisic acid of RNA synthesis in discs from healthy and tobacco mosaic virus-infected tobacco leaves. Planta 150:349-353

Whenham RJ, Fraser RSS (1981) Effect of systemic and local-lesion-forming strains of tobacco mosaic virus on abscisic acid concentration in tobacco leaves: consequences for the control of leaf growth. Physiol Plant Pathol 18:267-278

Whenham RJ, Fraser RSS, Snow A (1985) Tobacco mosaic virus-induced increase in abscisic acid concentration in tobacco leaves: intracellular location and relationship to symptom severity and to extent of virus multiplication. Physiol Plant Pathol 26:379-387

Whenham RJ, Fraser RSS, Brown, LP, Payne JA (1986) Tobacco mosaic virus-induced increase in abscisic-acid concentration in tobacco leaves: intracellular location in light and dark-green areas, and relationship to symptom development. Planta 168:592-598

Wright STC, Hiron RWP (1969) (+)-Abscisic acid, the growth inhibitor induced in detached wheat leaves by a period of wilting. Nature 224:719-720

Yerkes WD (1960) Interaction of potassium gibberellate and a stunting bean virus on beans, *Phaseolus vulgaris*. Phytopathology 50:525-527

Zaitlin M (1979) How viruses and viroids induce disease. In: Horsfall JG, Cowling EB (eds) Plant disease, an advanced treatise. Academic Press, New York, p 257

Zeevaart JAD, Creelman RA (1988) Metabolism and physiology of abscisic acid. Annu Rev Plant Physiol 39:439-473

RECOGNITION AND DISEASE DEVELOPMENT IN THE TOBACCO-TMV SYSTEM

L C van Loon, O van Kooten[1], E G A Linders, C Meurs and
M M G Wijdeveld
Department of Plant Physiology
Agricultural University
Arboretumlaan 4
6703 BD Wageningen
The Netherlands

ABSTRACT

In tobacco mosaic virus-infected tobacco symptoms develop as a result of recognition between any of the virus-coded proteins and host constituents. In systemically infected plants photosynthetic electron transport is reduced; photosystem II activity is affected most in the lower leaves that do not show symptoms. In the upper, mosaic-diseased leaves, the delineation of light-green and dark-green areas can be masked but not altered by plant growth regulators. The occurrence of viral 126 kDa protein in X-bodies near nuclei is associated with severe mosaic symptoms.

RECOGNITION BETWEEN VIRUS-CODED PROTEINS AND HOST FACTORS

Simple RNA viruses of plants code for a small number of proteins. In cells infected with tobacco mosaic virus (TMV) four virus-specified polypeptides of 183, 126, 30 and 17.5 kDa have been detected.

The 183 kDa protein is a readthrough product of the 126 kDa gene. Both polypeptides appear early during infection and have been implicated in virus replication (Scalla *et al.*, 1976; Meshi *et al.*, 1987). It has been established that when the amber codon of the 126 kDa gene is replaced by a tyrosine codon, viral multiplication is reduced but not abolished, whereas mutants with an intact 126 kDa but a defective 183 kDa gene are not infectious (Ishikawa *et al.,* 1986). These observations are

[1] Department of Plant Physiological Research, Agricultural University, Generaal Foulkesweg 72, 6703 BW Wageningen, The Netherlands

NATO ASI Series, Vol H 41
Recognition and Response in Plant-Virus Interactions
Edited by R S S Fraser

consistent with the view that the 183 kDa protein functions as a replicase, in accordance with the conserved RNA-polymerase domain in the 57 kDa readthrough part (Cornelissen and Bol, 1984; Haseloff *et al.*, 1984; Kamer and Argos, 1984). The 126 kDa protein appears to have an accessory function.

The 30 kDa protein occurs transiently in association with membranes, shortly after infection, but accumulates more stably in cell walls later (Moser *et al.*, 1988). It functions in enabling the virus to spread from cell to cell (Deom *et al.*, 1987; Meshi *et al.*, 1987). Strains and mutants defective in the 30 kDa protein accumulate to normal levels in those cells that receive virus during mechanical inoculation, but give rise to a subliminal infection, in which the virus is unable to move from the original infection centres.

The 17.5 kDa protein is the coat protein. It accumulates to high levels in the later stages of infection, both as the protein encapsidating the viral RNA in TMV rods, and as a pool of free trimeric aggregates (A-protein) (Fraenkel-Conrat, 1957). By site-directed mutagenesis of cDNA clones, it has been inferred that the coat protein is also required for long-distance transport within the plant (Dorokhov *et al.*, 1984; Takamatsu *et al.*, 1987), even though functional encapsidation may not occur (Jockusch, 1968; Dawson *et al.*, 1988b).

This small number of proteins is much less than the variety of types of symptom induced by TMV on different hosts or displayed by various mutants on a single host. Viruses do not seem to have the capacity to code for specific symptom-inducing proteins. Rather, symptoms seem to be the consequence of the interaction between viral proteins and host constituents (Van Loon, 1987). Compatibility between viral proteins necessary for multiplication and spread, and unknown host factors, is required for systemic infection to occur. Any of the viral functions may be involved in the recognition processes required. For instance, the loss of the ability to multiply in protoplasts might indicate incompatibility between the viral polymerase and host factors required for replication. Similarly, mutations in the transport or coat protein might be expected to affect the interaction with cellular components involved in cell-to-cell and long-distance transport, respectively.

In tobacco, all natural varieties are sensitive to TMV (Holmes, 1960) and develop systemic mosaic symptoms after infection with the common strain. Tobacco species, varieties, or cultivars that possess the dominant gene *N'* localize all strains of TMV except those resembling the common one (Valleau, 1952; Siegel and Wildman, 1954).

Table 1. *Genetic interactions between tobacco and TMV*

TMV Strain	Host gene		
	N	*N'*	*n'*
"common", e.g. U1	R	S	S
"mild", e.g. U2	R	R	S
"defective" mutants	R	R	R

S = susceptible reaction; R = resistance reaction

Virus mutants are easily generated that have lost the ability to spread systemically in all host genetic backgrounds, even in those that do not seem to possess genes for resistance (*n'* plants). The introduction, through crossings with *Nicotiana glutinosa*, of the gene *N* into *N. tabacum* created cultivars that are resistant to all strains of TMV. With the exception of a pepper strain of tomato mosaic virus that causes local lesions on *N. sylvestris* and *N. tabacum* cv. Xanthi-nc or *N. glutinosa* as well as systemic latent infections on the latter (Csilléry *et al.*, 1983), no virus strains or mutants are known which induce a resistance reaction in *n'* plants but cause systemic infection in *N'* plants, or which induce resistance in *N'* but are systemic in *N* plants. Given these relationships, TMV strains and mutants can be arranged in such a way that they are progressively less able to spread in hosts carrying less effective genes for resistance (Table 1). This scheme is consistent with viral functions being required for resistance to be evaded and systemic infection to occur (Van Loon, 1987). The triggering of resistance reactions then results from incompatibility between a viral and one or more host factors, and depends on the genetic constitution of both partners.

Variant strains or mutants usually exhibit reduced viral multiplication and/or spread. However, symptom expression is also often affected. Mutation of the 30 kDa gene has been linked to the induction of a hypersensitive reaction in *N'* tobacco (Taliansky *et al.*, 1982; Zimmern and Hunter, 1983). Virus isolates and mutants with temperature-sensitive coat protein were found to induce yellow mosaic symptoms

(Jockusch, 1966), whereas the introduction of various alterations in the coat protein gene by site-directed mutagenesis of cDNA clones gives rise in sensitive tobaccos to widely varying phenotypes, ranging from necrosis, marked yellowing, herringbone netting and transitory yellowing to no symptoms at all (Saito *et al.*, 1987; Knorr and Dawson, 1988; Dawson *et al.*, 1988a,b).

Since both these proteins are involved in virus transport, it seems that the induction of necrosis may be linked to restriction of virus movement. Other factors can also be involved, however, because the replacement of a single nucleotide in the 126 kDa gene of the tomato strain enabled the resulting mutant to overcome resistance conferred by the *Tm-1* gene in tomato (Meshi *et al.*, 1988). Perhaps the mutant strain had become more effective in its rate of multiplication and, indirectly, in its ability to spread.

When *N*-tobacco is infected at temperatures above 28°C, the *N* gene is not functional and a systemic disease develops. If a few days after inoculation the temperature is lowered to 20°C, hypersensitive necrosis ensues only in those cells which the virus had just reached. Cells in which the virus had already multiplied do not necrose. Thus, apart from genetic differences affecting recognition, conditions favouring rapid multiplication and spread of the virus may suppress the resistance reaction by not allowing host cells sufficient time to initiate defence reactions. This hypothesis is in line with observations that the virulence of several strains of TMV is related to the rate of synthesis of the 30 kDa protein in infected cells, which may in turn be influenced by either the 126 kDa or the 183 kDa protein (Watanabe *et al.*, 1987).

THE DEVELOPMENT OF MOSAIC SYMPTOMS

Virus multiplication and spread do not necessarily result in visible symptoms. In sensitive tobaccos systemic infection can give rise to symptoms varying from a severe yellow-green mosaic, such as caused by TMV *flavum*, to a barely detectable mottling in a few leaves only, as seen upon infection with the masked strain. Whereas young, developing leaves are invariably most affected, leaves present at the time of inoculation may develop no or only slight symptoms, even though the virus multiplies to similar extents. Upon inoculation of fully-grown leaves of young Samsun plants with TMV W U1, symptoms of vein clearing become apparent about 7 days later on the third and upper leaves above the inoculated ones. The third and fourth leaf then develop mottling symptoms, whereas on the younger leaves, which were 2 cm or less at the time of inoculation, light-green/dark-green mosaic symptoms and malformations appear. The mosaic breaks down into a yellow-green flecking without malformations at about the ninth leaf, when growth of the plant is slowing down, but well before flower buds are observable. After infection with the U1 strain, the inoculated leaves never develop symptoms, but during senescence they do yellow somewhat earlier than leaves from non-infected plants. After infection with various strains of TMV a similar sequence of variations in symptom expression is evident in leaves at different positions along the stem, indicating that in mosaic disease symptom expression is developmentally controlled.

Chlorophyll content in the light-green areas of mosaic-diseased leaves is decreased, but this loss is largely compensated for by a higher chlorophyll content in the dark-green areas (Peterson and McKinney, 1938; Whenham *et al.*, 1986). Malformations arise because growth in dark-green parts is not, or is only slightly inhibited as compared to that in light-green parts. Besides leaf area, stem length is also reduced; both effects can be attributed to an increased content of the hormone abscisic acid (ABA) (Whenham and Fraser, 1981). Whereas only about 1% of the ABA was found outside the chloroplast in control plants, up to 78% appeared to be located in the cytoplasm in severely mosaic-diseased leaves (Whenham *et al.,* 1985). These observations are suggestive of a direct effect of TMV on chloroplast properties. In mosaic-diseased leaves the number of chloroplast ribosomes, and the rate of chloroplast ribosomal RNA synthesis are reduced (Fraser, 1969; Oxelfelt, 1971). Chloroplast protein synthesis (Hirai and Wildman, 1969; Doke and Hirai, 1970) and the amount of ribulosebisphosphate carboxylase (Rubisco) (Van Loon and Van Kammen, 1970; Makovcová and Šindelář, 1978) are also decreased, even though the capacity for protein synthesis does not seem to be impaired (Reinero and Beachy, 1989).

Although it is difficult to separate the light from the dark areas, it appears that in the light-green/yellow parts the virus interferes with chloroplast development. Yet, on a chlorophyll basis chloroplast preparations isolated 1-4 days after infection from inoculated (asymptomatic) and 15-18 days after inoculation from systemically infected (mosaic symptom) leaves did not exhibit any change in the rates of NADP and ferricyanide photoreduction, although less oxygen photoreduction was observed in chloroplasts from leaves with systemic mosaic. NADH-malate dehydrogenase, glutamate-oxalacetate transaminase, superoxide dismutase and phosphoenolpyruvate carboxylase were either unchanged or little affected. In contrast, Rubisco activity was decreased by about 30%, suggesting that the dark processes of carbon assimilation were mainly affected (Montalbini and Lupattelli, 1989).

Coat protein has been reported to accumulate in both chloroplast stroma and thylakoid membranes of systemically-infected leaves (Reinero and Beachy, 1989). Coat protein of a TMV strain inducing a pale-white/green chlorosis accumulated at a rapid rate in the chloroplasts of the first two systemically infected leaves located immediately above the inoculated leaves. Coat protein of the masked strain was present in chloroplasts at lower levels and little of it was associated with the thylakoids. Under these conditions electron transport through photosystem II (PS II) was inhibited in plants infected with the aggressive strain but no reduction was observed in plants infected with the mild strain (Reinero and Beachy, 1989). These results were supported by analysis of *in vivo* chlorophyll fluorescence induction kinetics, showing that the ratio between the variable fluorescence (F_v) and the initial level of chlorophyll fluorescence after dark adaptation (F_o) was progressively reduced in leaves infected with the chlorosis-inducing strain but not after infection with the virtually symptomless strain. The accumulation of coat protein in both stroma and in thylakoid membranes was suggested to be responsible for the inhibition of PS II activity (Reinero and Beachy, 1989).

In similar types of experiments we determined the electron transport rate and calculated the light quantum yield Φ_{pc} of PS II according to the equation

$$\Phi_{pc} = (F_v)_m/F_m = (F_m - F_o)/F_m$$

where $(F_v)_m$ is the maximal variable fluorescence and F_m is the maximum level of fluorescence (Björkman and Demmig, 1987), for leaves at different insertion levels on non-infected and TMV W U1-infected Samsun plants. In non-infected plants the rate of electron transport, as determined from our measurements by the method proposed by Genty *et al.* (1989), decreased progressively with leaf age (Fig. 1A). This was independent of effects of leaf age on chlorophyll content. In plants infected for 4 weeks and expressing symptoms of varying severity at different positions up the stem, electron transport through PS II was always substantially reduced. However, F_v/F_m was reduced significantly and strongly only in leaves that had already expanded at the time of inoculation and showed no or only slight symptoms (Fig. 1B). The decrease in the older infected leaves is indicative of photochemical destruction of PS II. In contrast, symptom-bearing leaves were not affected by the reduced electron transport capacity.

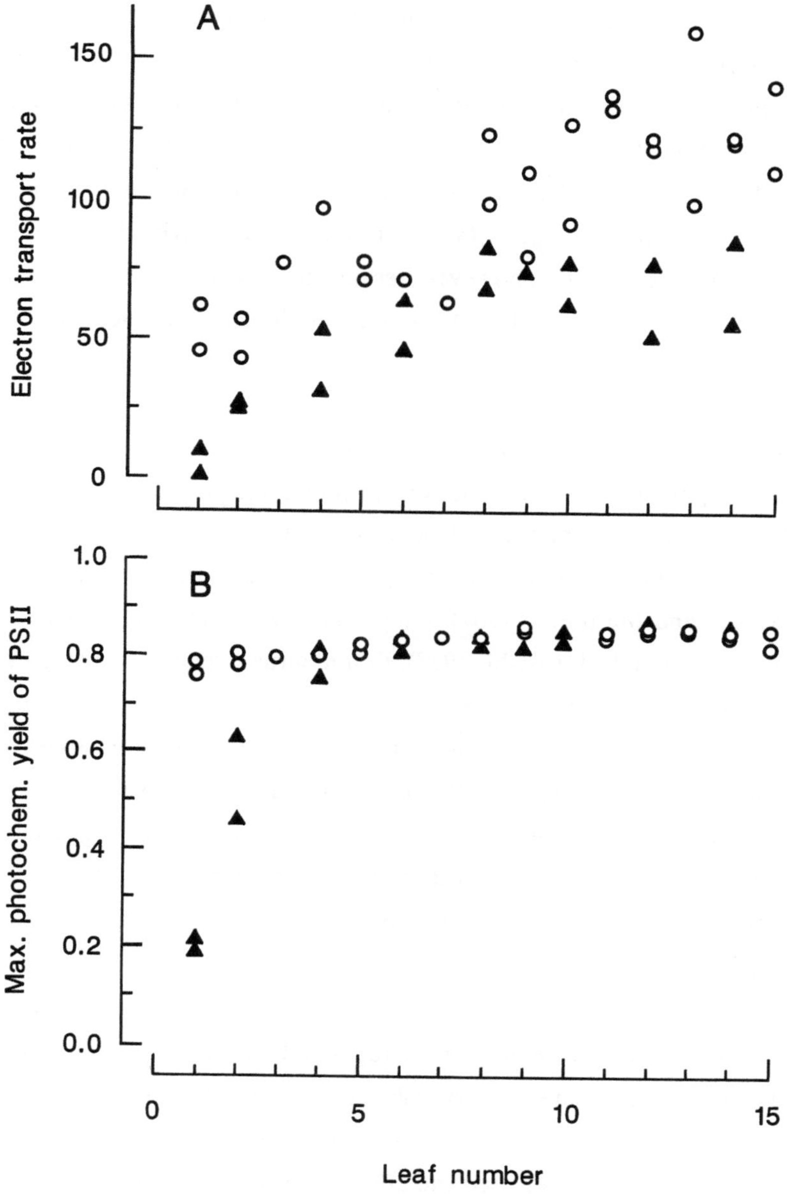

Figure 1. (A) Rates of electron transport through photosystem II at an irradiance of 340 μmoles m^{-2} s^{-1} (PAR) and (B) maximum quantum yield (Φ_{pc}) of photosystem II in individual leaves from two plants of (o) non-infected or (▲) mosaic-diseased Samsun tobacco. The infected plants had been inoculated with TMV W U1 28 days previously on three expanded leaves. Leaf 1 is the youngest inoculated leaf.

Similar effects are seen with other TMV strains (O van Kooten, C Meurs and L C van Loon, in preparation). Thus, mosaic-diseased leaves seem to compensate for reduced electron transport by adapting through energy-dissipating mechanisms in PS II. Since the mosaic comprises both light-green and dark-green areas, it cannot be excluded that reductions in PS II activity in light-green parts are compensated for by stimulation in dark-green areas. The influence of TMV infection on PS II activity does not appear to be related to symptom expression, however, because the severest effects occurred in the inoculated leaves and in those directly above the inoculated ones, which remain essentially symptomless.

The delineation of light-green and dark-green areas

The development of systemic mosaic symptoms results from the occurrence of clusters of leaf cells that either support or do not support virus multiplication. TMV accumulates to high levels in the light-green parts, whereas dark-green areas are devoid of virus (Atkinson and Matthews, 1967). Only very few of such "green islands" have a clonal origin (Carlson and Murakishi, 1978). Often, virus may at some stage penetrate into the dark-green areas, start multiplying and initiate additional yellowing. This is particularly evident in leaves at high insertion levels of plants with established mosaic symptoms. These observations support the notion that a diffusible substance is involved in the origin and maintenance of green islands. It has been suggested that in mosaic-diseased leaves an inhibitor of virus multiplication is present in dark-green areas, induced under the influence of the light-green parts (Carlson and Murakishi, 1978; Whenham *et al.*, 1986).

As shown previously, the formation of mosaic symptoms is dependent on leaf developmental stage. Leaf development is regulated by the interaction of different hormones, and hormones themselves, or substances induced through their action, can maintain specific physiological states (Trewavas, 1986). To see how far the nature of the mosaic can be altered, we investigated the effect of various plant growth regulators on symptom expression in Samsun tobacco after infection with TMV W U1. Starting 3 days after inoculation of a single fully-grown leaf, the growing shoot tip and developing leaves were sprayed with solutions containing various hormones every 2-3 days during the period of systemic infection. Virus was first detectable in the top of the plant 4 days after inoculation (Van Telgen *et al.*, 1985b), well after the treatment had started. Symptoms on the 12-15 leaves which developed after the plants had been inoculated

Figure 2. Consecutive leaves of a Samsun tobacco plant inoculated with TMV W U1 25 days previously. The largest leaf (bottom right) was 2 cm long at the time the plant was inoculated on one expanded leaf. The growing shoot tip was sprayed every 2-3 days with (A) water or (B) 0.1 mM naphthylacetic acid.

were followed up to 25 days. To determine the effect of the treatments on virus multiplication, the second leaf above the leaf 2 cm in length at the time of inoculation was harvested on day 10, and virus content was assessed by local lesion assay on *N. glutinosa*.

Starting from the leaf 2 cm in length at the time of inoculation, mosaic patterns with progressively larger patches of dark-green areas developed in the third to tenth newly formed leaves above. Later, starting between the seventh and tenth leaf, yellow spots developed in the dark-green areas and the last leaves formed exhibited yellow-green flecking. This sequence is evident in Fig. 2A, which depicts the symptoms in successive leaves 25 days after inoculation.

Of the various hormones tested, the auxins naphthylacetic acid (NAA) and indoleacetic acid (IAA) reduced virus accumulation, whereas 2,4-dichlorophenoxyacetic acid increased virus multiplication (Table 2), effects similar to those seen in hypersensitively reacting tobacco (Van Loon, 1979). Gibberellic acid (GA_3) increased virus content only marginally at a concentration of 1 μM, but by 283% at 0.1 mM. The cytokinins zeatin and kinetin strongly stimulated virus multiplication, whereas ABA tended to reduce it, and the ethylene precursor 1-aminocyclopropane-1-carboxylic acid (ACC) had no effect (Table 2).

Although cytokinins are known to stimulate chloroplast development and retard leaf senescence (Horgan, 1984), application of either zeatin or kinetin increased the severity of systemic mosaic symptoms. Particularly after treatment with zeatin, light-green areas turned into conspicuous yellow flecks, but the sequence of symptoms and the nature of the mosaic pattern was unchanged. GA_3 showed a similar effect when applied at 0.1 mM, suggesting that the increase in symptom severity was linked to the increase in virus multiplication. Prior to flower bud formation GA_3 stimulated stem elongation and retarded senescence of the lower leaves, but the delineation of the mosaic was not altered. ABA caused some growth retardation of stem and leaves (Whenham and Fraser, 1981) but, likewise, did not affect the mosaic pattern. ACC accelerated senescence of the lower leaves. The light-green areas on the systemically infected leaves were more chlorotic, yielding a more conspicuous mosaic, without any effect on its nature. Finally, the auxin NAA had little effect, except at concentrations above 10 μM. At 0.1 mM it reduced shoot development and caused leaf epinasty; the leaves were dark green and mosaic symptoms were almost fully masked (Fig. 2B). The mosaic was still present, but the light areas were only marginally less green than the

Table 2. *Effects on virus content of spraying TMV W U1-infected Samsun tobacco with 1 μM of various growth regulators.*

Growth regulator	Virus content (Percentage of water control)
1-Napthylacetic acid	77
2,4-Dichlorophenoxyacetic acid	234
Gibberellic acid (GA_3)	112
Zeatin	225
Kinetin	279
Abscisic acid	73
1-Aminocyclopropane-1-carboxylic acid	97

dark ones. The auxins indolebutyric acid and IAA had similar but less marked effects, whereas the auxin transport inhibitor triiodobenzoic acid tended to counteract the auxin-induced reduction in virus multiplication and symptom severity. Methylbenzimidazole-2-yl carbamate (carbendazim), reported to cause a similar masking of mosaic symptoms (Tomlinson *et al.*, 1976; Fraser and Whenham, 1978) reduced symptom severity to a lesser extent than NAA, even though it was effective in stimulating plant growth.

Combinations of the various growth regulators did not alter symptom expression beyond their additive effects. Thus, although some of the growth-regulating substances affected shoot morphology, as well as virus multiplication and symptom severity, no changes in the delineation of the mosaic pattern were detected. These observations indicate that the virus must have become established and the mosaic pattern fixed at a very early stage of leaf development. Symptoms might then arise in the areas where the virus is actively multiplying as a consequence of the interaction between a viral product and a host constituent.

The relationship between viral 126 kDa protein and symptom expression

The reduction in stem and leaf growth, the formation of mosaic patterns as a result of local interference with chloroplast development, and leaf malformation are all symptoms of disturbances of plant growth and development. Since growth and development are the result of selective gene expression under the control of the proteins complexed with the DNA in the chromatin (Cartwright *et al.*, 1982), we investigated whether changes in chromatin-associated proteins occur in mosaic-diseased tobacco leaves. Gel electrophoresis of purified chromatin protein preparations revealed that leaves systemically infected with TMV W U1 contained two additional polypeptides with apparent masses of 116 and 20 kDa, respectively. The 20 kDa protein was identified serologically as the viral coat protein (Van Telgen *et al.*, 1985a). It was only loosely associated with the chromatin and preferentially released upon incubation in 6 M urea. In contrast, the 116 kDa protein remained tightly bound under these conditions and was dissociated together with all other chromatin-associated proteins only in the presence of salt (Van Telgen *et al.*, 1985b). By protease V8 mapping it was established that this protein was identical to the virally-coded 126 kDa protein (Van Telgen *et al.*, 1985c). No other differences between the chromatin protein profiles of healthy and TMV-infected leaves could be detected, suggesting that the 126 kDa protein preferentially is associated with the chromatin.

To study this association further, an antiserum was raised against a fusion protein of ß-galactosidase and part of the 126 kDa protein, and used to localize the protein by immunoelectron microscopy in thin sections of mosaic-diseased leaves. Under these conditions no labelling of nuclei occurred. Instead, labelling was associated exclusively with X-bodies (Fig. 3). These amorphous inclusions appeared to be cytoplasmic domains without a surrounding membrane, and characterized by the presence of tubular structures. X-bodies were also present and specifically labelled in purified nuclear preparations from systemically-infected leaves. These results indicate that the presence of the 126 kDa protein among the chromatin-associated proteins is caused by the adherence of the X-bodies to the nuclei during chromatin purification. Indeed, by light microscopy of infected tissues X-bodies were almost always seen associated with nuclei, leaving the possibility that the X-bodies influence nuclear activity. Viral coat protein was detected exclusively in virus crystals in the cytoplasm; no specific association with either nuclei or chloroplasts was apparent (Wijdeveld *et al.*, 1989).

To follow the accumulation of the 126 kDa protein in X-bodies, plants were

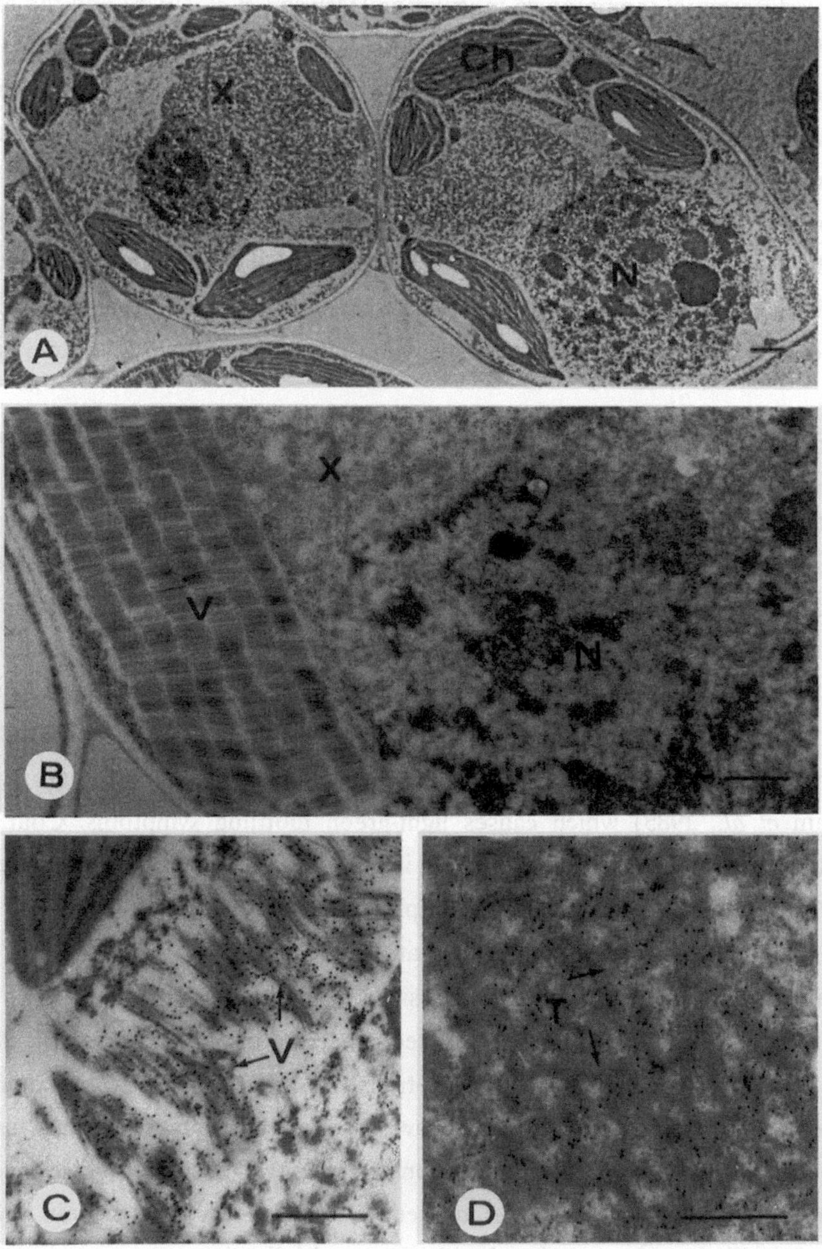

Figure 3. Electron micrographs of systemically infected leaf tissues, untreated (A,B), or immunogold-labelled with antibodies against TMV coat protein (C) or against the viral 126 kDa protein (D). Note the location of the X-body against the nucleus in (A) and (B), the labelling of the coat protein in a virus crystal in (C) and part of an X-body containing 126 kDa protein in (D). N = nucleus, Ch = chloroplast, V = virus crystal, X = X-body, T = tubules. Bar represents 1 μm (A) or 500 nm (B-D).

infected by differential temperature inoculation (Dawson *et al.*, 1975) and the amounts of the 126 kDa protein and coat protein as a measure of virus content were determined at intervals after transfer of the plants from 4 to 25°C. Both proteins were detected in extracts from systemically infected leaves by 39 h and increased in parallel up to 7 days after transfer. In thin sections some virus crystals and occasional X-bodies were already present by 24 h; the latter were detected by labelling with antiserum. The average size of the X-bodies, as well as their content of 126 kDa protein increased with time. At first, the small X-bodies were seen free in the cytoplasm, but by about 4 days, they had associated with nuclei, well before the first symptoms of vein clearing became apparent on day 5.

X-bodies similar to those present in cells infected with TMV W U1 were found in leaves infected with TMV *flavum*, which causes a severe yellow/dark-green mosaic. In contrast, in cells infected with the masked strain only very small X-bodies, not associated with nuclei, were detectable as the sites of the 126 kDa protein. It has been suggested that the aggregation of gene products into large inclusion bodies may be a method of removing proteins that, in large quantities, may be detrimental to the functioning of the cell. The inclusion bodies might also be associated with viral replication complexes and/or be the sites of virus assembly (Zaitlin and Hull, 1987), However, X-bodies were not detectable in plants infected with the tomato strain V4 (Whenham *et al.*, 1985) which causes light-green mottling symptoms. Coat protein accumulated and virus crystals were present in cells infected with any of the virus strains but no 126 kDa protein was detectable in plants infected with TMV V4. The amount of virus accumulated was about half of that found upon infection with TMV W U1. Either far less 126 kDa protein was synthesized in plants infected with V4, or the 126 kDa protein of V4 must have been degraded, whereas that of W U1 is not. Since the 126 kDa protein of V4 produced by *in vitro* translation was not significantly more prone to degradation by extracts from tobacco leaves, the latter possibility appears less likely. It is also difficult to envisage that replication of TMV W U1 would require much more 126 kDa protein than TMV V4 or masked. Thus, the 126 kDa protein present in X-bodies is unlikely to be associated with virus multiplication.

The 126 kDa proteins of the mild mosaic-inducing strains U2 and U5 were not recognized by the antisera against the protein of W U1. Although after infection with the former strains substantial amounts of virus accumulated in the cytoplasm and in chloroplasts, respectively, no 126 kDa protein accumulated to levels detectable by gel electrophoresis, and no X-bodies were apparent, similar to the situation after infection with TMV V4. This leaves the possibility that the excess of 126 kDa protein produced in W U1 or *flavum* infections, and its accumulation in X-bodies near nuclei is associated with the more severe type of mosaic after infection with these strains.

In systemically infected leaves virus starts multiplying well before symptoms are visible. By that time recognition events have occurred and various changes affecting leaf development have been initiated. This makes it difficult to distinguish primary events from secondary consequences. Of the proteins coded for by TMV, coat protein is made in large excess and alterations in its sequence affect symptom type. In severe strains, such as W U1 and *flavum*, the 126 kDa protein also appears to be produced in excess, and might be implicated in symptom severity.

REFERENCES

Atkinson PH, Matthews REF (1967) Distribution of tobacco mosaic virus in systemically infected tobacco leaves. Virology 32:171-173

Björkman O, Demmig B (1987) Photon yield of O_2 evolution and chlorophyll fluorescence characteristics at 77 K among vascular plants of diverse origins. Planta 170:489-504

Carlson PS, Murakishi HH (1978) Evidence on the clonal versus non-clonal origin of dark green islands in virus infected tobacco leaves. Plant Sci Lett 13:377-381

Cartwright IL, Abmayr SM, Fleischmann G, Lowenhaupt K, Elgin SCR, Keene MA, Howard GC (1982) Chromatin structure and gene activity: the role of non-histone chromosomal proteins. Crit Rev Biochem 13:1-86

Cornelissen BJC, Bol JF (1984) Homology between the proteins encoded by tobacco mosaic virus and two tricornaviruses. Plant Mol Biol 3:379-384

Csilléry G, Tóbiás I, Ruskó (1983) A new pepper strain of tomato mosaic virus. Acta Phytopathol Acad Sci Hung 18:195-200

Dawson WO, Schlegel DE, Lung MCY (1975) Synthesis of tobacco mosaic virus in intact tobacco leaves systemically inoculated by differential temperature treatment. Virology 65:565-573

Dawson WO, Knorr DA, Bubrick P (1988a) Elicitor active coat protein mutations in cDNA clones of tobacco mosaic virus. In: Keen NT, Kosuge T, Walling LL (eds) Physiology and biochemistry of plant-microbial interactions. American Society of Plant Physiologists, Rockville MD, p 139

Dawson WO, Bubrick P, Grantham GL (1988b) Modification of the tobacco mosaic virus coat protein gene affecting replication, movement, and symptomatology. Phytopathology 78:783-789

Deom CM, Oliver MJ, Beachy RN (1987) The 30-kilodalton gene product of tobacco mosaic virus potentiates virus movement. Science 237:389-394

Doke N, Hirai T (1970) Effects of tobacco mosaic virus infection on photosynthetic CO_2 fixation and $^{14}CO_2$ incorporation into protein in tobacco leaves. Virology 42:68-77

Dorokhov YL, Alexandrova NM, Miroshnichenko NA, Atabekov JG (1984) The informosome-like virus-specific ribonucleoprotein (vRNP) may be involved in the

transport of tobacco mosaic virus infection. Virology 137:127-134
Fraenkel-Conrat H (1957) Degradation of tobacco mosaic virus with acetic acid. Virology 4:1-4
Fraser RSS (1969) Effects of two TMV strains on the synthesis and stability of chloroplast ribosomal RNA in tobacco leaves. Mol Gen Genet 106:73-79
Fraser RSS, Whenham RJ (1978) Inhibition of the multiplication of tobacco mosaic virus by methyl benzimidazol-2-yl carbamate. J Gen Virol 39:191-194
Genty B, Briantais JM, Baker NR (1989) The relationship between the quantum yield of photosynthetic electron transport and quenching of chlorophyll fluorescence. Biochim Biophys Acta 990:87-92
Haseloff J, Goelet P, Zimmern D, Ahlquist P, Dasgupta R, Kaesberg P (1984) Striking similarities in amino acid sequence among nonstructural proteins encoded by RNA viruses that have dissimilar genomic organization. Proc Natl Acad Sci USA 81:4358-4362
Hirai A, Wildman SG (1969) Effect of TMV multiplication on RNA and protein synthesis in tobacco chloroplasts. Virology 38:73-82
Holmes FO (1960) Inheritance in tobacco of an improved resistance to infection by tobacco mosaic virus. Virology 12:59-67
Horgan R (1984) Cytokinins. In: Wilkins MB (ed) Advanced plant physiology. Pitman, London, p 53
Ishikawa M, Meshi T, Motoyoshi F, Takamatsu N, Okada Y (1986) *In vitro* mutagenesis of the putative replicase genes of tobacco mosaic virus. Nucleic Acids Res 14:8291-8305
Jockusch H (1966) Temperatursensitive Mutanten des Tabakmosaikvirus I. *In vivo*-Verhalten. Z Vererbungsl 98:320-343
Jockusch H (1968) Two mutants of tobacco mosaic virus temperature-sensitive in two different functions. Virology 35:94-101
Kamer G, Argos P (1984) Primary structural comparison of RNA-dependent polymerases from plant, animal and bacterial viruses. Nucleic Acids Res 12:7269-7282
Knorr DA, Dawson WO (1988) A point mutation in the tobacco mosaic virus capsid protein gene induces hypersensitivity in *Nicotiana sylvestris*. Proc Natl Acad Sci USA 85:170-174
Makovková O, Šindelář L (1978) Changes in phosphoenolpyruvate carboxylase and ribulosebisphosphate carboxylase activities in tobacco plants infected with tobacco mosaic virus. Biol Plant 20:135-137
Meshi T, Watanabe Y, Saito T, Sugimoto A, Maeda T, Okada Y (1987) Function of the 30 kd protein of tobacco mosaic virus: involvement in cell-to-cell movement and dispensability for replication. EMBO J 6:2557-2563
Meshi T, Motoyoshi F, Adachi A, Watanabe Y, Takamatsu N, Okada Y (1988) Two concomitant base substitutions in the putative replicase genes of tobacco mosaic virus confer the ability to overcome the effects of a tomato resistance gene, *Tm-1*. EMBO J 7:1575-1581
Montalbini P, Lupattelli M (1989) Effect of localized and systemic tobacco mosaic virus infection on some photochemical and enzymatic activities of isolated tobacco chloroplasts. Physiol Mol Plant Pathol 34:147-162
Moser O, Gagey MJ, Godefroy-Colburn T, Stussi-Garaud C, Ellwart-Tschürtz M, Nitschko H, Mundry KW (1988) The fate of the transport protein of tobacco mosaic virus in systemic and hypersensitive tobacco hosts. J Gen Virol 69:1367-1373
Oxelfelt P (1971) Development of systemic tobacco mosaic virus infection II. RNA

metabolism in systemically infected leaves. Phytopathol Z 71:247-256
Peterson PD, McKinney HH (1938) The influence of four mosaic diseases on the plastid pigments and chlorophyllase in tobacco leaves. Phytopathology 28:329-342
Reinero A, Beachy RN (1989) Reduced photosystem II activity and accumulation of viral coat protein in chloroplasts of leaves infected with tobacco mosaic virus. Plant Physiol 89:111-116
Saito T, Meshi T, Takamatsu N, Okada Y (1987) Coat protein gene sequence of tobacco mosaic virus encodes a host response determinant. Proc Natl Acad Sci USA 84:6074-6077
Scalla R, Boudon E, Rigaud J (1976) Sodium dodecyl sulphate-polyacrylamide-gel electrophoretic detection of two high molecular weight proteins associated with tobacco mosaic virus infection in tobacco. Virology 69:339-345
Siegel A, Wildman SG (1954) Some natural relationships among strains of tobacco mosaic virus. Phytopathology 44:277-282
Takamatsu N, Ishikawa M, Meshi T, Okada Y (1987) Expression of bacterial chloramphenicol acetyltransferase gene in tobacco plants mediated by TMV-RNA. EMBO J 6:307-311
Taliansky ME, Atabekova TI, Kaplan IB, Morozov SY, Malyshenko SI, Atabekov JG (1982) A study of TMV *ts* mutant Ni2519 I. Complementation experiments. Virology 118:301-308
Tomlinson JA, Faithfull EM, Ward CM (1976) Chemical suppression of the symptoms of two virus diseases. Ann Appl Biol 84, 31-41
Trewavas A (1986) Understanding the control of plant development and the role of growth substances. Austr J Plant Physiol 13:447-457
Valleau WD (1952) The evolution of susceptibility to tobacco mosaic virus in Nicotiana and the origin of the tobacco mosaic virus. Phytopathology 42:40-42
Van Loon LC (1979) Effects of auxin on the localization of tobacco mosaic virus in hypersensitively reacting tobacco. Physiol Plant Pathol 14:213-226
Van Loon LC (1987) Disease induction by plant viruses. Adv Virus Res 33:205-255
Van Loon LC, Van Kammen A (1970) Polyacrylamide disc electrophoresis of the soluble leaf proteins from *Nicotiana tabacum* var. 'Samsun' and 'Samsun NN' II. Changes in protein constitution after infection with tobacco mosaic virus. Virology 40:199-211
Van Telgen HJ, Van der Zaal EJ, Van Loon LC (1985a) Evidence for an association between viral coat protein and host chromatin in mosaic-diseased tobacco leaves. Physiol Plant Pathol 26:83-98
Van Telgen HJ, Van der Zaal EJ, Van Loon LC (1985b) Some characteristics of the association of the 116 kD protein with host chromatin in tobacco leaves infected with tobacco mosaic virus. Physiol Plant Pathol 26:99-109
Van Telgen HJ, Goldbach RW, Van Loon LC (1985c) The 126,000 molecular weight protein of tobacco mosaic virus is associated with host chromatin in mosaic-diseased tobacco plants. Virology 143:612-616
Watanabe Y, Morita N, Nishiguchi M, Okada Y (1987) Attenuated strains of tobacco mosaic virus. Reduced synthesis of a viral protein with a cell-to-cell movement function. J Mol Biol 194:699-704
Whenham RJ, Fraser RSS (1981) Effect of systemic and local-lesion-forming strains of tobacco mosaic virus on abscisic acid concentration in tobacco leaves: consequences for the control of leaf growth. Physiol Plant Pathol 18:267-278
Whenham RJ, Fraser RSS, Snow A (1985) Tobacco mosaic virus-induced increase in abscisic acid concentration in tobacco leaves: intracellular location and relationship to symptom severity and to extent of virus multiplication. Physiol

Plant Pathol 26:379-387
Whenham RJ, Fraser RSS, Brown LP, Payne JA (1986) Tobacco mosaic virus-induced increase in abscisic acid concentration in tobacco leaves: intracellular location in light and dark-green areas, and relationship to symptom development. Planta 168:592-598
Wijdeveld MGG, Goldbach RW, Verduin BJM, Van Loon LC (1989) Association of viral 126 kDa protein-containing X-bodies with nuclei in mosaic-diseased tobacco leaves. Arch Virol 104:225-239
Zaitlin M, Hull R (1987) Plant virus-host interactions. Annu Rev Plant Physiol 38:291-315
Zimmern D, Hunter T (1983) Point mutation in the 30-K open reading frame of TMV implicated in temperature-sensitive assembly and local lesion spreading of mutant Ni 2519. EMBO J 2:1893-1900

ENTRY OF TOBACCO MOSAIC VIRUS RNA INTO TOBACCO CHLOROPLASTS *IN VIVO* AND *IN VITRO*

Milton Zaitlin, James E Schoelz[1], Rosemary Grzesik and José O Gaspar[2]
Department of Plant Pathology
Cornell University
Ithaca
NY 14853
USA

ABSTRACT

Chloroplasts may be adversely affected in tobacco mosaic virus (TMV)-infected plants. Part of the perturbation could result from the known accumulation of virus and viral-coded components within that organelle. Our studies have been directed at ascertaining how these virus-related components gain entry into the chloroplasts. We prepared chloroplasts from either directly-inoculated or systemically-infected leaves of TMV-infected tobacco, and from uninfected controls. Intact chloroplasts, isolated from Percoll gradients, were treated with both pancreatic ribonuclease and thermolysin to destroy any TMV particles and RNA, either bound to or copurifying with the chloroplasts. RNA isolated from such chloroplasts was probed on Northern blots with ^{32}P-labelled cDNA prepared from TMV RNA. Genomic, but not sub-genomic TMV RNA was detected within the chloroplasts of directly-inoculated and systemically-invaded leaves, indicating that the uptake into the chloroplast is selective. Evidence showing that TMV RNA *per se* (as opposed to virions) could enter is based on the observation that the temperature-sensitive mutant Ts 38, in which no virions are formed at a restrictive temperature of 35°, still allows for entry of the viral RNA at that temperature.

An *in vitro* system has been developed in which isolated chloroplasts are incubated with labelled TMV RNA, or other RNAs as controls. As in the leaf, the uptake is specific to the genomic length TMV RNA, with subgenomic and non-viral RNAs being excluded. The parameters of this system are being optimized, and the sequences of genomic TMV RNA essential for uptake are being determined.

---

[1]Department of Plant Pathology, University of Missouri, Columbia, Missouri 65211 USA

[2]Centro de Biologia Molecular, Instituto de Biologia, Universidade Estadual de Campinas, 13081 Campinas, SP, Brasil

NATO ASI Series, Vol H 41
Recognition and Response in Plant-Virus Interactions
Edited by R S S Fraser

INTRODUCTION

Consequences of viral infection on chloroplast structure and function

Plant virus infections frequently have adverse affects on chloroplasts, resulting in many of the symptoms seen on the plant. A summary of those consequences for a number of viruses and hosts is given in Goodman *et al.* (1986). Effects may be seen on specific photosynthetic processes, chloroplast structure, chlorophyll content, and both accumulation and breakdown of starch. For instance, the TMV coat protein which accumulates in the chloroplast is membrane associated (Reinero and Beachy, 1986) and has a specific inhibitory effect on photosystem II (Hodgson *et al.*, 1989). Moreover, TMV infection has been shown to have an adverse affect on chloroplast ribosomal RNA synthesis, effectively shutting it down and causing degradation (Fraser, 1969; Hirai and Wildman, 1969).

In addition to TMV, other viruses have been shown to have an association with the chloroplast. Turnip yellow mosaic virus (Chalcroft and Matthews, 1966; Garnier *et al.*, 1966) and barley stripe mosaic virus (Carroll, 1970) are found in invaginations at the chloroplast periphery; they are not transported across the chloroplast outer membrane. Electron microscopic studies indicate that beet yellows virus may be present within sugar beet chloroplasts (Leyon, 1953; Cronshaw *et al.*, 1966). Most recently, the DNA of the geminivirus, abutilon mosaic virus has been isolated from intact chloroplasts (Groning *et al.*, 1987).

The association of TMV and its components with chloroplasts

Viruses or viral components have been shown to be associated with chloroplasts of TMV infected plants, and of plants infected with a few other viruses. These are enumerated in Schoelz and Zaitlin (1989). Virus particles, viral RNA, coat protein, and pseudovirions have all been observed within the tobacco leaf chloroplast. Fig. 1 shows a chloroplast from a tobacco leaf in which TMV particles are seen both in cross section and in longitudinal section (Esau and Cronshaw, 1967). Because the grouped virions were not bounded by membranes, they were considered to be within the chloroplast. Subsequent work in other laboratories employing different technologies has confirmed the presence of virus particles within the chloroplast.

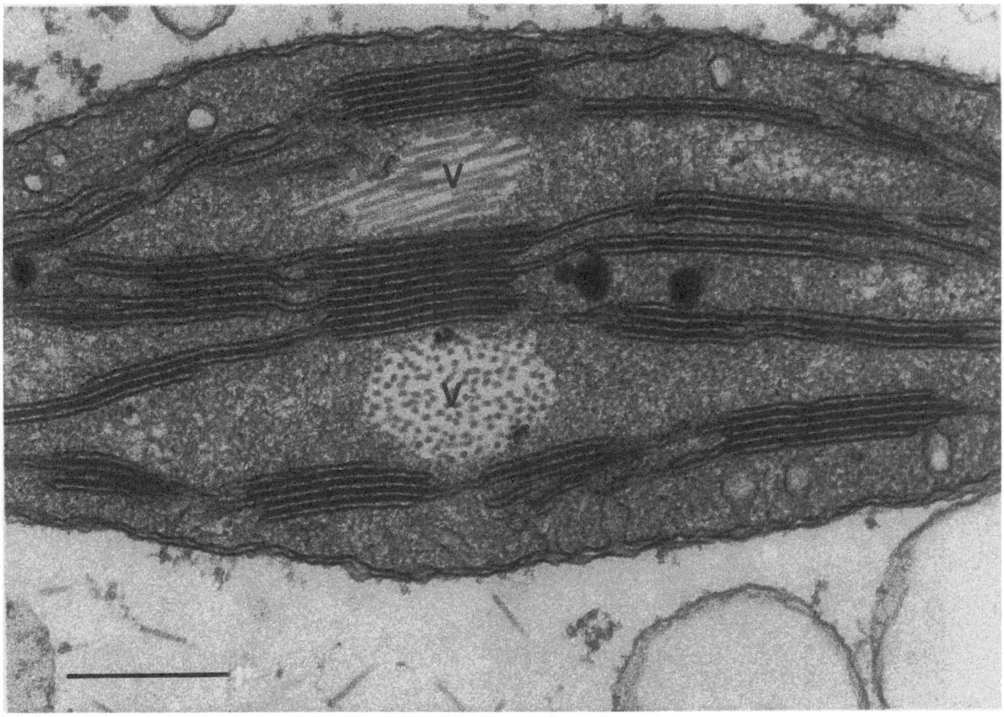

Figure 1. An electron micrograph of a chloroplast from a TMV-infected *Nicotiana tabacum* plant, showing virions (V), both in transverse and in cross section. The scale bar indicates 300 nm. Electron micrograph by courtesy of Professor Katherine Esau, University of California, Santa Barbara. Reproduced from the Journal of Cell Biology (1967) 33:665-678 by copyright permission of the Rockefeller University Press.

EXPERIMENTS ON TMV RNA AND CHLOROPLASTS

In vivo studies

Recent work in our laboratory has shown that TMV RNA can enter the chloroplast *in vivo* (Schoelz and Zaitlin, 1989). In these experiments, chloroplasts were isolated from TMV-infected leaves, and RNA was extracted; this RNA was analyzed on Northern blots with ^{32}P-cDNA of TMV RNA as a hybridization probe. The essential finding of these studies is shown in Fig. 2. Chloroplasts were prepared on Percoll gradients, and before their nucleic acids were extracted, they were treated with RNase and thermolysin to eliminate any possible contamination from cytoplasmic RNAs and/or TMV RNA and virions. Lanes 5 and 5' show RNA extracted from the chloroplasts of infected leaves. Hybridization of the probe was observed to genomic-length TMV RNA

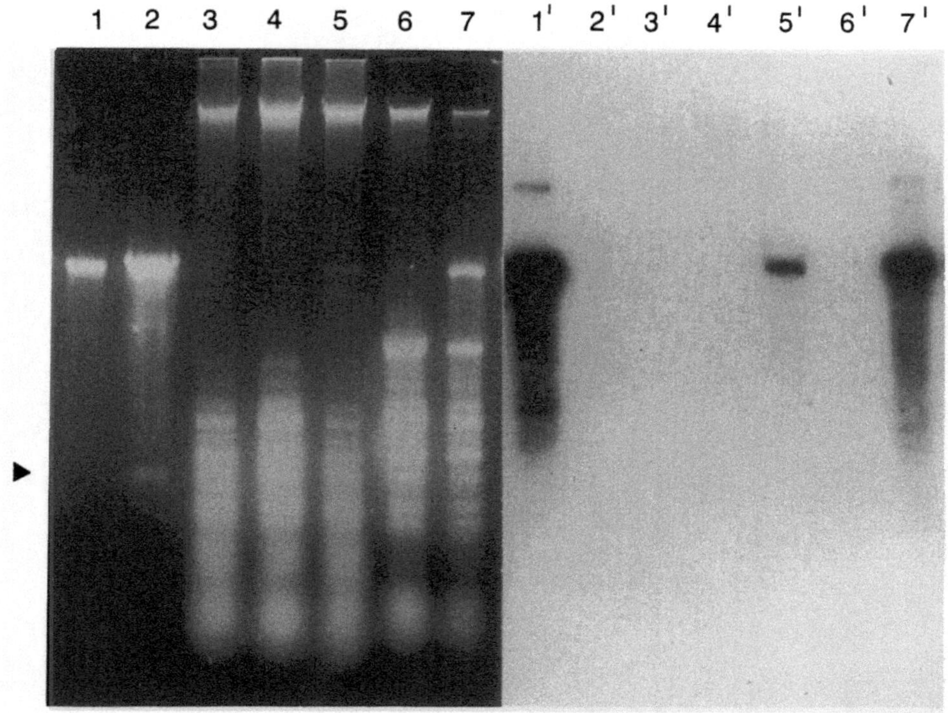

Figure 2. Analysis of RNA extracted from chloroplasts of TMV-infected and uninfected *Nicotiana tabacum* plants. Left panel, ethidium bromide-stained, non-denaturing 1.8% agarose gel. Right panel, a Northern blot of the transfer of the nucleic acids from the gel to nitrocellulose, followed by a probe with ^{32}P-labelled cDNA prepared to TMV Strain U2 RNA. Lanes 1 and 1', RNA from Strain U2 virions; 2 and 2', RNA from virions of the cowpea strain of TMV (the arrow marks the position of migration of the subgenomic coat protein mRNA); 3 and 3', nucleic acid from chloroplasts of an uninfected leaf; 4 and 4', chloroplast nucleic acid from the reconstruction experiment described in the text; 5 and 5', chloroplast nucleic acid from Strain U2-infected leaves; 6 and 6', nucleic acid from the post-chloroplast supernatant of uninfected leaves; 7 and 7', nucleic acid extracted from the post-chloroplast supernatant of Strain U2-infected leaves. From Schoelz and Zaitlin (1989).

in lane 5'. The lane from uninfected leaves (3') showed no such response. In addition, lanes 4 and 4' represent a reconstruction experiment in which a post-chloroplast supernatant containing TMV virions and RNA was added to chloroplasts from uninfected leaves, followed by chloroplast reisolation. Here too, the RNA extracted from those chloroplasts did not react with the TMV cDNA, validating the purity of the chloroplasts.

One characteristic feature of the hybridization was that only the genomic, and not the subgenomic TMV RNAs were found in the chloroplast, indicating that uptake is very sequence or structure specific (compare lane 1' with lane 5'). Studies were also undertaken to determine whether the RNA entered the chloroplast as free RNA or as virions. To ascertain this, we employed the temperature-sensitive TMV mutant Ts 38, which does not produce virions at restrictive temperature (Hariharasubramanian *et al.*, 1970). We found that TMV RNA entered chloroplasts at the restrictive temperature, indicating that it was the molecule involved. This finding has been confirmed in our experiments *in vitro*.

In vitro studies

More recently, it has been shown that chloroplasts isolated from either infected or uninfected tobacco leaves are capable of taking up TMV RNA from the external medium *in vitro* (R Grzesik, JO Gaspar, M Zaitlin, unpublished data). Genomic TMV RNA was taken up, while subgenomic RNAs, ribosomal RNAs, and transcripts representing non-5' regions of the TMV genome were tested and found to be excluded. These results demonstrate that a sequence at the 5' end of TMV RNA is necessary for transfer across the chloroplast double membrane. The parameters of the system are still being developed, but several factors are already established. Uptake is not dependent on active photosynthesis, nor on light. Active metabolism is essential, however, as uptake is inhibited at 4°C. Chloroplasts can be from either infected or uninfected leaves. Currently, the sequences at the 5' end of TMV RNA essential for uptake are being investigated by the use of transcripts from full length clones of TMV, generously provided by Dr WO Dawson of the University of California, Riverside.

Viral coat protein in the chloroplast

As stated above, chloroplasts from infected plants contain TMV coat protein, which can encapsidate either viral RNA to form virions (Zaitlin and Boardman, 1958), or chloroplast transcripts to form pseudovirions (Siegel, 1971; Rochon and Siegel, 1984). Coat protein might be accumulated within the chloroplast by the operation of two mechanisms. It could either be synthesized within the chloroplast from a viral RNA template, or it could be transported from the cytoplasm. There is supporting evidence for both mechanisms.

One possibility is synthesis of coat protein by translation of TMV RNA on the chloroplastic (prokaryotic) 70 S ribosomes, thereby precluding the necessity for the viral sub-genomic mRNAs normally necessary for the synthesis of several TMV-encoded proteins on 80 S ribosomes. Two studies *in vitro* have demonstrated that TMV coat protein can be synthesized on 70 S ribosomes, either from *Escherichia coli* (Glover and Wilson, 1982) or spinach chloroplasts (Camerino *et al.,* 1982), using *genomic* TMV RNA as messenger. Glover and Wilson (1982) pointed out that TMV RNA contains sequences for prokaryotic ribosome association (Shine and Dalgarno, 1975) upstream from the start of the coat protein cistron. This would suggest that it is possible that chloroplast ribosomes could translate coat protein mRNA directly without the intercession of sub-genomic coat protein mRNA. However, we have been unable to demonstrate coat protein synthesis within the chloroplast directly, using an active "*in organello*" synthesis system patterned after Fish and Jagendorf (1982). Similar experiments by Reinero and Beachy (1989) also failed to detect coat protein synthesis in the chloroplasts.

An experiment which might resolve these contradictory results involves determining if the coat protein of the TMV tomato strain (L) is present in tobacco chloroplasts. An examination of the L strain RNA sequences upstream from the coat protein start codon (Ohno *et al.,* 1984) has shown that the Shine-Dalgarno ribosome entry sequence has not been conserved as well as in the common strain (Glover and Wilson, 1982). If the L strain coat protein is present within chloroplasts, it would suggest that the coat protein is transported into chloroplasts. Conversely, the absence of the L strain coat protein in chloroplasts would indicate that the coat protein of the other strains is synthesized within the chloroplast, and that the Shine-Dalgarno sequence upstream from the coat protein cistron is necessary for synthesis.

The second possibility, that TMV coat protein enters the chloroplast from the cytoplasm, also has mixed supporting evidence. On the one hand, we see no sequences in the coat protein which might be transit peptides for transport into the chloroplast, analogous to those found by Karlin-Neumann and Tobin (1986) for nuclear-encoded chloroplast proteins. On the other hand, plants transformed with the coat protein of TMV express some of that protein in the chloroplast as small rodlets (Osbourn *et al.,* 1989). This implies that the protein enters the chloroplast and accumulates there. Alternatively, if the mRNA for the coat protein were transported rather than the coat protein itself, it would have to be translated on the 70 S chloroplast ribosomes. This mystery is unresolved!

ACKNOWLEDGEMENTS

The *in vivo* studies from our laboratory were supported in part by Grant 87-CRCR-1-2549 from the Competitive Grants Program of the United States Department of Agriculture. Additional support is currently being provided by Grant DCB-8904884 from the National Science Foundation. JO Gaspar was supported by a fellowship from the Brazilian Research Council - CNPg.

REFERENCES

Camarino G, Savi A, Ciferri O (1982) A chloroplast system capable of translating heterologous mRNAs. FEBS Lett 150:94-98

Carroll TW (1970) Relation of barley stripe mosaic to plastids. Virology 42:1015-1022

Chalcroft J, Matthews REF (1966) Cytological changes induced by turnip yellow mosaic virus infection. Virology 28:555-562

Cronshaw J, Hoefert L, Esau K (1966) Ultrastructural features of *Beta* leaves infected with beet yellows virus. J Cell Biol 31:429-443

Esau K, Cronshaw J (1967) Relation of tobacco mosaic virus to the host cells. J Cell Biol 33:665-678

Fish LE, Jagendorf AT (1982) High rates of protein synthesis by isolated chloroplasts. Plant Physiol 70:1107-1114

Fraser RSS (1969) Effects of two TMV strains on the synthesis and stability of chloroplast ribosomal RNA in tobacco leaves. Mol Gen Genet 106:73-79

Garnier M, Candresse T, Bové JM (1986) Immunocytological localization of TYMV-coded structural and nonstructural proteins by the protein A-gold technique. Virology 151:100-109

Glover JF, Wilson TMA (1982) Efficient translation of the coat protein cistron of tobacco mosaic virus in a cell-free system from *Escherichia coli*. Eur J Biochem 122:485-492

Goodman RN, Kiraly Z, Wood KR (1986) The biochemistry and physiology of plant disease. University of Missouri Press, Columbia

Groning BR, Abouzid A, Jeske H (1987) Single-stranded DNA from abutilon mosaic virus is present in the plastids of infected *Abutilon sellovianum*. Proc Natl Acad Sci USA 84:8996-9000

Hariharasubramanian V, Zaitlin M, Siegel A (1970) A temperature-sensitive mutant of TMV with unstable coat protein. Virology 40:579-589

Hirai A, Wildman SG (1969) Effect of TMV multiplication on RNA and protein synthesis in tobacco chloroplasts. Virology 38:73-82

Hodgson RAJ, Beachy RN, Pakrasi HB (1989) Selective inhibition of photosystem II in spinach by tobacco mosaic virus: an effect of the viral coat protein. FEBS Lett 245:267-270

Karlin-Neumann GA, Tobin EM (1986) Transit peptides of nuclear-encoded chloroplast proteins share a common amino acid framework. EMBO J 5:9-13

Leyon H (1953) Virus formation in chloroplasts. Exp Cell Res 4:362-370

Ohno T, Aoyagi M, Yamanashi Y, Saito H, Ikawa S, Meshi T, Okada Y (1984)

Nucleotide sequence of the tobacco mosaic virus (tomato strain) genome and comparison with the common strain genome. J Biochem 96:1915-1923
Osbourn JK, Plaskitt KA, Watts JW, Wilson TMA (1989) Tobacco mosaic virus coat protein and reporter gene transcripts containing the TMV origin-of-assembly sequence do not interact in double-transgenic tobacco plants: implications for coat protein-mediated protection. Mol Plant-Microbe Interact, in press
Reinero A, Beachy RN (1986) Association of TMV coat protein with chloroplast membranes in virus-infected leaves. Plant Mol Biol 6:291-301
Reinero A, Beachy RN (1989) Reduced photosystem II activity and accumulation of viral coat protein in chloroplasts of leaves infected with tobacco mosaic virus. Plant Physiol 89:111-116
Rochon D, Siegel A (1984) Chloroplast DNA transcripts are encapsidated by tobacco mosaic virus coat protein. Proc Natl Acad Sci USA 81:1719-1723
Schoelz JE, Zaitlin M (1989) Tobacco mosaic virus RNA enters chloroplasts *in vivo*. Proc Natl Acad Sci USA 86:4496-4500
Shine J, Dalgarno L (1975) Determinant of cistron specificity in bacterial ribosomes. Nature 254:34-38
Siegel A (1971) Pseudovirions of tobacco mosaic virus. Virology 46:50-59
Zaitlin M, Boardman NK (1958) The association of tobacco mosaic virus with plastids I. Isolation of virus from the chloroplast fraction of diseased-leaf homogenates. Virology 6:743-757

MODIFICATIONS OF THE COAT PROTEIN GENE OF TOBACCO MOSAIC VIRUS RESULTING IN THE INDUCTION OF NECROSIS

J N Culver and W O Dawson
Department of Plant Physiology
University of California
Riverside
CA 92521
USA

ABSTRACT

We are examining the interaction of tobacco mosaic virus (TMV) with *Nicotiana* species relative to the induction of necrosis. A number of point mutations in the coat protein gene have been found to induce a hypersensitive reaction (HR) in plants with the *N'N'* genotype. Several of these coat protein point mutations also induce a systemic necrosis on *nn* genotype plants. Partial deletions of the coat protein gene have been found to induce a generalized necrosis that appears not to be specific to any one host gene. Removal of the entire TMV coat protein gene does not effect the induction of HR in *NN* genotype plants. Thus, a viral component other than the coat protein gene can elicit HR in *NN* genotype plants.

INTRODUCTION

Little is known about how plant viruses interact with their hosts to cause disease or to induce resistance responses. Moreover, the understanding of the mechanisms involved in virus-host interactions and host recognition of viral components is a first step in the development of new methods to control virus diseases of plants. Recently, we have developed the ability to modify the genomes of plant viruses and to observe the resulting effect on the virus-host interaction (Ahlquist and Janda, 1984; Dawson *et al.*, 1986; Meshi *et al.*, 1986). This development has enabled us to begin a systematic examination of specific viral sequences associated with specific host responses. We describe here the results of modifications of the tobacco mosaic virus (TMV) coat protein gene that lead to the development of necrosis in different host plants.

NATO ASI Series, Vol H 41
Recognition and Response in Plant-Virus Interactions
Edited by R S S Fraser

The TMV genome is comprised of a single-stranded RNA, 6395 nucleotides in length, containing four open reading frames (Goelet *et al.*, 1982). Viral proteins consist of the 126 kDa and 183 kDa proteins presumably involved in viral RNA replication (Ishikawa *et al.*, 1986), a 30 kDa protein associated with cell-to-cell movement (Meshi *et al.*, 1987; Deom *et al.*, 1987), and a 17.5 kDa coat protein. Numerous coat protein mutants of TMV have been correlated with altered host responses, suggesting that the coat protein not only serves a structural role but can also affect the pattern of virus-host interaction.

One type of virus-induced host response is the development of necrosis, and this may occur in several forms. Often, the virus is localized around the site of infection, and a necrotic lesion develops. This is referred to as the hypersensitive reaction (HR). In other cases, the virus can spread systemically through the infected plant, causing either sporadic patches of necrosis or necrosis of specific tissues, leading to severe disease and death of the plant (Knorr *et al.*, 1983). It is not clear if all of these necrotic responses are genetically and physiologically distinct, or whether some of the apparently different virus-induced necrotic responses are variations of the same type of response. It is likely that some variations in the development of virus-induced necrotic reactions are due to the ability of the plant to recognize and then respond to a specific virus elicitor. Therefore, slower systemic necrotic responses might be induced by weaker virus elicitors and faster HR responses might be induced by stronger virus elicitors of the same mechanism.

INDUCTION OF THE HYPERSENSITIVE REACTION

Introduction of a gene for the hypersensitive reaction to a particular virus has been one of the most important means of controlling virus diseases of plants. Recently, workers in Okada's laboratory identified the coat protein gene of TMV as a determinant for the induction of HR in *Nicotiana sylvestris* (*N'N'* genotype) (Saito *et al.*, 1987). This was demonstrated through the production of hybrids of TMV-L, which induces HR in *N. sylvestris*, and TMV-OM, which does not induce HR on this host. When the coat protein gene of TMV-L was replaced by the coat protein gene of TMV-OM, HR was not induced.

Work in our laboratory has identified a single point mutation in the coat protein

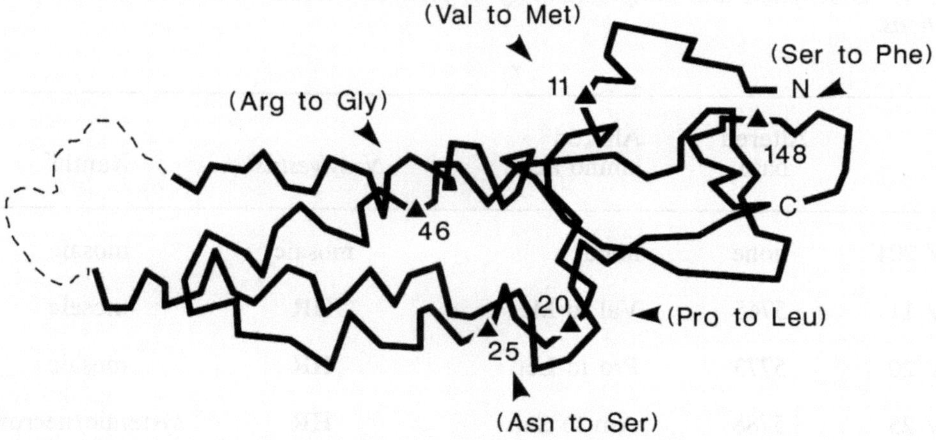

Figure 1. Schematic representation of a tobacco mosaic virus coat protein subunit (derived from Bloomer *et al.*, 1978). Five amino acid substitutions produced by five different point mutations in the coat protein gene are indicated.

gene of TMV as being responsible for the induction of HR in *N'N'* plants (Knorr and Dawson, 1988). This was shown by cloning segments of nitrous acid mutants, obtained from the non-HR inducing U1 strain of TMV, that were selected for their ability to induce HR in *N. sylvestris*. cDNA segments of these mutants were substituted into the original non-HR inducing infectious cDNA genome of TMV U1 to identify segments responsible for HR induction. Sequencing revealed that a single point mutation at nucleotide 6157 (cytosine to uracil) in the coat protein gene of mutant TMV DK34 was responsible for the induction of HR. The strategy of direct comparison of mutants of the same virus strain allowed precise assignment of sequence alterations to function.

This point mutation resulted in an amino acid substitution of phenylalanine for serine at amino acid position 148 in the coat protein of TMV. By coincidence this amino acid substitution had previously been identified in a TMV mutant selected for the ability to induce HR in *N. sylvestris* (Funatsu and Fraenkel-Conrat, 1964). However, the technology of the time did not allow exclusion of possible mutations in the rest of the genome and the assignment of specific coat protein mutations to induction of HR. Moreover, several other mutants which were selected for the ability to induce necrosis in *N. sylvestris* were also reported to contain amino acid substitutions in other areas of the coat protein.

Table 1. *Description and symptomatology of tobacco mosaic virus coat protein point mutations.*

	Altered[a] base	Altered amino acid	*N. sylvestris*[b]	Xanthi[c]
TMV 204	none	none	mosaic	mosaic
TMV 11	5745	Val to Met	HR	mosaic
TMV 20	5773	Pro to Leu	HR	mosaic
TMV 25	5788	Asn to Ser	HR	systemic/necrosis
TMV 46	5850	Arg to Gly	HR	systemic/necrosis
TMV DK32	6157	Ser to Phe	HR	systemic/necrosis

[a] Nucleotide position of coat protein point mutation.
[b] Induction of hypersensitive reaction (HR) on *N'N'* genotype.
[c] Systemic symptoms on *N. tabacum* cv. *nn* genotype.

To determine if other mutations in the coat protein gene of TMV might also lead to the induction of HR on *N. sylvestris* we created four independent point mutations in the gene (Culver and Dawson, 1989). Point mutations were selected so as to alter specific amino acids throughout the coat protein. Each of the selected amino acid substitutions had previously been described (Funatsu and Fraenkel-Conrat, 1964) (Fig. 1). Site-directed mutagenesis was utilized to create point mutations in the coat protein gene of the parental cDNA viral clone, TMV 204 (Dawson *et al.*, 1986) as shown in Table 1.

All four of the selected mutants were infectious and induced HR in *N. sylvestris*. However, the timing and development of HR was distinctly different for each mutant. Mutant TMV 11 produced necrotic local lesions that developed four to five days after inoculation and would often spread from the initial point of infection, leading to the collapse of large portions of the inoculated leaves. Mutants TMV 20 and TMV 25 both produced distinct necrotic local lesions at three days post-inoculation that did not spread further. TMV 46 produced necrotic local lesions at 4 days post-inoculation that only occasionally spread from the initial point of infection. All four mutants gave symptom responses identical to the parental virus, TMV 204, on the local lesion assay host *N. tabacum* cv. Xanthi-nc (*NN* genotype) as well as on a number of other hosts, demonstrating that these point mutations were specific for the induction of HR in the *N'N'* genotype of *N. sylvestris*.

At this time we cannot conclude that the coat protein itself is the active elicitor of HR. It is possible that the alteration in RNA sequence may play this role. However, the identification of several different mutations that can induce HR demonstrates that there are several coat protein gene sequences or coat protein structural configurations that may be responsible for the induction of HR. Also, the frequency of mutants that induce HR in *N. sylvestris* is relatively high (10^{-3} to 10^{-2}), while reversion mutation of HR inducing strains to non-HR inducing strains is extremely infrequent (Aldaoud, 1987). This suggests that the sequence of the TMV coat protein gene, responsible for the systemic infection by the virus, is a relatively rare sequence and that many alterations from this sequence may lead to induction of HR in *N'N'* plants.

In contrast to the induction of HR in *N'N'* genotype plants, the coat protein gene of TMV does not appear to play a role in the induction of HR in *NN* genotype plants. Infectious RNA from mutants having the entire coat protein gene deleted induced HR in Xanthi-nc (*NN* genotype) (Takamatsu *et al.,* 1987; Dawson *et al.,* 1988). Thus, a viral product other than the coat protein gene of TMV is the elicitor of HR in *NN* plants.

Induction of other plant necrotic reactions

We have previously described a number of coat protein deletion mutants and their effect on host responses (Dawson *et al.,* 1988). Of these mutants, cp 10 and cp 27 were shown to induce distinct necrotic lesions on a wide range of host plants, such as *N. tabacum* cv. Xanthi, which normally are not resistant to TMV and develop a systemic infection. Mutants cp 10 and cp 27 were created by Bal31 nuclease treatment to remove the coat protein gene sequences 6042-6085 nt, resulting in a 3' deletion of 45 amino acids, and sequences 5948-6080 nt, resulting in an internal deletion also of 45 amino acids, respectively. It is difficult to correlate the different coat protein deletions with their ability to induce necrosis. Although the deletions in mutants cp 10 and cp 27 included a common region of the coat protein (amino acids 109 to 123), several additional mutants having deletions in this same area did not induce this form of necrosis.

The ability of mutants cp 10 and cp 27 to induce necrosis in a number of different hosts, which normally are not resistant to TMV, suggests that this form of necrotic response is non-specific and does not rely on the interaction of specific host resistance genes with viral products. Therefore, the development of necrosis after infection by these two mutants may be due to the induction of a general, non-specific response as has been shown in other host-pathogen interactions.

Another type of necrotic reaction is the development of systemic necrosis. This was observed on the normally systemic host *N. tabacum* cv. Xanthi (*nn* genotype) for the point mutations of TMV DK32, TMV 25, and TMV 46 described previously (Table 1). Each of these mutants produced sporadic patches of necrosis on both inoculated and systemically-infected leaves, that lead to severe disease and stunting of the plant. However, necrosis occurred in only a small proportion of the infected tissues. The ability of these mutants, which normally induce HR in *N. sylvestris* (*N'N'* genotype), to produce necrosis on Xanthi might be due to a relationship of the interaction with *N'* and *n* genes. The induction of necrosis in *nn* genotype plants by these mutants might be a slower variation of the HR that occurs in *N'N'* plants.

CONCLUSION AND FUTURE EXPERIMENTS

We have shown that a number of virus-induced necrotic responses in plants can be mediated by alterations in the coat protein gene of TMV U1. The inducers of these necrotic responses include coat protein gene alterations that interact with specific host genotypes, as well as alterations that appear to be non-specific and are able to induce necrosis on a wide variety of host genotypes. This demonstrates that the coat protein gene plays a multifunctional role in determining the pattern of virus-host interaction.

Currently, we are trying to discover whether changes created in the coat protein or in its RNA act as the determinants responsible for the induction of HR. If the protein is the active molecule, what effect does each of the described mutations have on the secondary configuration of the protein that cause it to act as an elicitor of HR? Another important question is the precise nature of the determinant responsible for HR induction. It is possible that only specific regions of the coat protein are involved, in a manner similar to that of epitopes in the induction of antibodies. Alternatively, the entire coat protein or even aggregations of coat protein subunits might be recognized by the host in the event that triggers the resistance response.

We have also begun to look at other types of host necrotic responses that are important during the development of disease. Differences in the development of both localized and/or systemic necrotic responses may be due to the efficiency and timing of recognition of specific viral determinants by host mechanisms. In addition, the ability of particular coat protein deletion mutants to induce the development of necrotic local lesions on host plants that are not normally resistant to TMV may indicate the presence of a non-specific mechanism for the development of necrosis which perhaps could be utilized to provide resistance in non-resistant hosts. Finally, understanding how viruses induce necrotic reactions should provide an interesting parallel to the induction of necrosis by extracellular pathogens.

REFERENCES

Ahlquist P, Janda M (1984) cDNA cloning and in vitro transcription of the complete brome mosaic virus genome. Mol Cell Biol 4:2876-2882

Aldaoud R (1987) Biological characteristics, variations, and factors affecting variation in cloned tobacco mosaic virus. PhD thesis, University of California, Riverside

Bloomer AC, Champness JN, Bricogne G, Staden R, Klug A (1978) Protein disk of tobacco mosaic virus at 2.8 A resolution showing the interactions within and between subunits. Nature 276:362-368

Culver JN, Dawson WO (1989) Point mutations in the coat protein gene of tobacco mosaic virus responsible for the induction of hypersensitivity in *Nicotiana sylvestris*. Mol Plant Microbe Interact 2:209-213

Deom CM, Oliver MJ, Beachy RN (1987) The 30-kilodalton gene product of tobacco mosaic virus potentiates virus movement. Science 237:389-393

Dawson WO, Beck DL, Knorr DA, Grantham GL (1986) cDNA cloning of the complete genome of tobacco mosaic virus and production of infectious transcripts. Proc Natl Acad Sci USA 83:1832-1836

Dawson WO, Bubrick P, Grantham GL (1988) Modification of the TMV coat protein gene affects replication, movement and symptomatology. Phytopathology 78:783-789

Funatsu G, Fraenkel-Conrat H (1964) Location of amino acid exchanges in chemically evoked mutants of tobacco mosaic virus. Biochemistry 3:1356-1361

Goelet P, Lomonossoff GP, Butler PJG, Akam ME, Gait MJ, Karn J (1982) Nucleotide sequence of tobacco mosaic virus RNA. Proc Natl Acad Sci USA 79:5818-5822

Ishikawa M, Meshi T, Motoyoshi F, Takamatsu N, Okada Y (1986) *In vitro* mutagenesis of the putative replicase genes of tobacco mosaic virus. Nucleic Acids Res 14:8291-8305

Knorr DA, Laemmlen FF, Dawson WO (1983) Effects of a necrotic strain of alfalfa mosaic virus on stand losses of tomato. Phytopathology 73:1554-1558

Knorr DA, Dawson WO (1988) A point mutation in the tobacco mosaic virus capsid protein gene induces hypersensitivity in *Nicotiana sylvestris*. Proc Natl Acad Sci

USA 85:170-174
Meshi T, Ishikawa M, Motoyoshi F, Semba K, Okada Y (1986) *In vitro* transcription of infectious RNAs from full-length cDNAs of tobacco mosaic virus. Proc Natl Acad Sci USA 83 5043-5047
Meshi T, Watanabe Y, Saito T, Sugimoto A, Maeda T, Okada Y (1987) Function of the 30kd protein of tobacco mosaic virus: involvement in cell-to-cell movement and dispensability for replication. EMBO J 6:2557-2563
Saito T, Meshi T, Takamatsu N, Okada Y (1987) Coat protein gene sequence of tobacco mosaic virus encodes host response determinant. Proc Natl Acad Sci USA 84:6074-6077
Takamatsu N, Masayuki I, Meshi T, Okada Y (1987) Expression of bacterial chloramphenicol acetyltransferase gene in tobacco plants mediated by TMV-RNA. EMBO J 6:307-311

HYPERSENSITIVITY TO TOBACCO MOSAIC VIRUS IN *N'*-GENE HOSTS: WHICH VIRAL GENES ARE INVOLVED?

K W Mundry, W Schaible, M Ellwart-Tschürtz, H Nitschko, C Hapke
Biological Institute
Stuttgart University
Pfaffenwaldring 57
D-7000 Stuttgart 80
Federal Republic of Germany

ABSTRACT

Mapping was attempted of sites on the genome of tobacco mosaic virus which are involved in triggering a hypersensitive response on *N'*-gene tobacco plants. Two experimental approaches were tried: 1. "stripping mapping" by nitrous acid mutagenesis with particles partially uncoated with SDS and mapping necrotic mutants relative to the degree of stripping; 2. nitrous acid mutagenesis under "single-hit" conditions, selection of local lesion mutants on *N'*-gene tobacco leaves, and nucleotide sequencing of their p30 and coat protein genes. Mutant loci related to necrotic phenotypes were found in the coat protein cistron only, and were found there to be grouped in two "hot regions" which seem to form "pathogenic domains" in the tertiary structure of the capsid. The stripping mapping experiments suggested that the p30 gene might also be involved in the sense that a malfunctioning transport protein might permit necrogenesis in *N'*-gene tobacco to take place while wild type p30 might be able to prevent a hypersensitive reaction. Mutants with a hit in the p30 gene were however inaccessible to nucleotide sequence analysis. This may have been because they were lost during the local lesion transfer procedure required to separate and purify them from non-mutated parental material.

INTRODUCTION

The response of a plant to an infection with a virus is determined by both the genome of the virus and that of the plant. Tobacco mosaic virus (TMV) is one of those

NATO ASI Series, Vol H 41
Recognition and Response in Plant-Virus Interactions
Edited by R S S Fraser

cases which exemplify this in particular detail. TMV not only has a very broad host range, but also exists in a great number of different strains and mutants which cause a great variety of host responses. Infections with TMV may spread systemically, become restricted to the primary site of infection with local lesion formation, or may be "subliminal infections" restricted to a single cell. Systemic symptoms range from practically invisible (masked) forms of disease, through all variations of green and yellow mosaics, up to severe systemic necrosis able to kill the whole plant.

Studying disease induction by plant viruses or, in other words, trying to correlate a particular host response with a particular viral trigger function is best attempted with a virus-host system in which one single type of host (with an admittedly complicated genome) is capable of reacting in many different ways to infection by a virus with a relatively simple genome. Fortunately, all the major types of host responses to viral infection can be obtained with various forms of TMV in only one single class of tobacco cultivars, the so-called *N'*-gene tobaccos. Using this system, one should be able to correlate particular host responses with particular gene functions or structural features of TMV or its RNA.

Like other plant viruses with small genomes, the genome of TMV provides genes to fulfil the three main requirements for the existence of all plant viruses: reproduction, cell-to-cell and long distance transport within the infected host individual, and ability for (or stability to allow) spread within a population of host organisms. A particular host response thus needs to be correlated with one or more of these basic viral functions.

In trying to correlate host responses with particular viral gene functions two different approaches are possible: one is to produce defined alterations of the viral genome and then look for their biological consequences (e.g. Culver and Dawson, this volume). The other is to produce mutations which statistically are likely to be "single-hit", select mutants of a particular phenotype on the base of symptomatology and then analyze the genome and define the mutated positions by nucleotide sequence analysis. For studies on TMV-induced hypersensitivity, we used the latter approach.

HYPERSENSITIVITY IN GENERAL AND IN TMV-INFECTED *N'*-GENE TOBACCO LEAVES

Hypersensitivity in plant virus infections means that cells of a host first replicate virus for some time without suffering much from it, but later they die rather suddenly as a consequence of the infection. In leaves reacting hypersensitively the tissue turns necrotic and the virus does not invade the vascular system of the plant. Thus localized necrotic lesions appear around the primary site of infection, and the pathological damage to the plant is kept at a minimum.

Hypersensitivity cannot be directly attributed to impaired virus replication, because Otsuki *et al.* (1972) showed that TMV replication is normal in isolated mesophyll protoplasts of tobacco plants carrying the dominant hypersensitivity gene *N*. This has been confirmed meanwhile for numerous other virus-host systems. So how is hypersensitivity triggered: by special features of the cell-to-cell transport mechanism, by the virus capsid, or by peculiar structures in the viral RNA?

It is conceivable that localization could result when virus movement fails to occur. Since virus localization is often phenotypically expressed by the induction of localized necrotic lesions, in other words via activation of a hypersensitive host response, one might argue whether virus-coded transport factors are in addition capable of circumventing virus-localizing host reactions, to avoid or prevent a hypersensitive host response. Such transport factors have become known recently as "movement factors" or "transport proteins". They are virus coded. How they function is still unclear (for review see Hull, 1989).

TMV is one example where restricted virus movement resulting in virus localization and the induction of the hypersensitive response appear functionally linked (for review see Van Loon, 1987). Since the 30 kDa protein of TMV (p30) has been shown to be involved in cell-to-cell transport of the virus (Leonard and Zaitlin, 1982; Meshi *et al.,* 1987), it should also be involved in triggering (or circumventing) the hypersensitive host response, if this hypothesis is correct.

Hypersensitivity is certainly an effective resistance mechanism, but how it works in the case of plant virus infections, how the cascade of biochemical events is triggered and led to culminate in the "suicide" of a group of cells for the benefit of the survival of the whole plant is still largely unknown. What is known is that host and viral genome functions are required. Thus the "type" strains (*vulgare*, U1) of TMV produce a non-necrotic green systemic mosaic on *N'*-gene tobaccos, but mutate easily to "local lesion variants". These mutants fail to cause systemic infections of tobacco varieties carrying the gene *N'* from *Nicotiana sylvestris* (*N'* is assumed to be an allele of the hypersensitivity gene *N* of *Nicotiana glutinosa*). Instead they cause the host to react hypersensitively with formation of localized necrotic lesions. This system was used years ago to demonstrate the possibility of *in vitro* mutagenesis with nitrous acid (Gierer and Mundry, 1958; Mundry and Gierer, 1958). Attempts to obtain the reverse mutation from local lesion strains to systemic ones by *in vitro* mutagenesis with nitrous acid failed completely (Mundry, 1960). Occasionally this reverse mutation occurs spontaneously (see Van Loon, 1987).

These observations support the view, that - at least in TMV - systemic spread and localized necrogenesis are functionally linked and inversely related. Based on this reasoning, one might expect that a defective p30 transport protein of TMV may define the local-lesion phenotype of TMV mutants. In this case a "local lesion gene", identified functionally on TMV-RNA by some means, should match the position of the p30 gene in the nucleotide sequence of TMV-RNA, and in local lesion mutants of TMV the p30 gene should be the one which bears the mutation(s) defining this phenotype.

In an early experiment, Kado and Knight (1966) found evidence for a "local lesion gene" on TMV-RNA which obviously was not the capsid protein gene. On the other hand, it has been demonstrated recently that a single amino acid exchange in the capsid protein of TMV may create a local-lesion mutant capable of triggering the hypersensitive response of *Nicotiana sylvestris*, an *N'*-gene host (Knorr and Dawson, 1988; Culver and Dawson, this volume).

It seemed therefore worthwhile to study the problem again and in more detail with the aim of settling the question, or at least gaining more insight into the problem, of which genes of TMV finally determine the "local lesion" character of TMV mutants and trigger the hypersensitive response of *N'*-gene tobaccos.

Localization of the viral necrogenic function by stripping mapping

TMV is a nucleoprotein rod in which 2130 capsid protein subunits encapsidate one single-stranded RNA molecule in a helical manner. The 6395 nucleotide residues of the TMV genome have been sequenced (Goelet *et al.*, 1982). They code for four major proteins: 183 and 126 kDa (probably involved in replication), 30 kDa (seemingly required for cell-to-cell spread of the infection), and the capsid protein of 17.5 kDa. This rod can be artificially uncoated ("stripped") with 1% sodium dodecyl sulfate (SDS) in a polar manner. Stripping starts at the 5' end and gradually approaches the 3' end (Wilson *et al.*, 1976).

TMV mutants can be produced easily and with great efficiency by *in vitro* mutagenesis with nitrous acid. Under standard conditions (1M $NaNO_2$ at pH 4.2 and 20°C) the time required to cause one "hit" (= mutagenic event) per TMV genome is about 23 min for treatment of complete virions. Of particular interest is the circumstance in which local lesion mutants are derived from the parental systemic strain *vulgare* by a single hit (Gierer and Mundry, 1958).

From published data (Mundry and Gierer, 1958) it can be concluded that naked TMV RNA reacts under standard conditions (see above) about five times faster with nitrous acid than encapsidated TMV RNA. Therefore particles with known extents of stripping can be used to determine gene locations because the response of a gene to standard mutagenic treatment should rise when it becomes uncoated.

Using this approach, Kado and Knight (1966) found local-lesion mutations grouped in a "local-lesion" gene. When one takes into account that stripping proceeds in the direction opposite to that assumed by Kado and Knight (Wilson *et al.,* 1976) it becomes evident that Kado and Knight's local lesion gene does not match one of the open reading frames of TMV: instead it overlaps partly the 3'-terminus of the 183 kDa cistron and the 5'-terminus of the 30 kDa gene. However, their data leave little doubt that the capsid protein gene is not primarily involved (Fig. 1).

We repeated this experiment using refined techniques. TMV partially stripped with 1% SDS (in 0.02 M sodium cacodylate; pH 7.5; 37°C; for 10-20 min) was fractionated by repeated sucrose density gradient centrifugation and the stripping degree determined by four independent methods, namely 1) A_{260}:A_{280} absorbance ratio; 2) density in Cs_2SO_4; 3) length of the remaining nucleoprotein structure in the electron microscope; 4) length of RNA extracted with phenol from RNase-treated aliquots, as measured on gel electrophoresis under denaturing conditions. All four methods yielded comparable values for the degree of stripping, indicating that partially stripped particles are stable under the different conditions of measurement. In accord with reports in the literature, we have never seen two-tailed particles under the electron microscope which would indicate that stripping might proceed from both ends of TMV. For routine purposes the spectrophotometric determination was preferred while the density in Cs_2SO_4 gave the most accurate results with a minimum of sample.

Fractions with nearly identical A_{260}:A_{280} ratios were pooled from different sucrose gradients and the particles' bouyant density in Cs_2SO_4 was determined in a small aliquot. The remaining particles were then mutagenized with HNO_2 and inoculated after suitable dilution in RNA-inoculation buffer onto leaves of the *N'*-gene tobacco cv. Java, for

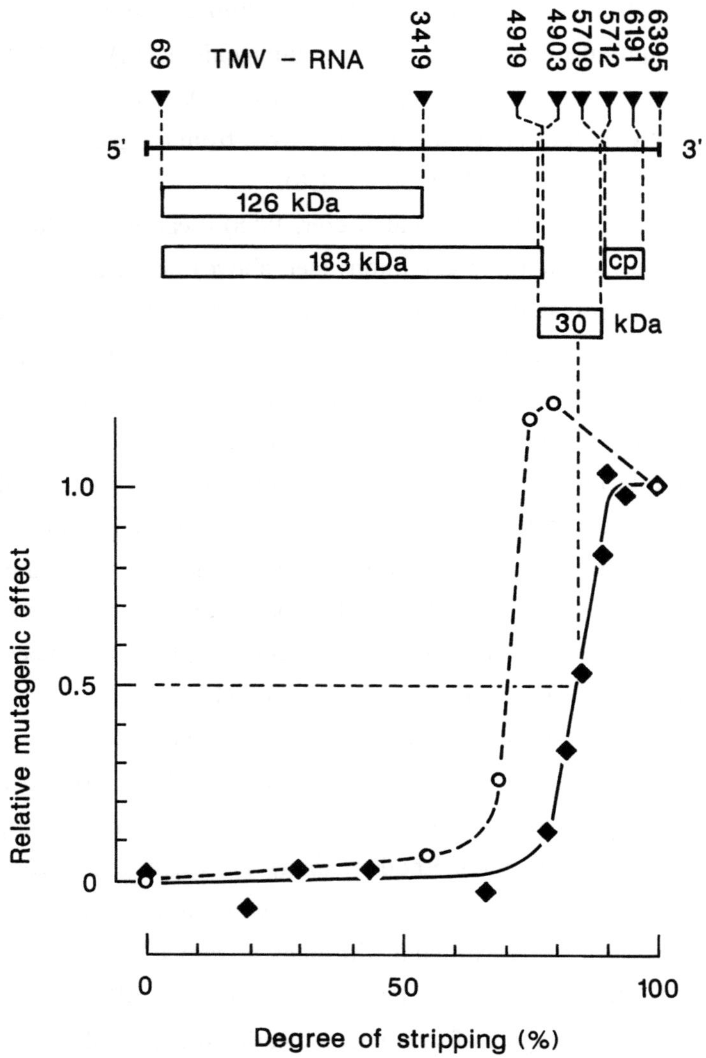

Figure 1. "Stripping mapping " of a local-lesion gene on TMV-RNA. New data (◆—◆) are shown in comparison with data from Kado and Knight (1965) (o---o). Partially uncoated virions were subjected to standard mutagenic treatment with HNO_2 and the frequency of mutants was determined as explained in the text. Relative mutagenic effect (ΔM) was set at 0 for results obtained with native TMV, and 1.0 for naked TMV-RNA. Each solid square (◆) represents one single experiment. The level for the base line at $\Delta M = 0$ is defined by the average of the mutant frequencies obtained in the first 5 experiments (referring to stripping degrees of 0, 20, 30, 44 and 66%). The TMV genome is schematically represented in the upper part of the Figure. The major gene products are shown, and the first and last nucleotide positions of their corresponding open reading frames are indicated.

quantification of local lesion mutants and onto *N. tabacum* cv. Xanthi nc leaves for total infectivity.

With a series of 12 independent experiments (one for each fraction) the revised curve shown in Fig. 1 was obtained. Particles with densities of about 1.300 g/cm^3 or less responded to standard HNO_2 treatment like native TMV (density: 1.2632 g/cm^3), while particles with densities of 1.3648 g/cm^3 and higher reacted like naked TMV RNA (density: 1.6527 g/cm^3). These densities are indicative of about 73 and 89% stripping, respectively (the position of the p30 gene is from 76.7 to 89.2% of genome length). 50% exposure of the local lesion gene as characterized by a relative mutagenic effect of 0.5 (see Fig. 1) corresponds to about 85% of the total genome stripped, which is close to the centre of the p30 gene at 83% genome length. Therefore, on the basis of these data and as far as one can rely on this technique, the p30 gene of TMV seems to be the local lesion gene. Considering that it is usually easier to impair gene function by statistical mutagenesis than to restore it, the mutated alleles may be thought to allow the hypersensitive response of *N'*-gene tobacco while the wild type allele is able to circumvent this host reaction.

However the surprise came when we sequenced the p30 gene of local lesion mutants.

Nucleotide sequence studies with local lesion mutants of TMV

Parallel with the stripping-mapping experiments we began sequencing the RNA of members of a "single-hit" population of nitrous acid-induced local lesion mutants. TMV *vulgare* was treated under conditions which would produce on the average 0.8 mutagenic hit/genome by reducing the time for incubation with HNO_2. Local lesion phenotypes were collected from inoculated Java tobacco leaves and purified biologically via repeated local lesion transfers on *N. tabacum* cv. Xanthi nc leaves with regular checks on Java tobacco that the local lesion phenotype had not been lost. However, after a minimum of 5 single lesion passages from 300 primary mutant isolates, only 37 survived as biologically clean local lesion variants which could be propagated sufficiently on Samsun tobacco. Java tobacco seedlings inoculated with any of these 37 mutant preparations remained free from systemically spreading "contaminants" until flowering. Finally, virus was isolated and the p30 and capsid protein genes were sequenced by primer extension with reverse transcriptase.

To our great surprise, 32 of the 33 mutants so far sequenced were wild type in

Table 1. *Amino acid substitutions in necrotic mutants of TMV* vulgare

n[a]	Position[b]	Identification code number[c]
	p30 Gene (NT.-position 4903 -5709):	
3	GLY_{223}:NT.-position 5571 : GLY→GLY	N119, N141. N277 (Silent mutations)
1	SER_{225}: NT.-position 5576 : ASN→SER	N72
	Coat protein gene (NT.-position 5712-6191):	
1	SER_{015}:NT.-position 5758 : SER→LEU	N277
9	PRO_{020}:NT.-position 5773 : PRO→LEU	N69, N73, N76, N91, N98, N122, N155, N272, N275
2	ILE_{021}: NT.-position 5776 : ILE→THR	N154, N246
1	VAL_{075}:NT.-position 5937 : VAL→ILE	N274
1	SER_{138}:NT.-position 6127 : SER→TYR	N274
13	SER_{138}:NT.-position 6127 : SER→PHE	N2, N27, N119, N141, N168, N178, N192, N202, N234, N251, N254, N259, N277
6(8)	SER_{148}:NT.-position 6157 :SER→PHE	N9, N22, N67, N72, N94, N145, Ni104; D

[a] n : number of independent observations

[b] position : position of amino acid in the putative translation product and of mutated nucleotide on the TMV genome

[c] identification code number : designates stock identification with prefix *N* for members of the "single-hit" mutant collection at Stuttgart University; *Ni*: member of the former mutant collection of the Max-Planck-Institute for Biology in Tübingen (now in stock in Stuttgart); *D*: see Knorr and Dawson (1988). Most isolates are single-hit mutants; double-hit mutants are identified by a dotted box, the (one) triple mutant by a solid box.

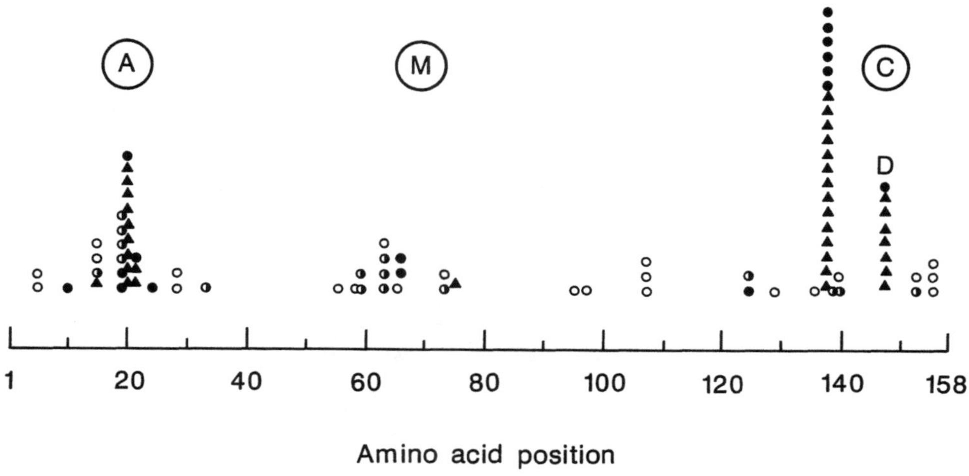

Figure 2. Location of the "hot regions" and amino acid positions in the polypeptide chain of TMV capsid protein, in relation to pathogenicity types of TMV on *N'* gene tobacco. The parent isolate caused dark green/light green systemic mosaic. ▲ indicates local lesion mutants obtained under single-hit conditions of mutagenesis, where the mutated locus was identified by nucleotide sequence analysis. ● indicates local lesion mutants obtained after multi-hit mutagenesis, and identified by amino acid sequence analysis. **o** and o represent multihit mutants identified by amino acid sequence analysis, giving yellow mosaic and mild green mosaic respectively. Data for ●, **o** and o are from Wittmann (1962, 1964), Wittmann and Wittmann-Liebold (1963) and Wittmann-Liebold *et al.* (1965). D is the mutant reported by Knorr and Dawson (1988).

respect to the peptide structure of the p30, and 29 of them were wild type in respect to the nucleotide sequence of their p30 genes. However, virtually all of these 33 local lesion mutants had suffered amino acid substitutions in the capsid protein (Table 1). Twenty nine mutants proved to be single-hit mutants, three had received two hits, and one was a triple-hit mutant. One of the two-hit mutants was found to have an amino acid exchange in its p30 (see above).

All mutations were found to be distributed over six of the 158 amino acid positions of the capsid peptide (positions 15, 20, 21, 75, 138 and 148). All except one (at position 75) were concentrated into two "hot regions" (Fig. 2).

To check whether these mutable positions on TMV RNA were not just generally selected during nitrous acid incubation, these findings were checked against earlier data from the literature on nitrous acid-induced amino acid substitutions in the coat protein of TMV (Wittmann 1962, 1964; Wittmann-Liebold and Wittmann, 1963; Wittmann-

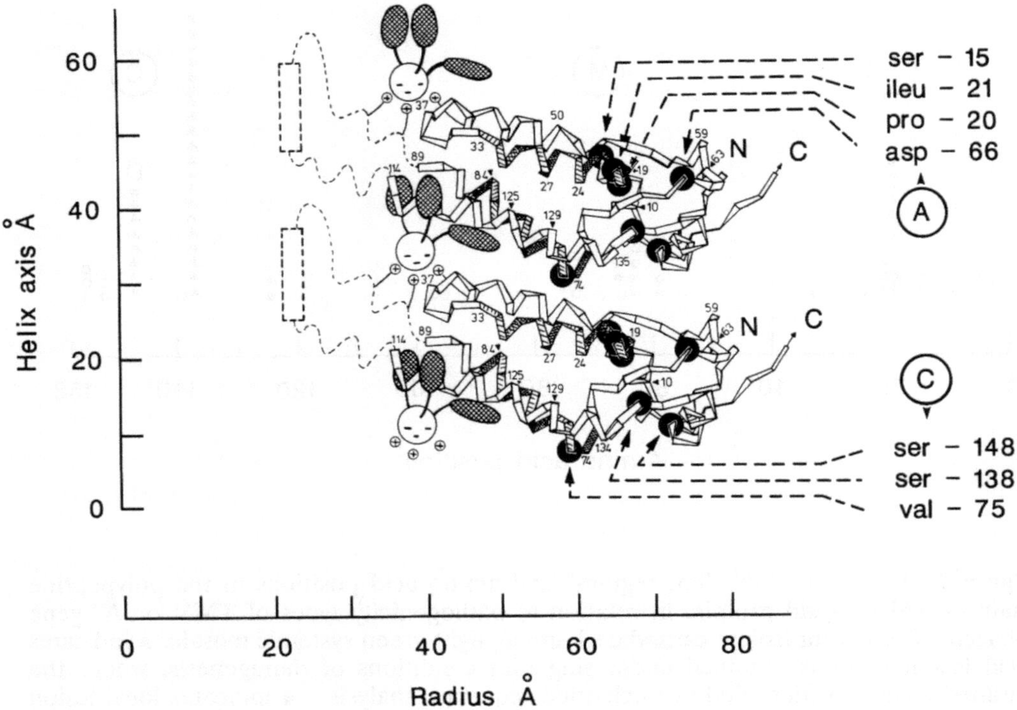

Figure 3. Amino acid positions within the 3-dimensional structure of TMV-vulgare wild type capsid protein (large black discs) where substitutions were found to yield the local-lesion pathotype able to trigger a hypersensitive response on *N'*-gene tobacco. The structure of two subunits of TMV capsid protein was redrawn from Bloomer *et al.* (1978) with the left slew and the right radial helices (positions 20 - 33, and 73 - 85, respectively) indicated by shading and the bases binding adjoining capsid protein subunits indicated by hatching. Hot regions A and C are as explained in Fig. 2 and the text.

Liebold *et al.,* 1965). Though all of these earlier results were obtained with "multiple-hit" conditions of HNO_2 treatment, it is clear from Fig. 2 that positions accessible to any nitrous acid mutagenesis are found all over the capsid protein gene but that mutations leading to local-lesion phenotypes are clustered (Fig. 2) within two (or perhaps three) "hot regions" in the primary structure of the polypeptide. One of these ("*A*") is situated close to the *amino* terminus and includes positions 15, 20 and 21; the other ("*C*") is close to the *carboxyl* terminus with the positions 138 and 148. There seems to be a less predominant third "hot region" ("*M*") near the *middle* of the primary structure of the capsid polypeptide with position 75 and - when the older multiple-hit mutants are included - positions 66 and 73. The other multiple hit mutants exhibiting the local lesion phenotype have amino acid substitutions within the coat protein in rather close vicinity to the two main hot regions *A* and *C* as they are deduced from the present single-hit experiments (positions 19, 21, 24, 125, 138). Knorr and Dawson (1988) have reported a local lesion phenotype due to a ser to phe replacement at position 148. This same amino acid exchange has been identified in the present study for seven independent cases (see Table 1 and Fig. 2).

Both hot regions *A* and *C* are situated at a radial distance from the virion axis of about 70 A and hence are not far from the surface of the virion. This appears from the tertiary structure of the protein (Fig. 3) (Bloomer *et al.,* 1978; Namba and Stubbs, 1986). Both regions seem accessible for interaction with the environment, which includes interactions with host components. Two facts are particularly interesting: hot regions *A* and *C* come close to each other, and hot region *M* perhaps "disappears" as such, when the capsid peptide assumes its tertiary structure, because position 66 may come close to hot region *A* (around position 20) while position 75 comes close to hot region *C* around position 148 (G Stubbs, personal communication).

A closer look at the tertiary structure of the capsid protein reveals further interesting details concerning the location of the "hot regions" and other features within their intramolecular vicinity. Thus with the proline at position 20 (pro_{20}) the left-slew helix starts; at ala_{74} the right radial helix begins, and the distal end of the left radial helix is at arg_{134}.

Putting everything together one could envisage recognition of *pathogenic "domains"* which arise when certain details of the wild type conformation of TMV capsid protein are disturbed. These "domains" are situated close to - though somewhat hidden from - the surface of the TMV particle. How they are recognized by the host and which details of their structure define the signalling properties which trigger the hypersensitive response of the host, and how virus localization is then effected, are fascinating problems for further study in molecular plant pathology.

Which genes of TMV trigger necrogenesis in N'-*gene tobacco?*

The findings reported here, old and new, seem to be at least partly in contradiction. Involvement of the p30 transport protein is indicated from stripping mapping of local lesion mutants, but when such mutants are isolated, purified, propagated and the nucleotide sequence of their p30 genes established, practically all of them turned out to be wild type in the p30 polypeptide. The only exception to this had, in common with all other mutants, an amino acid substitution in the coat protein. Its capsid protein had suffered a replacement of ser by phe in position 148, which had already been shown by others (and additionally in seven independent new cases in the present paper) to be one of the possible determinants for necrogenicity. A further 27 cases of amino acid substitutions regarding five other positions in the capsid polypeptide have been found. A total of 14 variants with a substitution of ser_{138} have arisen. All these mutants have in common the ability to trigger (or inability to avoid) local lesion formation on *N'*-gene hosts. All mutations are grouped together in a manner suggesting the existence of perhaps two pathogenic domains on the surface of the virus particle.

In trying to resolve this dilemma, one should recall our reproducible early finding, that about 90% of all local lesion mutants originally detectable on leaves of *N'*-gene tobaccos were lost during consecutive local lesion transfers on *N. tabacum* cv. Xanthi nc (or Samsun *NN*) when their biological purification was attempted (Ellwart-Tschürtz, 1987). It seems reasonable to speculate that they were lost because they lacked the transport function. It has been shown by Atabekov and coworkers that missing or malfunctioning transport proteins may be functionally complemented by transport functions not only from related (Dorokhov *et al.,* 1984) but also from unrelated viruses (Taliansky *et al.,* 1982). It has also been shown that transgenic tobacco plants expressing the wild type p30 of TMV support systemic spread of temperature-sensitive transport mutants of TMV at non-permissive temperatures when mutated transport protein is not functioning (Deom *et al.,* 1987).

Hence one may assume that artificial mutants of TMV with defective transport functions will remain detectable as long as they are plated in the presence of a sufficient surplus of virions with functional (i.e. wild type) p30. This is exactly what happens when a type strain of TMV (vulgare or U1) which has been treated with a low dose of mutagen is first tested on leaves of an *N'*-gene host. But when it is attempted to free these mutants from accompanying wild type helper virus by successive single lesion transfers they are likely to get lost because of eventual restriction to a subliminal (single cell) infection. However, mutants which are necrotic because of some other features which are critical for their interaction with a potentially hypersensitive host can be isolated, propagated, and analyzed as long as their p30 is functioning. Thus only one of possibly two classes of mutants were detected: those which have suffered changes in "pathogenic domains" in their capsid protein which are involved in triggering hypersensitivity. An extremely powerful but unanticipated selection pressure may have excluded p30 mutants which trigger hypersensitivity in the type of experiment used.

CONCLUDING REMARKS

Though final proof is still required we tend to conclude that necrogenesis may be triggered by viral mutants in different ways, or may be avoided, or suppressed by wild type TMV in multiple ways. For TMV and *N'*-gene hosts malfunctional p30 may be one and critical surface structures of the viral capsid may constitute another signal. It appears that TMV is virulent to (i.e. able to infect systemically) *N'*-gene hosts only when certain wild type features of these two proteins are maintained and critical "signal properties" triggering a hypersensitive response are avoided. That "signalling properties" triggering the hypersensitive response can be ascribed to certain surface properties of the TMV capsid has been shown in this study. Moreover, because silent mutations have not been found so far within the coat protein gene of our single-hit local-lesion mutants, and only three times within the p30 gene among 33 mutants sequenced, a direct involvement of a peculiar structure of the viral RNA in triggering the hypersensitive host response in *N'*-gene tobaccos seems thus far to be excluded.

How and with which cellular components the host recognizes these signals and responds to them by triggering the protective "suicide for survival" resulting in localized necrogenesis, and how wild type virus manages to escape localization, remain major problems which will be further investigated.

ACKNOWLEDGEMENTS

This work was supported by the Deutsche Forschungsgemeinschaft (grants Mu 124/7-1;2;3).

REFERENCES

Bloomer AC, Champness JN, Bricogne G, Butler PJG, Klug A (1978) Protein disk of tobacco mosaic virus at 2.8 A resolution showing the interactions within and between subunits. Nature 276:362-368

Deom CM, Shaw MJ, Beachy RN (1987) The 30-kilodalton gene product of tobacco mosaic virus potentiates virus movement. Science 237:389-394

Dorokhov YL, Alexandrova NM, Miroshnichenko NA, Atabekov JG (1984) The informosome-like virus-specific ribonucleoprotein (vRNP) may be involved in the transport of tobacco mosaic virus infection. Virology 137:127-134

Ellwart-Tschürtz M (1987) Charakterisierung der p30-Genprodukte nitrit-induzierter Lokalläsions-Mutanten des Tabakmosaik-Virus. PhD thesis, University of Stuttgart, FRG

Gierer A, Mundry K-W (1958) Production of mutants of tobacco mosaic virus by chemical alteration of its ribonucleic acid *in vitro*. Nature 182:1457-1458

Goelet P, Lomonossoff GP, Butler PJG, Akam ME, Gait MJ, Karn J (1982) Nucleotide sequence of tobacco mosaic virus RNA. Proc Natl Acad Sci USA 79:5818-5822

Hull R (1989) The movement of viruses in plants. Annu Rev Phytopathol 27:213-240

Kado C, Knight CA (1966) Location of a local lesion gene in tobacco mosaic virus RNA. Proc Natl Acad Sci USA 55:1276-1283

Knorr DA, Dawson WO (1988) A point mutation in the tobacco mosaic virus capsid protein gene induces hypersensitivity in *Nicotiana sylvestris*. Proc Natl Acad Sci USA 85:170-174

Leonard DA, Zaitlin M (1982) A temperature sensitive strain of tobacco mosaic virus defective in cell-to-cell movement generates an altered virus-coded protein. Virology 117:416-424

Meshi T, Watanabe Y, Saito T, Sugimoto A, Maeda T, Okada Y (1987) Function of the 30kd protein of tobacco mosaic virus: involvement in cell-to-cell movement and dispensibility for replication. EMBO J 67:1257-1271

Mundry K-W (1960) Mutationsuntersuchungen am Tabakmosaikvirus *in vitro*. I. Die Abhängigkeit des Erfolgs der Mutagenese mit HNO_2 von der Mutationsrichtung. Z Vererbungsl 91:81-86

Mundry K-W, Gierer A (1958) Die Erzeugung von Mutationen des Tabakmosaikvirus durch chemische Behandlung seiner Nukleinsäure *in vitro*. Z Vererbungsl 89:614-630

Namba K, Stubbs G (1986) Structure of tobacco mosaic virus at 3.6 A resolution: implications for assembly. Science 231:1401-1406

Otsuki Y, Shimomura T, Takebe I (1972) Tobacco mosaic virus multiplication and expression of the N gene in necrotic responding tobacco varieties. Virology 50:45-50

Taliansky ME, Atabekova TI, Kaplan IB, Morozov SY, Malyshenko SI, Atabekov JG (1982) A study of TMV ts mutant NI 2519. I. Complementation studies. Virology 118:301-308

Van Loon LC (1987) Disease induction by plant viruses. Adv Virus Res 33:205-255

Wilson TMA, Perham RN, Finch JT, Butler PJG (1976) Polarity of the RNA in the tobacco mosaic virus particle and the direction of protein stripping in sodium dodecyl sulphate. FEBS Lett 64:285-289

Wittmann HG (1962) Proteinuntersuchungen an Mutanten des Tabakmosaikvirus als Beitrag zum Problem des genetischen Codes. Z Vererbungsl 93:491-530

Wittmann HG (1964) Proteinanalysen von chemisch induzierten Mutanten des Tabakmosaikvirus. Z Vererbungsl 95:333-344

Wittmann HG, Wittman-Liebold B (1963) Tobacco mosaic virus mutants and the genetic coding problem. Cold Spring Harbor Symp Quant Biol 28:589-595
Wittmann-Liebold B, Jauregui-Adell J, Wittmann HG (1965) Die primäre Proteinstructur temperatursensitivier Mutanten des Tabakmosaikvirus. II. Chemisch induzierte Mutanten. Z Naturforsch Teil B 20:1235-1249

INDUCTION OF PLANT GENES BY COMPATIBLE AND INCOMPATIBLE VIRUS-PLANT INTERACTIONS

Huub JM Linthorst, Ben JC Cornelissen[1], Jan AL van Kan, Miranda van de Rhee, Ralph LJ Meuwissen, Maite T González Jaén and John F Bol
Gorlaeus Laboratories
Leiden University
Einsteinweg 5
2333 CC Leiden
The Netherlands

ABSTRACT

cDNAs were cloned to one host mRNA induced by a compatible virus-plant interaction and several host mRNAs induced by an incompatible virus-plant interaction. Regulatory sequences in two of the corresponding genes were analysed and three of the inducible genes were constitutively expressed in transgenic tobacco to test a putative role of the encoded proteins in defence mechanisms.

INTRODUCTION

Because of their limited genetic information, plant viruses rely heavily for their replication on components of the host cell. Host ribosomes are responsible for translation of viral RNAs but may also be involved in co-translational disassembly of some viruses (Wilson, 1985). The role of host proteins in viral RNA replication has been documented in most detail for turnip yellow mosaic virus (TYMV). In addition to a virus-specific 115 kDa protein, the purified TYMV replicase was found to contain a 45 kDa subunit of host origin (Candresse *et al.*, 1986).

The interplay between host and virus results in the development of symptoms that can range from a mild chlorosis with no major alteration in cellular metabolism (compatible interaction) to severe necrosis and cell death (incompatible interaction).

[1]MOGEN Int., Einsteinweg 97, 2333 CB Leiden, The Netherlands

NATO ASI Series, Vol H 41
Recognition and Response in Plant-Virus Interactions
Edited by R.S S Fraser

The first type of interaction usually results in high levels of virus multiplication and sometimes several grams of virus can be isolated from a kilogram of infected leaf material. It could be surmized that this massive replication would require the amplification of host components that are necessary for this process. However, little is known about the induction of host genes during a compatible virus-plant interaction. The activity of a host enzyme which *in vitro* acts as an RNA-dependent RNA-polymerase (RdRp) is known to increase after the systemic invasion of plants by viruses (see Fraenkel-Conrat, 1983). Available evidence indicates that this enzyme is not involved in virus replication (Dorssers *et al.*, 1983). The role of this enzyme in healthy and diseased plants is still enigmatic.

In contrast to the compatible interaction, the incompatible interaction between virus and plant is known to trigger the induction of dozens of host genes. These include genes encoding enzymes for the aromatic biosynthetic pathway, cell wall proteins and the "pathogenesis-related" (PR) proteins (for recent reviews see Bol and Van Kan, 1988; Van Loon, 1988; and Bol *et al.*, 1989). The accumulation of these proteins is parallelled by the induction of a systemic resistance of the plant against a broad range of pathogens. Available evidence on the function of plant proteins that are induced by the hypersensitive response to virus infection supports the notion that these proteins are involved in various defence mechanisms. Studies on genes encoding enzymes from the phenylpropanoid pathway and hydroxyproline-rich cell wall proteins have been carried out mainly with plants reacting hypersensitively to fungal infection or with plant cells treated with elicitors.

Although these genes are also induced by the incompatible interaction of plants with viruses (Van Loon, 1982; Memelink, 1988), studies on virus-inducible plant genes have mainly focussed on the genes encoding PR proteins. These proteins are characterized by their low isoelectric point, their resistance to proteases and their extracellular location. Initially, these proteins accumulate around local lesions in the inoculated leaf but after several days they are also systemically induced in the virus-free parts of the plant. Interestingly, the host RdRp is systemically induced both by compatible and incompatible interactions of the plant with viruses (JAL van Kan, unpublished results) and by wounding of leaves (Zaitlin and Hull, 1987). Thus, it could be that the induction pathways activated by compatible and incompatible virus-plant interactions partially overlap.

We have studied the induction of host mRNAs in virus-infected plants by a differential screening of cDNA libraries with probes made to poly(A) RNA from either infected or healthy leaves. Here we report on the cloning of mRNAs induced by the compatible interaction of cucumber with cucumber mosaic virus (CMV) and the incompatible interaction of Samsun *NN* tobacco to tobacco mosaic virus (TMV). The tobacco cDNA clones have been used as probes to isolate the corresponding genes from a Samsun *NN* genomic library. These genomic clones were used to study regulatory sequences involved in the induction of the genes by TMV infection or by abiotic elicitors, and to construct transgenic tobacco plants that constitutively express virus-inducible proteins. The role of these proteins in possible defence mechanisms has been tested by assaying the transgenic plants for their susceptibility to virus infection and insect attack.

HOST GENES INDUCED BY A COMPATIBLE VIRUS-PLANT INTERACTION

The infection of cucumber with CMV has been reported to result in a 50- to 100-fold increase in the activity of the host RdRp (Kumarasamy and Symons, 1979). In an attempt to clone the mRNA of this 110 kDa protein, we made a cDNA library to poly(A) RNA which had been isolated from primarily-infected leaves of cucumber, seven days after inoculation. The poly(A) RNA was size-fractionated by sucrose-gradient centrifugation in order to enrich for mRNA larger than 2500 nt. An extensive screening of the cDNA library by a differential hybridization procedure yielded one single clone corresponding to a CMV-induced host mRNA (Van Kan, 1988). Fig. 1 shows a Northern blot probed with this clone, which was called HL5. The clone did not hybridize to CMV RNAs (lane 1) or RNAs from healthy cucumber (lane 2) but hybridized strongly to a 0.9 kb mRNA present in CMV-infected cucumber (lane 3). In view of its size, this mRNA cannot encode the host RdRp. Possibly, it is an abundant mRNA in infected plants, explaining its occurrence in the size-selected RNA fraction that was used as a template for cDNA cloning.

Clone HL5 contained an incomplete copy of the 0.9 kb mRNA. Sequence studies showed that the insert was 343 nucleotides long and did not contain a poly(A) tail. The plus-strand in the insert was identified by hybridizing strand-specific probes, subcloned in M13, to the blot of Fig. 1. The plus-strand was found to contain two overlapping reading frames represented by the 5'-terminal 231 and 261 nucleotides of the insert. The

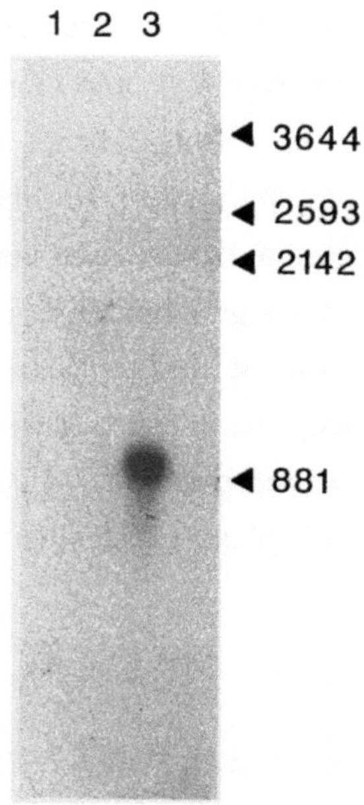

Figure 1. Induction of host mRNA after infection of cucumber with CMV. A Northern blot was run with 2 μg of CMV RNA (lane 1), 2.5 μg of poly(A)-RNA from healthy cucumber plants (lane 2), or 2.5 μg of poly(A)-RNA from CMV-infected cucumber plants (lane 3). The blot was probed with ^{32}P-labelled DNA of clone HL5. The length of size markers (AlMV RNAs) is given in nucleotides in the right margin.

encoded polypeptide sequences did not show any homology to protein sequences stored in the GENBANK or EMBL databanks.

To our knowledge, these results represent the first demonstration of the induction of a host mRNA by a compatible virus-plant interaction. The function of the induction of the 0.9 kb mRNA after infection of cucumber with CMV, if any, is as yet unknown.

HOST GENES INDUCED BY AN INCOMPATIBLE VIRUS-PLANT INTERACTION

The approach used to analyse a compatible interaction, as described in the previous paragraph, has also been applied to Samsun *NN* tobacco plants reacting hypersensitively to TMV- infection. Differential screening of a cDNA library made to poly(A) RNA from infected tobacco yielded groups of clones corresponding to seven

TMV-induced host mRNAs. These groups were called "clusters A to G" (Hooft van Huijsduijnen *et al.*, 1986b; Cornelissen *et al.*, 1987). Three of these clusters were found to correspond to known acidic PR proteins of tobacco. A classification of tobacco PRs into five groups has been proposed by Van Loon *et al.* (1987). Group 1 contains PR proteins 1a, 1b and 1c, all of unknown function. Group 2 contains PR proteins 2, N and O, which were shown to be ß-1,3-glucanases (Kauffmann *et al.*, 1987). Group 3 consists of PR proteins P and Q, which were found to have chitinase activity (Legrand *et al.*, 1987). Group 4 contains PR-R of unknown function. Finally, group 5 contains one or more proteins collectively known as PR-S in the nomenclature of Van Loon *et al.* (1987), which show homology to the sweet-tasting protein thaumatin and a bifunctional maize protein that inhibits α-amylase and proteases in the digestive tract of insects (Cornelissen *et al.*, 1986; Pierpoint *et al.*, 1987; Richardson *et al.*, 1987). Available evidence suggests that the homology between proteins within one group is about 90%. In addition to these acidic proteins, basic homologues with an approximately 65% homology to their acidic counterparts have been observed for proteins of groups 1, 2, 3 and 5 (for references see Bol *et al.*, 1989). In contrast to the acidic PRs, the basic isoforms accumulate largely intracellularly (Bol *et al.*, 1989) and are constitutively expressed in the roots of healthy plants (Memelink, 1988).

Our cDNA clones were found to correspond to acidic and basic PR-1 proteins (clusters B and G, respectively), acidic and basic chitinases (clusters D and F, respectively) and the acidic thaumatin-like protein PR-S (cluster E) (Hooft van Huijsduijnen *et al.*, 1986b; Hooft van Huijsduijnen *et al.*, 1987; Cornelissen *et al.*, 1986). cDNA clones from clusters A and C corresponded to 40 kDa and 12 kDa proteins that did not react with an antiserum to PR proteins. The 12 kDa protein was found to be a glycine-rich protein (GRP) of 109 amino acids including a putative signal peptide of 26 amino acids (Van Kan *et al.*, 1988). By analogy to other stress-induced glycine-rich proteins the GRP may be a cell wall component (Keller *et al.*, 1988).

STRUCTURE OF VIRUS-INDUCIBLE GENES

The seven clusters of cDNA clones isolated in our laboratory have been used as probes to estimate the number of corresponding genes from genomic blots of Samsun *NN* DNA, and to isolate corresponding genes from a genomic library of Samsun *NN* tobacco. At present genes encoding acidic PR-1, GRP and acidic PR-S proteins have been completely sequenced. Fig. 2 shows a schematic drawing of a representative gene from each gene family.

The tobacco genome was estimated to contain about eight genes for acidic PR-1 proteins, three of which are expressed after TMV infection. Of the three PR-1 genes that were cloned and sequenced, one was found to encode PR protein 1a whereas the two others probably represent pseudo genes (Cornelissen *et al.,* 1987). There are also approximately eight GRP genes in the tobacco genome, at least half of which are expressed after TMV infection. One expressed gene (GRP-8, Fig. 2) with an intron of 555 bp and one putatively silent gene with an intron of 1954 bp were sequenced (Van Kan *et al.*, 1988). The PR-S genomic clones were found to correspond to two homologous genes called E2 and E22 (Van Kan *et al.*, 1989). Genomic blots indicate that there may be no additional genes for acidic thaumatin-like proteins in the tobacco genome. Both PR-S genes are expressed after TMV- infection as corresponding mRNAs and proteins have been identified (Cornelissen *et al.*, 1986; Pierpoint *et al.*, 1987; Payne *et al.*, 1988).

REGULATORY SEQUENCES OF VIRUS-INDUCIBLE GENES

In primary inoculated leaves the mRNAs corresponding to cDNAs A to F reached maximum levels of accumulation three to four days after TMV-infection. About eight days after inoculation induction of these mRNAs became detectable in the non-inoculated virus-free leaves of the tobacco plant (Hooft van Huijsduijnen *et al.*, 1986b). This indicates that one or more long-distance signals migrated from the site of infection to the systemically induced leaves. Spraying of tobacco plants with a 5 mM salicylate solution resulted in a rapid induction of the PR-1 and GRP mRNAs whereas there was only a low induction of other PR mRNAs under this condition (Hooft van Huijsduijnen *et al.*, 1986a). The differences in tissue-specific expression of acidic and basic PR proteins and their differential response to various stress conditions (Memelink, 1988) indicate that the promoter regions of PR genes are a complex mosaic of cis-acting elements reacting with different trans-acting factors. A comparison of the flanking sequences of the PR-1, GRP and PR-S genes showed only limited homologies within the first 80 bp of the upstream sequences (Van Kan *et al.,* 1989).

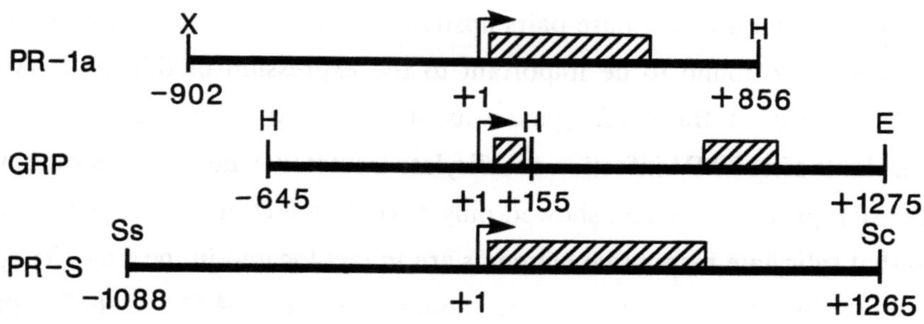

Figure 2. Schematic representation of the PR-1a gene, the GRP-8 gene and the PR-S E22 gene. The transcription start site is numbered as +1. Hatched boxes show the coding sequences of the genes. Restriction sites: X, *Xho*I; H, *Hind*III; E, *Eco*RI, Ss, *Sst*I; Sc, *Sca*I.

Regulatory elements involved in the induction of PR-1 and GRP genes were analysed by fusing the upstream sequences of these genes to reporter genes. Bal31 digestion starting from the *Hind*III at position +856 was used to remove the coding sequence of the PR-1a gene up to position +29. An *Xho*I site was used to generate a fragment of -902 to +29 of the PR-1a upstream sequence that was fused to the GUS reporter gene. In addition, Bal31 digestion from the *Xho*I site was used to create 5-deletions of the PR-1a promoter fragment ranging from position - 902 to -287. The coding sequence of the GRP gene was removed up to position +8 by Bal31 digestion starting at the *Hind*III site in the intron of GRP-8 at position +155. The upstream sequence of the GRP gene from -645 to +8 and truncated derivatives thereof were fused to the CAT reporter gene. The PR-1a/GUS and the GRP/CAT constructs were used to transform tobacco by an *Agrobacterium tumefaciens*-mediated transformation procedure. Transgenic tobacco plants thus obtained were assayed for the expression of the reporter gene after TMV infection or salicylate treatment of the plants (MD van de Rhee, JAL van Kan, MT González Jaén and JF Bol, manuscript in preparation).

The 902 bp upstream sequence of the PR-1a gene was sufficient to induce the GUS-gene after TMV-infection or salicylate treatment of the transgenic plants. In the infected plants the GUS-gene was induced in the primary inoculated leaves as well as in the virus-free non-inoculated leaves while mock-inoculation had no effect. Reduction of the upstream sequence to 689 bp had no effect on the induction of GUS by TMV or salicylate, whereas a further reduction to 633 bp abolished the induction. Similarly, sequences several hundreds of base pairs upstream of the transcription initiation site of the GRP-gene were found to be important to the expression of this gene. A 645 bp upstream sequence of the GRP gene was sufficient to induce the CAT-gene in transgenic plants after TMV-infection or salicylate treatment whereas constructs with 400 bp of the GRP-promoter or less showed only a weak induction by salicylate. The data indicated that salicylate responsive elements are in fact located in the first 135 bp of the GRP-promoter but that an enhancer-like sequence between -400 and -645 is probably required for an efficient induction of the gene by salicylate (MD van de Rhee, JAL van Kan, MT González Jaén and JF Bol, manuscript in preparation).

SUSCEPTIBILITY TO VIRUS INFECTION AND INSECT ATTACK OF TRANSGENIC PLANTS EXPRESSING PR GENES

Spraying of tobacco plants with salicylate results in an induction of a subset of PR proteins. When these plants are subsequently infected with alfalfa mosaic virus (AlMV) the replication of the virus is reduced by 90% compared to the controls. Moreover, addition of 1 mM of salicylate to the incubation medium was sufficient to inhibit the replication of AlMV in cowpea protoplasts by more than 99% while there was no detectable inhibition of host metabolism (Hooft van Huijsduijnen *et al.*, 1986a). Salicylate or salicylate-induced components interfered with the synthesis of both viral plus- and minus-strand RNA. In tobacco, PR-1 and GRP were strongly induced by salicylate and it is conceivable that these proteins play a role in the salicylate-induced inhibition of virus multiplication. Another tobacco PR protein that has been implicated in a defence mechanism is PR-S. Its homology to a maize inhibitor of digestive enzymes of insects has led to the suggestion that PR-S is involved in a defence against insect pests (Richardson *et al.*, 1987). To test the putative role of PR proteins in defence mechanisms, we have constructed transgenic tobacco plants expressing PR-1, GRP and PR-S genes constitutively. These plants have been assayed for their susceptibility to virus infection and insect attack.

Sequences of the PR-1a gene (nucleotides +30 to +856), the GRP-8 gene (nucleotides -14 to +1275) and the PR-S E2 gene (nucleotides +31 to +885) were fused to the CaMV 35S promoter and these constructs were used to transform Samsun *NN* tobacco by an *Agrobacterium tumefaciens*-mediated transformation procedure. Although

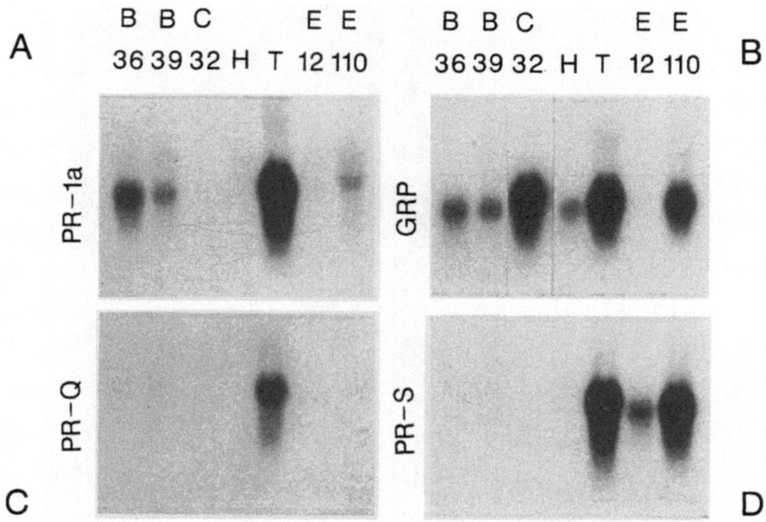

Figure 3. Expression of chimeric genes in transgenic tobacco. Four identical Northern blots containing total RNA from healthy (H) and TMV-infected (T) non-transformed tobacco, and transgenic tobacco plants with chimeric genes for PR-1a (B36, B39), GRP (C32) and PR-S (E12, E110), were hybridized with cDNA probes for PR-1a (panel A), GRP (panel B), PR-Q (panel C) and PR-S (panel D). Reprinted from Linthorst *et al.* (1989), with permission.

the various transformants differed considerably in their expression levels, for each construct a number of plants were identified that selectively expressed the inserted genes at levels comparable to those in TMV-infected plants (Linthorst *et al.*, 1989). The Northern blot in Fig. 3 shows the expression at the RNA level in several transgenic plants. Plants B36 and B39 were transformed with the PR-1a gene, plant C32 was transformed with the GRP gene and plants E12 and E110 were transformed with the PR-S gene. Lanes H and T were loaded with RNA from healthy and TMV-infected tobacco, respectively. Panels A, B and D were probed with cDNA to PR-1a, GRP and PR-S, respectively. As a control, panel C was probed with cDNA to the acidic chitinase PR-Q. Compared to the non-transformed healthy plants (lanes H), plant E110 showed an increased level of PR-1 and GRP mRNA which must have been expressed from the corresponding endogenous genes. The other transgenic plants selectively expressed the gene that was put under the control of the 35S promoter. The plants were also analysed by Western blotting for expression at the protein level. Plants expressing PR-1, GRP and

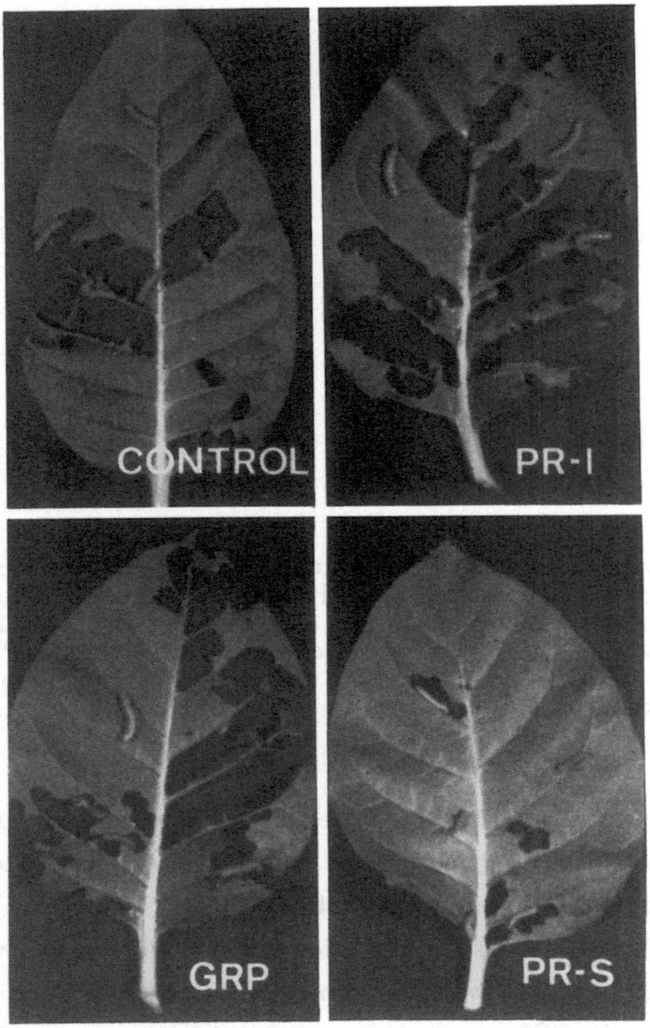

Figure 4. Susceptibility of transgenic tobacco plants to insect attack. Leaves from non-transformed tobacco plants (control) and transgenic plants expressing PR-1a (plant B21), GRP (plant C32) and PR-S (plant E12) were fed to larvae of *Spodoptera exigua.*

PR-S at levels comparable to that of TMV-infected plants were tested for a possible resistance to pathogens (i.e. plants C32, E12 and plant B21, not shown in Fig. 3).

The PR-1a, GRP and PR-S expressing plants were infected with TMV and AlMV. With TMV no significant difference in the number and size of the lesions was observed compared to control plants. Moreover, there was no detectable inhibition of the replication of AlMV in the transgenic plants (Linthorst *et al.*, 1989). Fig. 4 shows the result of an experiment in which leaves of the transgenic plants were fed to larvae of *Spodoptera exigua*. Leaves of the transgenic plants and non-transformed controls were eaten by the insect larvae at comparable speeds. At the time that the photographs of Fig. 4 were taken, the leaf of the PR-S plant was somewhat less affected but the next day it was completely consumed by the larvae. The results of these assays indicate that PR-1a, GRP and PR-S alone are not able to confer resistance in the plants to virus infection or insect attack.

CONCLUSIONS

Compared to the incompatible virus-plant interaction, relatively few efforts have been made to identify host genes that are induced by a compatible virus-plant interaction. Our demonstration that the systemic infection of cucumber with CMV induces the synthesis of a 0.9 kb host mRNA may be a first step in this direction. On the other hand, a number of genes induced by the incompatible virus-plant interaction have now been characterized at the molecular level. The TMV-induced ß-1,3-glucanases and chitinases of tobacco are believed to play a role in the TMV-induced resistance of the plant to fungal infection (Boller, 1988; Gianinazzi, 1983). Our results so far do not support the assumption that PR proteins with an as yet unknown function play a role in the induced resistance to virus infection or in a resistance to insect attack. Further research will be required to identify the components involved in these resistance mechanisms and to shed light on the function of PR-1, GRP and PR-S proteins in diseased plants.

REFERENCES

Bol JF, Van Kan JAL (1988) The synthesis and possible functions of virus-induced proteins in plants. Microbiol Sci 5:47-52

Bol JF, Linthorst HJM, Meuwissen RLJ (1989) Molecular responses of tobacco to virus infection. In: Smith CJ (ed) Biochemistry and molecular biology of plant-pathogen interactions. Oxford University Press, Oxford, in press

Boller T (1988) Ethylene and the regulation of antifungal hydrolases in plants. In: Miflin

BJ (ed) Oxford surveys of plant molecular and cell biology, Vol. 5, Oxford University Press, Oxford, in press

Candresse T, Mouchès C, Bové JM (1986) Characterization of the virus encoded subunit of turnip yellow mosaic virus RNA replicase. Virology 152:322-330

Cornelissen BJC, Hooft van Huijsduijnen RAM, Bol JF (1986) A tobacco mosaic virus-induced tobacco protein is homologous to the sweet-tasting protein thaumatin. Nature 321:531-532

Cornelissen BJC, Horowitz J, Van Kan JAL, Goldberg RB, Bol JF (1987) Structure of tobacco genes encoding pathogenesis-related proteins from the PR-1 group. Nucleic Acids Res 15:6799-6811

Dorssers L, Van der Meer J, Van Kammen A, Zabel P (1983) The cowpea mosaic virus RNA replication complex and the host-encoded RNA-dependent RNA polymerase-template complex are functionally different. Virology 125:155-174

Fraenkel-Conrat H (1983) RNA-dependent RNA polymerases of plants. Proc Natl Acad Sci USA 80:422-424

Gianinazzi S (1983) Genetic and molecular aspects of resistance induced by infection or chemicals. In: Nester EW, Kosuge, T (eds) Plant-microbe interactions; molecular and genetic perspectives, Vol. I. Macmillan, New York, p 321

Hooft van Huijsduijnen RAM, Alblas SW, De Rijk RH, Bol JF (1986a) Induction by salicylic acid of pathogenesis-related proteins and resistance to alfalfa mosaic virus infection in various plant species. J Gen Virol 67:2135-2143

Hooft van Huijsduijnen RAM, Van Loon LC, Bol JF (1986b) cDNA cloning of six mRNAs induced by TMV infection of tobacco and a characterization of their translation products. EMBO J 5:2057-2061

Hooft van Huijsduijnen RAM, Kauffmann S, Brederode FT, Cornelissen BJC, Legrand M, Fritig B, Bol JF (1987) Homology between chitinases that are induced by TMV infection of tobacco. Plant Mol Biol 9: 411-420

Kauffmann S, Legrand M, Geoffroy P, Fritig B (1987) Biological function of "pathogenesis-related" proteins. Four PR-proteins of tobacco have 1,3-ß-glucanase activity. EMBO J 6:3209-3212

Keller B, Sauer N, Lamb CJ (1988) Glycine-rich cell wall proteins in bean: gene structure and association of the protein with the vascular system. EMBO J 7: 3625-3633

Kumarasamy R, Symons RH (1979) Extensive purification of the cucumber mosaic virus-induced RNA replicase. Virology 96:622-632

Legrand M, Kauffmann S, Geoffroy P, Fritig B (1987) Biological function of "pathogenesis-related" proteins: four tobacco PR-proteins are chitinases. Proc Natl Acad Sci USA 84:6750-6754

Linthorst HJM, Meuwissen RLJ, Kauffmann S, Bol JF (1989) Constitutive expression of pathogenesis-related proteins PR-1, GRP and PR-S in tobacco has no effect on virus infection. The Plant Cell 1:285-291

Memelink J (1988) Altered gene expression in T-DNA transformed tobacco tissues. PhD thesis, Leiden University

Payne G, Middlesteadt W, Williams S, Desai N, Parks TD, Dincher S, Carnes M and Ryals J (1988) Isolation and nucleotide sequence of a novel cDNA clone encoding the major form of pathogenesis-related protein R. Plant Mol Biol 11:223-224

Pierpoint WS, Tatham AS, Pappin DJC (1987) Identification of the virus-induced protein of tobacco leaves that resembles the sweet protein thaumatin. Physiol Mol Plant Pathol 31:291-298

Richardson M, Valdes-Rodriguez S, Blanci-Labra A (1987) A possible function for thaumatin and a TMV-induced protein suggested by homology to a maize inhibitor. Nature 327:432-434

Van Kan JAL (1988) Structure and expression of virus-inducible plant genes. PhD thesis, Leiden University

Van Kan JAL, Cornelissen BJC, Bol JF (1988) A virus-inducible tobacco gene encoding a glycine-rich protein shares putative regulatory elements with the ribulose bisphosphate carboxylase small subunit gene. Mol Plant-Microbe Interact 1:107-112

Van Kan JAL, Van de Rhee MD, Zuidema D, Cornelissen BJC, Bol JF (1989) Structure of tobacco genes encoding thaumatin-like proteins. Plant Mol Biol 12:153-155

Van Loon LC (1982) Regulation of changes in proteins and enzymes associated with active defence against virus infection. In: Wood RKS (ed) Active defence mechanisms in plants. Plenum Press, New York, p 247

Van Loon LC (1988) Stress proteins in infected plants. In: Kosuge T, Nester EW (eds) Plant-microbe interactions; molecular and genetic perspectives, Vol. 3. Macmillan, New York, in press

Van Loon LC, Gerritsen YAM, Ritter CE (1987) Identification, purification and characterization of pathogenesis-related proteins from virus-infected Samsun *NN* tobacco leaves. Plant Mol Biol 9:593-609

Wilson TMA (1985) Nucleocapsid disassembly and early gene expression by positive-strand RNA viruses. J Gen Virol 66:1201-1207

Zaitlin M, Hull R (1987) Plant virus-host interactions. Annu Rev Plant Physiol 38:291-315

DEFENCE PROTEINS, GLYCANHYDROLASES AND OLIGOSACCHARIDE SIGNALS IN PLANT-VIRUS INTERACTIONS

B Fritig, S Kauffmann, J Rouster, B Dumas, P Geoffroy, M Kopp and M Legrand
Institut de Biologie Moléculaire des Plantes du CNRS
12 rue du Général Zimmer
67000 Strasbourg
France

ABSTRACT

Most defence proteins and enzymes that are typical of incompatible plant-fungi interactions are also induced or stimulated in hypersensitive plant-virus interactions. Examples include "pathogenesis-related" (PR) proteins, proteases, inhibitors of proteases, hydrolases (chitinases and 1,3-ß-glucanases), enzymes of ethylene biosynthesis, of aromatic metabolism, of phytoalexin biosynthesis and oxidative enzymes. At least 10 PR proteins are glycanhydrolases (4 chitinases and 6 glucanases) with differential specific activity. They are direct anti-microbial defence enzymes and are supposed to cause other defence responses through the release of oligosaccharidic elicitors of pathogen and/or plant origin. Oligosaccharides are shown to protect plants from infection by viruses by a host-mediated process that may not be a classical defence mechanism.

INTRODUCTION

When plants are infected by parasites a variety of situations are encountered, ranging from very weak or no defence to very intense and active defence ultimately leading to resistance. The hypersensitive response is one of the most efficient natural mechanisms of defence that is induced by the infection itself. It has two main characteristics: necrosis at and around each point at which the leaf tissue was infected; and localization of the parasite to the region of each infection initiated. The cells surrounding the necrotic area undergo marked metabolic changes which are believed to cause, or at least to contribute to, the resistance observed.

In this paper we shall compare aspects of the hypersensitive response of plants

NATO ASI Series, Vol H 41
Recognition and Response in Plant-Virus Interactions
Edited by R S S Fraser

to viruses, with features of incompatible interactions between plants and fungi. We shall show that, even though the triggering of the hypersensitivity reaction results from a very specific gene-for-gene recognition between plant and pathogen, what the plant does thereafter to defend itself appears to be unspecific and consists of a set of metabolic alterations that are specific only to the host. This suggests that intermediate signals of host origin must be operating in all cases and that these are responsible for the induction of the metabolic changes.

Hypersensitivity to plant viruses provides a simple model system in which to search for and identify such signals. Their identification as well as the identification of the host enzymes responsible for their release might then lead to strategies that could be used to engineer a given plant and confer resistance, not only to a particular virus, but also to a wide range of pathogens. Thus, in the present paper special emphasis has been devoted to the search for defence-related direct gene products: detection of host proteins highly induced during the defence response and characterization of their biological function.

Since several of these defence proteins have been found to exhibit glycanhydrolase activity, they might be involved in the release of oligosaccharidic signals from the pathogen (in the case of fungi) or from the host. Such signals would necessarily be of host origin in the case of virus infections. Different types of oligosaccharides, known to be elicitors of defence reactions in several plant species, were assayed in plant-virus interactions. Surprisingly, they did not modulate the defence response of hypersensitive reactions, but interfered with the establishment of virus infection, even in the case of systemic non-necrotic infections.

THE ONSET OF RESISTANCE TO VIRUS REPLICATION AND SPREAD IN THE HYPERSENSITIVE REACTION

In most plant-pathogen interactions leading to hypersensitivity (also called incompatibility in the case of fungal and bacterial pathogens), the mechanism of hypersensitivity is believed to consist of recognition at two successive levels of signalling, followed by the induction of metabolic alterations responsible for resistance (Fritig *et al.*, 1987).

Genetic analysis of plant-pathogen interactions has shown that in most cases active defence related to incompatibility is under the control of only one host gene, the resistance gene. A parasite gene, called the "avirulence" gene, corresponds to this resistance gene. An interaction between the products of these genes is thought to trigger the hypersensitivity response, which is usually accompanied by a "stress" reaction of the plant and by major metabolic alterations. No host resistance gene has been characterized yet at the molecular level. Probably the most accessible component of the resistance induction pathway is the parasite's avirulence gene. Very significant progress has been made recently in the identification of a number of bacterial avirulence genes (Staskawicz *et al.*, 1988; Keen and Staskawicz, 1988; reviewed by Vivian, this volume). In the case of viruses, there has also been recent significant progress in the localization on the tobacco mosaic virus (TMV) genome of the avirulence function corresponding to the *N'* gene of tobacco that confers hypersensitive resistance against only a proportion of TMV strains (Saito *et al.*, 1987; Knorr and Dawson, 1988; Culver and Dawson, this volume). The genetics of plant-virus interactions and the avirulence function in viruses are extensively discussed in other chapters.

We investigated the exact time of appearance of resistance to virus multiplication in the hypersensitive combination of *Nicotiana tabacum* cv. Samsun *NN* with TMV. Was the resistance mechanism already operating shortly after virus inoculation, or was it induced later in the infection process? If the latter, did the onset of resistance precede or follow the stress evidenced by necrotic cell death? To answer these questions we followed virus multiplication in the hypersensitive host at intervals after inoculation and compared it with multiplication in a reference system consisting of a systemic infection, by the same TMV-strain (U1), of the almost isogenic line *N. tabacum* cv. Samsun.

In this type of comparison the measurements should be made on the same number of infection sites in the two hosts. However, classical mechanical inoculation creates infection sites at random on the leaves. Therefore, the exact location and number of points of virus entry can never be known for the systemically reacting host and are known only after the appearance of local lesions for the hypersensitively reacting host. To overcome this difficulty we used a microinoculation procedure (Konate and Fritig, 1984) which yielded infection sites at predetermined locations on the leaves of the two hosts.

At various intervals after the microinoculations the individual infection sites were assayed for virus content by an enzyme-linked immunosorbent assay (ELISA). We found large fluctuations in virus content from one infection site to another, even on the same leaf, and whatever the host or the infection period. In our comparative studies of virus multiplication we therefore analysed more than 3000 individual infection sites. This enabled us to express the results as the frequency of sites containing a given amount of virus. To summarize the results (which will be described in detail elsewhere), up to 30 hours after microinoculation, the frequency distribution curves obtained for the cultivars Samsun and Samsun *NN* were very similar. These results resemble those mentioned by Fraser (1988) for the accumulation of viral RNA in the case of leaves of White Burley tobacco inoculated either by a necrotic (avirulent) virus strain of TMV, forming local lesions, or by a systemically-spreading strain (virulent). In our system where one virus strain was inoculated to two genetically related tobacco lines, we observed that at the time (33-36 hours after inoculation) when visible necrosis appeared, the rates of virus multiplication did not differ significantly between the two hosts. This observation using individual infection sites confirmed our previous results, obtained by assaying whole leaves inoculated by a classical mechanical procedure (Konate *et al.*, 1982).

Once the lesions were visible on the Samsun *NN* leaves, the virus titres continued to increase in the infection sites of this host but at a progressively reducing rate compared to that in Samsun infection sites. The differences between the rates of virus multiplication were demonstrated by injecting radioactive uridine around all infection sites at the end of the infection period and assaying them individually for incorporation of label into the viral RNA (Konate and Fritig, 1983). The relative differences in the rates of virus multiplication during the last few hours of a two-day infection period were much greater than the relative differences between the amount of virus accumulated since the beginning of this infection period.

We therefore concluded that hypersensitive resistance to virus infection does not preexist but is induced by the infection itself. Hypersensitive resistance to virus multiplication appears at about the same time as the necrotic symptoms, at least in the Samsun *NN*-TMV interaction, and its efficiency increases with time after infection. Does this progressively increased resistance to virus multiplication result from the production and activity of specific anti-viral factors? This possibility will be reviewed elsewhere in this book. Another possibility, also discussed in this volume, would be an inhibition of cell-to-cell movement of the infection by interaction between altered host metabolism and the normal functioning of the virus-coded movement protein. Altered host metabolism during the hypersensitive reaction occurs preferentially in a ring of cells (1.0 - 1.5 mm in width in the case of tobacco Samsun *NN* leaves) surrounding the necrotic lesions.

METABOLIC CHANGES CHARACTERISTIC OF INCOMPATIBLE (HYPERSENSITIVE) PLANT-PATHOGEN INTERACTIONS

The recognition between the resistance gene and the avirulence gene products apparently induces a stress reaction (necrosis) itself associated with the induction of intense metabolic alterations believed to contribute to the resistance observed. Those metabolic alterations that have been most frequently reported in relation to active defence during incompatible plant-pathogen interactions have been reviewed in the last decade by several authors (Bailey and Deverall, 1983; Darvill and Albersheim, 1984; Fraser, 1985; Goodman *et al* 1986; Ponz and Bruening, 1986; Collinge and Slusarenko, 1987; Fritig *et al.*, 1987; Lamb *et al.*, 1989).

The best known and most studied host response is undoubtedly the production of phytoalexins. These substances were discovered 45 years ago and are low molecular weight, antimicrobial compounds which plants synthesize and accumulate after exposure to microorganisms (Bailey and Mansfield, 1982; Darvill and Albersheim, 1984). For a long time they were thought to be the major active principles responsible for resistance. Phytoalexins have also been found in plants reacting hypersensitively to viruses (Bailey and Mansfield, 1982), but there is no evidence that any phytoalexin could be directly or indirectly responsible for anti-viral activity. Even in the case of incompatible plant-fungus interactions, and though good correlations have been reported between levels of phytoalexins and the degrees of resistance to some fungi, it is now generally believed that resistance ultimately results not only from phytoalexins but from the superimposition of various other metabolic changes. These include the production of mechanical barriers and defence proteins, including defence enzymes.

Among the mechanical barriers are aromatic macromolecules such as lignins, suberins and other phenolic polymers that are produced in large amounts and incorporated into the host cell wall. Evidence for an involvement of these aromatic polymers in virus localization has been obtained from approaches using competitive inhibitors of an early enzyme in the aromatic pathway (Massala *et al.*, 1980, 1987). Polysaccharides such as callose, a ß-1,3-polyglucan, are also actively synthesized and found in several cellular compartments (cell wall, intercellular spaces, plasmodesmata).

It has been reported that callose surrounds local lesions in some viral infections where it may help to prevent spread of the virus (Shimomura and Dijkstra, 1975; Bell, 1981).

Another biochemical change contributing to mechanical barriers is the increased synthesis of hydroxyproline-rich glycoproteins (HRGPs) which become insoluble soon after their secretion into the cell wall (McNeil *et al.*, 1984; Varner and Lin, 1989). These proteins may function in defence simply by forming a denser, impenetrable cell wall barrier or by providing nucleation sites for the deposition of lignin, again resulting in a more protective barrier against potential pathogens. Since they possess the ability to agglutinate bacteria, these glycoproteins may also function in defence by immobilizing pathogens in the wall. There are not many reports of changes in the production of HRGPs during virus infections, but a slight increase has been found in tobacco reacting hypersensitively to TMV (Mazau and Esquerre-Tugaye, 1986).

The proteins believed to participate in active defence are listed in Table 1. In addition to the cell wall HRGP, there are some other major proteins, the best known being the famous "pathogenesis-related" (PR) proteins. Until recently, no biological function was known for PR proteins despite their characteristic properties and the large quantities sometimes accumulated during active defence. The PR proteins will be discussed in more detail later in this paper.

Proteases are known to be widespread and abundant in plants even in the absence of disease. With one exception to be discussed later, their possible participation in active defence is not well documented. Production of inhibitors of proteases and polygalacturonases enables the plant to limit the action of enzymes that play a key role in the virulence of various pathogens.

Defence enzymes are another important category of host response. They can be subdivided into two classes: enzymes that catalyse the increased production of various metabolites participating in resistance (ethylene, phytoalexins, aromatic compounds, oxidized metabolites); and direct defence enzymes (hydrolases such as chitinases and glucanases). Chitin and glucans are major structural components of many important pests, such as insects and fungi (Bartnicki-Garcia, 1968; Cabib, 1987). The production by the plant of enzymes able to degrade these polymers is therefore of interest to plant pathologists.

The different metabolic changes associated with defence mentioned above may occur sequentially and are not all independent from each other; some are closely related. For instance, some of the phytoalexins are derivatives of aromatic metabolism.

Table 1. *Host-encoded proteins believed to be involved in defence against various pathogens*[a]

HRGP ("hydoxy-proline rich glycoproteins")
PR-proteins ("pathogenesis-related")
Proteases
Inhibitors of proteases
Inhibitors of polygalacturonases
Hydrolases (chitinases, glucanases)
Enzymes of ethylene biosynthesis
Enzymes of aromatic metabolism
Enzymes involved in the biosynthesis of plant antibiotics ("phytoalexins")
Oxidative enzymes (polyphenoloxidases and peroxidases)

[a]Proteins that have been found in the case of hypersensitivity to viruses are underlined with continuous lines; dotted lines indicate preliminary results.

Also, the mechanical barrier provided by the deposition of lignin involves both the production of aromatic metabolites (substituted cinnamyl alcohols) and their polymerization by oxidative enzymes (specific peroxidases).

PR PROTEINS AND THEIR BIOLOGICAL FUNCTIONS

Some of the defence enzymes cited above, for instance those of aromatic metabolism, are highly stimulated during the hypersensitive reaction (Legrand *et al.*, 1976, 1978). But since they exhibit very high specific catalytic activity, they occur in rather low amounts, even in the diseased plants (Dumas *et al.*, 1988).

Until very recently, this contrasted strikingly with the situation of the PR proteins. When plants are infected with viruses, viroids, fungi or bacteria, the development of symptoms is accompanied by the accumulation of soluble host-encoded proteins. Such proteins were first detected in tobacco cultivars showing hypersensitivity to TMV (Van Loon and Van Kammen, 1970; Gianinazzi *et al.*, 1970) but have now been found in about 20 plant species (Van Loon, 1985) in various circumstances. Since their appearance could at first only be related to pathological conditions, they were named "pathogenesis-related" proteins. They have characteristic properties which aid in their detection: they are selectively extractable at low pH; are highly resistant to proteolytic enzymes; localized predominantly in the intercellular spaces; and are easily resolved by electrophoresis on polyacrylamide gels under native conditions. Despite the great wealth of information on their characteristic properties and their occurrence in large amounts under various circumstances, no biological function was found for any of these proteins until long after their discovery.

The highest number of PR proteins was found in solanaceaeous plants such as tomato (Granell *et al.*, 1987) and tobacco (Van Loon, 1985). In Samsun *NN* tobacco, ten major PR proteins could be detected (Van Loon, 1982) and were referred to as PR proteins 1a, 1b, 1c, 2, N, O, P, Q, R and S in order of decreasing mobility on electrophoresis in native gels. The three more mobile PR proteins, 1a, 1b and 1c, have been purified to homogeneity (Antoniw *et al.*, 1980) and shown to be serologically related (Antoniw and White, 1983 and Bol *et al.*, this volume).

Since very little information was available on the physicochemical properties of the other PR proteins, we designed a preparative purification procedure using a combination of conventional and high-performance liquid chromatography. The procedure, which is summarized in Fig. 1, enabled us to purify to homogeneity all known PR proteins (Jamet and Fritig, 1986; Kauffmann, 1988). We also were able to detect new PR proteins with acidic isoelectric points, designated as Q', r_1 and r_2, s_1 and s_2, because of similar migrations upon electrophoresis on native basic gels with the known PR proteins Q, R and S, respectively (Kauffmann, 1988). By using antibodies raised against the proteins we confirmed the serological relationships between proteins 1a, 1b and 1c, and proteins PR-2, -N and -O (Fortin *et al.*, 1985), and demonstrated new groups of serologically-related proteins: PR-P and -Q (Hooft van Huijsduijnen *et al.*, 1987), and PR-R and -S (Kauffmann, 1988).

To try to identify the function of some of these PR-proteins we looked at other metabolic changes during Samsun *NN* tobacco-TMV interaction, namely changes in hydrolases such as chitinases and glucanases. Previous reports have mentioned highly stimulated activities of chitinase in cucumber leaves infected by tobacco necrosis virus

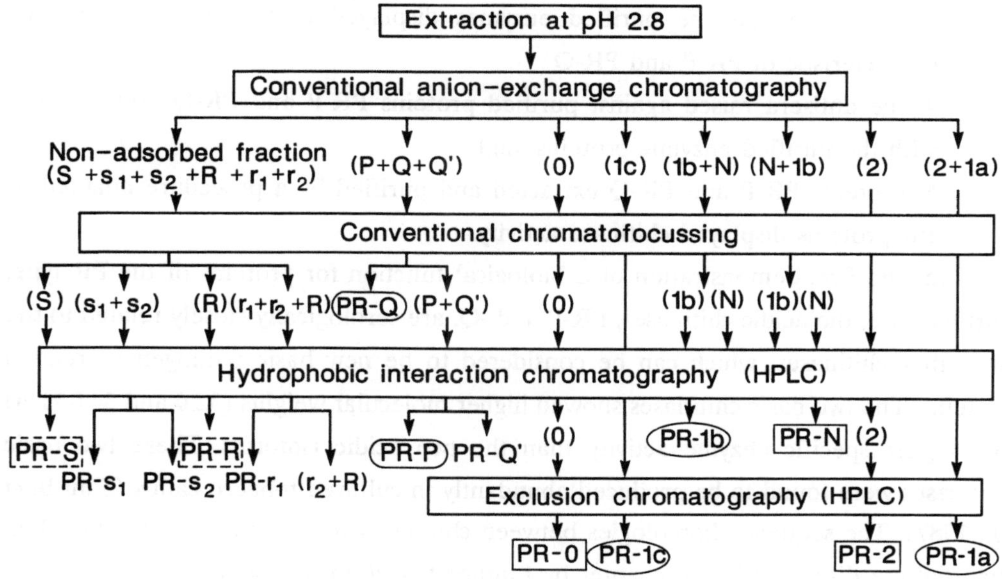

Figure 1. Purification procedure for tobacco acidic PR proteins. Serologically-related proteins are enclosed by boxes of similar shape.

(Métraux and Boller, 1986) and of 1,3-ß-glucanase in *Nicotiana glutinosa* infected by TMV (Moore and Stone, 1972). We found that these enzyme activities were also strongly enhanced in the Samsun *NN*-TMV combination, with time course curves that were strikingly similar to those for the induction of several PR proteins (Legrand *et al.*, 1987). It turned out that four chitinases - two acidic forms with low isoelectric points and two basic ones with high pI values - were produced upon infection. These four enzymic proteins have been purified to homogeneity and characterized. They are separable by electrophoresis under native conditions at suitable pH and have different electrophoretic mobilities on sodium dodecyl sulphate-polyacrylamide gels. Their relative molecular weights were estimated as 27.5 and 28.5 kDa for the acidic isoforms and 32.0 and 34.0 kDa for the basic isoforms.

Several lines of evidence demonstrated that the two acidic chitinases from tobacco were, in fact, proteins PR-P and PR-Q:

1. the acidic chitinases and the two PR proteins displayed the same chromatographic behaviour on various supports
2. the estimated M_r values for the enzymic proteins purified to homogeneity were the same as those measured for the two purified PR proteins

3. on native gels the purified enzymes displayed electrophoretic mobilities characteristic of PR-P and PR-Q

4. the antisera raised against purified proteins PR-P and PR-Q cross-reacted with the purified enzymic proteins, and

5. proteins PR-P and PR-Q extracted and purified by a procedure adapted to PR proteins displayed chitinase activity.

This was the first demonstration of a biological function for proteins of the PR type. Furthermore, the acidic chitinases, PR-P and -Q, are serologically closely related to the two other chitinases, which can be considered to be new basic pathogenesis-related proteins. The two basic chitinases showed higher molecular weights (32.0 and 34.0 kDa) and higher specific enzyme activity than the two acidic isoforms. These two basic chitinases were found to be produced abundantly in cultured tobacco callus (Shinshi *et al.*, 1987). The sequence homologies between chitinases from tobacco and other plant material are discussed in the chapter by Linthorst *et al.* in this book.

A similar approach was used to identify the proteins that had 1,3-ß-glucanase activity. Enzymatic activity was found associated with both basic and acidic proteins. A major basic 1,3-ß-glucanase of molecular weight 33 kDa was identified whose features resemble those described for a 1,3-ß-glucanase produced in tobacco tissues cultured on auxin-containing medium (Felix and Meins, 1985). Among the acidic glucanases, 3 were identified as the well known PR proteins 2, N and O (Kauffmann *et al.*, 1987). Another acidic glucanase was identified as protein Q' that has behaviour typical of a PR protein (Kauffmann, 1988). A sixth glucanase, called O' since it migrates on native gels like PR protein O, is readily detectable in healthy leaves, is less induced upon TMV infection than the other glucanases and PR proteins, is apparently a dimer rather than a monomer, and therefore does not have all of the properties typical of PR proteins. All acidic glucanases, except O', are serologically related to the basic 33.0 kDa glucanase that can be considered as a new tobacco PR protein with a basic isoelectric point.

At almost the same time as several PR proteins induced in tobacco leaves reacting hypersensitively to TMV were identified as chitinases (Legrand *et al.*, 1987) and 1,3-ß-glucanases (Kauffmann *et al.*, 1987), a similar finding was made in potato leaves during the incompatible interaction with *Phytophthora infestans* or after treatment with fungal elicitor (Kombrink *et al.*, 1988).

The biological functions of the other well known PR proteins, 1a, 1b, 1c, R and S are not known yet. The proteins of the PR-1 group have been suspected of involvement in localized and/or systemic resistance to virus infection, but there are also arguments difficult to reconcile with such a rôle (Van Loon, 1985). Recently, transgenic Samsun *NN* tobacco plants were obtained that expressed either PR-1a or PR-S constitutively, and it was shown that their susceptibility to infection by TMV or alfalfa mosaic virus was not altered (Linthorst *et al.*, 1989). Proteins R and S are serologically related (Kauffmann, 1988) and show sequence homology to the sweet-tasting protein thaumatin (Cornelissen *et al.*, 1986; Pierpoint *et al.*, 1987) and to a bifunctional enzyme (amylase and proteinase) inhibitor from maize (Richardson *et al.*, 1987). We found no significant α-amylase or proteinase inhibitor activity in tobacco PR proteins R and S, or in a serologically-related 24.0 kDa basic protein (S Kauffmann, M Legrand, B Fritig, unpublished results).

In tobacco, none of the known PR proteins (Van Loon, 1985) or of the newly identified PR proteins (Kauffmann, 1988) has proteolytic activity. This contrasts with tomato where one of the PR proteins induced by infection with citrus exocortis viroid, P-69, was shown to be an alkaline endoproteinase (Vera and Conejero, 1988). The rôle of P-69 in pathogenesis and defence is still unclear.

Polypeptide inhibitors of proteases are widely distributed in all plant tissues and are thought to have a role in defence against herbivores, since they tend to be active against animal, but not against endogenous plant proteases (Ryan, 1981). Microorganisms produce serine proteinases resembling animal proteinases and, therefore, proteinase inhibitors produced by plants are thought also to be involved in defence against attacking microbial pathogens. Indeed, the fact that inhibitors from plants are able to inhibit proteinases from bacteria and fungi is well documented. In melon plants infected by *Colletotrichum lagenarium*, an increase of activity inhibiting the protease produced by the fungus has been demonstrated (Roby *et al.*, 1987). It is one of the rare reports of an increase of proteinase inhibitory activity in plants infected by pathogens. Concerning virus infections, there is a report of induction of proteinase inhibitor activity in tobacco reacting hypersensitively to TMV (Pierpoint *et al.*, 1981), but this activity does not correspond to PR proteins 1a, 1b and 1c (Pierpoint *et al.*, 1981) or to any of the other well known or recently isolated PR proteins (P Geoffroy, M Legrand, B Fritig, unpublished results) listed in Table 2. The Samsun *NN* tobacco inhibitors of serine proteinases are currently being isolated and their role in active defence will be examined. A protease inhibitor, probably unrelated to the ones mentioned above, is apparently involved in the resistance of cowpea to cowpea mosaic virus. This most likely operates by interfering with the processing of the viral polyprotein (Bruening *et al.*, 1987; Ponz *et al.*, 1988).

GLYCANHYDROLASES AND ACTIVE DEFENCE

The lack of specificity of the metabolic alterations which confer resistance contrasts with the high specificity (gene-for-gene recognition) required to initiate the hypersensitive response. The defence proteins and enzymes that are typical of incompatible interactions between plants and fungi, and which have also been found in hypersensitive plant-virus interactions, are listed in Table 1. As shown in Table 2, a significant proportion of major proteins induced in hypersensitively reacting tobacco are glycanhydrolases, and similar results have been found in potato (Kombrink *et al.*, 1988). The glycanhydrolases may represent very efficient "direct" defence enzymes against fungi and bacteria (Boller *et* al., 1983; Schlumbaum *et al.*, 1986; Métraux *et al.*, 1989) and a combination of chitinases and 1,3-ß-glucanases may increase their efficiency (Mauch *et al.*, 1988).

But in addition to their role as direct anti-microbial enzymes, glycanhydrolases might also be involved in the production of chemical signals of pathogen and/or plant origin. The discrepancy between the very specific initial recognition between plant and pathogen, and the rather unspecific host response, has led many plant pathologists to search for intermediary chemical signals that would mimic incompatible plant-pathogen interactions and induce the same metabolic changes in the absence of any infection. Such regulatory molecules, called "elicitors", were obtained first from various pathogens (for a review see Darvill and Albersheim, 1984) and, later, from plant cell walls (for reviews see Darvill and Albersheim, 1984; McNeil *et al.*, 1984). They are usually carbohydrates or contain carbohydrates. Synergistic effects between elicitors of pathogen or host origin were observed in some cases. All these observations led to a hypothetical scheme involving these regulatory molecules in plant-fungi interactions (Albersheim and Darvill, 1985; Fritig *et al.*, 1987; Lamb *et al.,* 1989; Hahn *et al.,* 1989). The stress response resulting from the gene-for-gene recognition would initiate the defence response by activating host hydrolases. Some of these hydrolases would be able to attack the host cell wall and release endogenous elicitors while others would attack the cell wall of the fungal pathogen and release exogenous elicitors which in turn could

Table 2. *The tobacco PR proteins*

PR proteins	Isoelectric point	Molecular weight[a] (x 10^{-3})	Biological activity
The well-known PR proteins (Van Loon 1982)			
1a	acidic	15.5	?
1b	"	15.5	?
1c	"	15	?
2	"	39.5	endo-ß-1,3 glucanase
N	"	40	endo-ß-1,3 glucanase
O	"	40.5	endo-ß-1,3 glucanase
P	"	28	endo-chitinase
Q	"	29	endo-chitinase
R	"	24	?
S	"	24	?
The newly characterized PR proteins (Kauffmann 1988)			
Q'	acidic	35	endo-ß-1,3 glucanase
s_1	"	14.5	?
s_2	"	13	?
r_1	"	14.5	?
r_2	"	13	?
gluc.b.	basic	33	endo-ß-1,3 glucanase
ch.32	"	32	endo-chitinase
ch.34	"	34	endo-chitinase
24K	"	24	?
16K	"	16	?
45K	"	45	?
O'[b]	acidic	25	endo-ß-1,3 glucanase

[a] determined on SDS - polyacrylamide gels

[b] O' does not have all of the typical properties of PR proteins. It is probably a dimer, and the molecular weight indicated is that of the enzyme subunit.

activate host hydrolases and thus amplify the host response. Furthermore, the pathogens themselves contain hydrolases that are likely to attack the host cell wall and also release endogenous elicitors. Therefore, it appears that a complex network of elicitors of both pathogen and host origin would act with synergistic or cascade effects and that this could explain the amplified efficiency of the defence responses.

OLIGOSACCHARIDE SIGNALS AND PLANT-VIRUS INTERACTIONS

Can carbohydrates interfere with a virus infection? There has been a report (Modderman *et al.*, 1985) suggesting that polysaccharides of plant origin may modulate hypersensitive resistance to virus infection. These authors partially digested isolated cell walls from tobacco leaves with a commercial fungal cellulase. The released carbohydrate material was fractionated by gel filtration and the resulting fractions were injected into the intercellular spaces of tobacco leaves reacting hypersensitively to TMV. A large reduction in mean lesion diameter was observed in the areas injected with carbohydrate-containing fractions.

It is certainly worthwhile to search for putative elicitors that might be involved in hypersensitivity to viruses. Plant viruses do not contain polysaccharidic cell wall material and hydrolases, so the scheme of elicitation (Albersheim and Darvill, 1985; Fritig *et al.*, 1987; Lamb *et al.*, 1989; Hahn *et al.*, 1989) will need to be simplified considerably, since it should involve exclusively elicitors of host origin. The occurrence and activity of such elicitors have been reviewed recently (Ryan 1987, 1988; Hahn *et al* 1989).

We used oligosaccharidic fractions, known to be elicitors of defence reactions, in an attempt to prestimulate typical defence pathways, and to examine whether the pretreatment would result in any interference with a subsequent virus infection. This approach led to very surprising results. A glucan preparation obtained from the mycelial walls of the fungus *Phytophthora megasperma* f. sp. *glycinea* and known as an active elicitor of phytoalexins in soybean was shown to be a very efficient inducer of resistance against viruses in tobacco (Kopp *et al.*, 1989). The glucan preparation protected against mechanically transmitted viral infections on the upper and lower leaf surfaces. Whether the glucan preparation was applied by injection, inoculation, or spraying, it protected the plants if applied before, at the same time as, or not later than 8 hours after virus inoculation. At concentrations ranging from 0.1 to 10 μg/ml, the glucan preparation induced protection ranging from 50 to 100% against both symptom production (necrotic local lesions, necrotic rings, or systemic mosaic) and virus accumulation in all *Nicotiana*-virus combinations examined. However, no significant protection against some of the same viruses was observed in bean or turnip. The host plants successfully protected included *N. tabacum* (9 different cultivars), *N. sylvestris*, *N. glutinosa*, and *N. clevelandii*. The viruses belonged to several taxonomic groups and included TMV, alfalfa mosaic virus, and tomato black ring virus. The glucan preparation did not act directly on the virus and did not interfere with virus disassembly; rather, it appeared to induce changes in the host plant that prevented infections from being initiated or recently established infections from enlarging.

Preliminary experiments aiming to find a relationship between the degree of protection and prestimulated defence reactions led also to surprising results (J Rouster, M Kopp, B Fritig, unpublished results): the glucan-mediated protection did not depend on induction of pathogenesis-related proteins or enzymes of the phenylpropanoid pathway. We believe that the induced protection results from a mechanism that has yet to be elucidated. Other oligosaccharides known as elicitors (Ryan, 1987; 1988) have similar protecting activity against virus infections with perhaps broader host specificities, which appears to be a new biological activity for these signalling molecules (J Rouster, M Kopp, B Fritig, unpublished results).

CONCLUSION

The hypersensitive reaction is switched on by the very specific recognition between the avirulence gene product of the pathogen and the resistance gene product of the host. However, active defence in plants is associated with metabolic alterations that are apparently only host specific and confer a universal type of resistance. It is noteworthy that some of the tobacco PR proteins, which were first detected in viral infections and were thought to be related to antiviral defence, are chitinases and glucanases and are therefore defence enzymes directed against insects and fungi. It follows that controlling the production of regulatory molecules such as elicitors would confer resistance to various pathogens. Hypersensitivity to viruses might be a simplified model in which to identify elicitors of host origin. These elicitors could then be used to induce metabolic alterations, and the elicitors that proved most effective in inducing resistance against challenge infection by various pathogens could be selected. The genes of the corresponding hydrolases releasing the most efficient elicitors would be useful for plant genetic engineering. If adequate regulatory sequences were used, the expression of these genes could be targeted to the sites most likely to be attacked by pathogens.

The finding that various oligosaccharides known as elicitors of defence reactions can protect plants against virus infections is intriguing. More has to be discovered about host specificity of the protection, efficiency against virus transmission by other than mechanical means, and relationship with defence or stress reactions, before it can be concluded that oligosaccharides may be of practical value in protecting plants against virus infections.

ACKNOWLEDGEMENTS

We acknowledge financial support from the Centre National de la Recherche Scientifique (Grant 88-60-N-5104) and from the Ministère de la Recherche et de l'Enseignement Supérieur (Grant 86-C-0952).

REFERENCES

Albersheim P, Darvill AG (1985) Oligosaccharins. Sci Am 253:44-50

Antoniw JF, White RF (1983) Biochemical properties of the PR proteins from tobacco. Neth J Plant Pathol 89:255-264

Antoniw JF, Ritter CE, Pierpoint WS, van Loon LC (1980) Comparison of three pathogenesis-related proteins from plants of two cultivars of tobacco infected with TMV. J Gen Virol 47:79-87

Bailey JA, Mansfield JW (eds) (1982) Phytoalexins. Blackie, Glasgow London

Bailey JA, Deverall BJ (eds) (1983) The dynamics of host defence. Academic Press, Sydney New York London

Bartnicki-Garcia S (1968) Cell wall chemistry, morphogenesis and taxonomy of fungi. Annu Rev Microbiol 22:87-108

Boller T, Gehri A, Mauch F, Vogeli U (1983) Chitinase in bean leaves: induction by ethylene, purification, properties, and possible function. Planta 157:22-31

Bruening G, Ponz F, Glascock C, Russel ML, Rowhani A, Chay C (1987) Resistance of cowpeas to cowpea mosaic virus and to tobacco ringspot virus. In: Evered D, Harnett S (eds) Ciba Foundation Symposium 133: Plant resistance to viruses. John Wiley & Sons, Chichester, p 23

Cabib E (1987) The synthesis and degradation of chitin. Adv Enzymol 59:59-101

Collinge DB, Slusarenko AK (1987) Plant gene expression in response to pathogens. Plant Mol Biol 9:389-410

Cornelissen BJC, Hooft van Huijsduijnen RAM, Bol JF (1986) A tobacco mosaic virus-induced tobacco protein is homologous to the sweet-tasting protein thaumatin. Nature 321:531-532

Darvill AG, Albersheim P (1984) Phytoalexins and their elicitors - a defense against microbial infection in plants. Annu Rev Plant Physiol 35:243-275

Dumas B, Legrand M, Geoffroy P, Fritig B (1988) Purification of tobacco O-methyltransferases by affinity chromatography and estimation of rate of synthesis of the enzymes during the hypersensitive reaction to virus infection. Planta 176:36-41

Felix G, Meins J (1985) Purification, immunoassay and characterization of an abundant, cytokinin-regulated polypeptide in cultured tobacco tissues. Evidence that the protein is a ß-1,3-glucanase. Planta 164:423-428

Fortin MG, Parent JG, Asselin A (1985) Comparative study of two groups of b proteins (pathogenesis-related) from the intercellular fluid of *Nicotiana* leaf tissue infected by TMV. Can J Bot 63:932-93

Fraser RSS (1985) Mechanisms of genetically controlled resistance and virulence: virus diseases. In: Fraser RSS (ed) Mechanisms of resistance to plant diseases. Nijhoff, Dordrecht, p 143

Fraser RSS (1988) Virus recognition and pathogenicity: implications for resistance mechanisms and breeding. Pestic Sci 23:267-275

Fritig B, Kauffmann S, Dumas B, Geoffroy P, Kopp M, Legrand M (1987) Mechanism of the hypersensitivity reaction of plants. In : Evered D, Harnett S (eds) Ciba Foundation Symposium 133: Plant resistance to viruses. John Wiley & Sons, Chichester, p 92

Gianinazzi S, Martin C, Vallée JC (1970) Hypersensibilité aux virus, température et protéines solubles chez le *Nicotiana* Xanthi nc. Apparition de nouvelles macromolécules lors de la répression de la synthèse virale. C R Acad Sci (Paris) D270:2383-2386

Goodman RN, Kiraly Z, Wood KR (1986) The biochemistry and physiology of plant disease. University of Missouri Press, Columbia, Missouri

Granell A, Bellés JM, Conejero V (1987) Induction of pathogenesis-related proteins in tomato by citrus exocortis viroid, silver ion and ethephon. Physiol Mol Plant Pathol 31:83-90

Hahn MG, Bucheli P, Cervone F, Doares SH, O'Neil RA, Darvill A, Albersheim P (1989) The roles of cell wall constituents in plant pathogen interactions. In: Nester E, Kosuge T (eds) Plant-Microbe Interactions, Vol. 3. Macmillan, London New York, in press.

Hooft van Huijsduijnen RAM, Kauffmann S, Brederode FT, Cornelissen BJC, Legrand M, Fritig B, Bol J (1987) Homology between chitinases that are induced by TMV infection of tobacco. Plant Mol Biol 9:411-420

Jamet E, Fritig B (1986) Purification and characterization of 8 of the pathogenesis-related proteins in tobacco leaves reacting hypersensitively to tobacco mosaic virus. Plant Mol Biol 6:69-80

Kauffmann S (1988) Les protéines PR (Pathogenesis-Related) du Tabac: Des protéines impliquées dans les réactions de défense aux agents phytopathogènes. Isolement, propriétés sérologiques et activités biologiques. Thèse de l'Université Louis Pasteur, Strasbourg

Kauffmann S, Legrand M, Geoffroy P, Fritig B (1987) Biological function of 'pathogenesis-related' proteins: four PR proteins of tobacco have 1,3-ß-glucanase

activity. EMBO J 6:3209-3212
Keen NT, Staskawicz B (1988) Host range determinants in plant pathogens and symbionts. Annu Rev Microbiol 42:421-440
Knorr DA, Dawson WO (1988) A point mutation in the tobacco mosaic virus capsid protein gene induces hypersensitivity in *Nicotiana sylvestris*. Proc Natl Acad Sci USA 85:170-174
Kombrink E, Schröder M, Hahlbrock K (1988) Several "pathogenesis-related" proteins in potato are 1,3-ß-glucanases and chitinases. Proc Natl Acad Sci USA 85:782-786
Konate G, Fritig B (1983) Extension of the ELISA method to the measurement of the specific radioactivity of viruses in crude cellular extracts. J Virol Meth 6:347-356
Konate G, Fritig B (1984) An efficient microinoculation procedure to study plant virus multiplication at pre-determined individual infection sites on the leaves. Phytopathol Z 109:131-138
Konate G, Kopp M, Fritig B (1982) Multiplication du virus de la mosaïque du tabac dans des hôtes à réponse systémique ou nécrotique: approche biochimique à l'étude de la résistance hypersensible aux virus. Phytopathol Z 105:214-225
Kopp M, Rouster J, Fritig B, Darvill A, Albersheim P (1989) Host-pathogen interactions XXXII. A fungal glucan preparation protects Nicotianae against infection by viruses. Plant Physiol 90:208-216
Lamb CJ, Lawton MA, Dron M, Dixon RA (1989) Signals and transduction mechanisms for activation of plant defenses against microbial attack. Cell 56:215-224
Legrand M, Fritig B, Hirth L (1976) Enzymes of the phenyl-propanoid pathway and the necrotic reaction of hypersensitive tobacco to tobacco mosaic virus. Phytochemistry 15:1353-1359
Legrand M, Fritig B, Hirth L (1978) o-diphenol O-methyltransferases of healthy and tobacco mosaic virus-infected hypersensitive tobacco. Planta 144:101-108
Legrand M, Kauffmann S, Geoffroy P, Fritig B (1987) Biological function of "pathogenesis-related" proteins: four tobacco PR-proteins are chitinases. Proc Natl Acad Sci USA 84:6750-6754
Linthorst HJM, Meuwissen RLJ, Kauffmann S, Bol J (1989) Constitutive expression of pathogenesis-related proteins PR-1, GRP, and PR-S in tobacco has no effect on virus infection. The Plant Cell 1:285-291
McNeil M, Darvill AG, Fry SC, Albersheim P (1984) Structure and function of the primary cell walls of plants. Annu Rev Biochem 53:625-663
Massala R, Legrand M, Fritig B (1980) Effect of α-aminooxyacetate, a competitive inhibitor of phenylalanine ammonia-lyase, on the hypersensitive resistance of tobacco to tobacco mosaic virus. Physiol Plant Pathol 16:213-226
Massala R, Legrand M, Fritig B (1987) Comparative effects of two competitive inhibitors of phenylalanine ammonia-lyase on the hypersensitive resistance of tobacco to tobacco mosaic virus. Plant Physiol Biochem 25:217-225
Mauch F, Mauch-Mani B, Boller T (1988) Antifungal hydrolase in pea tissue. II. Inhibition of fungal growth by combinations of chitinase and ß-1,3-glucanase. Plant Physiol 88:936-942
Mazau D, Esquerre-Tugaye M-T (1986) Hydroxyproline-rich glycoprotein accumulation in the cell walls of plants infected by various pathogens. Physiol Mol Plant Pathol 29:147-157
Métraux JP, Boller T (1986) Local and systemic induction of chitinase in cucumber plants in response to viral, bacterial and fungal infections. Physiol Mol Plant Pathol. 28:161-169
Métraux JP, Burkhart W, Moyer M, Dincher S, Middlesteadt W, Williams S, Payne G, Carnes M, Ryals J (1989) Isolation of a complementary DNA encoding a

chitinase with structural homology to a bifunctional lysozyme/chitinase. Proc Natl Acad Sci USA 86:896-900
Modderman PW, Schot CP, Klis FM, Wieringa-Brants DH (1985) Acquired resistance in hypersensitive tobacco against tobacco mosaic virus, induced by plant cell wall components. Phytopathol Z 113:165-170
Moore AE, Stone BA (1972) Effect of infection with TMV and other viruses on the level of a ß-1,3-glucan hydrolase in leaves of *Nicotiana glutinosa*. Virology 50:791-798
Pierpoint WS, Robinson NP, Leason MB (1981) The pathogenesis-related proteins of tobacco: their induction by viruses in intact plants and their induction by chemicals in detached leaves. Physiol Plant Pathol 19:85-97
Pierpoint WS, Tatham AS, Pappin DJC (1987) Identification of the virus-induced protein of tobacco leaves that resembles the sweet-protein thaumatin. Physiol Mol Plant Pathol 31:291-298
Ponz F, Bruening G (1986) Mechanisms of resistance to plant viruses. Annu Rev Phytopathol 24:355-381
Ponz F, Glascock CB, Bruening G (1988) An inhibitor of polyprotein processing with the characteristics of a natural virus resistance factor. Mol Plant-Microbe Interact 1:25-31
Richardson M, Valdes-Rodriguez S, Blanco-Labra A (1987) A possible function for thaumatin and a TMV-induced protein suggested by homology to a maize inhibitor. Nature 327:432-434
Roby D, Toppan A, Esquerre-Tugaye MT (1987) Cell surfaces in plant-microorganism interactions. VIII Increased proteinase inhibitor activity in melon plants in response to infection by *Colletotrichum lagenarium* or to treatment with an elicitor fraction from this fungus. Physiol Mol Plant Pathol 30:453-460
Ryan CA (1981) Proteinase inhibitors. In: Stumpf PK, Conn EE (eds) Biochemistry of plants, Vol. 6. Academic Press, London, p 351
Ryan CA (1987) Oligosaccharide signalling in plants. Annu Rev Cell Biol 3:295-317
Ryan CA (1988) Oligosaccharides as recognition signals for the expression of defensive genes in plants. Biochemistry 27:8879-8883
Saito T, Meshi T, Takamatsu N, Okada Y (1987) Coat protein gene sequence of tobacco mosaic virus encodes a host response determinant. Proc Natl Acad Sci USA 84:6074-6077
Schlumbaum A, Mauch F, Vogeli U, Boller T (1986) Plant chitinases are potent inhibitors of fungal growth. Nature 324:365-367
Shimomura T, Dijkstra J (1975) The occurrence of callose during the progress of local lesion formation. Neth J Plant Pathol 81:107-121
Shinshi H, Mohnen D, Meins F (1987) Regulation of a plant pathogenesis-related enzyme: inhibition of chitinase and chitinase mRNA accumulation in cultured tobacco tissues by auxin and cytokinin. Proc Natl Acad Sci USA 84:89-93
Staskawicz B, Bonas U, Dahlbeck D, Huynh T, Kearny B, Ronald P, Whalen M (1988) Molecular determinants of specificity in plant-bacterial interactions. In: Keen NT, Kosuge T, Walling LL (eds) Physiology and biochemistry of plant-microbial interactions. American Society of Plant Physiologists, Rockville, p 124
Van Loon LC (1982) Regulation of changes in proteins and enzymes associated with the active defence against virus infection. In: Wood RKS (ed) Active defence mechanisms in plants. Plenum Press, New York, p 247
Van Loon LC (1985) Pathogenesis-related proteins. Plant Mol Biol 4:111-116
Van Loon LC, van Kammen A (1970) Polyacrylamide disc electrophoresis of the soluble leaf proteins from *Nicotiana tabacum* var. "Samsun" and "Samsun NN". Changes

in protein constitution after infection with TMV. Virology 40:199-201
Varner JE, Lin LL (1989) Plant cell wall structure. Cell 56:231-239
Vera P, Conejero V (1988) Pathogenesis-related proteins of tomato. P-69 as an alkaline endoproteinase. Plant Physiol 87:58-63

INHIBITOR OF VIRUS REPLICATION ASSOCIATED WITH RESISTANCE RESPONSES

G Loebenstein and A Gera
Department of Virology
Agricultural Research Organization
Volcani Center
Bet Dagan 50-250
Israel

ABSTRACT

Protoplasts from tobacco plants containing the *N* gene for resistance to tobacco mosaic virus (TMV), after infection with TMV, release a 23 kDa protein that inhibits replication of several plant viruses. The protein, known as inhibitor of virus replication (IVR), was also obtained from inoculated and induced-resistant leaf tissue. Poly- and monoclonal antibodies to IVR have been prepared which will assist in the future cloning of the IVR gene.

INTRODUCTION

Resistance can be defined as the ability of the plant to prevent or restrict virus multiplication. We may distinguish between *passive preformed defences* prepared in advance of infections, and *active defences* which generally develop after infection. The former type may include examples such as mechanical barriers, or lack of metabolites required for virus multiplication (Loebenstein *et al.*, 1984). Another likely candidate for preformed resistance could be the inability of host- and virus-coded subunits to form a functional replicase.

Active host resistance generally develops after infection, and probably requires induction of transcription of a gene or genes in the host. This may lead to structural changes forming physical barriers to virus spread, or to functional changes that interfere with viral replication or translocation (Loebenstein and Stein, 1985). Active resistance mechanisms are known in systemic infections, as in cucumbers resistant to cucumber

NATO ASI Series, Vol H 41
Recognition and Response in Plant-Virus Interactions
Edited by R S S Fraser

mosaic virus (CMV) (Nachman *et al.*, 1971; Levy *et al.*, 1974) and in local lesion-responding plants (Loebenstein *et al.,* 1984). Systemically effective resistance in cucumber plants depends on three genes (Kooistra, 1969), while localization in most cases depends on one dominant gene (Fraser, 1987).

The local lesion response is probably the most notable active resistance phenomenon, whereby after inoculation the virus invades and multiplies in several hundred cells but does not spread to other tissues. The following types of localized infections are known (Loebenstein *et al.,* 1982):

1. necrotic lesions of the nonlimited type
2. self-limiting necrotic local lesions
3. chlorotic local lesions
4. ring-like patterns or ringspots
5. starch lesions
6. microlesions, and
7. subliminal infections

Resistance is also induced in uninoculated parts of a hypersensitive host by inoculating other parts of the plant with necrotic local lesion-producing viruses, fungi (Loebenstein, 1972; Kuć, 1983) and bacteria (Jenns *et al.*, 1979). For example, inoculation of the lower leaves of tobacco plants containing the *N* gene for tobacco mosaic virus (TMV) resistance or with tobacco necrosis virus (TNV) induced resistance in the upper uninoculated leaves; inoculation of a primary leaf of Pinto beans with TMV or TNV induced resistance in the opposite leaf. Lesions developing after challenge inoculation of the resistant tissue were consistently smaller, and usually fewer in number, than those formed on previously uninoculated control plants. In tobacco *NN* plants, induced resistance, as measured by reduction in number and size of lesions, was found to be closely correlated with virus concentration as determined by ELISA (Stein *et al.*, 1985). This indicates that virus replication, and not only the development of necrotic local lesions, was suppressed in the "resistant" tissues. However, there have been other reports indicating that virus multiplication is not suppressed in leaves with induced resistance (Fraser, 1979; Coutts and Wagih, 1983).

HYPOTHESES EXPLAINING LOCALIZATION

Some possible mechanisms

Several hypotheses have been raised to explain localization. Details have been discussed previously (Loebenstein *et al.*, 1982; Loebenstein *et al.*, 1984; Loebenstein and Stein, 1985); therefore only some major points are presented here. It has been suggested that the death of cells may localize or inactivate the virus. This explanation is not satisfactory even in necrotic local lesion hosts and even less so in a chlorotic or starch-lesion host, as virus particles are found in apparently viable cells outside the necrotic area (Milne, 1966).

Several barrier substances such as lignin and callose, in cells surrounding the local lesion, have been implicated as possible factors preventing virus movement. However, callose depositions were also observed in infections which produce systemic necrosis, where the virus does not remain localized, as with tobacco infected with tomato spotted wilt and tobacco rattle virus (Shimomura and Dijkstra, 1975). On the other hand, no callose deposition was observed around TMV-induced starch lesions on cucumber cotyledons, where the infection remains localized without necrosis (Cohen and Loebenstein, 1975). It seems, therefore, that barrier substances could be a response to necrotization and not necessarily responsible for localization.

Another possible mechanism of localization is the inactivation of a "transport protein". Virus spread is thought to require the help of a virus-coded protein. For TMV, this factor has been identified as a 30 kDa non-structural protein (Leonard and Zaitlin, 1982). Using immunogold cytochemistry, it was shown that the 30 kDa protein accumulated in plasmodesmata, an observation consistent with a role in virus spread (Tomenius *et al.*, 1987). Similarly, immunogold labelling has shown the association of a non-structural protein (P3) of alfalfa mosaic virus, considered to be involved in cell-to-cell spread, with the middle lamella of cell walls (Stussi-Garaud *et al.*, 1987 and Godefroy-Colbourn *et al.*, this volume). Plasmodesmata have been identified as subcellular locations of the P1 protein product of cauliflower mosaic virus (CaMV) gene I, suggesting its involvement in the systemic spread of CaMV (Linstead *et* al., 1988). Recently, it was reported that the amount of the 30 kDa protein in the cell wall fraction of TMV-infected Samsun *NN* tobacco plants decreased sharply as soon as necrosis was visible, compared with that in the systemic host Samsun *nn*. It was suggested that this may explain why TMV infection becomes localized (Moser *et al.*,

1988). Whether the decrease of transport protein is responsible for localization or is a result of the necrosis remains to be seen, especially as virus particles are found outside the necrotic area (Milne, 1966). In this respect, it may be of interest to study the kinetics of the transport protein after transfer of plants with necrotic lesions to higher temperatures, where localization fails, as well as in chlorotic- or starch-lesion infections.

Finally, localization might be due to reduced virus multiplication caused by a substance or substances that inhibit virus multiplication. Indications of a host-mediated process that suppresses virus multiplication in a starch-lesion host were obtained from studies using antimetabolites (Loebenstein *et al.*, 1969; Sela *et al.*, 1969) and ultraviolet (UV) irradiation (Loebenstein *et al.*, 1970). Virus concentration in cucumber cotyledons increased markedly when cotyledons were treated with actinomycin D, chloramphenicol, or UV one day after inoculation with TMV. The antimetabolites also stimulated the biosynthesis of TNV in *Chenopodium amaranticolor* (Faccioli and Rubies-Autonell, 1976). It was suggested that the localizing mechanism operates mainly via inhibition of virus multiplication (Loebenstein *et al.*, 1984).

An inhibitor of virus replication (IVR) from TMV-infected protoplasts and from intact tissue of a local lesion-responding tobacco.

During recent years, we have collected data supporting the association of a protein, termed "inhibitor of virus replication" (IVR) with the local lesion response in tobacco varieties containing the *N* gene (Loebenstein and Gera, 1981; Gera *et al.*, 1983; Loebenstein and Gera, 1988). This inducible protein is a true inhibitor of virus replication, and not an inhibitor of infection, as it affects virus titres in protoplasts when applied up to 18 hours after inoculation. IVR was released into the medium from TMV-infected protoplasts of Samsun *NN* tobacco, a cultivar in which infection in the intact plant is localized. It inhibited virus replication in protoplasts from both local lesion-responding resistant Samsun *NN* and systemically responding, susceptible Samsun *nn* plants. IVR was not produced in protoplasts from susceptible Samsun plants, or from non-inoculated protoplasts of the resistant cultivar. IVR did not affect TMV directly. IVR was found to be sensitive to trypsin and chymotrypsin but not to RNase, and its activity was abolished by incubation at 60°C for 10 minutes, suggesting that it has a proteinaceous nature (Gera and Loebenstein, 1983). A specific protein band with an approximate molecular mass of 23 kDa was regularly observed in polyacrylamide gel electrophoresis of crude preparations of IVR (Fig. 1a; Spiegel *et al.*, 1989; Gera *et al.*,

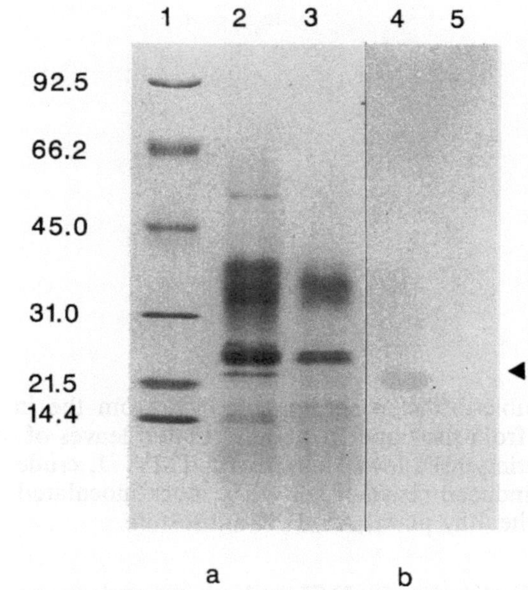

Figure 1. Polyacrylamide gel electrophoresis of crude IVR (a) and immunoblotting of IVR recovered after electroelution from the 23 kDa band (b). Polypeptide markers (lane 1); crude IVR (10 units) (lane 2); control preparation (10 units) (lane 3); IVR (10 units) after electroelution (lane 4); control after electroelution (lane 5).

1990). Electroelution of protein from this band recovered a biologically active fraction with a twentyfive-fold increase in specific activity over that of the crude preparation. This fraction revealed only one band at 23 kDa in polyacrylamide gel electrophoresis, providing evidence that the 23 kDa band is IVR purified to a high degree. This band gave a specific reaction in immunoblots with IVR antiserum (Fig. 1b) (see below; Gera *et al.*, 1990).

Polyclonal (Gera and Loebenstein, 1989; Gera *et al.*, 1990) and monoclonal (Gera *et al.*, 1989) antibodies were prepared, which neutralized the antiviral activity of IVR and enabled its detection in immunoblots.

IVR was found also to inhibit TMV replication in Samsun *nn* leaf disks, with inhibition rates ranging between 60% and 90%, as determined by both local lesion assay and ELISA. Similar results were obtained when IVR was applied to TMV-infected tomato leaf disks. IVR also inhibited replication of CMV, potato virus X (Gera and Loebenstein, 1983), and potato virus Y (Loebenstein and Gera, 1988) in different host tissues, thereby indicating that it is neither virus- nor host-specific. The

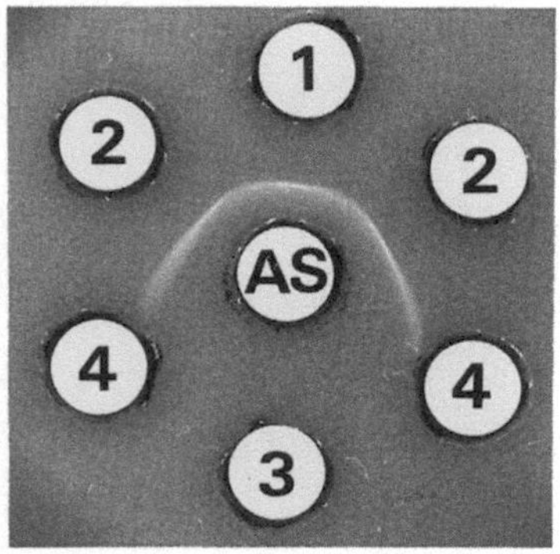

Figure 2. Agar-gel double-diffusion serology of IVR from the intercellular fluid of induced-resistant tissue from the upper two uninoculated leaves of tobacco *NN* plants, seven days after inoculating three lower leaves with TMV. 1, crude protoplast IVR; 2, crude tissue IVR from induced-resistant leaves; 3, mock-inoculated protoplast IVR; 4, intercellular fluid from healthy plant; AS, IVR antiserum.

inhibition of TMV multiplication by IVR in both protoplasts and leaf tissue disks showed a clear dose-response effect (Gera *et al.*, 1986).

IVR was also obtained from the intercellular fluid of Samsun *NN* tobacco leaves infected with TMV (Spiegel *et al.*, 1989). This facilitates the bulk recovery of IVR.

IVR could also be recovered from the intercellular fluid of induced-resistant, non-inoculated tissues of *NN* tobacco (Spiegel *et al.*, 1989). Extracts from induced-resistant tissue between TMV-inoculated strips had a higher inhibition rate than those from upper leaves with induced systemic resistance. In agar-gel diffusion tests, clear precipitation lines were obtained which fused without spurs with the precipitation line obtained between crude protoplast IVR and the antiserum (Fig. 2).

The interspecific hybrid of *N. glutinosa* x *N. debneyi* is highly resistant to infection with TMV and TNV (Ahl and Gianinazzi, 1982) and on inoculation, produced only a small number of tiny lesions. When the rootstock of this hybrid was grafted to other *Nicotiana* species, the scions expressed acquired resistance (Gianinazzi and Ahl, 1983). IVR was obtained from the intercellular fluid of uninoculated leaves of this hybrid. Inoculation with TMV somewhat increased the potency of the extract. A specific protein band was observed on polyacrylamide gel electrophoresis and had the same electrophoretic mobility as the putative IVR from protoplasts. In this connection, it may be apposite to note that this hybrid constitutively produced a PR-1 type protein (Ahl and Gianinazzi, 1982).

CONCLUDING REMARKS

The recovery of IVR from protoplasts and intact tissue of a resistant cultivar; its purification and production of poly- and monoclonal antisera, may be the bases for the isolation and cloning the gene coding for the IVR. This gene might then be useful in transforming susceptible plants.

ACKNOWLEDGEMENT

This research was supported in part by the United States-Israel (Binational) Agricultural Research and Development Fund (BARD).

REFERENCES

Ahl P, Gianinazzi S (1982) b-Proteins as a constitutive component in highly (TMV) resistant interspecific hybrids of *Nicotiana glutinosa* x *Nicotiana debneyi*. Plant Sci Lett 26:173-181

Cohen J, Loebenstein G (1975) An electron microscope study of starch lesions in cucumber cotyledons infected with tobacco mosaic virus. Phytopathology 65:32-39

Coutts RHA, Waigh EE (1983) Induced resistance to viral infection and soluble protein alterations in cucumber and cowpea plants. Phytopathol Z 107:57-69

Faccioli G, Rubies-Autonell C (1976) Action of antimetabolites on the biosynthesis of tobacco necrosis virus in locally-infected cells of *Chenopodium amaranticolor*. Phytopathol Z 87:48-56

Fraser RSS (1979) Systemic consequences of the local lesion reaction to tobacco mosaic virus in a tobacco variety lacking the *N* gene for hypersensitivity. Physiol Plant Pathol 14:383-394

Fraser RSS (1987) Genetics of plant resistance to viruses. In: Evered D, Harnett S (eds) Plant resistance to viruses, Ciba Foundation Symposium 133. John Wiley & Sons, Chichester, p 6

Gera A, Loebenstein G (1983) Further studies of an inhibitor of virus replication from tobacco mosaic virus-infected protoplasts of a local lesion-responding tobacco cultivar. Phytopathology 73:111-115

Gera A, Loebenstein G (1989) Evaluation of antisera to an inhibitor of virus replication. J Phytopathol 124:366-371

Gera A, Loebenstein G, Salomon R, Frank A (1990) An inhibitor of virus replication from tobacco mosaic virus infected protoplasts of a hypersensitive tobacco cultivar is associated with a 23K protein species. Phytopathology (in press)

Gera A, Loebenstein G, Shabtai S (1983) Enhanced tobacco mosaic virus production and suppressed synthesis of a virus inhibitor in protoplasts exposed to antibiotics. Virology 127:475-478

Gera A, Sadka A, Spiegel S, Salomon R, Smorodinsky NI (1989) Use of monoclonal antibodies in the purification of an inhibitor of virus replication by affinity chromatography. J Gen Virol 70:1293-1296

Gera A, Spiegel S, Loebenstein G (1986) Production, preparation and assay of an antiviral substance from plant cells. Methods Ezymol 119:729-733

Gianinazzi S, Ahl P (1983) The genetic and molecular basis of b-proteins in the genus *Nicotiana*. Neth J Plant Pathol 89:275-281

Jenns AE, Caruso FL, Kuć J (1979) Non-specific resistance to pathogens induced systemically by local infection of cucumber with tobacco necrosis virus, *Colletotrichum lagenarium* or *Pseudomonas lachrymans*. Phytopathol Mediterr 18:129-134

Kooistra E (1969) The inheritance of resistance to cucumis virus I in cucumber (*Cucumis sativus* L.). Euphytica 18:326-332

Kuć J (1983) Induced systemic resistance in plants to diseases caused by fungi and bacteria. In: Bailey JA, Deverall BJ (eds) Dynamics of host defense. Academic Press, Sydney, New York, p 191

Leonard DA, Zaitlin M (1982) A temperature sensitive strain of tobacco mosaic virus defective in cell-to-cell movement generates an altered viral-coded protein. Virology 117: 416-424

Linstead PJ, Hills GJ, Plaskitt KA, Wilson IG, Harker CL, Maule AJ (1988) The subcellular location of the gene 1 product of cauliflower mosaic virus is consistent with a function associated with virus spread. J Gen Virol 69:1809-1818

Levy A, Loebenstein G, Smookler M, Drori T (1974) Partial suppression by UV irradiation of the mechanism of resistance to cucumber mosaic virus in a resistant cucumber cultivar. Virology 60:37-44

Loebenstein G (1972) Localization and induced resistance in virus-infected plants. Annu Rev Phytopathol 10:177-208

Loebenstein G, Chazan R, Eisenberg M (1970) Partial suppression of the localizing mechanism to tobacco mosaic virus by UV irradiation. Virology 41:373-376

Loebenstein G, Gera A (1981) Inhibitor of virus replication released from tobacco mosaic virus infected protoplasts of a local lesion-responding tobacco cultivar. Virology 114:132-139

Loebenstein G, Gera A (1988) Resistance responses of plants to virus infections. Acta Hortic 234:403-411

Loebenstein G, Gera A, Stein A (1984) Plant defense mechanisms to viral infections. In: Kurstak E (ed) Control of virus diseases. Marcel Dekker, New York, p 375

Loebenstein G, Sela B, van Praagh T (1969) Increase of tobacco mosaic local lesion size and virus multiplication in hypersensitive hosts in the presence of actinomycin D. Virology 37:42-48

Loebenstein G, Spiegel S, Gera A (1982) Localized resistance and barrier substances. In: Wood RKS (ed) Active defense mechanisms in plants. Plenum, New York, p 211

Loebenstein G, Stein A (1985) Plant defense responses to viral infections. In: Key JL, Kosuge T (eds) Cellular and molecular biology of plant stress. Alan R. Liss, New York, p 413

Milne RG (1966) Electron microscopy of tobacco mosaic virus in leaves of *Nicotiana glutinosa*. Virology 28:527-532

Moser O, Gagey MJ, Godefroy-Colburn T, Stussi-Garaud C, Ellwart-Tschurtz M, Nitschko H, Mundry KW (1988) The fate of the transport protein of tobacco mosaic virus in systemic and hypersensitive tobacco hosts. J Gen Virol 69:1367-1373

Nachman I, Loebenstein G, van Praagh T, Zelcer A (1971) Increased multiplication of cucumber mosaic virus in a resistant cucumber cultivar caused by actinomycin D. Physiol Plant Pathol 1:67-72

Sela B, Loebenstein G, van Praagh T (1969) Increase of tobacco mosaic virus multiplication and lesion size in hypersensitive hosts in the presence of chloramphenicol. Virology 39:260-264

Shimomura T, Dijkstra J (1975) The occurrence of callose during the process of local lesion formation. Neth J Plant Pathol 81:107-121

Spiegel S, Gera A, Salomon R, Ahl P, Harlap S, Loebenstein G (1989) Recovery of an inhibitor of virus replication from the intercellular fluid of hypersensitive tobacco infected with tobacco mosaic virus and from uninfected induced-resistant tissue. Phytopathology 79:258-262

Stein A, Spiegel S, Loebenstein G (1985) Studies on induced resistance to tobacco mosaic virus in Samsun *NN* tobacco and changes in ribosomal fractions. Phytopathol Z 114:295-300

Stussi-Garaud C, Garaud JR, Berna A, Godefroy-Colburn T (1987) *In situ* location of an alfalfa mosaic virus non-structural protein in plant cell walls: correlation with virus transport. J Gen Virol 68:1779-1784

Tomenius K, Clapham D, Meshi T (1987) Localization by immunogold cytochemistry of the virus-coded 30K protein in plasmodesmata of leaves infected with tobacco mosaic virus. Virology 160:363-371

INTERCELLULAR SPREAD OF POTATO LEAFROLL LUTEOVIRUS: EFFECTS OF CO-INFECTION AND PLANT RESISTANCE

B D Harrison, H Barker and P M Derrick
Scottish Crop Research Institute
Invergowrie
Dundee
DD2 5DA
UK

ABSTRACT

Potato leafroll luteovirus (PLRV) moves inefficiently from cell to cell and, in intact plants, is virtually confined to phloem tissue. Its spread to non-phloem tissue is enhanced by co-infection with some sap-transmissible viruses whereas others have no effect. The effective sap-transmissible viruses seem not to increase plasmodesmatal permeability to small peptides. Viruses which can aid intercellular movement of PLRV may themselves pass through virus-modified plasmodesmata as virus nucleic acid whereas some other viruses probably pass from cell to cell, as virus nucleoprotein particles, through virus-induced tubules which lead to plasmodesmata.

TISSUE LOCALIZATION OF POTATO LEAFROLL LUTEOVIRUS (PLRV)

The luteovirus group includes several widely disseminated plant pathogens, of which PLRV is one of the most important economically. The viruses are transmitted by aphids in the persistent manner but not by inoculation with sap, and in their plant hosts they remain virtually confined to, and can damage, phloem tissue (Waterhouse *et al.*, 1988). In plants infected with PLRV, virus-like particles have been detected by ultrastructural studies in the phloem, notably in phloem parenchyma and companion cells (Arai *et al.*, 1969; Kojima *et al.*, 1969), and PLRV particle antigen has been found, especially in companion cells, by fluorescent antibody staining (Barker and Harrison, 1986). Indeed Barker and Harrison (1986) suggested that the virus might multiply mainly in companion cells, with the sieve tubes serving to convey it from one companion cell to another. However, despite the apparent restriction of PLRV to phloem tissue in intact plants, mesophyll protoplasts can readily be infected when suspensions of them are inoculated (Kubo and Takanami, 1979). Evidently mesophyll cells are infectible, although they do not become infected in intact leaves. Thus, except for those in phloem tissue, most cells in intact plants are resistant to infection with PLRV.

These findings focus attention on mechanisms of spread of PLRV in intact tissue. In this paper we are concerned not with the long distance transport of PLRV within plants but with possible reasons for the inefficiency of its cell-to-cell spread. After considering the results of recent work, we also discuss the mechanisms and recognition events which may be involved in intercellular movement of a range of other plant viruses.

Possible existence of a movement-protein gene in PLRV RNA

Nucleotide sequencing of PLRV-RNA has revealed six open reading frames with coding capacities for proteins of molecular mass of more than 10 kDa (Mayo *et al.*, 1989). These open reading frames are conserved in beet western yellows luteovirus (BWYV; Veidt *et al.*, 1988), and all but that nearest the 5'-end also occur in barley yellow dwarf luteovirus (Miller *et al.*, 1988). Of the genes that are conserved in all three luteoviruses, one is thought to encode a viral polymerase and another encodes the particle protein (Mayo *et al.*, 1989). However, none of the putative gene products of PLRV has an amino-acid sequence with obvious similarities to that of the movement protein of tobacco mosaic, tobacco rattle or alfalfa mosaic viruses (MA Mayo, unpublished results). Such evidence suggests that PLRV RNA does not contain an open reading frame which is an obvious candidate for a viral movement-protein gene, and we tentatively conclude that there may not be one. This could explain the inefficiency of intercellular movement of PLRV.

Effects of co-infection with other viruses

To explore the possibility that other viruses can provide a movement function which PLRV lacks, the effects of co-infection with unrelated viruses were examined. In earlier work Atabekov *et al.* (1984) showed that PLRV concentration is greater in *Datura stramonium* plants co-infected with potato X potexvirus than in singly infected plants, and that PLRV accumulates in some parenchyma cells of the doubly infected leaves. In further experiments, in which *Nicotiana clevelandii* plants were infected with PLRV + tobacco rattle tobravirus, a virus which has a putative movement-protein gene

Table 1. *Effects of co-infection with unrelated viruses on the concentration of luteoviruses in systemically infected leaves*[a]

Luteovirus	Host[b]	Co-infecting virus	Effect on luteovirus concentration
Potato leafroll	NC	Alfalfa mosaic virus	< x 2
		Cucumber mosaic cucumovirus	< x 2
		Parsnip yellow fleck virus	< x 2
		Tomato black ring nepovirus	< x 2
		Carrot mottle umbravirus	x 9
		Narcissus mosaic potexvirus	x 6
		Pea early-browning tobravirus	x 7
		Pea early-browning tobravirus (NM form)	x 10
		Potato V potyvirus	x 10
Beet western yellows	NC	Alfalfa mosaic virus	< x 2
		Carrot mottle umbravirus	x 9
		Pea early-browning tobravirus	x 5
		Potato V potyvirus	x 9
Potato leafroll	NB	Alfalfa mosaic virus	< x 2
		Narcissus mosaic potexvirus	x 12

[a] Data are from Barker (1989)
[b] NB = *Nicotiana benthamiana*, NC = *N. clevelandii*

(Boccara *et al.*, 1986), or with PLRV + potato Y potyvirus, many more cells were found to become infected with PLRV than in plants infected with PLRV alone, and the PLRV concentration of leaves was increased considerably (Barker, 1987; Barker, 1989). Of the protoplasts from these systemically infected plants that could be stained with fluorescent antibody to PLRV particles, most contained chloroplasts and closely resembled mesophyll protoplasts. Fluorescent antibody staining of sections showed that the additional PLRV-infected cells in the doubly infected plants were in the phloem or immediately adjàcent to it (H Barker, unpublished results). Thus although co-infection with these viruses resulted in more extensive invasion of plants by PLRV, the infection was still far from being generalized.

Further work has shown that co-infection with sap-transmissible viruses in four taxonomic groups can increase the concentration of PLRV, presumably by aiding its intercellular movement, but that viruses in several other groups have little or no effect (Table 1). Similar results were obtained in *N. clevelandii* and *N. benthamiana*, and also when BWYV was substituted for PLRV (Barker, 1989). Where examined, the concentration of the sap-transmissible virus was not affected by concurrent infection with PLRV. Cell-to-cell spread of luteoviruses therefore is enhanced in dual infections with a virus from a restricted number of other virus groups.

In co-infections with tobraviruses, the NM isolates of pea early browning (Table 1) and tobacco rattle viruses, which lack the smaller genome segment (RNA-2) were at least as effective in increasing invasion by PLRV as were the particle-producing M isolates. This rules out a possible role for the particle protein of tobraviruses in aiding PLRV invasion, because this protein is encoded by RNA-2. Moreover RNA-1, which constitutes the genome of NM isolates, contains four genes, one of which encodes the tobravirus putative movement protein (Boccara *et al.*, 1986).

Effects of helper viruses on plasmodesmatal conductance

Research on the intercellular movement of materials through plasmodesmata of plants such as *Elodea* has shown that there is little transport of compounds with molecular weights (M_r) exceeding 750 (Goodwin, 1983). This suggests that intercellular movement of plant viruses must involve a mechanism different from that operating for smaller molecules and/or that plasmodesmatal permeability is altered by virus infection. To ascertain whether plasmodesmatal permeability to smaller molecules is affected, fluorescent amino acids and peptides of different molecular weights, such as FITC-Gly, FITC-Glu-Glu and FITC-Ala_6, were injected into the symplast of epidermal cells of control and virus-infected leaves of *N. clevelandii*. In the control leaves, molecules of M_r less than 680 were transported remarkably quickly and within 5 minutes had usually entered at least 3 additional cells at each injection point. In contrast, there was no transport of molecules of M_r 839 in 30 minutes (Derrick *et al.*, 1989). The size at which transport ceased was between M_r 750 and 839, a value close to that quoted by Goodwin (1983) for *Elodea*. When the tests were repeated with *N. clevelandii* leaves infected with tobacco rattle tobravirus, potato Y potyvirus or tomato black ring nepovirus, the results were the same as for virus-free leaves. Infection with these viruses, which include one that has a putative movement protein, therefore does not increase plasmodesmatal permeability non-specifically.

A different result was obtained in tests on *N. clevelandii* infected with carrot mottle umbravirus. In leaves of these plants, plasmodesmatal permeability was decreased (Derrick *et al.*, 1989) despite the fact that this virus aids invasion of leaf cells by PLRV (Barker, 1989). This decrease in permeability can probably be attributed to the modifications of plasmodesmatal structure described by Murant *et al.* (1973), and which result in the formation of outgrowths from the plasmodesmata into the cytoplasm. That these changes seem to enhance the spread of PLRV while decreasing that of small molecules is particularly noteworthy, and again points to the specificity of the mechanism of virus transport. The frequency of intercellular virus movement clearly does not reflect plasmodesmatal conductance in general.

Restriction of PLRV spread in genetically resistant potato plants

PLRV typically reaches a concentration of 1-2 μg/g leaf tissue of susceptible potato cultivars, but only about 5% of this value in several resistant cultivars (Barker and Harrison, 1985). This form of virus resistance is expressed most strongly in leaf tissue and least strongly in the basal parts of stems. In leaves, the lower virus content is parallelled by a decrease in the number of phloem companion cells which can be stained with fluorescent antibody to PLRV particles. Moreover, the few stained cells found in resistant genotypes are nearly all confined to the internal phloem whereas stained cells occur in both internal and external phloem of susceptible potato genotypes (Barker and Harrison, 1986). Evidently the virus accumulates to similar extents in individual cells of resistant and susceptible genotypes but it infects many fewer cells in the resistant plants. This has the practical result that infected leaves of resistant genotypes are poor sources of virus for vector aphids, and that relatively little spread of PLRV is therefore expected to occur within crops of resistant potato cultivars (Barker and Harrison, 1986). Resistance seems to be caused by the partial inhibition of cell-to-cell spread of PLRV in phloem tissue.

In recent work to determine the pattern of inheritance of this resistance to PLRV, seedling progenies of crosses between resistant and susceptible parents segregated into two populations which respectively resembled one or other parent in the ability of PLRV to accumulate in them (H Barker, RM Solomon and GR Mackay, unpublished results). This suggests that intercellular spread of PLRV is under simple genetic control by the host, and therefore that one or a few host proteins play a crucial role in the processes involved in spread.

In tests of other kinds, potato genotypes identified as being resistant to infection with PLRV were shown to lose their resistance when they were infected with potato virus X or Y (Anon, 1988). Although it is uncertain whether the resistance that broke down in these experiments is of the same type as that studied by Barker and Harrison (1986), it is tempting to speculate that the two are the same, and that the effects of potato viruses X and Y mirror those of co-infection of *N. clevelandii* with these viruses.

EVIDENCE FOR DIFFERENT TYPES OF INTERCELLULAR VIRUS MOVEMENT

The experiments on mixed infections of either PLRV or BWYV in combination with one of a range of sap-transmissible viruses show that these sap-transmissible viruses fall into two categories: those which do, and those which do not, enhance cell-to-cell spread of luteoviruses. This implies that individual sap-transmissible viruses move from cell to cell by at least two alternative mechanisms. Support for this suggestion is provided by comparing some of the other properties of several of the sap-transmissible viruses. However, of these properties (Table 2), not all have been determined for each of the viruses given as examples, so that some of the assignments are tentative.

Viruses with properties of Type A (Table 2) can exist as variants which do not produce nucleoprotein particles but can nevertheless spread from cell to cell (Cadman and Harrison, 1959; Siegel *et al.*, 1962; Kassanis and Welkie, 1963). Moreover, variants that lack any form of coat protein (Sänger, 1968) can spread in inoculated leaves at least as fast as particle-producing strains, in some instances apparently faster (Frost *et al.*, 1967). This suggests that not only do all these variants spread from cell to cell as infective RNA but also that the particle-producing strains of the viruses spread in the same way. Possibly the entity that spreads is double-stranded RNA or replicative-intermediate RNA, which would be more resistant to inactivation by nucleases than single-stranded RNA. Some of these viruses are also known to have a movement protein which, in cells infected with tobacco mosaic tobamovirus (TMV), becomes associated with plasmodesmata (Tomenius *et al.*, 1987), and so may function by facilitating the passage of viral RNA through them. One may also speculate that the other viruses which assist the intercellular spread of PLRV, including potexviruses,

Table 2. *Characteristics of viruses with two apparently contrasting types of cell to cell movement*

Characteristic	Type A	Type B
Cell to cell spread by mutants which do not produce virus particles	+	0
Virus particles in plasmodesmata	0	+
Virus particles in tubules	0	+
Virus-coded protein in plasmodesmata	+	0
Virus-coded protein in tubules	0	+
Assist transport of PLRV	+	0
Possible examples	tobacco mosaic tobacco necrosis tobacco rattle	cowpea mosaic parsnip yellow fleck tomato black ring

potyviruses and umbraviruses, likewise are transported from cell to cell as RNA, and that they assist PLRV to spread in this form. Indeed, the observation that potato X potexvirus can assist the spread of TMV in TMV-resistant tomato plants which carry gene *Tm-2* (Taliansky *et al.*, 1982), suggests that these two viruses have similar intercellular movement mechanisms. Spread in the form of RNA is also plausible for carrot mottle umbravirus because no typical virus-like particle has been found for this virus, and it relies for transmission by vector aphids on its RNA being packaged in particles which have coats consisting of the protein of carrot red leaf luteovirus (Waterhouse and Murant, 1983).

In some other virus groups there is evidence for a second kind of movement (Table 2, Type B). These viruses do not aid the intercellular spread of PLRV and they induce the formation of tubules, which occur in the cytoplasm, contain virus particles, and which may extend from or through plasmodesmata (Roberts and Harrison, 1970). Moreover, recent work with one such virus, cowpea mosaic comovirus, indicates that the tubules contain a virus-coded protein (Goldbach *et al.*, this volume). Their function may be to facilitate the cell-to-cell movement of virus particles. If these viruses can only spread intercellularly as nucleoprotein particles, the lack of examples of variants with defective particle proteins is explained, because such variants could not be maintained in infected plants.

Viruses in yet other groups seem not to have either of the above categories of intercellular movement, and other mechanisms may well exist. Candidates for viruses that have such systems include alfalfa mosaic virus (Stussi-Garaud *et al.*, 1987), bromoviruses, cucumoviruses, and probably several other virus groups.

Possible recognition events involved in intercellular virus movement

The hypotheses outlined in the previous section imply that several kinds of signal recognition are needed for cell-to-cell spread of plant viruses to occur and that these recognition events differ in the different mechanisms proposed. For example, Type A movement may involve recognition of viral RNA by a virus-coded movement protein, and of the virus-coded movement protein by host plasmodesmatal material. In Type B movement, there may be recognition of virus coat protein by a virus-coded tubule protein, and of the latter by host material. In such a system the viral coat protein, which must also have the abilities to assemble to form particles that protect viral RNA from inactivation, and to confer vector transmissibility, would therefore have a third function. Perhaps this triple role is the reason why the coat proteins of tubule-inducing viruses are larger in size (nepoviruses) or greater in species complexity (comoviruses, parsnip yellow fleck virus) than those of many other plant viruses.

In neither of the systems proposed is there direct physical evidence for the occurrence of host-coded components of the virus-movement system, but the conclusion that individual, or a few, host genes seem able to control the cell-to-cell movement of TMV (Nishiguchi and Motoyoshi, 1987), cowpea mosaic comovirus (Beier *et al.*, 1979) and PLRV (H Barker, RM Solomon and GR Mackay, unpublished results) strongly encourages the view that they exist.

CONCLUSIONS

The results and hypotheses outlined above can be summed up as follows. Luteoviruses such as PLRV are normally confined to phloem tissue and their cell-to-cell spread is inefficient. Spread within phloem and into non-phloem tissue is enhanced by co-infection of plants with members of specific groups of sap-transmissible viruses whereas viruses in other groups have no such effect. Neither category of sap-transmissible virus increases epidermal plasmodesmatal permeability to molecules of M_r 600-900. It is suggested that (a) the viruses which assist PLRV spread do so by encoding a movement protein which helps their own RNA, and that of PLRV, to pass through plasmodesmata; (b) the function of the cytoplasmic, virus-containing tubules, which are produced by infection with several groups of sap-transmissible viruses that do not assist PLRV spread, is to enable particles of the homologous virus to pass through plasmodesmata; and (c) different sets of recognition signals are involved in these two systems. The involvement of specific host components in such systems is implied by evidence for major host genes which control cell-to-cell spread of viruses such as TMV and PLRV.

REFERENCES

Anon (1988) Interaction of potato viruses. Annual Report of the International Potato Center, Lima, Peru, 1987-1988, p 80

Arai K, Doi Y, Yora K, Asuyama H (1969) Electron microscopy of the potato leafroll virus in leaves of three kinds of host plants and the partial purification of the virus. Ann Phytopathol Soc Jpn 35:10-15

Atebekov JG, Taliansky ME, Drampyan AH, Kaplan IB, Turka IE (1984) Systemic infection by a phloem-restricted virus in parenchyma cells in a mixed infection (in Russian). Biol Nauki 10:28-31

Barker H (1987) Invasion of non-phloem tissue in *Nicotiana clevelandii* by potato leafroll virus is enhanced in plants also infected with potato Y potyvirus. J Gen Virol 68:1223-1227

Barker H (1989) Specificity of the effect of sap-transmissible viruses in increasing the accumulation of luteoviruses in co-infected plants. Ann Appl Biol 155:in press.

Barker H, Harrison BD (1985) Restricted multiplication of potato leafroll virus in resistant potato genotypes. Ann Appl Biol 107:205-212

Barker H, Harrison BD (1986) Restricted distribution of potato leafroll virus antigen in resistant potato genotypes and its effect on transmission of the virus by aphids. Ann Appl Biol 109:595-604

Beier H, Bruening G, Russell ML, Tucker CL (1979) Replication of cowpea mosaic virus in protoplasts isolated from immune lines of cowpeas. Virology 95:165-175

Boccara M, Hamilton WDO, Baulcombe DC (1986) The organisation and interviral homologies of genes at the 3' end of tobacco rattle virus RNA1. EMBO J 5:223-229

Cadman CH, Harrison BD (1959) Studies on the properties of soil-borne viruses of the tobacco-rattle type occurring in Scotland. Ann Appl Biol 47:542-556

Derrick PM, Barker H, Oparka KJ (1989) Effects of virus infection on plasmodesmatal permeability. Rep Scott Crop Res Inst for 1988, in press

Frost RR, Harrison BD, Woods RD (1967) Apparent symbiotic interaction between particles of tobacco rattle virus. J Gen Virol 1:57-70

Goodwin PB (1983) Molecular size limit for movement in the symplast of the *Elodea* leaf. Planta 157:124-130

Kassanis B, Welkie GW (1963) The nature and behaviour of unstable variants of tobacco necrosis virus. Virology 21:540-550

Kojima M, Shikata E, Sugawara M, Murayama D (1969) Purification and electron microscopy of potato leafroll virus. Virology 39:162-174

Kubo S, Takanami Y (1979) Infection of tobacco mesophyll protoplasts with tobacco necrotic dwarf virus, a phloem-limited virus. J Gen Virol 42:387-398

Mayo MA, Robinson DJ, Jolly CA, Hyman L (1989) Nucleotide sequence of potato leafroll luteovirus RNA. J Gen Virol 70:1037-1051

Miller WA, Waterhouse PM, Gerlach WL (1988) Sequence and organization of barley yellow dwarf virus genomic RNA. Nucleic Acids Res 16:6097-6111

Murant AF, Roberts IM, Goold RA (1973) Cytopathological changes and extractable infectivity in *Nicotiana clevelandii* leaves infected with carrot mottle virus. J Gen Virol 21:269-283

Nishiguchi M, Motoyoshi F (1987) Resistance mechanisms of tobacco mosaic virus strains in tomato and tobacco. In: Evered D, Harnett S (eds) Plant resistance to viruses. Ciba Foundation Symposium 133. John Wiley & Sons, Chichester, New York p 38

Roberts IM, Harrison BD (1970) Inclusion bodies and tubular structures in *Chenopodium amaranticolor* plants infected with strawberry latent ringspot virus. J Gen Virol 7:47-54

Sänger H (1968) Characteristics of tobacco rattle virus. I. Evidence that its two particles are functionally defective and mutually complementing. Mol Gen Genet 101:346-367

Siegel A, Zaitlin M, Sehgal OP (1962) The isolation of defective tobacco mosaic virus strains. Proc Natl Acad Sci USA 48:1845-1851

Stussi-Garaud C, Garaud J-C, Berna A, Godefroy-Colburn T (1987) *In situ* location of an alfalfa mosaic virus non-structural protein in plant cell walls: correlation with virus transport. J Gen Virol 68:1779-1784

Taliansky ME, Malyshenko SI, Pshennikova ES, Atabekov JG (1982) Plant virus-specific transport function. II. A factor controlling virus host range. Virology 122:327-331

Tomenius K, Clapham D, Meshi T (1987) Localization by immunogold cytochemistry of the virus-coded 30K protein in plasmodesmata of leaves infected with tobacco mosaic virus. Virology 160:363-371

Veidt I, Lot H, Leiser M, Scheidecker D, Guilley H, Richards K, Jonard G (1988) Nucleotide sequence of beet western yellows virus RNA. Nucleic Acids Res 16:9917-9932

Waterhouse PM, Murant AF (1983) Further evidence on the nature of the dependence of carrot mottle virus on carrot red leaf virus for transmission by aphids. Ann Appl Biol 103:455-464

Waterhouse PM, Gildow FE, Johnstone GR (1988) Luteovirus group. AAB Descriptions of Plant Viruses No 339

EXAMINATION OF MECHANISMS OF CROSS PROTECTION WITH NON-TRANSGENIC PLANTS

L A Urban, J L Sherwood[1], J A M Rezende[2] and U Melcher[3]
Department of Plant Pathology
Oklahoma State University
Stillwater
Oklahoma 74078
USA

ABSTRACT

Cross protection, the activity of a virus in a plant preventing the expression of a challenge virus, has been used to control only a few virus diseases. Utilization of host metabolites or structures, sequestering of the nucleic acid of the challenge virus, inhibition of uncoating of the challenge virus, and prevention of systemic spread of the challenge virus have been proposed to be involved in cross protection. The host may play an important role in the events that occur in cross protection. Strains of tobacco mosaic virus that reciprocally cross protect and *Arabidopsis thaliana* can be used to study virus-host interactions in cross protection.

INTRODUCTION

Cross protection, the activity of a virus in a plant preventing the expression of a subsequent challenge virus (Dodds, 1982), was first described in the late 1920s (McKinney, 1929; Wingard, 1928). Yet some sixty years later, cross protection is used to control only a few virus diseases. The mechanisms of cross protection in non-transgenic (Fulton, 1980; Hamilton, 1980; Ponz and Bruening, 1986; Zaitlin and

---

[1] To whom correspondence should be sent.
[2] Present address: Department of Plant Pathology, Oklahoma State University. Permanent address: Seção de Virologia, Instituto Agronômico de Campinas, 13020 Campinas, SP, Brazil.
[3] Department of Biochemistry, Oklahoma State University.

NATO ASI Series, Vol H 41
Recognition and Response in Plant-Virus Interactions
Edited by R S S Fraser

Hull, 1987; Sherwood, 1987a) as well as in transgenic plants (see Beachy *et al.,* this volume) have been the subject of much research, speculation, and review. However,the events that occur in the host that control the outcome of the interaction between virus strains have yet to be fully explained. Cross protection involves an interaction or interactions between related virus strains, and may therefore encompass recognition and signalling events between viral RNAs or virus-coded proteins. But evidence to be discussed in this chapter also shows that host-encoded molecules can influence the outcome of the interaction. Understanding of this three-way signalling is necessary if cross-protection is to be fully exploited in practical crop protection. This paper focusses on studies utilizing non-transgenic plant systems and the contributions of these studies to an understanding of the phenomenon of cross protection.

CROSS PROTECTION AND VIRUS DISEASE CONTROL

Although Salaman (1937) and Johnson (1937) suggested the use of cross protection for disease control over 50 years ago, there are few examples of the use of cross protection for control of plant virus diseases. Research emphasis has focussed on the mechanism of cross protection rather than its application to control virus diseases. Cacao swollen shoot in Africa (Posnette and Todd, 1955) and passion fruit woodiness in Queensland, Australia (Simmonds, 1959) were the first virus diseases in which plants protected with mild strains showed good development and yield increase under field conditions. Programmes for the control of these diseases by cross protection have not continued, and the diseases are still of economic importance. Citrus tristeza, tomato mosaic, and papaya ringspot diseases appear to be the only examples in which cross protection has been earnestly investigated for disease control.

Grant and Costa (1951) demonstrated that cross protection could be used to control citrus tristeza virus (CTV) in Brazil. In 1961, Müller and Costa (1977) initiated a research program to utilize cross protection to control CTV in São Paulo State. The isolation of suitable mild strains from the field required an immense effort. The identification of isolates by mild symptom expression was not sufficient; each isolate had then to be checked in each type of citrus to be protected. From some 45 mild isolates, each checked on Pêra sweet orange, Galego lime, and Ruby Red grapefruit, which had been budded on three different rootstocks, only 6 isolates were satisfactory. Three isolates could be used for Pêra sweet orange, two isolates for Galego lime, and one isolate for Ruby Red grapefruit. An isolate that worked in one type of citrus did not work in another.

By 1987, a total of 50 million Pêra sweet orange trees protected with a mild strain had been planted in Brazil (Salibe, 1987). Control of citrus tristeza through cross protection has also had a great impact on the citrus industry in Australia, India, Israel, Japan, South Africa and the USA (Hamilton, 1985).

Cross protection was also widely applied for control of tomato mosaic, caused by tobacco mosaic virus (TMV) (Rast, 1975; Fletcher and Rowe, 1975; Broadbent, 1976; Channon *et al.*, 1978). However, difficulties have developed in trying to control TMV using cross protection. Mild strains generally protect less effectively at 25-30°C than at lower temperatures. The mild strain MII-16, produced by nitrous acid mutation of strain 1 (Rast, 1975), does not protect tomatoes against all other strains of the virus. In some areas where the MII-16 strain has been used, there has been an increase in the prevalence of strain 1. Today growers rely primarily on resistant varieties to control the disease (Fulton, 1986).

The use of cross protection to control papaya ringspot disease has been investigated in Taiwan and Hawaii (Lin, 1980, Yeh and Gonsalves, 1984; Yeh *et al.*, 1988), and Brazil (Rezende *et al.*, 1981; Rezende, 1985). A mild strain of papaya ringspot virus (PRV), produced by nitrous acid treatment, was used in field experiments in Taiwan. Papaya plants protected by the mild strain had 82% greater fruit yield than unprotected trees, resulting in a 111% increase in grower income. Protection was effective when protected plants were planted in solid blocks, and disease pressure within the test orchard was minimized by roguing severely infected plants once every 10 days up to flowering. However, protection waned under high disease pressure by other strains from areas near the protected trees. Due to the success of these experiments, more than one million papaya seedlings inoculated with the mild strain of PRV were planted in the field in 1986 (Yeh *et al.*, 1988).

In Brazil, in spite of very satisfactory results of protection in greenhouse and field experiments, protected papaya plants in the field showed mild symptoms for only 6-8 months. Afterwards the symptoms increased in severity. Symptom intensification occurred in a synchronized manner in all plants. The change of symptoms was not considered a breakdown of protection, but rather a change in the mild strains selected from the field due to mutation and selective competition (Rezende and Costa, 1987).

The success of cross protection as a method of controlling papaya ringspot in Brazil is therefore dependent upon finding more stable mild strains.

Although cross protection has been commercially applied to control few diseases, it may be an effective option for control of many other diseases. Cross protection has been proposed for control of cauliflower mosaic virus (CaMV) in Brussels sprouts (Tomlinson and Shepherd, 1978), cucumber mosaic virus (CMV) in tomato (Dodds, 1982; Yoshida *et al.,* 1985) and pepper (Tien and Chiang, 1983), TMV in sweet pepper (Goto *et al.,* 1984), potato virus Y (PVY) in tobacco (Latorre and Flores, 1985), tomato aspermy virus in tomato (Kuti and Moline, 1986), and grapevine fanleaf virus in grapevines (Huss *et al.,* 1989). Fulton (1986) suggested that avocado sun blotch, concave gum and psoriasis of citrus, and some stone and pome fruit viruses may be effectively controlled by cross protection. Greater attention to the practical application of cross protection could result in control of more virus diseases by this method. However, the decision to use cross protection to control diseases must be well thought out and precautions should be taken to prevent additional problems (Fulton, 1986). The anxiety about potential problems could be the reason for the scarcity of examples where cross protection has been used for disease control in the field.

THEORIES ON THE MECHANISM OF CROSS PROTECTION

Because viruses are obligate parasites they require a suitable host for replication. Most mechanisms proposed to account for specificity in cross protection assume that the events that control the outcome of the interaction of virus strains are relatively independent of the host. However, the host may play as important a role as the virus strains in the events that occur in cross protection.

Many theories have been proposed to account for specificity in cross protection. Development of experiments to test some of these theories has not been possible. However, this does not provide grounds for their dismissal as a possible basis for the events in cross protection. Four theories proposed to explain cross protection are summarized here. The theories are listed separately, but some combination of each may contribute to the events in cross protection.

Utilization and depletion of host metabolites or structures

Ross (1974) hypothesized that the initial virus would utilize the available ribosomes by a rapid increase in mRNA. This would leave the challenge virus nucleic acid unbound and therefore suceptible to degradation before it could be expressed. Another possible host constituent with a role in cross protection could be a protein that is required as part of the RNA-dependent RNA polymerase of an RNA virus (Ponz and Bruening, 1986). Depletion of this component by the first virus could prevent the challenge virus from replicating.

Specific sequestering of nucleic acid

Palukaitis and Zaitlin (1984) suggested that for (+)RNA viruses, infection by the challenging virus would be reduced or prevented by inhibiting its ability to synthesize (+)RNA. This could occur when the nascent (-)RNA of the challenge virus became hybridized to the excess (+)RNA of the protecting strain. If a mechanism is to account for the cross protection that occurs between strains of viroids as well as between strains of viruses it will most likely involve the regulation of replication of nucleic acid.

Involvement of coat protein

DeZoeten and Fulton (1975) suggested a model to explain cross protection in which the coat protein already present in the cell would encapsidate the viral RNA of the challenge strain before it could be expressed. Zaitlin (1976) tested their hypothesis using a coat protein mutant of TMV (PM1), which produces insoluble coat protein that does not encapsidate TMV RNA. When plants inoculated with the coat protein mutant were challenge-inoculated with TMV (U1), less of the U1 strain was recovered than in the controls. This suggests that protection was conferred by the coat protein mutant and that encapsidation of the challenge RNA was not involved in protection in this system.

Horikoshi *et al.* (1987) found that the coat protein of brome mosaic virus, *in vitro*, blocked the binding site of the replicase thereby interfering with RNA synthesis. They suggested that the regulation of replication by coat protein may be the basis of cross protection: this would obviously not depend on complete encapsidation of the RNA of the challenge virus.

The inability of the challenge virus to uncoat when inoculated onto plants containing a protecting strain was proposed as the basis for cross protection with TMV in *Nicotiana sylvestris* (Sherwood and Fulton, 1982). A mosaic of dark and light green areas develops in the leaves of *N. sylvestris* after inoculation with the common strain of TMV (TMV-C). When a leaf with mosaic is inoculated with a strain of TMV that produces local lesions in *N. sylvestris*, the lesions are confined to the dark green areas of the mosaic (Fulton, 1951; Sherwood and Fulton, 1982). When a leaf with mosaic symptoms is inoculated with RNA from a strain of TMV that produces local lesions, the lesions are produced in both the light green and the dark green areas (Sherwood and Fulton, 1982). This suggests that the lesion producing strains are unable to uncoat in the light green areas where the concentration of TMV is greatest. DeZoeten and Gaard (1984) reported that 2 to 7.5 times more TMV antigen was detectable in cell walls of light green areas than in dark green areas. The first encounter between the challenge virus and the protected host may be at the cell wall.

Other supporting evidence for the involvement of coat protein in cross protection was found by Dodds *et al.* (1985) in tomato protected with CMV. As with TMV in *N. sylvestris*, a breakdown of cross protection occurred when the leaves were challenged with viral RNA but not with intact virions.

Work with coat protein-free mutants has created evidence for and against the specific involvement of coat protein in cross protection. Sherwood (1987b) inoculated leaves of *N. sylvestris* with the DT-1G coat protein-free strain of TMV (Sarkar and Smitamana, 1981) or TMV-C. The leaves were then challenged with turnip mosaic virus (TuMV) to test non-specific protection, or with a necrotizing strain of TMV. The leaves inoculated with strain DT-1G had equal susceptibility to TuMV and the necrotizing strain of TMV. Infection was less with the necrotizing strain than with TuMV when TMV-C was the protecting virus.

Gerber and Sarkar (1989) found that the DT-1G mutant of TMV induced protection which was up to 90% effective against the U1 strain of TMV in "Samsun" tobacco. These conflicting reports support the idea that the mechanisms of cross protection may be different in different hosts.

Coat protein may not be the only factor involved in cross protection. Zinnen and Fulton (1986) compared the cross protection effects of two tobamoviruses; sunn-hemp mosaic virus (SHMV) and TMV-C. Cowpeas systemically infected with SHMV were challenged with either virions or RNA of a necrotic lesion strain of SHMV which had been produced by nitrous acid mutagenesis. In both cases protection occurred. However, when SHMV infected plants were challenged by TMV-C RNA encapsidated in SHMV coat protein, or TMV-C intact virions, the plants inoculated with

SHMV-encapsidated TMV-C RNA produced 5 to 27-fold fewer lesions than those inoculated with TMV-C.

Prevention of systemic spread of the challenge virus

The initial virus infection establishes a mechanism for virus spread throughout the plant. The challenge virus may be limited in its ability to spread from the site of initial inoculation due to saturation or changes in the putative receptors involved in cell to cell movement of virus. This specific involvement of the control of virus movement in cross protection is supported by the work of Dodds (1982) and Dodds *et al.* (1985). Using tomato as the host and two strains of CMV they showed that the challenge strain was able to increase only in the inoculated leaves. They also demonstrated that when the protecting strain was not well established systemically, an increase in the challenge strain also occurred in the upper leaves.

Work with two serologically distinct strains of TMV that reciprocally cross protect in *Arabidopsis thaliana* supports this theory. The challenge strain multiplied in inoculated leaves to concentrations detectable by ELISA, but it did not move systemically in the plant (Urban *et al.*, 1988). This suggests an impairment of systemic movement of the challenge virus and a possible interaction of the 30 kDa movement protein with a host component.

Blum *et al.* (1989) suggested a virus-induced, host specific inhibitor of viral transport. They hypothesized that the challenge virus cannot synthesize its movement protein because the protecting strain induces the host to synthesize an inhibitor of 30 kDa synthesis. They further suggested that if the challenge virus cannot use the movement system of the protecting strain or the protein has already dissipated, the challenge virus will be unable to spread. Additional support for the interaction of the 30 kDa protein with host proteins comes from Moser *et al.* (1988). In a TMV local lesion host, *N. tabacum* cv. Samsun *NN*, the amount of movement protein in cell wall fractions decreased when necrosis became visible and the production of coat protein ceased.

The pathogenesis related (PR) proteins are possible host components that may interact with the movement protein to stop the spread of the challenge virus. Good candidates might be the group C encoded proteins. Bol (1988) reported that this group is strongly induced by TMV infection and salicyclic acid treatment. He speculated that members of this group might be cell wall components because of their high glycine content, and that they might alter wall structure and subsequent virus movement. PR proteins have not been detected in *N. sylvestris* systemically infected with TMV (Sherwood, 1985), but can be found in healthy tobacco when it begins to flower and senesce (Fraser, 1981). The specific role PR proteins play in resistance to plant virus infection is still unclear (see Fritig *et al.* and Linthorst *et al.*, this volume).

SOME FUTURE PROSPECTS

We are using *Arabidopsis thaliana*, a member of the Brassicaceae, to study host-virus interaction in cross protection. *Arabidopsis* can be infected by a number of plant viruses including TMV, CMV, turnip yellow mosaic virus, tomato black ring virus, potato virus X, PVY (Sosnova and Polak, 1975), and CaMV (Balázs and Lebeurier, 1981). Cross protection has been shown to occur in *Arabidopsis* using two strains of TMV. The protection is reciprocal and results in non-movement of the challenge strain (Urban *et al.*, 1988). When plants previously inoculated with TMV are challenge inoculated with CMV, CMV can later be found throughout the plant indicating the specificity of the protection.

The merits of *Arabidopsis* as a model system for plant molecular biology have been discussed by Pang and Meyerowitz (1987). In short, it is small in size, has a small genome with little repetitive DNA, a short generation time, and well defined classical genetics. Genetic maps of many phenotypic traits and restriction fragment length polymorphisms (RFLPs) are available. The plants produce abundant seed and the seed can be mutagenized. Some of these mutations have been mapped. A number of genes have been isolated from *Arabidopsis* using cDNA clones from *Arabidopsis* RNA or heterologous probes from other plants or yeast. Some examples are genes for alcohol dehydrogenase, nitrate reductase, chalcone synthase, and heat shock proteins. *Arabidopsis* has also been transformed using the *Agrobacterium* Ti-plasmid system (Lloyd *et al.*, 1986) and the genome modified by T-DNA insertion mutagenesis (Feldman *et al.*, 1989). This plant may serve as a model for study of other pathogen-host interactions for other members of the Brassicacae such as broccoli, cauliflower and turnip.

CONCLUSION

There are many events that occur in cross protection, and the regulation of this phenomenon probably involves events in addition to the ones discussed here. The use of transgenic plants to identify the function of specific genes and gene products has and will add a new dimension to this ongoing research. Further research into areas such as PR proteins, antiviral factors, and the hypersensitive response will add key information to our understanding of cross protection. The role the host plays in the events that occur in the cross protection phenomenon must not be overlooked. A concurrent development of an understanding of the biology of plant viruses, of hosts of plant viruses, and of the interaction that occurs between the two will be necessary to utilize cross protection fully as a control measure for virus diseases.

ACKNOWLEDGEMENT

This work was supported in part by NSF grant NSF-RII-OK-8610676-7 and the Oklahoma Agricultural Experiment Station. Journal Article No. J-5614, Oklahoma Agricultural Experiment Station, Oklahoma State University.

REFERENCES

Balázs E, Lebeurier G (1981) *Arabidopsis* is a host of cauliflower mosaic virus. *Arabidopsis* Inf Serv 18:130-134

Blum H, Gross HJ, Beier H (1989) The expression of the TMV-specific 30-kDa protein in tobacco protoplasts is strongly and selectively enhanced by actinomycin. Virology 169:51-61

Bol JF (1988) Structure and expression of plant genes encoding pathogenesis-related proteins. In: Verma DPS, Goldberg RB (eds) Temporal and spatial regulation of plant genes. Springer-Verlag, Vienna, p 201

Broadbent L (1976) Epidemiology and control of tomato mosaic virus. Annu Rev Phytopathol 14:75-96

Channon AG, Cheffins NJ, Hitchon GM, Barker J (1978) The effect of inoculation with an attenuated mutant strain of tobacco mosaic virus on the growth and yield of early glasshouse tomato crops. Ann Appl Biol 88:121-130

DeZoeten GA, Fulton RW (1975) Understanding generates possibilities. Phytopathology 65:221-222

DeZoeten GA, Gaard G (1984) The presence of viral antigen in the apoplast of systemically virus-infected plants. Virus Res 1:713-725

Dodds JA (1982) Cross-protection and interference between electrophoretically distinct strains of cucumber mosaic virus in tomato. Virology 118:235-240

Dodds JA, Lee SQ, Tiffany M (1985) Cross protection between strains of cucumber mosaic virus: effect of host and type of inoculum on accumulation of virions and double-stranded RNA of the challenge strain. Virology 144:301-309

Feldman KA, Marks MD, Christianson ML, Quatrano RS (1989) A dwarf mutant of *Arabidopsis* generated by T-DNA insertion mutagenesis. Science 243:1351-1354

Fletcher JT, Rowe JM (1975) Observations and experiments on the use of an avirulent mutant strain of tobacco mosaic virus as a means of controlling tomato mosaic. Ann Appl Biol 81:171-179

Fraser RSS (1981) Evidence for the occurrence of the "pathogenesis-related" proteins in leaves of healthy tobacco plants during flowering. Physiol Plant Pathol 19:69-76

Fulton RW (1951) Superinfection by strains of tobacco mosaic virus. Phytopathology 41:579-592

Fulton RW (1980) The protective effects of systemic virus infection. In: Wood RKS (ed) Active defense mechanisms in plants. Plenum, New York, London p 231

Fulton RW (1986) Practices and precautions in the use of cross protection for plant virus disease control. Annu Rev Phytopathol 24:67-81

Gerber M, Sarkar S (1989) The coat protein of tobacco mosaic virus does not play a significant role for cross-protection. J Phytopathol 124:323-331

Goto T, Iizuka N, Komochi S (1984) Selection and utilization of an attenuated isolate of pepper strain of tobacco mosaic virus. Ann Phytopathol Soc Jpn 50:221-228

Grant TJ, Costa AS (1951) A mild strain of the tristeza virus of citrus. Phytopathology 41:114-122

Hamilton RI (1980) Defenses triggered by previous invaders: viruses. In: Horsfall JG, Cowling EB (eds) Plant disease: an advanced treatise, Vol V. Academic Press, New York, p 279

Hamilton RI (1985) Using plant viruses for disease control. HortScience 20:848-852

Horikoshi M, Nakayama M, Yamaoka N, Furosawa I, Shishiyama J (1987) Brome mosaic virus coat protein inhibits viral RNA synthesis *in vitro*. Virology 158:15-19

Huss B, Walter B, Fuchs M (1989) Cross-protection between arabis mosaic virus and grapevine fanleaf virus isolates in *Chenopodium quinoa*. Ann Appl Biol 114:45-60

Johnson J (1937) An acquired partial immunity to the tobacco streak disease. Trans Wis Acad Sci Arts Lett 30:27

Kuti JO, Moline HE (1986) Effects of inoculation with a mild strain of tomato aspermy virus on the growth and yield of tomatoes and the potential for cross-protection. J Phytopathol 115:56-60

Latorre BA, Flores V (1985) Strain identification and cross protection of potato virus Y affecting tobacco in Chile. Plant Dis 69:930-932

Lin CC (1980) Strains of papaya ringspot virus and their cross protection. PhD Thesis, National University of Taiwan

Lloyd AM, Barneson AR, Rogers SG, Byrne MC, Fraley RT, Horsch RB (1986) Transformation of *Arabidopsis* with *Agrobacterium tumefaciens*. Science 234:464-466

McKinney HH (1929) Mosaic diseases in the Canary Islands, West Africa, and Gibraltar. J Agric Res 39:557-578

Moser O, Gagey M-J, Godefroy-Colburn T, Stussi-Garaud C, Ellwart-Tschürtz M, Nitschko H, Mundry K-W (1988) The fate of the transport protein of tobacco mosaic virus in systemic and hypersensitive tobacco hosts. J Gen Virol 69:1367-1373

Müller GW, Costa AS (1977) Tristeza control in Brazil by preimmunization with mild

strains. Proc Int Soc Citricult 3:868-872
Palukaitis P, Zaitlin M (1984) A model to explain the cross-protection phenomenon shown by plant viruses and viroids. In: Kosuge T, Nester EW (eds) Plant-microbe interactions: molecular and genetic perspectives. MacMillan, New York, p 420
Pang PP, Meyerowitz EM (1987) *Arabidopsis thaliana*: a model system for plant molecular biology. Bio/Technology 5:1177-1181
Ponz F, Bruening G (1986) Mechanisms of resistance to plant viruses. Annu Rev Phytopathol 24:355-381
Posnette AF, Todd JMcA (1955) Virus diseases of cacao in West Africa. IX. Strain variation and interference in virus 1A. Ann Appl Biol 43:433-453
Rast ATB (1975) Variability of tobacco mosaic virus in relation to control of tomato mosaic in glasshouse tomato crops by resistance breeding and cross protection. Agric Res Rep No 834. Wageningen, Netherlands
Rezende JAM (1985) Tentativas de premunização para o controle do mosaico do mamoeiro. MS thesis, University of São Paulo, Brazil
Rezende JAM, Costa AS (1987) Intensificação dos sintomas dificulta o controle do mosaico do mamoeiro por premunização. Fitopatol Bras 12:100-101
Rezende JAM, Costa AS, Soares NB (1981) Ocorrência de um isolado fraco do vírus do mosaico do mamoeiro *Carica papaya* L. Fitopatol Bras 6:534
Ross AF (1974) Interaction of viruses in the host. Acta Hortic 36:247-260
Salaman RN (1937) Acquired immunity against the "Y" potato virus. Nature 139:924-925
Salibe AA (1987) Clones nucelares de citros no Estado de São Paulo. Laranja 2:443-466
Sarkar S, Smitamana P (1981) A proteinless mutant of tobacco mosaic virus: evidence against the role of a viral coat protein for interference. Mol Gen Genet 184:158-159
Sherwood JL (1985) The association of "pathogenesis-related" proteins with viral induced necrosis in *Nicotiana sylvestris*. J Phytopathol 112:48-55
Sherwood JL (1987a) Mechanisms of cross-protection between plant virus strains. In: Evered D, Harnett S (eds) Plant resistance to viruses. John Wiley & Sons, Chichester, New York, p 136
Sherwood JL (1987b) Demonstration of the specific involvement of coat protein in tobacco mosaic virus (TMV) cross protection using a TMV coat protein mutant. J Phytopathol 118:358-362
Sherwood JL, Fulton RW (1982) The specific involvement of coat protein in tobacco mosaic virus cross protection. Virology 119:150-158
Simmonds JH (1959) Mild strain protection as a means of reducing losses from the Queensland woodiness virus in the Passion vine. Queensl J Agric Sci 16:371-380
Sosnova V, Polak Z (1975) Susceptibility of *Arabidopsis thaliana* (L) Heynh to infection with some plant viruses. Biol Plant 17:156-158
Tien P, Chiang XH (1983) Control of two seed-borne virus diseases in China by the use of protective inoculation. Seed Sci Technol 11:969-972
Tomlinson JA, Shepherd RJ (1978) Studies on mutagenesis and cross-protection of cauliflower mosaic virus. Ann Appl Biol 90:223-231
Urban LA, Sherwood JL, Melcher U (1988) The use of *Arabidopsis thaliana* cv. Columbia in plant virus interaction studies. Phytopathology 78:1586
Wingard SA (1928) Hosts and symptoms of ringspot, a virus disease of plants. J Agric Res 37:127-153
Yeh S-D, Gonsalves D (1984) Evaluation of induced mutants of papaya ringspot virus

for control by cross protection. Phytopathology 74:1086-1091
Yeh S-D, Gonsalves D, Wang H-L, Namba R, Chiu R-J (1988) Control of papaya ringspot virus by cross protection. Plant Dis 72:375-380
Yoshida K, Goto T, Iizuka N (1985) Attenuated isolates of cucumber mosaic virus produced by satellite RNA and cross protection between attenuated isolates and virulent ones. Ann Phytopathol Soc Jpn 51:238-242
Zaitlin M (1976) Viral cross-protection: more understanding is needed. Phytopathology 66:382-383
Zaitlin M, Hull R (1987) Plant virus-host interactions. Annu Rev Plant Physiol 38:291-315
Zinnen TM, Fulton RW (1986) Cross protection between sunn-hemp mosaic and tobacco mosaic viruses. J Gen Virol 67:1679-1687

COAT PROTEIN-MEDIATED PROTECTION AGAINST VIRUS INFECTION[1]

Richard S Nelson,
Samuel Roberts Noble Foundation
Plant Biology Division
PO Box 2180
Ardmore, OK 73402
USA

Patricia A Powell[2] and Roger N Beachy[3]
Washington University
Plant Biology Program, Department of Biology
Box 1137
St Louis, MO 63130
USA

ABSTRACT

Research on cross protection between virus strains has suggested that more than one type of mechanism may be involved. The ability to produce transgenic plants expressing virus coat proteins has offered a means of investigating these mechanisms, and raises other interesting questions. Coat protein-mediated protection has now been reported for seven families of single-stranded RNA nonenveloped plant viruses, and evidence of efficacy under field conditions has been obtained. Protection appears to operate against primary establishment of infection, and separately, against a later stage of virus spread or multiplication. The mechanism(s) of protection remain largely unclear, however, the coat protein and not the coat protein transcript *per se* is required for protection in transgenic plants expressing tobacco mosaic virus coat protein.

COAT PROTEIN-MEDIATED PROTECTION AND CROSS PROTECTION

The discovery that expression of viral coat protein (CP) in transgenic plants protects these plants against virus infection is, in some ways, a natural extension of work by previous researchers on cross protection. This phenomenon, first reported by McKinney (1929), is defined in its simplest form as the protection of a plant against

---

[1] A similar contribution is to be published in "Genetic Engineering of Crop Plants", eds GW Lycett and D Grierson, Butterworth Scientific Ltd., London, UK

[2] Present address: Salk Institute for Biological Studies, Plant Biology Laboratory, 10010 N Torrey Pines Road, La Jolla, CA 92037, USA

[3] To whom correspondence should be sent

NATO ASI Series, Vol H 41
Recognition and Response in Plant-Virus Interactions
Edited by R S S Fraser

invasion by a pathogen because of prior inoculation with another pathogen. Cross protection was first observed between related plant viruses (see reviews by Fulton, 1986; Ponz and Bruening, 1986; Urban *et al.,* this volume), but similar observations have been made for viroids (Fernow, 1967; Niblett *et al.,* 1978). The inclusion of the protection between viroids under the umbrella of cross protection presents a dilemma since viroids do not code for the synthesis of a capsid protein and there is evidence, to be discussed below, which implicates the CP in cross protection between viruses (Sherwood and Fulton, 1982). It is clear, therefore, that cross protection is defined more by the final result than by the mechanism of protection.

With regard to plant viruses and cross protection, much research has been directed towards determining whether the CP, the RNA, or both from the protecting virus provide protection. Zaitlin (1976) and Sarkar and Smitamana (1981), using protecting tobacco mosaic virus (TMV) strains which produced a defective CP or no CP at all, respectively, reported that protection occurred when plants harbouring these strains were challenged with a related virus. These results appear to be in conflict with those of Sherwood and Fulton (1982), who showed that the CP of the protecting virus was important for protection. Sherwood (1987) resolved much of the controversy by showing that the CP-minus TMV strain used by Sarkar and Smitamana (1981) did indeed provide protection against superinfection by a normally encapsidated TMV strain, but that greater protection was observed when a CP-plus strain was used as the protecting virus. The protection afforded by the CP-minus strain was non-specific since infection by turnip mosaic potyvirus (TuMV) was also inhibited. The protection provided by the CP-expressing strain was specific for related viruses, since there was an approximately 5-fold greater level of protection against a second TMV strain compared with that observed against TuMV. This and other work (Zinnen and Fulton, 1986) suggests, therefore, that cross protection may be composed of two or more mechanisms.

One problem with interpreting the results from cross protection experiments is that replication of a viral genome is involved. Viral replication and its regulation are complicated processes which are poorly understood. Researchers are left to identify mutants which lack a function associated with a specific gene product and then to correlate this loss with a loss of protection. Mutants which have lost the ability to replicate cannot, of course, be analyzed for cross protection. In contrast, the tools of molecular biology and genetic transformation allow the researcher to express fragments of the viral genome throughout transgenic plants or in specific tissues and then to evaluate the ability of these fragments to provide protection in the absence of replication of the protecting virus.

COAT PROTEIN-MEDIATED PROTECTION

First proposals for genetically engineered protection

Hamilton (1980) was one of the earliest researchers to propose that expressing viral sequences in transgenic plants could be a method for protecting plants against virus infections. Others followed with more refined ideas as to which viral sequences could potentiate protection (Sequeira, 1984; Beachy *et al.*, 1985; Sanford and Johnston, 1985). On the basis of earlier models of cross protection proposed by de Zoeten and Fulton (1975) and Sherwood and Fulton (1982), Sequeira (1984) suggested that insertion and expression of the CP gene in plants might result in protection. Beachy *et al.* (1985), in addition to discussing the use of a CP gene for protection, discussed the potential for using genes that encode antisense viral sequences. Sanford and Johnston (1985) suggested that the replicase of the protecting virus might bind to a "replicase attachment site" but not replicate the challenge virus RNA, thereby preventing the challenge virus replicase from binding. All of these ideas were based on possible explanations for cross protection put forth previously by plant pathologists.

Transgenic plants and coat protein-mediated protection

By the early to mid 1980s researchers in several laboratories were cloning plant viruses and inserting portions of the viral genome into plants under the control of heterologous promoters, to test whether expression of the inserted sequences could provide protection against virus infection. Bevan *et al.* (1985) produced transgenic tobacco plants that expressed the CP of TMV but expression levels were low (approximately 0.17μg/gram fresh weight) and no results were presented indicating protection of these plants after virus challenge. Beachy *et al.* (1985) reported expression of TMV-CP in transgenic tobacco plants at levels greater than ten times those reported by Bevan *et al.* (1985). Powell Abel *et al.* (1986) showed that these plants were protected after challenge with TMV. Many reports quickly followed showing the usefulness of CP gene expression in providing protection in a number of other plant-virus combinations (Loesch-Fries *et al.*, 1987; Tumer *et al.*, 1987; van Dun *et al.*, 1987;

Table 1: *Compilation of Results and Conclusions in the Literature Concerning CP-Mediated Protection in Plants*

Issue Addressed	System Studied[a]	Phenotype of CP-Mediated Protection
Site of protection	T/T/Tb A/A/Tb TS/TS/Tb P/P/Tb	Decrease in lesion numbers (chlorotic, necrotic, starch) on inoculated leaves after challenge with virus from which the CP was isolated.
	T/T/Tb	Decrease in virus accumulation in systemically infected leaves in spite of virus accumulation in inoculated leaves equal to that found in controls.
Mechanism of protection	T/T/Tb A/A/Tb A/A/Al	Decrease in viral CP and (+) and (-) strand viral RNA accumulation in challenged leaves and protoplasts.
	T/T/Tb A/A/Tb P/P/Tb	Less protection in inoculated leaves and protoplasts challenged with viral RNA or pH 8.0 treated virions except for studies involving PVX.
	A/A/Tb T/T/Tb	Protection is correlated with CP levels and not CP-RNA levels; a threshold level of CP expression is apparently required for protection.

[a] virus from which the CP gene was derived/challenge virus/host plant
T = tobacco mosaic virus; A = alfalfa mosaic virus; TS = tobacco streak virus; P = potato virus X; C = cucumber mosaic virus; S = soybean mosaic virus; TR = tobacco rattle virus; - = various viruses.
Tb = tobacco; Al = Alfalfa; Tm = tomato.

Conclusions Drawn	References
Protection manifested at the primary infection site.	Loesch-Fries *et al.* (1987); Nelson *et al.* (1987); Tumer *et al.* (1987); Van Dun *et al.* (1987); Hemenway *et al.* (1988); Van Dun *et al.* (1988)
Protection manifested after primary infection and inhibiting systemic spread of the infection.	Wisniewski LA, Powell P, Nelson RS, Beachy RN (submitted manuscript); Nelson *et al.* (this review)
Primary infection blocked early in infection; for TMV, blockage is at or before virus replication.	Register and Beachy (1988); Halk *et al.* (1989)
Interaction between the challenge CP and its viral RNA is required for maximum protection. Indirectly indicates that RNA:RNA interactions between the transgenic CP transcript and the challenge virus RNA are not providing much protection against primary infection. Results for PVX indicate that mechanisms other than those working in TMV and AlMV systems may provide significant protection against PVX-RNA.	Loesch-Fries *et al.* (1987); Nelson *et al.* (1987); Van Dun *et al.* (1987); Hemenway *et al.* (1988); Register and Beachy (1988); Van Dun *et al.* (1988)
Protection requires interaction between the CP expressed by the transgenic plants and the challenge virus. More than a low level of biologically active CP is required for protection.	Loesch-Fries *et al.* (1987); Van Dun *et al.* (1988); Powell *et al.* (1990)

Continued overleaf

Table 1. (continued)

Issue Addressed	System Studied[a]	Phenotype of CP-Mediated Protection
Mechanism of protection	T/T/Tb	CP expressed in transgenic plants forms aggregates; aggregates of CP provided greater protection in protoplasts than non-aggregated CP.
	A/-/Tb T/-/Tb T/-/Tm S/-/Tb C/-/Tb TR/-/Tb	Protection generally declines as the amino acid sequence similarity of the challenge virus CP with the CP expressed in the transformant declines. For TMV- and SMV-CP expressing plants, protection is observed against viruses with CPs having relatively low amino acid sequence similarities with the protecting CPs.
	T/T/Tb T/T/Tm	Continuous high temperatures result in a decline in CP accumulation but not CP-mRNA. A decline in protection in tobacco but not tomato plants expressing TMV-CP was observed under these conditions.
	T/T/Tb	Grafted stem and leaf sections of CP(+) plants can block systemic spread of virus in CP(-) tissue after challenge; primary infection not inhibited in non-expressing tissue.
	T/T/Tb	Expression of the entire viral genome in plants provides greater protection than CP expression alone; resistance against challenge with viral RNA.

Conclusions Drawn	References
Aggregation state of CP (i.e. its quaternary structure) plays an important role in primary protection.	Register and Beachy (1989); Wilson (1989)
Protection relies on an undetermined degree of amino acid sequence similarity and it is clear that at least some of the protection is related to tertiary and/or quaternary structures.	Loesch-Fries *et al.* (1987); Nelson *et al.* (1988); Van Dun and Bol (1988); Van Dun *et al.* (1988); Anderson *et al.* (1989); Quemada *et al.* (1989); Register and Beachy (1989); Stark *et al.* (1990); Stark and Beachy (1989)
Implicates the role of the CP and not the mRNA in protection. The difference between responses of tobacco and tomato plants indicates potential differences in host response.	Nejidat and Beachy (1989)
Implies that protection against systemic spread is effected by expression in tissue through which the challenge virus must move.	Wisniewski LA, Powell P, Nelson RS, Beachy RN (submitted manuscript)
Expression of the entire genome may utilize antisense inhibition or other mechanism(s) suggested for cross protection to a greater extent than plants expressing the CP gene only.	Yamada *et al.* (1988)

Cuozzo *et al.,* 1988; Hemenway *et al.,* 1988; Stark and Beachy, 1989). Thus, of the 28 plant virus families consisting of ssRNA without envelopes (see Wilson (1989) for a recent collation of plant virus families), CPs from viruses within seven of the families have now been shown to provide protection against virus infection in transgenic plants. The prospects for practical application of this technology are promising: transgenic tomato plants that expressed the TMV-CP gene were protected in field experiments under conditions of high infection pressure. The benefits of protection were reflected in the fruit yields, which were 1.4 to 3 times greater than those from inoculated plants not expressing the CP gene (Nelson *et al.,* 1988; Gasser and Fraley, 1989). Recently, results have been presented showing that expression of CP from potato virus Y in transgenic potato plants provides protection against aphid transmission of that virus (Lawson *et al.,* 1989).

SITES AND POTENTIAL MECHANISMS OF CP-MEDIATED PROTECTION

While it is clear that CP-mediated protection is moving towards application in a commercial setting, it is less clear where and how the protection is effected in transgenic plants. However, information is beginning to appear which addresses these issues (Table 1). Many of the studies cited were conducted to identify the extent of the protection and its applicability in practice, and not necessarily to determine the mechanism(s) of protection *per se*. However, even these results often give clues exposing the underlying mechanism(s).

Site of protection

Protection by CP expression appears to exert its effect at several sites. For combinations of plants and viruses which result in a localization response after inoculation, a decrease in lesion numbers on inoculated leaves of CP(+) plants has been found in every system tested to date. Thus, one site of protection is at the primary infection site (see Table 1 for references). For TMV, there is a decrease in the apparent number of primary infections in transgenic CP(+) plants that are local lesion (*Nicotiana tabacum* cv. Xanthi nc) or systemic (cv. Xanthi) hosts (Nelson *et al.,* 1987). The chlorotic spots ("lesions") which form on inoculated leaves of systemic hosts are taken to be equivalent to necrotic lesions as indicators of successful infection events.

Although there is a trend indicating that the decrease in "lesion" numbers is greater in the local lesion host than in the systemic host, additional experiments are required to validate these observations. Thus, any suggestion of an interaction between the local lesion gene *N* and the CP to enhance protection requires further proof.

To determine whether expression of the TMV CP gene stimulates synthesis of potentially protective pathogenesis-related (PR) proteins, TMV CP(+) local lesion and systemic hosts have been analyzed (Carr *et al.,* 1989). Expression of PR proteins is known to be induced by pathogens and other antagonists (see Fritig *et al.,* this volume) and although debated, the expression of the PR1 group has been shown to be correlated with protection against spread of TMV in tobacco (see van Loon, 1985; and references therein). PR1 protein accumulation was not enhanced by expression of the TMV CP gene in the systemic host (Carr *et al.,* 1989). It was enhanced in the transgenic local lesion host compared with controls but the level was at least 25-fold lower than in an untransformed local lesion host infected with TMV. Whether this increase is sufficient to increase protection in the CP(+) local lesion host above that of the systemic host requires not only proof that the hypersensitive host is more resistant to challenge than the systemic host, but also that PR1 expression in CP-expressing transgenic plants increases protection against virus infection. The effect of CP gene expression on the expression of other PR proteins has not been analyzed. Linthorst *et al.* (1989) and Cutt *et al.* (1989) have recently shown that constitutive expression of PR1a or PR1b, respectively, in transgenic tobacco plants does not protect these plants against virus infection.

Prevention of systemic spread of virus by CP-mediated protection has been reported in several examples of CP-mediated resistance but has been studied in detail only for the TMV/tobacco system (LA Wisniewski, P Powell, RS Nelson, RN Beachy, submitted manuscript). In this study, CP(+) or CP(-) systemic host plants were inoculated with TMV or TMV-RNA and infection was followed in the inoculated and systemically infected leaves. Previously, Nelson *et al.* (1987) showed that inoculation with TMV-RNA effectively overcame much of the protection at the primary infection site in terms of lesion numbers. A follow-up experiment was conducted to determine whether virus accumulation within established primary infection sites differed between leaves from CP(+) and CP(-) tobacco plants (Fig. 1). The concentration of virus inoculum was varied in this experiment to ensure both similar "lesion" numbers on CP(+) and CP(-) leaves and a range of "lesion" numbers for each plant line for linear

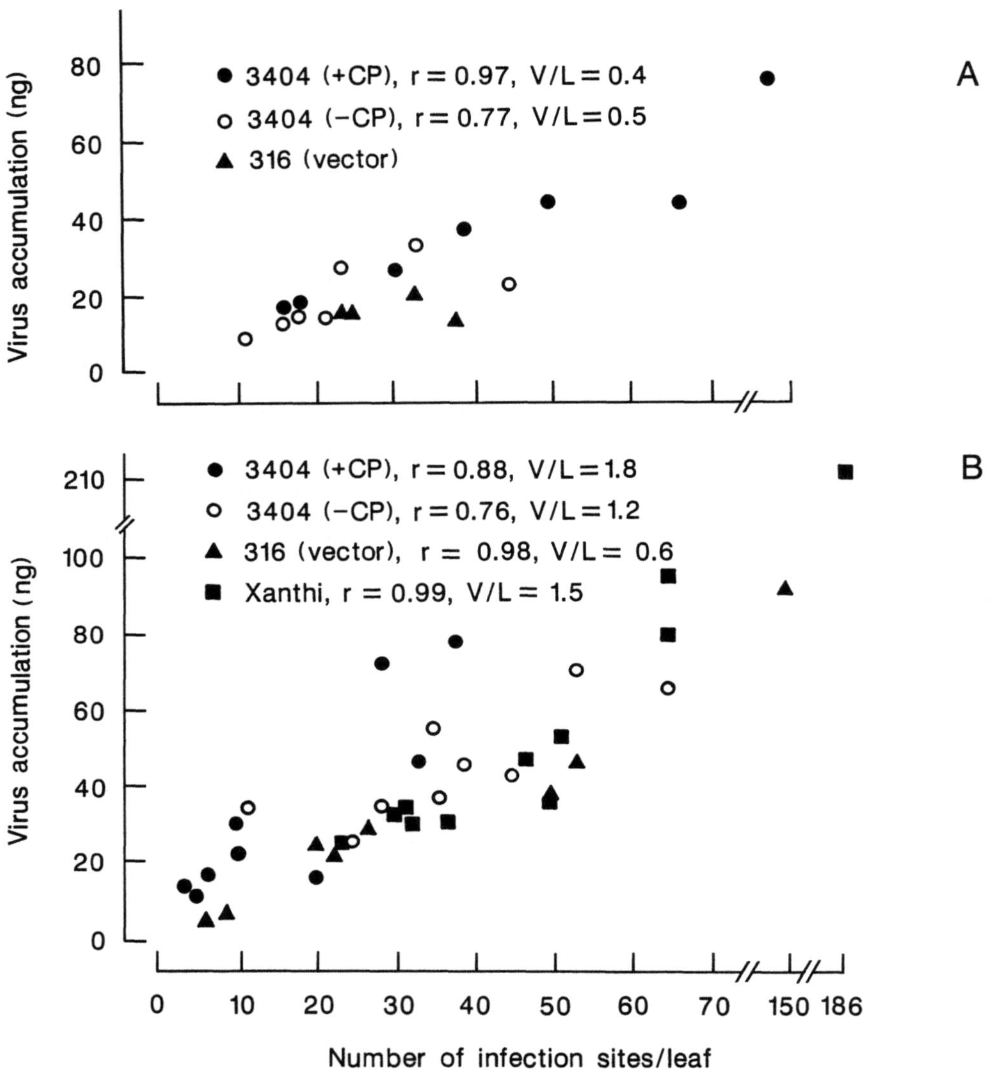

Figure 1. TMV accumulation versus visible chlorotic "lesion" numbers (presumed primary infection sites) on leaves from CP(+) and CP(-) tobacco plants (*Nicotiana tabacum* cv. Xanthi) after inoculation with TMV and TMV-RNA (strain PV230). Plants were greenhouse grown and inoculation (Nelson *et al.*, 1987) took place on fully expanded leaves 51 days after planting. Inoculum concentrations ranged from 0.1 to 10 μg/ml for TMV and 1 to 10 μg/ml for TMV-RNA. Leaf tissue was harvested 5 days post inoculation and extracts analyzed for TMV by ELISA (A): plants inoculated with TMV. (B): plants inoculated with TMV-RNA. 3404(+CP): CP-expressors, R_1 generation; 3404(-CP): non-expressors, R_1 generation; 316: vector control (i.e. plants transformed with vector not containing the TMV-CP coding sequence); Xanthi: untransformed parental line; r, linear correlation coefficient; V/L, virus accumulation/"lesion". Each point represents a value from a single plant. No V/L ratio is given for the 316 (vector) leaf samples shown in A due to the small sample size (4 plants) and lack of slope.

correlation analysis. Results from this work (Fig. 1) and from Wisniewski *et al.* (submitted manuscript) show that virus accumulation within primary "lesions" was essentially the same in both CP(+) and CP(-) plants (note V/L values in Fig. 1). In spite of similar "lesion" numbers and amounts of virus per "lesion" in inoculated leaves, spread, as determined by immunoassay of tissue sections and extracts from systemic leaves and stems, was decreased in CP(+) plants compared with CP(-) plants (Wisniewski *et al.*, submitted manuscript). These results clearly indicate that a second site exists in CP(+) plants, beyond the primary site, providing further protection against virus multiplication or spread.

Mechanism(s) of protection

Intensive research efforts during the last two years are gradually revealing the mechanism(s) of CP-mediated protection. However, much of the data gathered to date must be evaluated cautiously, because many of the published results address protection against primary and general infection and not specifically against systemic infection; the general conclusions drawn in Table 1 often depend upon findings from a limited number of reports; and reports on plants expressing TMV or alfalfa mosaic virus (AlMV)-CP predominate in the literature and the mechanism of protection in these systems may or may not reflect the mechanism(s) of protection in other systems. Thus for example, plants expressing potato virus X (PVX)-CP and inoculated with viral RNA are still protected; see Table 1 for reference.

With these precautions in mind the information presented in Table 1 for plants that express the TMV-CP can be summarized as follows.

1. There is a strong indication that there is a blockage of the infection process prior to the production of the (-) strand viral RNA (Register and Beachy, 1988).
2. Maximum protection is much greater against TMV than against TMV-RNA and involves the CP and not the CP gene transcript in the transgenic plant (Nelson *et al.*, 1987; Register and Beachy 1988; Nejidat and Beachy 1989; Powell *et al.*, 1990). This also applies to the protection against A1MV observed for AlMV CP-expressing plants (Loesch-Fries *et al.*, 1987; Van Dun *et al.*, 1987; Van Dun *et al.*, 1988).
3. Register and Beachy (1989) have shown that not only is CP expression important but that the state of aggregation of the CP can affect the extent of protection in a protoplast transient assay system. The hypothesis that the

quaternary structure of the protecting CP is important in protection is supported by the finding that plants that expressed the CP gene of TMV were resistant to challenge by tobacco mild green mosaic virus which has only 67% amino acid sequence similarity compared with the protecting CP (Stark *et al.* 1990; A Nejidat and RN Beachy, unpublished results). Similar results were obtained in CP-mediated resistance against potyviruses (Stark and Beachy, 1989).

4. Wisniewski *et al.* (submitted manuscript) demonstrated by grafting TMV CP(+) stem and leaf tissue onto CP(-) root and shoot stocks, that CP expression prevented systemic spread but not primary infection by the challenge virus. This systemic protection appears to involve the CP and not the CP transcript since Powell *et al.* (1990) observed no protection either at the primary infection site or systemically when only TMV CP-transcript was synthesized.

Gene constructs that express RNA complementary to CP mRNA (i.e. antisense RNA) might be expected to provide greater protection than (+) sense CP transcripts since they could, in theory, hybridize to both (+) sense genomic and subgenomic viral RNAs. However, antisense gene constructs gave only minimal protection in the PVX, cucumber mosaic virus (CMV), and TMV systems (Cuozzo *et al.*, 1988; Hemenway *et al.*, 1988; Powell *et al.*, 1989). Powell *et al.* (1989) were able to show that only sequences complementary to the 3' untranslated region of TMV RNA, rather than the CP-coding sequence, were able to confer protection. Although protection via antisense RNA is readily overcome, antisense protection may have some ability to limit secondary infections under field situations. Other researchers have investigated the use of antisense sequences against the 5' or 3' ends of viral genomes to provide protection (Beachy *et al.*, 1985; Crum *et al.*, 1988; Valle *et al.*, 1989) and this work is continuing.

CONCLUSIONS

The research results summarized here extend our understanding of how CP-mediated protection is effected. They may also shed light on the mechanism of cross protection, as it is suggested that CP-mediated protection is part of cross protection. It is of interest that expression of the entire TMV viral genome in transgenic tobacco plants, which would closely resemble cross protection, results in a highly effective protection against TMV infection (Yamada *et al.*, 1988). Since these plants are nearly completely protected against challenge by TMV-RNA it is likely that the effectiveness is due to more than the simple presence of higher levels of CP in the leaves prior to challenge inoculation. Thus for these plants and for plants that are cross-protected by classical methods, anti-sense inhibition or other mechanism(s) may be involved to a greater extent than in plants protected by CP gene expression. For a better understanding of the implications of the results of Yamada *et al.* (1988), there should be further work involving expression of portions of the TMV genome other than the CP and 3' untranslated region.

Future work on CP-mediated protection should address how the CP expressed in the transgenic plant prevents or retards virus replication. Site-directed mutations of the CP and further studies on virus strains known to have high structural similarity but low sequence similarity with the transgenic CP could further define the protein conformation necessary for maximum protection. Namba *et al.* (1989) have recently determined the structure of intact TMV by X-ray fibre diffraction methods. They, as well as others (Register *et al.*, 1989), postulate that co-translational disassembly (Shaw *et al.*, 1986; Wilson *et al.*, this volume) may be initiated by a change in Ca^{2+} concentrations and pH values within the cellular milieu. In the Namba report at least two potential Ca^{2+} binding sites were identified in subunit interfaces. They postulate that upon virus entry low Ca^{2+} concentration and high pH within the cell removes calcium and protons from the virion leading to a general destabilization and potentially to ribosome binding and co-translational disassembly. Experiments with CP-expressing protoplasts in the presence of varying Ca^{2+} concentrations may provide evidence linking co-translational disassembly and CP-mediated protection.

In addition to defining the mechanism of protection at the primary infection site, future work should address both the mechanism and specific site of protection against systemic spread of virus. Analysis of cells in the vascular region of CP(+) tissue prior to inoculation might show whether CP accumulates at high levels in this area and thus leads to greater protection. Analysis of cells around the vascular system of inoculated leaves from both CP(+) and CP(-) plants would provide evidence for or against the hypothesis that transgenic CP interferes in some way with virus entry into the phloem of protected plants. These are but a few of the many areas which should be addressed to provide further understanding of CP-mediated protection in plants.

ACKNOWLEDGEMENTS

The authors wish to thank Allyson Wilkins for preparation of the manuscript and Susan Martin for graphics on Figure 1. Roger N Beachy acknowledges support by a grant from the Monsanto Company.

REFERENCES

Anderson EJ, Stark DM, Nelson RS, Powell PA, Tumer NE, Beachy RN (1989) Transgenic plants that express the coat protein genes of TMV and A1MV interfere with disease development of some non-related viruses. Phytopathology (in press)

Bevan MW, Mason SE, Goelet P (1985) Expression of tobacco mosaic virus coat protein by a cauliflower mosaic virus promoter in plants transformed by *Agrobacterium*. EMBO J 4:1921-1926

Beachy RN, Abel P, Oliver MJ, De B, Fraley RT, Rogers SG, Horsch RB (1985) Potential for applying genetic transformation to studies of viral pathogenesis and cross-protection. In: Zaitlin M, Day P, Hollaender A (eds) Biotechnology in plant science: relevance to agriculture in the nineteen eighties. Academic Press, New York, p 265

Carr JP, Beachy RN, Klessig DF (1989) Are the PR1 proteins of tobacco involved in genetically engineered resistance to TMV? Virology 169:470-473

Crum C, Johnson JD, Nelson A, Roth D (1988) Complementary oligodeoxynucleotide mediated inhibition of tobacco mosaic virus RNA translation *in vitro*. Nucleic Acids Res 16:4569-4581

Cuozzo M, O'Connell KM, Kaniewski W, Fang R, Chua N, Tumer NE (1988) Viral protection in transgenic tobacco plants expressing the cucumber mosaic virus coat protein or its antisense RNA. Bio/Technology 6:549-557

Cutt JR, Harpster MH, Dixon DC, Carr JP, Dunsmuir P, Klessig DF (1989) Disease response to tobacco mosaic virus in transgenic tobacco plants that constitutively express the pathogenesis-related PR1b gene. Virology 173:89-97

de Zoeten GA, Fulton RW (1975) Understanding generates possibilities. Phytopathology 65:221-222

Fernow KH (1967) Tomato as a test plant for detecting mild strains of potato spindle tuber virus. Phytopathology 57:1347-1352

Fulton RW (1986) Practices and precautions in the use of cross protection for plant virus disease control. Annu Rev Phytopathol 24:67-81

Gasser CS, Fraley RT (1989) Genetically engineering plants for crop improvement. Science 244:1293-1299

Halk EL, Merlo DJ, Liao LW, Jarvis NP, Nelson SE, Krahn KJ, Hill KK, Rashka KE, Loesch-Fries LS (1989) Resistance to alfalfa mosaic virus in transgenic tobacco and alfalfa. In: Staskawicz BJ, Ahlquist P, Yoder O (eds) Molecular biology of plant-pathogen interactions. Alan R Liss, New York, p 283

Hamilton RI (1980) Defenses triggered by previous invaders - viruses In: Horsfall JG, Cowling EB (eds) Plant disease: an advanced treatise. Vol 5. Academic Press, New York, p 279

Hemenway C, Fang R, Kaniewski WK, Chua N, Tumer NE (1988) Analysis of the mechanism of protection in transgenic plants expressing the potato virus X coat protein or its antisense RNA. EMBO J 7:1273-1280

Lawson C, Kaniewski W, Haley L, Rozman R, Newell C, Tumer N (1989) Expression analysis and field performance of russet burbank potato genetically engineered for resistance to potato virus X and potato virus Y. Horticultural Biotechnol. Symposium Abstr p 42

Linthorst HJM, Meuwissen RLJ, Kauffmann S, Bol JF (1989) Constitutive expression of pathogenesis-related proteins PR-1, GRP, and PR-S in tobacco has no effect on virus infections. Plant Cell 1:285-291

Loesch-Fries LS, Merlo D, Zinnen T, Burhop L, Hill K, Krahn K, Jarvis N, Nelson S, Halk E (1987) Expression of alfalfa mosaic virus RNA 4 in transgenic plants confers virus resistance. EMBO J 6:1845-1851

McKinney HH (1929) Mosaic diseases in the Canary Islands, West Africa, and Gibraltar. J Agric Res 39:557-578

Namba K, Pattanayek R, Stubbs G (1989) Visualization of protein-nucleic acid interactions in a virus: refined structure of intact tobacco mosaic virus at 2.9 A resolution by X-ray fiber diffraction. J Mol Biol 208:307-325

Nejidat A, Beachy RN (1989) Decreased levels of TMV coat protein in transgenic tobacco plants at elevated temperatures reduces resistance to TMV infection. Virology (in press)

Nelson RS, Powell Abel P, Beachy RN (1987) Lesions and virus accumulation in inoculated transgenic tobacco plants expressing the coat protein gene of tobacco mosaic virus. Virology 158:126-132

Nelson RS, McCormick SM, Delannay X, Dubé P, Layton J, Anderson EJ, Kaniewska M, Proksch RK, Horsch RB, Rogers SG, Fraley RT, Beachy RN (1988) Virus tolerance, plant growth, and field performance of transgenic tomato plants expressing coat protein from tobacco mosaic virus. Bio/Technology 6:403-409

Niblett CL, Dickson E, Fernow KH, Horst RK, Zaitlin M (1978) Cross protection among four viroids. Virology 91:198-203

Ponz F, Bruening G (1986) Mechanism of resistance to plant viruses. Annu Rev Phytopathol 24:355-381

Powell-Abel P, Nelson RS, De B, Hoffman N, Rogers SG, Fraley RT, Beachy RN (1986) Delay of disease development in transgenic plants that express the tobacco mosaic virus coat protein gene. Science 232:738-743

Powell P, Sanders PR, Tumer N, Fraley RT, Beachy RN (1990) Protection against tobacco mosaic virus infection in transgenic plants requires accumulation of coat protein rather than coat protein RNA sequences. Virology (in press)

Powell P, Stark DM, Sanders PR, Beachy RN (1989) Protection against tobacco mosaic virus in transgenic plants that express TMV antisense RNA. Proc Natl Acad Sci USA 86:6949-6952

Quemada H, Slightom JL, Gonsalves D (1989) The ability of CMV-C coat protein to protect against other strains of CMV in transgenic plants. J Cell Biochem Suppl 13D:341

Register III JC, Beachy RN (1988) Resistance to TMV in transgenic plants results from interference with an early event in infection. Virology 166:524-532

Register III JC, Beachy RN (1989) A transient protoplast assay for capsid protein-mediated protection: effect of capsid protein aggregation state on protection against tobacco mosaic virus. Virology (in press)

Register III JC, Powell P, Nelson RS, Beachy RN (1989) Genetically engineered cross protection against TMV interferes with initial infection and long distance spread of the virus. In: Staskawicz BJ, Ahlquist P, Yoder O (eds) Molecular biology of plant-pathogen interactions. Alan R Liss, New York, p 269

Sanford JC, Johnston SA (1985) The concept of parasite-derived resistance: deriving resistance genes from the parasite's own genome. J Theor Biol 113:395-405

Sarkar S, Smitamana P (1981) A proteinless mutant of tobacco mosaic virus: evidence against the role of a viral coat protein for interference. Mol Gen Genet 184:158-159

Sequeira L (1984) Cross protection and induced resistance: their potential for plant disease control. Trends Biotechnol 2:25-29

Shaw JG, Plaskitt KA, Wilson TMA (1986) Evidence that tobacco mosaic virus particles disassemble cotranslationally *in vivo*. Virology 148:326-336

Sherwood JL (1987) Demonstration of the specific involvement of coat protein in tobacco mosaic virus (TMV) cross protection using a TMV coat protein mutant. J Phytopathol 118:358-362

Sherwood JL, Fulton RW (1982) The specific involvement of coat protein in tobacco mosaic virus cross protection. Virology 119:150-158

Stark DM, Beachy RN (1989) Protection against potyvirus infection in transgenic plants: evidence for broad spectrum resistance. Bio/Technology (in press)

Stark DM, Register III JC, Nejidat A, Beachy RN (1990) Toward a better understanding of coat protein mediated protection. In: Lamb CJ, Beachy RN (eds) Plant gene transfer. Alan R Liss, New York, (in press)

Tumer NE, O'Connell KM, Nelson RS, Sanders PR, Beachy RN, Fraley RT, Shah DM (1987) Expression of alfalfa mosaic virus coat protein gene confers cross-protection in transgenic tobacco and tomato plants. EMBO J 6:1181-1188

Valle RPC, Lambert M, Joshi RL, Morch MD, Haenni AL (1989) Strategies for interfering *in vivo* with the replication of turnip yellow mosaic virus. J Cell Biochem Suppl 13D:345

Van Dun CMP, Bol JF (1988) Transgenic tobacco plants accumulating tobacco rattle virus coat protein resist infection with tobacco rattle virus and pea early browning virus. Virology 167:649-652

Van Dun CMP, Bol JF, van Vloten-Doting L (1987) Expression of alfalfa mosaic virus and tobacco rattle virus coat protein genes in transgenic tobacco plants. Virology 159:299-305

Van Dun CMP, Overduin B, van Vloten-Doting L, Bol JF (1988) Transgenic tobacco expressing tobacco streak virus or mutated alfalfa mosaic virus coat protein does not cross-protect against alfalfa mosaic virus infection. Virology 164:383-389

Van Loon LC (1985) Pathogenesis-related proteins. Plant Mol Biol 4:111-116

Wilson TMA (1989) Plant viruses: A tool-box for genetic engineering and crop protection. BioEssays 10:179-186

Yamaya J, Yoshioka M, Meshi T, Okada Y, Ohno T (1988) Cross protection in transgenic tobacco plants expressing a mild strain of tobacco mosaic virus. Mol Gen Genet 215:173-175

Zaitlin M (1976) Viral cross protection: More understanding is needed. Phytopathology 66:382-383

Zinnen TM, Fulton RW (1986) Cross-protection between sunn-hemp mosaic and tobacco mosaic viruses. J Gen Virol 67:1679-1687

THE USE AND MISUSE OF VIRUSES IN CLONING AND EXPRESSION IN PLANTS

Roger Hull
John Innes Institute
AFRC Institute of Plant Science Research
Colney Lane
Norwich
NR4 7UH
UK

ABSTRACT

Plant viruses have attracted considerable attention as possible vectors for introducing "foreign" genes into plants. Two basic systems have been studied: integration of viral sequences into the host chromosome, and use as episomal vectors. Integrated promoter sequences from DNA viruses, primarily the cauliflower mosaic virus 35S promoter, have been used successfully. Episomal vectors have been constructed both from DNA and RNA viral sequences for both the transient and the constitutive expression of genes; some major problems remain to be fully overcome. The use of viral sequences in plants, whether as gene vectors or to confer non-conventional forms of resistance, raises the possibility that there may be risks on the field release of modified plants.

INTRODUCTION

The recent advances in molecular biology and in plant transformation have led to the consideration of how plant viral sequences, which are normally detrimental to plants, can be used for crop improvement. There are now two basic areas in which the potential of plant viral sequences as beneficial agents is being explored. The concepts of using viral or virus-related sequences to confer resistance or immunity to viruses, first encapsulated by Sandford and Johnston (1985), are discussed in other chapters in this volume. The potential of viral sequences as gene vectors in plants has attracted much research, some of the main facets of which are reviewed in this chapter. The use of these sequences to confer resistance characters or to express genes in plants raises questions about possible risks on the field release of such genetically manipulated plants; this will also be discussed in this chapter.

VIRUSES AND VECTORS

It is primarily the ability of most DNA and RNA viruses to replicate efficiently and express in a manner which is not dependent upon the cell cycle that has led to attempts to use them as gene vectors. This has inspired much research which has enhanced our knowledge of the molecular biology of several viruses, especially those with DNA genomes. There have been several previous reviews on the subject including those by Buck and Coutts (1983), Davies and Stanley (1989); Davies *et al.* (1987); Gardner (1983); Gronenborn (1984, 1987); Hohn and Hohn (1982); Hohn *et al.* (1982); Howell (1982); Hull (1978, 1980, 1981, 1983, 1985); Hull and Davies (1983); van Vloten-Doting (1983) and Wilson (1989); most of these reviews deal with specific groups of viruses. In this chapter I do not intend to repeat all the information covered in the previous reviews but I wish to take a general view of the situation. For more detailed information, the reader is referred to these previous reviews.

There are two basic types of vector system in which viral sequences are involved: integrative vectors and episomal vectors.

Integrative vectors

No plant virus is known specifically to integrate into the host genome and thus viral sequences cannot be used to direct integration specifically. The sequences which are used are those which enhance the expression of chimeric vectors with the integrative capacity often being provided by *Agrobacterium* Ti or Ri plasmid sequences; in other cases constructs are transferred directly into the plant genome by integration (see Paszkowski *et al.*, 1984; Shillito *et al.*, 1985). The viral sequence which has been used most is the 35S promoter from cauliflower mosaic virus (CaMV), which has been shown to be among the strongest of plant promoters (Dandekar *et al.*, 1987; Ecker and Davies, 1986; Ellis *et al.*, 1987; Morelli et *al.*, 1985; Nagy *et al.*, 1987; Sanders *et al.*, 1987). Duplication of the 35S promoter enhances its activity and the duplicated region also enhances heterologous promoters (Kay *et al.*, 1987).

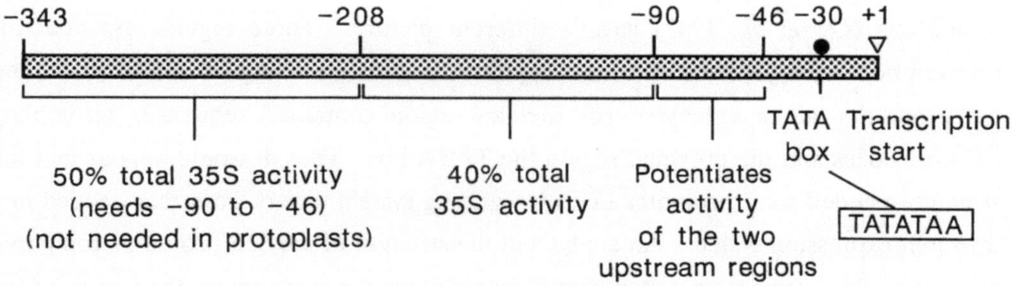

Figure 1. The structure of the promoter region of the CaMV 35S RNA. The box represents the 343 nucleotides 5' of the transcription start site, with regions which control transcription activity being indicated. Based on information from Fang *et al.* (1989).

The structure of the promoter has recently been determined. The TATA box, which is thought to establish the polarity of the promoter and to position the transcription start site, is at 31 nucleotides (-31) from the initiation site. The majority of the promoter strength lies in the sequence of 343 nucleotides upstream of the initiation site (Odell *et al.*, 1985). The studies of Fang *et al.* (1989) on integrated constructs showed that this sequence is made up of three functional regions (Fig. 1). The -343 to -208 and the -208 to -90 regions are responsible for about 50 and 40% of the promoter activity respectively. The -90 to -46 region has little activity on its own but enhances the transcriptional activity of the two upstream regions. Very recently Benfrey *et al.* (1989) showed tissue specificity for different regions of the CaMV 35S promoter. Expression from the -90 to +8 region is strongest in the radicle of the embryo, the radicle pole of the endosperm and in the root tissues of seedlings and mature tobacco plants. The region from -343 to -90 gives strongest expression in the cotyledons and adjacent endosperm cells in seeds, in the cotyledons of seedlings and in the leaves and stems of mature plants. These studies were done with tobacco which is a non-host of CaMV; it is not known if these regions act in the same way in a host of the virus. Other factors which affect the overall promoter activity include the chromosomal location of the insert in the transgenic plant.

A major point that comes from the work of Fang *et al.* (1989) is that promoters need to be well defined. They should not just be sequences upstream of transcripts which have convenient restriction endonuclease sites.

An analysis of the 35S promoter structure in a transient expression system in protoplasts (Ow *et al.*, 1987) gave a different picture. Three regions essential for transcription activity were identified. The distal region extending from -148 to -89 had the properties of an enhancer; the median region contained sequences resembling CCAAT boxes and the proximal region the TATA box. Thus, it would appear that the structure needed for a promoter in an integrating system differs from that needed in a transient expression system. An analysis of the strength of the 35S promoter compared with other plant promoters, noted above, should also take into account the factors of the promoter construct, transient or integrated expression (see Michael, 1988).

The CaMV 19S promoter has also been used to direct transcription of genes introduced into both di- and monocotyledons (see Paszkowski *et al.*, 1984; Potrykus *et al.*, 1985). In a comparison of the two CaMV promoters, Ow *et al.* (1987) showed that the 35S promoter was 80 times more active than the 19S promoter in transient expression systems; comparison of the two promoters in integrative systems is very difficult because of positional effects of the sites of integration. Furthermore, the detailed structure of the 19S promoter has not yet been determined and it is possible that the "easy to obtain" restriction endonuclease fragment omits some essential regions.

Episomal vectors

As viruses are *per se* episomal expressing sequences their potential for the episomal expression of genes is continually being explored. Viral sequences are used primarily in two different systems, namely for transient expression or for constitutive expression.

Transient expression is mostly used to analyse sequences which control transcription or translation. The viral sequences are attached to a reporter gene, frequently chloramphenicol acetyl transferase (CAT), ß-glucuronidase (GUS), neomycin phosphotransferase (NPT) II, or luciferase, and introduced into protoplasts by polyethylene glycol treatment or by electroporation. The activity of the expressed enzyme is then measured. As noted above, this approach has been used in experiments to define the CaMV 35S promoter.

A transient assay using the CAT gene enabled the promoter transcribing the mRNA for the coat protein of maize steak virus (MSV) to be defined and studied (Fenoll *et al.*, 1988). A 122 bp sequence upstream of the mRNA initiation site enhances the promoter activity in an orientation- but not position-independent manner. This region is interchangeable with the distal region (-148 to -89) of the CaMV 35S promoter described above.

In what was termed a transient assay, the coat protein region of the geminivirus wheat dwarf (WDV) was replaced by the NPT II gene and the constructs imbibed into dry wheat embryos (Topfer *et al.*, 1989). Constructs with dimers of WDV were effective in expressing NPT II activity. Similarly, constructs of more-than-monomer MSV genome, in which either the coat protein gene or an open reading frame predicted to encode a 10.9 kDa protein were replaced by the CAT gene, expressed CAT activity on agroinfection of maize (Lazarowitz *et al.*, 1989). For both these viral constructs, it is debatable whether true transient expression was occurring, or whether it was constitutive expression with the virus replicating but not spreading from cell to cell.

Two types of viral sequence have been used in transient expression vectors based on RNA viruses. The 5'-leader RNA sequence of tobacco mosaic virus (TMV), termed the Ω fragment, enhances the expression of reporter gene mRNAs to which it is attached (Gallie *et al.*, 1987a,b; Sleat *et al.*, 1987; 1988). The evidence suggests that this is because the Ω fragment remains free from secondary structure and can recruit ribosomes with great facility (Sleat *et al.*, 1988). The 5'-leader sequence of alfalfa mosaic virus RNA 4 also stimulates the expression of reporter genes (Joblin and Gehrke, 1987; Joblin *et al.*, 1988) though not to the extent of the TMV fragment (Wilson, 1989). The other type of viral sequence that is used is that which initiates the assembly of TMV particles. The origin of assembly sequence (OAS) has been attached to various constructs, the transcripts of which were packaged in TMV coat protein to form pseudovirions (Sleat *et al.*, 1986; Gallie *et al.*, 1987c). When pseudovirions comprising TMV OAS and Ω fragment together with a reporter gene were introduced into protoplasts the gene was expressed transiently (see Wilson, 1989).

The efficiency of transient expression depends upon various factors as well as the strength of the promoter and the efficiency of the leader sequence. For instance, the conditions for electroporation of the constructs into protoplasts, such as the concentration and quality of the DNA preparations and the number of pulses and field strength of the electric discharge had considerable effects on the transient expression of CAT in tobacco protoplasts (Michael, 1988).

The aim of most of the attempts to use viruses as *constitutive expression* vectors has been to try to insert "foreign" genes into the viral sequence without inhibiting the normal viral replication. Most of these studies have been on viruses with DNA genomes. There have been numerous attempts to find "non-lethal" sites for gene insertion in CaMV DNA. Infectivity was retained after small insertions (12 bp) in any of several sites but only two could tolerate larger insertions (see Hull, 1985). Insertions into one of these sites, the gene VII region, have proved to be unstable. The gene II region is dispensible for viral replication and constructs have been made with some or most of it removed. This region has then been replaced by the genes coding for dihydrofolate reductase (Brisson *et al.*, 1984) and for interferon (Gronenborn, 1987) and, in both cases, active gene products have been expressed. Great care has to be taken in the constructs as features which affect the translational polarity of CaMV abolish viral replication (see Gronenborn, 1987).

Coat protein deletion mutants of whitefly-transmitted geminiviruses retain their infectivity (Etessami *et al.*, 1988; Gardiner *et al.*, 1988; Hayes *et al.*, 1988a; Ward *et al.*, 1988a). Such mutants of tomato golden mosaic virus (TGMV) elicit very attenuated symptoms (Gardiner *et al.*, 1988; Hayes et *al.*, 1988a); however, it is likely that this is due to a large reduction in viral replication and expression. Coat protein deletion mutants of African cassava mosaic virus (ACMV) give normal symptoms but analysis of the progeny showed that the viral genome had reverted to the normal size of ACMV DNA (Etessami *et al.*, 1988); this was due to duplication of viral sequences. As was pointed out by Davies and Stanley (1989), different experimental approaches have been used to analyse these two viruses; ACMV mutants were inoculated directly to plants whereas those of TGMV were agroinfected. This could explain the differences in the results.

The ACMV coat protein gene could be replaced by genes of similar size (e.g. the CAT gene) and the inserted gene was expressed (Ward *et al.*, 1988a), whereas larger inserts were deleted to give a genome of viral size (Stanley and Townsend, 1986); these recombinant ACMV infections were made by mechanical inoculation. Using agroinfection it was possible to maintain larger inserts (e.g. the GUS and NPT II genes) in coat protein-deleted versions of TGMV (Hanley-Bowdoin *et al.*, 1988; Hayes *et al.*, 1988b; 1989). However, in a proportion of the plants deletions occurred giving molecules with a range of sizes down to that of the viral genome. Thus, even after agroinfection, deletions take place, presumably during the constitutive replication of the virus. The 76 nucleotide region upstream of the TGMV coat protein initiation codon was shown to be essential to the expression of the inserted gene and could be replaced by a CaMV 35S promoter fragment with increased expression (Hayes *et al.*, 1989).

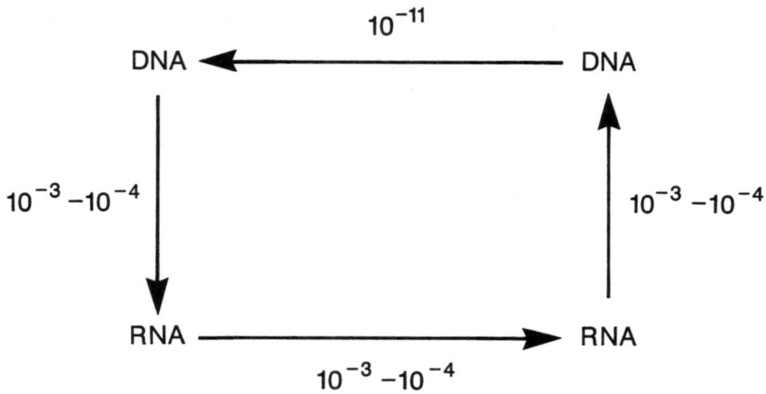

Figure 2. Representation of the error rates in transfer of genetic information between DNA or RNA molecules and from DNA to RNA and *vice versa*. The numbers are the mis-incorporated nucleotides per number of nucleotides transcribed. Thus on DNA to DNA there is one mis-incorporation per 10^{11} nucleotides incorporated. Based on data from Reanney (1984), Leider *et al.* (1988), and Ward *et al.* (1988b).

Much less attention has been paid to the use of RNA viruses as constitutive episomal vectors. The coat protein of brome mosaic virus has been replaced by the CAT gene which was then expressed in barley protoplasts (French *et al.* 1986). Takamatsu *et al.* (1987) replaced the coat protein gene of TMV with the CAT gene and showed that biologically active CAT was produced in inoculated tobacco leaves; however, there was no systemic spread of the recombinant RNA presumably because the coat protein was lacking.

Problems with episomal vectors

From the results so far obtained it would appear that viruses have only limited potential as vectors to transfer and express foreign genes in crop plants. This is for a variety of reasons, the main one being that the replication and expression of viral genomes are highly coordinated. The two results of this are that there is difficulty in finding insertion sites which do not interfere with replication and that it is often impossible to dissect symptom production from replication functions. Thus, the aim of developing a vector which has the replication, expression and short- and long-distance movement properties of a virus but is not symptom inducing is very difficult to achieve. Even if it were easy, most viral replication systems, with the possible exception of those of geminiviruses, do not involve proof reading of newly incorporated nucleotides. This leads to a high error rate (Fig. 2). If genes are under selection, as is the case for those of viruses, this is not a problem because lethal errors will not be propagated. However, in most cases inserted "foreign" genes will not be under continuous selection and thus might be ineffective when they are needed.

Even though it is unlikely that viruses or viral sequences will be used as episomal vectors for the genetic engineering of crop plants, it is certain that various viral functions will be incorporated into chimeric vectors. This leads to the need to assess the risks that might occur on the field release of plants containing such sequences.

RISKS ON FIELD RELEASE

As noted above there are likely to be two reasons for incorporating viral sequences into plants: for non-conventional resistance and as parts of gene vectors. The main question that has to be answered in assessing the risks on field release of such plants is: what is the risk of other viruses interacting with viral genes or sequences which are genetically engineered into plants for use as vectors or to confer resistance?

I will consider this question in relation to coat protein-mediated resistance, which is likely to be the first form of non-conventional resistance to be available for general field release, and to viral sequences forming parts of vectors. Some aspects of these risks are discussed by Hull (1989).

Coat protein

There are three major areas of risk associated with the introduction of coat protein genes into plant genomes; that owing to recombination is discussed below. Of the other two the main one is the possibility of an infecting heterologous virus genome being encapsidated in the expressed viral coat protein. There is also the possibility of synergistic interactions between the heterologous virus and the expressed coat protein.

The encapsidation of the genome of one virus in the capsid protein of another has been shown in *in vitro* experiments with several viruses (see Bancroft, 1970; Hull, 1970). These experiments indicated that there were fewer constraints on the encapsidation of nucleic acid (or other polycations) by the coat proteins of certain viruses with isometric particles. Studies on the assembly of some viruses with rod-shaped particles, and especially TMV, have shown that specific interactions between the viral coat protein and nucleic acid are needed to initiate particle synthesis. Thus, viral coat proteins can be considered to be either "catholic" or "conservative" in their ability to encapsidate other nucleic acids. Heterologous encapsidation (transcapsidation) has been demonstrated *in vivo* between strains of barley yellow dwarf virus (Rochow, 1970) and has been suggested as the mechanism of the helper activity in the insect transmission of other viruses (Falk *et al.*, 1989; Waterhouse and Murant, 1983). Even the "conservative" TMV coat protein has been found forming particles (pseudovirions) around chloroplast RNA (Rochon and Siegel, 1984). It is likely that, in normal virus infection, replication of the genome and encapsidation of progeny nucleic acid are co-ordinated and thus relatively little coat protein would be available for transcapsidation. In transgenic plants, however, coat protein is being produced constitutively and thus, although it might provide protection against the homologous virus, it could be available for the encapsidation of heterologous viral sequences.

As well as protecting the viral genome the coat protein in many cases is involved in determining the specificity of transmission of viruses, for example by insects (see Harrison and Murant, 1984, and Murant, this volume). This raises the possibility of the vector specificity of the virus infecting the transgenic plant being changed by transcapsidation. Although this would be for only one transmission passage (in the non-transgenic host it would be encapsidated by its own coat protein and thus revert to its normal means of transmission) the virus might be taken to ecological situations to which its normal vector would not usually introduce it. This could change its epidemiology.

A potentially more serious risk is that of making the infecting heterologous virus seed-transmitted in the transgenic host. A high rate of seed transmission would effectively take the virus into the germ line. For many seed-transmitted viruses there is variation between strains in the efficiency of transmission (Carroll, 1972; Haber and Hamilton, 1980; Hananda and Harrison, 1977; Harrison *et al.*, 1974). This indicates that, at least in part, the efficiency of seed transmission is controlled by viral gene product(s) but there are no data as to which is involved.

Another situation which could arise from infection of a coat protein transgenic plant by another virus is that of synergism (Holmes, 1956). Several combinations of viruses, e.g. tomato mosaic and potato virus X in tomato (causing tomato streak), and maize chlorotic mottle and maize dwarf mosaic viruses (giving corn lethal necrosis) cause much more severe symptoms than those induced by the individual viruses. There is some evidence that coat protein is involved in symptom production (see Dawson *et al.*, 1988) but there are no firm indications as to whether it contributes to synergistic responses.

Recombination

In a plant which is transgenic in viral sequences, there are several ways in which there could be recombination between nucleic acid sequences which could lead to problems. There can be homologous or heterologous recombination (see King, 1988; Zimmern, 1988; for reviews). There can be DNA/DNA or RNA/RNA recombinations. Recombinations can take place in the nucleus or cytoplasm, or nucleic acid sequences can move from the nucleus to the cytoplasm or from the cytoplasm to the nucleus, and then recombine.

Homologous recombination has been reported for plant viruses with DNA and RNA genomes. The recombination in the DNA virus CaMV (Choe *et al.*, 1985; Dixon *et al.*, 1986; Geldereich *et al.*, 1986; Grimsley *et al.*, 1986a; Howell *et al.*, 1983; Lebeurier *et al.*, 1982; Walden and Howell, 1982) most probably takes place in the RNA to DNA phase of the replication cycle.

Recently, evidence for homologous recombination in geminiviruses has become available. The "infilling" of deletions in ACMV DNA has been mentioned above. The agroinfection of plants with geminiviruses (or with CaMV) requires that more-than-monomer constructs are introduced into the plant (see Boulton *et al.*, 1989a,b; Grimsley *et al.*, 1986b, 1987; Hanley-Bowdoin *et al.*, 1988; Hayes *et al.*, 1988a; Hohn *et* al., 1987; Woolston *et al.*, 1988). It is considered that, at least in the case of the geminiviruses, the viral genome excises by intramolecular recombination. Boulton *et al.* (1989b) present evidence for intermolecular recombination.

The evidence for homologous recombination among RNA plant viruses comes from observations on viral sequences (Bujarski and Kaesberg, 1986; Robinson *et al.*, 1987). Although there are only a few records of recombination in plant RNA viruses, the observations on animal RNA viruses (King, 1988) would suggest that homologous recombination is a relatively common event. It is much harder to obtain information on heterologous recombination. However, the hypothesis that viruses have a modular construction (Hull and Covey 1986; Zimmern, 1988) with genes to adapt to specific host situations is dependent upon heterologous recombination. Whether the transcript of a constitutively expressed viral coat protein gene could recombine into the genome of a heterologous virus is unknown. But, if the modular hypothesis is correct there is a possibility that this event could occur.

A recent report on targetting of genes in plants (Paszkowski *et al.*, 1988) raises the possibility that recombinational events could lead to the integration of extra viral sequences into the host chromosome.

Thus, the results of the molecular approaches to virology are raising questions for which at present there are no answers. This should, however, stimulate further research and thus increase our knowledge of how viruses function.

ACKNOWLEDGEMENTS

I thank Drs JW Davies and J Stanley for their helpful comments on this manuscript and Mr M Harvey for preparing the figures.

REFERENCES

Bancroft JB (1970) The self-assembly of spherical plant viruses. Adv Virus Res 16:99-134

Benfrey PN, Ren L, Chua N-H (1989) The CaMV 35S enhancer contains at least two domains which can confer differential developmental and tissue-specific expression patterns. EMBO J 8:2195-2202

Boulton MI, Buchholz WG, Marks MS, Markham PG, Davies JW (1989a) Specificity of *Agrobacterium*-mediated delivery of maize streak virus DNA to members of the Gramineae. Plant Mol Biol 12:31-40

Boulton MI, Steinkeller H, Donson J, Markham PG, King DI, Davies JW (1989b) Mutational analysis of the virion-sense genes of maize streak virus. J Gen Virol (in press)

Brisson N, Paszkowski J, Penswick JR, Gronenborn B, Potrykus L, Hohn T (1984) Expression of a bacterial gene in plants by using a viral vector. Nature 310:511-514

Buck KW, Coutts RHA (1983) The potential of geminiviruses in the construction of plant host vector systems. Plant Mol Biol 2:351-357

Bujarski JJ, Kaesberg P (1986) Genetic recombination between RNA components of a multipartite plant virus. Nature 321:528-531

Carroll TC (1972) Seed transmissibility of two strains of barley stripe mosaic virus. Virology 48:323-336

Choe IS, Melcher U, Lebeurier G, Essenberg RC (1985) Recombination between mutant cauliflower mosaic virus DNAs. Plant Mol Biol 5:281-289

Dandekar AM, Gupta PK, Durzan DJ, Knauf V (1987) Transformation and foreign

gene expression in micropropagated Douglas fir (*Pseudotsuga menzieri*). Bio/Technology 5:587-590
Davies JW, Stanley J (1989) Geminivirus genes and vectors. Trends Genet 5:77-81
Davies JW, Townsend R, Stanley J (1987) The structure, expression, functions and possible exploitation of geminivirus genomes. In: Hohn T, Schell J (eds) Plant DNA infectious agents. Springer-Verlag, Vienna, p 31.
Dawson WDO, Bubrick P, Grantham GL (1988) Modifications of the tobacco mosaic virus coat protein gene affecting replication, movement and symptomatology. Phytopathology 78:783-789
Dixon L, Nyffenegger T, Delley G, Martinez-Izquierdo J, Hohn T (1986) Evidence for replicative recombination in cauliflower mosaic virus. Virology 150:463-468
Ecker JR, Davies RW (1986) Inhibition of gene expression in plant cells by expression of antisense RNA. Proc Natl Acad Sci USA 83:5372-5376
Ellis JG, Llewellyn DJ, Dennis ES, Peacock WJ (1987) Maize *Adh*-1 promoter sequences control anaerobic regulation: addition of upstream promoter elements from constitutive genes is necessary for expression in tobacco. EMBO J. 6:11-16.
Etessami P, Watts J, Stanley J (1988) Size reversion of African cassava mosaic virus coat protein gene deletion mutants during infection of *Nicotiana benthamiana*. J Gen Virol 70:277-289
Falk BW, Duffus JE, Morris TJ (1979) Transmission, host range and serological properties of the viruses that cause lettuce speckles disease. Phytopathology 69:612-617
Fang R-X, Nagy F, Sivasubramaniam S, Chua N-H (1989) Multiple *cis* regulatory elements for maximal expression of the cauliflower mosaic virus 35S promoter in transgenic plants. The Plant Cell 1:141-150
Fenoll C, Black DM, Howell SH (1988) The intergenic region of maize streak virus contains promoter elements involved in rightward transcription of the viral genome. EMBO J. 7:1589-1596
French R, Janda M, Ahlquist P (1986) Bacterial gene inserted in an engineered RNA virus: Efficient expression in monocotyledonous plant cells. Science 231:1294-1297.
Gallie DR, Sleat DE, Watts JW, Turner PC, Wilson TMA (1987a) The 5'-leader sequence of tobacco mosaic virus RNA enhances the expression of foreign gene transcripts *in vitro* and *in vivo*. Nucleic Acids Res 15:3257-3273
Gallie DR, Sleat DE, Watts JW, Turner PC, Wilson TMA (1987b) A comparison of eukaryotic viral 5'-leader sequences as enhancers of mRNA expression *in vivo*. Nucleic Acids Res 15:8693-8711
Gallie DR, Sleat DE, Watts JW, Turner PC, Wilson TMA (1987c) *In vivo* uncoating and efficient expression of foreign mRNAs packaged in TMV-like particles. Science 236:1122-1124
Gardiner W, Sunter G, Brand L, Elmer JS, Rogers SG, Bisaro DM (1988) Genetic analysis of tomato golden mosaic virus: the coat protein is not required for systemic spread or symptom development. EMBO J. 7:899-904
Gardner RC (1983) Plant viral vectors: CaMV as an experimental tool. In: Kosuge T, Meredith CP, Hollaender A (eds) Genetic engineering in plants. Plenum, New York, p 21
Geldereich A, Lebeurier G, Hirth L (1986) In vivo dimerization of cauliflower mosaic virus DNA can explain recombination. Gene 48:277-286
Grimsley N, Hohn T, Hohn B (1986a) Recombination in a plant virus: template-switching in cauliflower mosaic virus. EMBO J 5:641-646
Grimsley N, Hohn B, Hohn T, Walden R (1986b) Agroinfection an alternative route

for viral infection of plants by using the Ti plasmid. Proc Natl Acad Sci USA 83:3282-3286
Grimsley N, Hohn T, Davies JW, Hohn B (1987) *Agrobacterium*-mediated delivery of infectious maize streak virus into maize plants. Nature 325:177-179
Gronenborn B (1984) Cauliflower mosaic virus a possible vector for genetic engineering in plants. Hoppe-Seyler's Z Physiol Chem 365:217-218
Gronenborn B (1987) The molecular biology of cauliflower mosaic virus and its application as plant gene vector. In: Hohn T, Schell J (eds) Plant DNA infectious agents. Springer-Verlag, Vienna, p 1
Haber S, Hamilton RI (1980) Distribution of determinants for symptom production, nucleoprotein component distribution and antigenicity of coat protein between the two RNA components of cherry leaf roll virus. J Gen Virol 50:377-389
Hanada K, Harrison BD (1977) Effects of virus genotype and temperature on seed transmission of nepoviruses. Ann Appl Biol 85:79-92
Hanley-Bowdoin L, Elmer JS, Rogers SF (1988) Transient expression of heterologous RNAs using tomato golden mosaic virus. Nucleic Acids Res 16:10511-10528
Harrison BD, Murant AF (1984) Involvement of virus-coded proteins in transmission of plant viruses by vectors. In: Mayo MA, Harrap KA (eds) Vectors in virus biology. Academic Press, London, p 1
Harrison BD, Murant AF, Mayo MA, Roberts IM (1974) Distribution of determinants for symptom production, host range and nematode transmissibility between the two RNA components of raspberry ringspot virus. J Gen Virol 22:233-247
Hayes RJ, Coutts RHA, Buck KW (1988a) Agroinfection of *Nicotiana* spp with cloned DNA of tomato golden mosaic virus. J Gen Virol 69:1487-1496
Hayes RJ, Petty IDT, Coutts RHA, Buck KW (1988b) Gene amplification and expression in plants by a replicating geminivirus vector. Nature 334:179-182
Hayes RJ, Coutts RHA, Buck KW (1989) Stability and expression of bacterial genes in replicating geminivirus vectors in plants. Nucleic Acids Res 17:2391-2403
Hohn B, Hohn T (1982) Cauliflower mosaic virus: a potential vector for plant genetic engineering. In: Kahl G, Schell JS (eds) Molecular biology of plant tumors. Academic Press, New York, p 549
Hohn T, Richards KE, Lebeurier G (1982) CaMV on its way to becoming a useful plant vector. Curr Top Microbiol Immunol 96:193-236
Hohn B, Hohn T, Boulton MI, Davies JW, Grimsley N (1987) Agroinfection of *Zea mays* with maize streak virus DNA. In: von Wettstein D, Chua N-H (eds), Plant molecular biology. Plenum, New York, p 459
Holmes FO (1956) A simultaneous-infection test for viral inter-relationships as applied to aspermy and other viruses. Virology 2:611-617
Howell SH (1982) Plant molecular vehicles: potential vectors for introducing foreign genes into plants. Annu Rev Plant Physiol 33:609-650
Howell SH, Walden RM, Marco Y (1983) Recombination and replication of cauliflower mosaic virus DNA. In: Goldberg RB (ed) Plant molecular biology. UCLA Symposium on molecular and cellular biology Vol. 12. Alan R Liss Inc, p 137
Hull R (1970) Studies on alfalfa mosaic virus. III. Reversible dissociation and reconstitution studies. Virology 40:34-47
Hull R (1978) The possible use of plant virus DNAs in genetic manipulation in plants. Trends Biochem Sci 3:254-256
Hull R (1980) Genetic engineering in plants: the possible use of cauliflower mosaic virus DNA as a vector. In: Sala F, Parisi B, Cella R, Ciferri O (eds) Plant cell cultures: Results and perspectives. Elsevier, Amsterdam, p 219
Hull R (1981) Cauliflower mosaic virus DNA as a possible gene vector for higher

plants. In: Panopoulos NJ (ed) Genetic engineering in plant sciences. Praeger Scientific, New York, p 99
Hull R (1983) The current status of plant viruses as potential DNA/RNA vector systems. In: Mantell SH, Smith H (eds) Plant biotechnology. Society for Experimental Biology Seminar Series Vol. 18. Cambridge University Press, Cambridge, p 299
Hull R (1985) Viruses as vectors for plant genes. In: Dodds JH (ed) Plant genetic engineering. Cambridge University Press, Cambridge, p 95
Hull R (1989) Non-conventional resistance to viruses in plants - concepts and risks. Proc 19th Stadler Conference (in press)
Hull R, Covey SN (1986) Genome organization and expression of reverse transcribing elements: variations and a theme. J Gen Virol 67:1751-1758
Hull R, Davies JW (1983) Genetic engineering with plant viruses and their potential as vectors. Adv Virus Res 28:1-31
Joblin SA, Gehrke L (1987) Enhanced translation of chimeric messenger RNAs containing a plant viral leader sequence. Nature 325:622-625
Joblin SA, Cuthbert CM, Rogers SG, Fraley RT, Gehrke L (1988) *In vitro* transcription and translational efficiency of chimeric SP6 messenger RNAs devoid of 5'-vector nucleotides. Nucleic Acids Res 16:4483-4498
Kay R, Chan A, Daly M, McPherson J (1987) Duplication of CaMV 35S promoter sequences creates a strong enhancer for plant genes. Science 236:1299-1302
King AMQ (1988) Genetic recombination in positive strand RNA viruses. In: Domingo E, Holland JJ, Ahlquist P (eds) RNA genetics Vol. 2. CRC Press, Boca Raton, p 149
Lazarowitz SG, Pinder AK, Damsteegt VD, Rogers SG (1989) Maize streak virus genes essential for systemic spread and symptom development. EMBO J. 8:1023-1032
Lebeurier G, Hirth L, Hohn B, Hohn T (1982) In vivo recombination of cauliflower mosaic virus DNA. Proc Natl Acad Sci USA 79:2932-2936
Leider JM, Palese P, Smith FI (1988) Determination of the mutation rate of a retrovirus. J Virol 62:3084-3091
Michael AJ (1988) Transcriptional promoters of cauliflower mosaic virus. PhD Thesis, University of East Anglia, UK
Morelli G, Nagy F, Fraley RT, Rogers SG, Chua N-H (1985) A short conserved sequence is involved in the light-inducibility of a gene encoding ribulose 1,5-bisphosphate carboxylase small subunit of pea. Nature 315:200-204
Nagy F, Boutry M, Hsu M-Y, Wong M, Chua N-H (1987) The 5'- proximal region of the wheat Cab-1 gene contains a 268-bp enhancer-like sequence for phytochrome response. EMBO J 6:2537-2542
Odell JT, Nagy F, Chua N-H (1985) Identification of the DNA sequences required for activity of the cauliflower mosaic virus 35S promoter. Nature 313:810-812
Ow DW, Jacobs JD, Howell SH (1987) Functional regions of the cauliflower mosaic virus 35S RNA promoter determined by the firefly luciferase gene as a reporter of promoter activity. Proc Natl Acad Sci USA 84:4870-4874
Paszkowski J, Shillito RD, Saul M, Mandak V, Hohn T, Hohn B, Potrykus I (1984) Direct gene transfer to plants. EMBO J 3:2717-2722
Paszkowski J, Baur M, Bogucki A, Potrykus I (1988) Gene targetting in plants. EMBO J 13:4021-4026
Potrykus I, Saul M, Petruska J, Paszkowski J, Shillito RD (1985) Direct gene transfer to cells of a graminaceous monocot. Mol Gen Genet 199:183-188
Reanney D (1984) Genetic noise in evolution. Nature 307:318-319
Robinson DJ, Hamilton WDO, Harrison BD, Baulcombe DC (1987) Two anomolous

tobamovirus isolates: evidence for RNA recombination in nature. J Gen Virol 68:2551-2561
Rochon D, Siegel A (1984) Chloroplast DNA transcripts are encapsidated by tobacco mosaic virus coat protein. Proc Natl Acad Sci USA 81:1719-1723
Rochow WF (1970) Barley yellow dwarf virus: phenotypic mixing and vector specificity. Science 167:875-878
Sanders PR, Winter JA, Barnason AR, Rogers SG, Fraley RT (1987) Comparison of cauliflower mosaic virus 35S and nopaline synthase promoters in transgenic plants. Nucleic Acids Res 15:1543-1558
Sandford JC and Johnston SA (1985) The concept of parasite-derived resistance - deriving resistance genes from the parasite's own genome. J Theor Biol 113:395-405
Shillito RD, Saul MW, Paszkowski J, Muller M, Potrykus I (1985) High efficiency direct gene transfer to plants. Bio/Technology 3:1099-1103
Sleat DE, Turner PC, Finch JT, Butler PJG, Wilson TMA (1986) Packaging of recombinant RNA molecules into pseudovirus particles directed by the origin-of-assembly sequence from tobacco mosaic virus RNA. Virology 155:299-308
Sleat DE, Gallie DR, Jefferson RA, Bevan MW, Turner PC, Wilson TMA (1987) Characterisation of the 5'-leader sequence of tobacco mosaic virus RNA as a general enhancer of translation *in vitro*. Gene 60:217-225
Sleat DE, Hull R, Turner PC, Wilson TMA (1988) Studies on the mechanism of translational enhancement by the 5'-leader sequence of tobacco mosaic virus RNA. Eur J Biochem 175:75-86
Stanley J, Townsend R (1986) Infectious mutants of cassava latent virus generated in vivo from intact recombinant DNA clones containing single copies of the genome. Nucleic Acids Res 14:5981-5988
Takamatsu N, Ishikawa M, Meshi T, Okada Y (1987) Expression of bacterial chloramphenicol acetyltransferase gene in tobacco plants mediated by TMV-RNA. EMBO J 6:307-311
Topfer R, Gronenborn B, Schell J, Steinbiss H-H (1989) Uptake and transient expression of chimeric genes in seed-derived embryos. The Plant Cell 1:133-139
van Vloten-Doting L (1983) Advantages of multipartite genomes of single-stranded RNA plant viruses in nature, for research and for genetic engineering. Plant Mol Biol Rep 1:55-60
Walden RM, Howell SH (1982) Intergenomic recombination events among pairs of defective cauliflower mosaic virus genomes. J Mol Appl Genet 1:447-456
Ward A, Etassami P, Stanley J (1988a) Expression of a bacterial gene in plants mediated by infectious geminivirus DNA. EMBO J 7:1583-1587
Ward CD, Stokes MAM, Flanegan JB (1988b) Direct measurement of the poliovirus RNA polymerase error frequency in vitro. J Virol 62:558-562
Waterhouse PM, Murant AF (1983) Further evidence on the nature of the dependence of carrot mottle virus on carrot red leaf virus for transmission by aphids. Ann Appl Biol 103:455-464
Wilson TMA (1989) Plant viruses: a tool-box for genetic engineering and crop protection. BioEssays 10:179-186
Woolston CJ, Barker R, Gunn H, Boulton MI, Mullineaux PM (1988) Agroinfection and nucleotide sequence of cloned wheat dwarf virus DNA. Plant Mol Biol 11:35-43
Zimmern D (1988) Evolution of RNA viruses. In: Domingo E, Holland JJ, Ahlquist P (eds) RNA genetics Vol. 2. CRC Press, Boca Raton, p 211

INDEX

abscisic acid (ABA), 244, 290-297, 300, 315, 320
AMCV, see African cassava mosaic virus
active defences, 395
acquired resistance, 72, 117, 396, 400
adaptability, 237
African cassava mosaic virus, 448
Ag^+, see silver
aggregation, 437
Agrobacterium, 264, 367-368, 422, 446
agroinfection, 227, 447
alfalfa mosaic virus, 170, 176, 209, 211, 212-216, 219-227, 368-371, 389, 397, 437
alpha-amanitin, 189
alphaviruses, 149-151, 171
1-aminocyclopropane-1-carboxylic acid (ACC), 243, 301, 320
amino vinyl glycine, 301
amorphous inclusions, 165
amplification, 43, 188, 239, 362, 388
AMV, see arabis mosaic virus
anthriscus yellows virus, 61
animal viruses, 148
antigenic conservation, 66
antigenic diversity, 66
antisense RNA, 12, 268, 429, 433, 438
 domain, 269
antiviral factors, 9-10, 378, 395-401
aphids, 53, 60, 101, 106, 411, 434
apical buds, 244
apoplast, 251
Arabidopsis thaliana, 421, 422
arabis mosaic virus, 56
artificial recombinants, 81
 - see also pseudorecombinants
avocado sunblotch viroid, 192-193
aspartyl endoproteinase, 250
assembly, 5, 324
attenuated virus, 266
auxin, 244, 302-303
avirulence genes, 19-23, 39, 41, 47, 377
avocado sunblotch viroid, 188
Avr genes, see avirulence genes
AYV, see anthriscus yellows virus

back mutation, 270
 - see also reversion
bacteriophages, 168
barley stripe mosaic virus, 100, 330
barley yellow dwarf virus, 62, 105, 151, 303-304, 406
basal lamina, 62
bean common mosaic virus, 77, 83
beetles, 53
beet cryptic virus, 275-284
beet curly top virus, 111-113
beet western yellows virus, 62, 151, 406
beet yellows virus, 330
benign sat RNA, 266-268
beta-glucan, 40
beta-glucuronidase, 137
beta-1,3-glucanase, 41, 250-255, 365, 382-384, 386-388
blasticidin S, 45
Botrytis
 cinerea, 33
 fabae, 38
Bremia lactucae, 42, 44-46, 95
brome mosaic virus, 170, 171, 174-175, 211 419, 449
BYDV, see barley yellow dwarf virus

CAAT box, 190, 446
cacao swollen shoot virus, 95, 416
calcium, 127, 242, 439
callose, 36, 379, 397
calmodulin, 127, 242
CaMV, see cauliflower mosaic virus
cap structures, 147
capsid morphology, 147
carnation cryptic virus, 275
carrot red leaf virus, 63-65
carrot mottle virus, 63, 409
cascade control, 8-11, 252, 347, 388
catalytic RNA, 193
cationic PR proteins, 249
cauliflower mosaic virus, 60, 211, 216, 220, 397, 344, 448, 452
 19S promoter, 446
 35S promoter, 444-446

CCMV, see cowpea chlorotic mottle virus
CCV, see carnation cryptic virus
cDNA
 clones, 156
 libaries, 363
cell
 cycle, 240
 degeneration, 251
 division, 238, 254, 276, 293, 297, 300
 expansion, 293
 -free translation, 127-130, 134
 membrane, 45
 -to-cell movement, 76, 78, 157, 186, 208-228, 378, 405-413
 -to-cell transport, 312, 347
 wall, 36, 126, 136, 160, 207-222, 245, 251-252, 379, 386, 420
central conserved region, 192, 196-198
CEVd, see citrus exocortis viroid
challenge virus, 415, 429
chimaeric
 constructs, 194
 RNA, 128, 45
 vectors, 450
chitinase, 41, 250-251, 365, 382-384, 386-388
chlorophyll, 300, 315
 fluorescence, 316
chloroplasts, 290, 295-297, 300, 305, 329-334
 protein synthesis, 315
 ribosomes, 315, 334
 ribosomal RNA, 330
chlorotic lesions, 302, 398
chromatin-associated proteins, 322
chrysanthemum stunt viroid, 190, 194, 241
circular RNA, 185, 195, 237
circulative viruses, 55, 61-65
citrus exocortis viroid, 190, 194, 196, 235, 241, 243-245, 248-255
citrus tristeza virus, 98, 416-418
CMV, see cucumber mosaic virus
Cladosporium fulvum, 32-35, 41, 248
CMotV, see carrot mottle virus
coat protein, 12, 57, 82, 116, 134, 155, 159, 211, 215, 264, 280, 296, 312, 315, 330, 333, 337-343, 412, 419-421, 450-452
 defective, 428
 -mediated protection, 264, 427-439, 450
 mutants, 420
 subgenomic RNA, 168-171
cocoa, see cacao
co-evolution, 4, 6, 94, 114, 239, 252
cognon/cognor, 39
co-infection, 406-409
Colletotrichum
 lagenarium, 385
 lindemuthianum, 33-35
colonization, 32-35
common ancestry, 149, 151
comoviruses, 148, 152-161
companion cells, 405
compatible interaction, 361-365
competitive inhibition, 288-290
complementation, 159
"confidence men", 38
constitutive expression, 446
continuous selection, 450
control, see cascade control
convergence, 236
copper resistance, 20
cosmid, 23
cotranslational disassembly, 127-135, 439
cowpea chlorotic mottle virus, 170, 175-176
cowpea mosaic virus, 87, 148, 152-161, 165, 172, 178, 217, 228 385, 411
CRLV, see carrot red leaf virus
crop
 improvement, 443
 protection, 11-13, 116, 416, 418
cross protection, 11, 135-141, 243, 255, 415-423, 427-429, 438
cryptic viruses, 7, 273-284
cryptotope, 133
crystalline structures, 164
CSSV, see cacao swollen shoot virus
CSVd, see chrysanthemum stunt viroid
CTV, see citrus tristeza virus
cucumber mosaic virus, 100, 170, 177-178, 211, 218, 263-271, 291, 297, 302, 304, 363-364, 395
cuticle, 126
cyclic AMP, 242
cysteine proteinase, 250
cytokinins, 297-301, 320
cytopathic modifications, 164-168, 245

dark green areas, see green islands
defeated resistance genes, 80, 86, 107
defence, 234
 active, 395
 preformed, 395
 proteins, 375-390
 reactions, 254
deletion mutants, 341
desmotubule, 208
detoxification, 38
developmental disease, 235, 246, 253-255
differentiation, 253
differential interaction, 18
dihydrophaseic acid, 291
disease
 control, see crop protection
 development, 311-325
 establishment, 73, 243
 expression, 245-255
 outbreaks, 95-100, 113
 pressure, 417
disassembly, 126-141
DNA-dependent
 RNA polymerase I, 189
 RNA polymerase II, 189-191, 198
dsRNA, 165, 168-171
 -dependent protein kinase, 244
 genome, 275

early interactions, 123-141
electron transport, 316
electroporation, 447
elicitors, 39-42, 233, 338, 341, 363, 376, 386, 388
encapsidation, 154, 419
epidemics, 115
epidemiology, 4, 93-117, 451
epidemiological competence, 102
epinasty, 302, 320
episomal vectors, 446-450
error rates, 449
ethephon, 243, 249, 255, 301
ethylene, 7, 235, 242-244, 249, 251-252, 301-302, 320
evolution, 5, 74, 88, 200, 278
 - see also co-evolution
exclusion limit, 208
expression vectors, 81
extensin, 252

field release, 450-453
flanking sequences, 366
flower bud, 320
fruit tree viruses, 98
fungus, biotrophic, 33

GDD motif, 152, 172-179
geminiviruses, 210, 330, 452
gene
 activity, 1
 clustering, 24
 induction, 361-371
 insertion, 448
 structure, 365
 vectors, 12, 443-450
gene-for-gene relationships, 19, 39, 83, 104, 313, 376, 386
 Capsicum annuum-TMV, 84
 potato-potato virus X, 84
 Phaseolus vulgaris-BCMV, 84
 tomato-TMV, 84
genetic mapping, 80-82, 148, 153, 337, 343, 345-357
genome-linked proteins, 147, 148, 153
genotype x environment interactions, 77
ghost effect, 86
gibberellins, 244, 303-304, 320
glucanases, see beta-1,3-glucanases
glycanhydrolases, see chitinase and beta-1,3-glucanase
glycine-rich protein (GRP), 365, 367
green islands, 287, 297, 305, 318-322, 420
green revolution, 108
groundnut rosette virus, 63-65
group I introns, 193
 II introns, 194
growth inhibition, 287-306, 320
growth rate, 289
GRV, see groundnut rosette virus
guiding sheath, 56
Gynura aurantiaca, 235, 243, 255

haemocoel, 62
hairpin, 156, 199
hammerhead RNA, 192, 194
harp genes, 19, 24-25
haustorium, 32-34, 42
heat-shock, 45

helper component, 58-60, 61, 66, 149
helper functions, 73-76
helper virus, 58, 209, 263, 266, 356
heptaglucoside elicitor, 40
herbivores, 385
heterologous encapsidation, 63, 451
histones, 241
HMV, see hop mosaic virus
homeostasis, 237
hormones, see plant growth regulators
hop mosaic virus, 99
host
 factors, 186, 200, 311-314
 range, 3, 18, 26, 71-77, 125
 response, 337, 346
 specificity, 25, 186
 variation, 94-100
hot regions, 353-355
HR, see hypersensitive response
HRGP, see hydroxyproline-rich glycoprotein
Hrp genes, see harp genes
hydroxyproline rich glycoproteins (HRGP), 35-38, 252, 380
hypersensitive response, 18, 35-38, 42-44, 74, 220-222, 301, 313, 345-357, 364-371, 375-381, 388-343, 396-405, 434

inclusion bodies, 60
incompatibility, 23, 361-371, 376
indole acetic acid (IAA), 303, 320
induced resistance, see acquired resistance
infection, 215
 pressure, 113
 sites, 377, 434
 structures, 32
infectious transcripts, 156-159
information processing, 237
informosome, 6, 210
inhibitors, 74
insect attack, 368-371
insecticides, 106, 113
integrative vectors, 444-446
interactions
 plant-bacterium, 17-27
 plant/fungus, 31-47
 plant-virus, 1-11
intercellular
 fluids, 41, 400
 spaces, 250
 washing fluids (IWF), 250
intermediate signals, 376, 386
introns, 366
 group I, 193
 group II, 194
inverted cascade, 11
irreversible membrane damage (IMD), 42-46

kinetin, 320

landraces, 113
latent
 infection, 188, 273
 period, 55
leader sequences, 128, 130-134, 447
leaf
 growth, 293
 hoppers, 53, 106
 malformation, 241, 243, 315
lesion, see hypersensitive response
lethal-necrotic domains, 269
lettuce downy mildew, see *Bremia lactucae*
lettuce mosaic virus, 101
light-green/dark-green mosaic, 315, 318-322
 - see also green islands
light quantum yield, 316
lignification, 37, 279, 397
lipoxygenase, 44
LMV, see lettuce mosaic virus
local lesion gene, 348
 - see also hypersensitive response
localization, 78, 396
 - see also hypersensitive response
long-distance signals, 366
Longidorus, 56
luteoviruses, 62-65, 152, 405-408

maize chlorotic dwarf virus, 61
 dwarf mosaic, 105
 rough dwarf virus, 96
 streak virus, 446
masked strain, 314, 324
MCDV, see maize chlorotic dwarf virus
membrane damage, 36, 39
membrane protein, 20, 24
microinoculation, 377

mild strains, 104, 116, 416
mites, 53
modifier genes, 77
modular construction, 452
modular evolution, 151
mosaic symptoms, 287, 292-294, 314-325, 340, 420
movement protein, 77, 81, 82, 207-228, 312, 347, 356, 378, 397, 406, 410, 421
MRDV, see maize rough dwarf virus
MSV, see maize streak virus
multiplication, see replication
mutation, 81, 82, 150-159, 199, 210, 313, 338, 339, 346, 348
mycoviruses, 278, 284

N.Co.310, 97
N gene, 79, 301, 313, 341, 396, 398, 435
N' gene, 81-82, 221, 312, 338-343, 345-357, 377
naphthylacetic acid (NAA), 319
necrosis, 9, 19, 295, 301, 305, 314, 337-343, 375
 - see also hypersensitive reaction
necrotrophic fungi, 33
Nectria haematococca, 38
nematodes, 53
neotope, 139
Nephotettix virescens, 108
nepoviruses - transmission, 55-57
Nilaparvata lugens, 106
non-circulative viruses, 55-61
non-competitive inhibition, 288-290
non-host, 239
 immunity, 71-77
 - see also host range
non-specific response, 342
NTP-binding motif, 149, 172-179
nucleolus, 241

odontostyle, 56
oesophagus, 56
oligogenic resistance, 77, 80
oligomeric RNAs, 186
oligosaccharide, 41
 signals, 376, 388-390
omega fragment, 128, 130-134, 447
origin of assembly sequence, 447
origin of replication, 190, 199

papaya ringspot virus, 416-418
papillae, 35
parsnip yellow fleck virus, 61
passion fruit woodiness virus, 416
pathogen pressure, 114
pathogenic domains, 355
pathogenic fitness, 85-86
pathogenicity, 3, 6, 82, 186, 245-255, 305
 functions, 73-75
 genes, 17-25
 mechanisms, 34
pathogenesis-related (PR) proteins, 8, 235, 243, 247-251, 362-371, 380-388, 421, 435
pea enation mosaic virus, 61-62, 168
pesticides, 115
PFWV, see passion fruit woodiness virus
phaseic acid, 291
phenolics, 75
phenotypic mixing, 63
phenoxyacetic acid (PAA), 303
phenylpropanoids, 47
phloem, 405
photosystem II, 316, 330
phytoalexins, 9, 19, 36-42, 379
Phytophthora
 infestans, 384
 megasperma, 40, 388
picornavirus-related plant viruses, 148-149
plant
 breeding, 77, 94, 104, 113
 growth, 287-300
 growth regulators, 287-306, 318
 - see also individual types
planthoppers, 53, 97, 106
plasmalemma, 242, 306
plasmids, 20
plasmodesmata, 6, 124, 160, 208-222, 397, 408, 410
PLRV, see potato leafroll virus
plum pox virus, 98
POJ 2878, 97
poliovirus, 172, 178
pollen transmission, 277
poly(A) RNA, 148, 363
polymorphisms, 94
polyphenols, 45
polyproteins, 148, 153, 155, 211

population dynamics, 93-117
potato leafroll virus, 151, 405-413
potato spindle tuber viroid, 188-191, 193-198, 241, 244, 248-250
potato virus C, 58
potato virus X, 76, 103, 128, 209, 399, 410, 437
potato virus Y, 58, 79, 101, 399, 408, 410
potyvirus transmission, 57-60
PPV, see plum pox virus
promoters, 12, 190, 366, 368, 429, 444
protease, 149, 153, 155, 380, 385
 inhibitors, 385
protein phosphorylation, 244
protoplasts, 45, 76, 128, 136, 153-154, 165, 179, 211, 219, 252, 312, 347, 368, 398, 406, 446
protoviroids, 240
PR protein, see pathogenesis-related proteins
PRV, see papaya ringspot virus
pseudo genes, 366
Pseudomonas syringae, 18-27
pseudorecombinant, 57, 81, 199
pseudovirions, 123, 125, 128, 133, 330, 333, 447
PSTVd, see potato spindle tuber viroid
PVC, see potato virus C
PVFV, see parsley yellow fleck virus
PVX, see potato virus X
PVY, see potato virus Y

quasi-genetic model, 75

raspberry ringspot virus, 56-57, 102
RCMV, see red clover mottle virus
RCNMV, see red clover necrotic mosaic virus
RCV, see ryegrass cryptic virus
receptors, 39, 75, 123, 228, 295
recognition, 1-13, 17, 80
 in transmission, 65-67, 72
 plant-bacterium, 3
 plant-fungus, 3
recombination, 5, 151, 452
red clover mottle virus, 161
red clover necrotic mosaic virus, 95
regulatory sequences, 156, 366-368
reoviruses, 275, 284
repetitive DNA, 22
replicase, 5, 75, 82, 153, 164-165 171-179, 189, 263, 266, 312, 395, 429
replication, 4, 148, 163-180, 311, 313, 347, 368, 428, 449
replicative
 forms, 168-171
 intermediates, 168-171, 186, 410
replicon, 185, 200
reporter gene, 47, 367, 446
reporter structures, 126, 137
resistance, 2, 8, 254, 264, 313, 362, 375, 443, 450
 cell-to-cell spread, 409-410
 defeated genes, 80, 86, 107
 dominant genes, 78
 durable, 79, 102
 genetics, 13, 71-88, 377
 gene-dosage dependent, 78
 mechanisms, 245, 337
 minor gene, 113
 non-host, 36, 40
 negative mechanisms, 73-76, 78
 oligogenic, 77, 80
 positive mechanisms, 73-76, 78
 race-specific, 36, 39
 recessive, 79
 targets, 73
resistance-breaking, 79, 103
 - see also virulence
response network, 252
restriction fragment length polymorphisms, 422
retention site, 56, 59
reversion, 200
 mutation, 341, 348
RGSV, see rice grassy stunt virus
ribosomal RNA, 296
ribulosebisphosphate carboxylase (Rubisco), 315
rice grassy stunt virus, 106
rice stripe virus, 101
rice tungro virus, 98, 108, 297
rice waika virus, 98
RMV, see ryegrass mosaic virus
RNA-dependent RNA polymerase, 149, 164, 172, 283, 362, 419
RNA ligase, 193
RNA-polymerase, 149, 152, 164, 172, 189
 - see also replicase

roguing, 101, 111, 417
rolling-circle mechanism, 187-189
RRV, see raspberry ringspot virus
RSV, see rice stripe virus
RTV, see rice tungro virus
RWV, see rice waika virus
ryegrass cryptic virus, 275, 279
ryegrass mosaic virus, 279

Saccharomyces cerevisiae, 223-227
salicylate, 366, 368
satellite RNAs, 7, 65, 116, 179, 192, 263-271
SBMV, see southern bean mosaic virus
secondary invaders, 10
secondary messengers, 41, 239, 242, 251
secondary metabolites, 39
seed transmission, 100, 102, 275-278, 451
selection, 200, 280, 357
self and non-self, 240
self-cleavage, 192, 194
senescence, 287, 300, 315, 320
sequencing, 20
sequence homology, 148, 178, 211, 280, 385, 438
sequestration, 419
Shine-Dalgarno sequence, 334
signalling, 72, 233
 cell-to-cell, 1-11
signal
 peptide, 365
 recognition particle, 199, 235, 241, 412
 sequences, 20
 transduction, 40, 46, 87
silver, 233, 235, 236, 243, 253, 254
sindbis virus, 149
sindbis-like plant viruses, 149-157
single lesion passages, 351
site-directed mutagenesis, 87, 199, 314, 340, 439
southern bean mosaic virus, 151
specificity, 32, 239, 418
 determinants, 42
Spodoptera exigua, 370
spread factor - see movement protein
starch, 330
 lesions, 397
stomata, 294
stress, 242, 245, 255, 302, 377
striposomes, 129
stripping mapping, 348-351
stunting, 244, 290, 303
suberization, 37
subgenomic RNA, 333-334
subliminal infections, 76, 312, 346, 356
sugar beet, 273
sugarcane
 C.P. cultivars, 110
 fijivirus, 97
 mosaic virus, 109
 N.Co.310, 110
 P.O.J., 110
superoxides, 44
suppressor, 38
 tRNA, 176
susceptibility genes, 75
symptom, 264, 273, 278-280, 304, 330, 340, 417
 amelioration, 266-268
 formation, 6, 268, 288, 312, 449
synergistic interactions, 279, 450-451
systemic spread, 208, 421, 435

TASVd, see tomato apical stunt viroid
TATA box, 190, 445
TAV, see tomato aspermy virus
TBRV, see tomato blackring virus
temporary biotrophy, 34
TEV, see tobacco etch virus
TGMV, see tomato golden mosaic virus
thaumatin, 365, 385
"thugs", 38
thylakoids, 316
Tipula irridescent virus, 124
tissue localization, 405-410
tissue specificity, 445
TMV, see tobacco mosaic virus
TMV 126 kDa polypeptide, 129, 133, 322-325
Tm-1 gene, 79, 82, 85, 101, 314
Tm-2 gene, 78, 82, 209, 411
Tm-2^2 gene, 78, 86
Tn5, 23, 25
TNV, see tobacco necrosis virus
tobacco, 264, 288
 etch virus, 58-60
 mosaic virus, 7, 76, 78, 79, 81-82, 85,

126-141, 169, 176-177, 209, 210, 216, 219-227, 248, 288-306, 311-325, 329-334, 337-343, 345-357, 363-371, 377-390, 396, 410, 417, 419, 428, 447
necrosis virus, 301, 396
rattle virus, 210-211, 228, 408
ringspot virus, 192, 264, 266, 297
vein mottling virus, 58
TobRV, see tobacco ringspot virus
tolerance, 73, 99, 240
tomato, 243, 250, 264, 288, 297
apical stunt viroid, 241
aspermy virus, 264, 266
black ring virus, 56-57, 408
golden mosaic virus, 448
mosaic virus, 101, 416-418
- see also tobacco mosaic virus
planta macho viroid, 241-248
tonoplast, 42, 242
toxins, 44
TPMVd, see tomato planta macho viroid
transcapsidation, 63, 451
transformation, 88, 265, 443
transgenic plants, 12, 135-141, 210, 264, 356, 367, 368-371, 385, 416, 427-439, 443-450
transient expression system, 446
transit peptides, 334
translation, 5, 154, 333-334
transmission, 3, 53-67
transport, 154, 159-161
protein, see movement protein
transposon mutagenesis, 19
transposon tagging, 47, 87
TRV, see tobacco rattle virus
tubules, 411
tulipalin, 36
turnip mosaic virus, 95, 428
turnip yellow mosaic virus, 170, 174, 330
TVMV, see tobacco vein mottling virus
TYMV, see turnip yellow mosaic virus

uncoating, 4, 75, 127, 419-420

vacuole, 250
varietal mixtures, 104
variety rotation, 109
VCV, see vicia cryptic virus
vectors, 3, 53-67, 93-117, 124, 234
biotypes, 54, 104-106
specificity, 54, 57, 451
transmission, 154
vein clearing, 324
velvet tobacco mosaic virus, 179
vicia cryptic virus, 275
viroids, 2, 135, 179, 428
DNA, 186, 195
pathogenesis, 233-256
RNA processing, 192-198
replication, 185-201
structure, 198-201
viroplasms, 166, 177
virulence, 8, 71-88, 102, 222, 337-343, 345-357
genetics, 79-86
-modulating (VM) region, 200, 241
virulent sat RNA, 266-268
virus
diseases, 95-100, 113, 243, 311-325, 337, 417
evolution, 5, 74, 88, 147-161, 200, 278
genetic organization, 80-82, 147-161, 337-343, 345-357
movement, 6, 207-228, 278, 406, 410, 421
replication, 4, 137, 148, 163-180, 288, 311, 347, 368, 377, 428, 449
strains, 100-104
supergroups, 147-152
taxonomy, 147-161
virusoids, 192
VPg, 147, 148, 153
VTMoV, see velvet tobacco mosaic virus

war games, 115
water stress, 295
WDV, see wheat dwarf virus
weather, 115
weed hosts, 113
wheat dwarf virus, 447
whiteflies, 53, 106

Xanthomonas campestris, 18
X-bodies, 166, 322
Xiphinema, 56

yeast, 223-227
yield, 417, 434

yellow mosaic symptoms, 267-270, 313

zeatin and derivatives, 298-301, 320
zinc, 155

NATO ASI Series H

Vol 1 **Biology and Molecular Biology of Plant-Pathogen Interactions.** Edited by J A Bailey 415 pages 1986

Vol 2· **Glial-Neuronal Communication in Development and Regeneration.** Edited by H H Althaus and W Seifert 865 pages 1987

Vol 3 **Nicotinic Acetylcholine Receptor: Structure and Function.** Edited by A Maelicke. 489 pages 1986

Vol 4· **Recognition in Microbe-Plant Symbiotic and Pathogenic Interactions.** Edited by B Lugtenberg 449 pages 1986

Vol 5· **Mesenchymal-Epithelial Interactions in Neural Development.** Edited by J R Wolff, J Sievers, and M Berry 428 pages 1987

Vol 6· **Molecular Mechanisms of Desensitization to Signal Molecules.** Edited by T M Konijn, P J M Van Haastert, H Van der Starre, H Van der Wel, and M D Houslay 336 pages. 1987

Vol 7 **Gangliosides and Modulation of Neuronal Functions.** Edited by H Rahmann 647 pages 1987

Vol 8· **Molecular and Cellular Aspects of Erythropoietin and Erythropoiesis.** Edited by I N Rich 460 pages 1987

Vol 9 **Modification of Cell to Cell Signals During Normal and Pathological Aging.** Edited by S Govoni and F Battaini 297 pages. 1987

Vol 10· **Plant Hormone Receptors.** Edited by D Klambt 319 pages 1987

Vol 11 **Host-Parasite Cellular and Molecular Interactions in Protozoal Infections.** Edited by K -P Chang and D Snary 425 pages 1987

Vol 12. **The Cell Surface in Signal Transduction.** Edited by E Wagner, H Greppin, and B Millet 243 pages 1987

Vol 13. **Toxicology of Pesticides: Experimental, Clinical and Regulatory Perspectives.** Edited by L G Costa, C L Galli, and S D Murphy 320 pages 1987

Vol 14· **Genetics of Translation. New Approaches.** Edited by M F Tuite, M Picard, and M Bolotin-Fukuhara 524 pages 1988

Vol 15· **Photosensitisation. Molecular, Cellular and Medical Aspects.** Edited by G Moreno, R H Pottier, and T G Truscott 521 pages 1988

Vol 16· **Membrane Biogenesis.** Edited by J A F Op den Kamp 477 pages 1988

Vol 17· **Cell to Cell Signals in Plant, Animal and Microbial Symbiosis.** Edited by S Scannerini, D Smith, P Bonfante-Fasolo, and V Gianinazzi-Pearson 414 pages 1988

Vol. 18 **Plant Cell Biotechnology.** Edited by M S S Pais, F Mavituna, and J M Novais 500 pages 1988

Vol 19 **Modulation of Synaptic Transmission and Plasticity in Nervous Systems.** Edited by G Hertting and H -C Spatz 457 pages 1988

Vol 20· **Amino Acid Availability and Brain Function in Health and Disease.** Edited by G Huether 487 pages 1988

NATO ASI Series H

Vol 21: **Cellular and Molecular Basis of Synaptic Transmission.** Edited by H Zimmermann 547 pages 1988

Vol. 22 **Neural Development and Regeneration. Cellular and Molecular Aspects.** Edited by A Gorio, J R Perez-Polo, J de Vellis, and B. Haber 711 pages 1988

Vol 23. **The Semiotics of Cellular Communication in the Immune System.** Edited by E E. Sercarz, F. Celada, N A Mitchison, and T Tada 326 pages 1988

Vol 24: **Bacteria, Complement and the Phagocytic Cell.** Edited by F C Cabello und C Pruzzo 372 pages 1988

Vol 25: **Nicotinic Acetylcholine Receptors in the Nervous System.** Edited by F. Clementi, C. Gotti, and E Sher 424 pages 1988

Vol 26: **Cell to Cell Signals in Mammalian Development.** Edited by S.W de Laat, J.G Bluemink, and C L Mummery 322 pages 1989

Vol 27: **Phytotoxins and Plant Pathogenesis.** Edited by A Graniti, R D Durbin, and A Ballio 508 pages 1989

Vol. 28: **Vascular Wilt Diseases of Plants. Basic Studies and Control.** Edited by E C Tjamos and C H Beckman 590 pages 1989

Vol 29. **Receptors, Membrane Transport and Signal Transduction.** Edited by A E Evangelopoulos, J P Changeux, L. Packer, T G Sotiroudis, and K W A Wirtz 387 pages 1989

Vol 30: **Effects of Mineral Dusts on Cells.** Edited by B T Mossman and R O Bégin 470 pages 1989

Vol 31: **Neurobiology of the Inner Retina.** Edited by R Weiler and N.N Osborne. 529 pages 1989

Vol 32: **Molecular Biology of Neuroreceptors and Ion Channels.** Edited by A Maelicke 675 pages 1989

Vol 33: **Regulatory Mechanisms of Neuron to Vessel Communication in Brain.** Edited by F. Battaini, S Govoni, M S Magnoni, and M Trabucchi 416 pages 1989

Vol 34: **Vectors as Tools for the Study of Normal and Abnormal Growth and Differentiation.** Edited by H Lother, R Dernick, and W Ostertag 477 pages 1989

Vol 35: **Cell Separation in Plants: Physiology, Biochemistry and Molecular Biology.** Edited by D J Osborne and M B Jackson 449 pages 1989

Vol 36: **Signal Molecules in Plants and Plant-Microbe Interactions.** Edited by B J J Lugtenberg 425 pages 1989

Vol 37 **Tin-Based Antitumour Drugs.** Edited by M Gielen 226 pages 1990

Vol 38 **The Molecular Biology of Autoimmune Disease.** Edited by A G Demaine, J-P Banga, and A M McGregor 404 pages 1990

Vol 39: **Chemosensory Information Processing.** Edited by D Schild 403 pages 1990

Vol 40: **Dynamics and Biogenesis of Membranes.** Edited by J A.F Op den Kamp 367 pages 1990

Vol 41 **Recognition and Response in Plant-Virus Interactions.** Edited by R S S Fraser 467 pages 1990

MIX
Papier aus verantwortungsvollen Quellen
Paper from responsible sources
FSC® C105338

If you have any concerns about our products,
you can contact us on
ProductSafety@springernature.com

In case Publisher is established outside the EU,
the EU authorized representative is:
Springer Nature Customer Service Center GmbH
Europaplatz 3, 69115 Heidelberg, Germany

Printed by Libri Plureos GmbH
in Hamburg, Germany